# *The Politics of Excellence*

# The Politics of Excellence

## Behind the Nobel Prize in Science

ROBERT MARC FRIEDMAN

A W. H. Freeman Book
Times Books
Henry Holt and Company
New York

Times Books
Henry Holt and Company, LLC
*Publishers since 1866*
115 West 18th Street
New York, New York 10011

Henry Holt® is a registered trademark of Henry Holt and Company, LLC.

Copyright © 2001 by Robert Marc Friedman
All rights reserved.

Library of Congress Cataloging-in-Publication Data
Friedman, Robert Marc, 1949–
   The politics of excellence : behind the Nobel Prize in science / by Robert Marc Friedman
      p.cm.
Includes bibliographic references
   ISBN: 0-7167-3103-7
   1. Nobel Prizes—History. 2. Physics—Awards—History. 3. Chemistry—Awards—History.
I. Title.
   QC49.F75 2001
   507.9—dc21                                                       2001002099

Henry Holt books are available for special promotions and premiums.
For details contact: Director, Special Markets.

First Edition 2001

Designed by Cambraia Fonseca Fernandes

Printed in the United States of America
10 9 8 7 6 5 4 3 2 1

For my parents,

*Lillian Kamlet Friedman*
*David Friedman*

# *C*ontents

| | | |
|---|---|---|
| *Preface* | | ix |
| *Author's Note and Acknowledgments* | | xi |
| Introduction | Legendary Excellence | 1 |

## Part I
### PERMANENT BATTLES WILL SURELY BE WAGED FOR EVERY PRIZE 11

| | | |
|---|---|---|
| One | The Stupidest Use of a Bequest That I Can Imagine! | 13 |
| Two | Coming Apart at the Seams | 26 |
| Three | Sympathy for an Area Closely Connected with My Own Specialty | 40 |
| Four | Each Nobel Prize Can Be Likened to a Swedish Flag | 54 |

## Part II
### HAS THE SWEDISH ACADEMY OF SCIENCES … SEEN NOTHING, HEARD NOTHING, AND UNDERSTOOD NOTHING? 69

| | | |
|---|---|---|
| Five | Should the Nobel Prize Be Awarded in Wartime? | 71 |
| Six | While the Sores Are Still Dripping Blood! | 93 |

## Part III
### SMALL POPES IN UPPSALA 117

| | | |
|---|---|---|
| Seven | Einstein Must Never Get a Nobel Prize | 119 |
| Eight | To Sit on a Nobel Committee Is Like Sitting on Quicksand | 141 |
| Nine | Clamor in the Academy | 163 |

## Part IV
### DON'T SHOOT THE PIANO PLAYER, HE'S DOING THE BEST HE CAN     177

Ten     It Can Happen That Pure Pettiness Enters     179

Eleven     One Ought to Think the Matter Over Twice     190

## Part V
### SCANDALOUS TRAFFIC     211

Twelve     Dazzling Dialects     213

Thirteen     Completely Lacking an Unambiguous, Objective Standard     225

Fourteen     The Knights Templar     251

*Further Reflections*     267

*Appendix A*     279

*Appendix B*     284

*Appendix C*     286

*Notes*     289

*Index*     361

# Preface

*J*ust what do we know about the Nobel Prize? As a symbol of excellence and a shorthand indicator for genius it occupies a prominent niche in popular and scientific cultures. Considerable media attention greets the annual announcements, from describing the achievements to flashing a box score of national prizes. It is routinely enrolled in advertising, in Swedish foreign relations, in defining a superelite, even in selling a California sperm bank. Some universities try to "buy" Nobel laureates; those that have them, flaunt them. Biographical dictionaries of Nobel winners regularly appear, creating a pantheon of scientific heroes. For some researchers, the prize is not only the *most* important thing, it is the *only* thing by which scientific success is gauged.

At the Nobel festivities in Stockholm, leaders of industry, government, science, and culture wine, dine, and dance through the night in the company of the Swedish royal family. At home, the rest of us watch the spectacle on television and hear the laureates propose solutions to the world's problems. For tourists in Stockholm, there is an exact replica of the Nobel dinner, for a price. And, just as with the Hollywood Oscars, or the European Melody Grand Prix, no matter how bewildered or disapproving, we come back every year.

As icon, myth, and ritual, the Nobel Prize is well secured. Yet the realities of nomination, evaluation, and selection remain obscure. This book breaks the illusion of the Nobel Prize as being an impartial, objective crowning of the "best" in physics and chemistry. Based on extensive research, using the archives of the Nobel Committees for Physics and Chemistry and the Royal Swedish Academy of Sciences, it explores the history of why and how various people used the Nobel Prize to further their own scientific, cultural, and personal agendas. "Excellence" is not an unambiguous concept, not even in science.

The Nobel Prize may well be international in scope, but since the prize's beginnings in 1901 the Royal Swedish Academy of Sciences has determined the outcome. In evaluating nominations, the five members each of the Academy's Nobel Committees for Physics and for Chemistry have relied upon their own judgment. Nominators, moreover, rarely provide clear mandates. Even if they do, committees seldom select the rare consensual or even majority candidate. Some candidates enjoying clear mandates, such as Henri Poincaré and Albert Einstein (for relativity theory), have been rejected; winners such as Arthur Harden and Harold Urey each received but one nomination. No juggling of statistics related to nominations—number, frequency, and origin—can explain the awards. Similarly, success or failure in winning a prize has not depended upon timeless, fixed standards of excellence. Rather, the changing priorities and agendas of committee members, as well as their comprehension of scientific accomplishment, have been critical.

The history of the Nobel Prize is more than a history of prizewinners. As an institution, the Nobel Prize developed and evolved: Interpretation of the bylaws, committee procedures and priorities, criteria for selecting winners, degree of reliance on nominators, as well as relations within the Academy and committees changed over time. Deliberations frequently became enmeshed in the process by which factions within the Swedish science community attempted to define the scope, methods, and priorities for physics and chemistry. The prize helped shape the growth of these scientific disciplines in Sweden and influenced developments abroad. Some committee members tried to be dispassionate; others championed their own agendas, some openly and some cunningly.

A politics of excellence goes beyond choosing winners. Speeches and ceremonies associated with the annual ritual have a charismatic quality; fields of research, national traditions, scientific ideals, and the prize itself are thereby extolled. Commentators around the world have diffused heroic visions of scientific genius. Although these shape the public image and self-understanding of science, they frequently misrepresent the actual practice of modern science.

Shrouded in secrecy and legend, the Nobel Prize first became an object for serious study in 1976 when the Nobel Foundation opened its fifty-plus-year-old archives to researchers. Although access to the archives is restricted to the years up to 1950, this limitation does not preclude some discussion of the more recent history of the prizes.

Looking behind the scenes in the committees and the Academy enables us to understand the working of the prize. But there are also larger questions involved. Through observation of the processes of selecting winners, we can also examine the changing value system of science and the role of prizes in modern society. Alfred Nobel stipulated that his prizes should be awarded to those who confer "the greatest benefit on mankind." What did Nobel intend? How did committee members interpret his intent? How ought we understand it today? Although often critical, this book is not an assault upon science or upon the Royal Swedish Academy of Sciences. Rather, it asks us to reflect upon the meaning of such prizes in a culture characterized by intense competition for resources, indecorous commercialism, and hype. As a new century dawns, and the scientific community adjusts to a post–cold war era, how should we rethink and reclaim Alfred Nobel's legacy?

# Author's Note and Acknowledgments

*B*eing one of the few professional historians of science researching modern Swedish science, I was invited in 1980 to work with the newly opened Nobel archives. My preliminary results appeared in 1981 in the British science journal *Nature*. For the first time, the Swedish givers of the prize came into focus; speculation and myth could be replaced with insight. Although many different questions can be asked with respect to the prize, it seemed that subsequent discussion should be based on an understanding of what actually went on in the committees and Academy. To make sense of the why and how of awarding prizes, I accepted that it was necessary to study the annual evaluations of candidates in the context of Swedish and international science.

Who the committee members were, how they saw their role in selecting winners, what they considered important in science, and what range of interests might influence their actions all seemed necessary components of such a history. Might it be possible to reconstruct what these men were up to as they went about their extremely difficult task of evaluating candidates and deliberating the annual decision?

These were the guiding questions I arrived at from my first encounters with the history of the prize. At that time, the history of Swedish physics and chemistry was almost virgin territory. The small but excellent history of the science community in Sweden had, up to that point, largely focused on developments before 1900. Little of more recent science had been studied. It also became clear that the names of most committee members were little known in the history of science. Few appear in such standard reference works as the *Dictionary of Scientific Biography*.

It might have been tempting to jump the gun and rush through the official committee reports to write a series of sensational articles spilling the gossip on why X did not get a prize, why Y had to wait so long, and why Z surprisingly became a laureate. Gossip is frequently misleading and inaccurate; those who have tried to create a quick sensation by reading these reports, or obtaining translations from the Swedish, have again and again missed the underlying processes and intentions. Some have got the story wrong. It is easy to be lulled into thinking the committee reports represent the full story of why and how committees assessed the candidates. In actuality, the reports were written to justify the committees' recommendations to the Royal Swedish Academy of Sciences. They are vital documents for any study of the prizes, but they cannot be used in isolation. They are misleading without an understanding of their authors and their intended audience. The challenge to historical scholarship is to pierce the smooth surface of the reports, revealing the messy, complex, and confusing reality of thought and process behind each decision.

My task entailed putting the committee members into context. What characterized Swedish physics and chemistry in the early 1900s? What in particular distinguished these men as scientists? It would be of no use to accept an idealized image of a scientist, nor would it be sufficient to rely upon amiable obituary notices written by appreciative disciples. As is the practice among professional historians of science, I turned to the published and archival traces of committee members' activities. Correspondence among themselves and with colleagues near and far, lecture notes, evaluations of candidates for university positions, annual reports and budget requests, and, of course, their own scientific publications—all offer insight into the life and times of the committee members. I embarked in the mid-1980s on the task of researching a history of Swedish physics and chemistry that could provide a context for analyzing committee actions.

Knowing the sensitivity of the history of the Nobel Prize, I subsequently published in specialized historical journals. I did not want to address a broader public until I could consider the whole rather than the many parts. In 1989, after a decade in Scandinavia, I returned home to the United States where I joined the tenured staff of the University of California at San Diego. Many obligations during the 1990s slowed work on the project. Recently, I moved from California to Norway. Having experienced differing academic cultures in Sweden, America, and Norway, I was prompted to ask questions in this book about the changing values pervading science and its institutions. At the University of Oslo, I was engaged to develop a unit for university history; here, too, came valuable stimulation for thinking about the big issues of what is truly important in academic life. The Nobel Prize, of course, intersects the broader histories of the sciences and universities.

In writing this book I aim to reach out to a general audience, satisfy the curiosity of many scientists, and earn the approval of my professional colleagues in the history of science. All authors must compromise, otherwise—as I began to fear—their ideal book must remain just that, a beautiful vision. In transforming many years of painstaking research into a readable book, I was forced to condense and simplify more than I would in an academic monograph; a multitude of information remains on note cards and databases. In particular, difficult choices were necessary in deciding on the amount of scientific detail: How could I keep the interest of a general reader without losing credibility with physicists and chemists? Equally important, how could I make the recent past a less familiar place without making the book frightfully long? Yesterday's scientists must be allowed to live in their own world, on their own terms. To satisfy all potential audiences would require a book many times this size; few would want to wade through so much detail. The text can be read as a self-contained narrative; endnotes provide details of the archival and published sources on which the analyses are built as well as further details that might interest specialists.

In this book, I focus on those persons responsible for giving out the prize and on the process of making choices. Readers interested in fuller biographical information on those who won prizes and their valuable work should consult one or more of the many celebratory histories published during the one hundredth anniversary observance in 2001. This book is not the last word on the subject; it is written to promote new reflection and debate on the prize, its meaning and significance.

My study of the Nobel Prize began twenty years ago; the list of people and institutions to which I am indebted is long. The brief and simple thanks expressed here cannot do justice to my very deeply felt appreciation for their confidence, generosity, and kindness.

Author's Note and Acknowledgments **xiii**

Primary support for research was made possible by three Scholars Grants from the National Science Foundation's Program in the Study of Science, Technology, and Society (earlier, the Program in the History and Philosophy of Science). These funds sustained me through difficult times and allowed me to work in Sweden for extended periods to study archival collections and to begin writing the results of my research. Additional opportunity for research was afforded by the Swedish Collegium for Advanced Studies in the Social Sciences, Swedish Council for Research on Universities, Norwegian Research Council for Science and the Humanities, Norwegian Institute for Studies on Research and Higher Education, and a sabbatical leave from the University of California at San Diego. Smaller grants allowed me to consult archival materials in Sweden and elsewhere; these included assistance from the Niels Bohr Center for History of Physics at the American Institute of Physics, Royal Swedish Academy of Sciences' Center for History of Science, University of California at San Diego's Academic Senate Research Fund, Rockefeller Archives Center, Maison des Sciences de l'Homme, and Norwegian Academy of Science and Letters' Fridtjof Nansen Fund. At a critical point in writing this book, I received a grant from the Norwegian Non-Fiction Writers and Translators Association.

While working on this project, I enjoyed hospitality from several institutions, sometimes for long periods. Professor Tore Frängsmyr and his Department of History of Science at Uppsala University provided me with an office, which in many respects was my home for almost four years in the 1980s. Then, and on many subsequent visits, I was treated with generosity and openness. The friendships developed and the affection for Swedish society and culture cultivated have made my appreciation yet deeper. Similarly, the Royal Swedish Academy of Sciences' library and archives, now incorporated in the Center for History of Science, has always been a warm and inviting base for research, on this and several other projects. Reaching back to my first visits to the Academy, I am indebted to Wilhelm Odelberg and Kai-Inge Hillerud; more recently to Urban Wråkberg, Julia Lindqvist, and Karl Grandin; and to Christer Wijkström, who has provided important assistance and a friendly face of continuity over twenty years. When I moved to Norway in 1996 and began writing without an institutional affiliation, Dr. Keith Smith generously allowed me to stay at his research unit, Studies in Technology, Innovation, and Economic Policy (S.T.E.P.), which made my landing a bit softer.

In addition to the people mentioned, I benefited from conversations and support from a large number of colleagues. Although I fear, after so many years, I may omit many names, I would like to acknowledge the following persons: Finn Aaserud, Anders Barany, Gunnar Broberg, Craig Calhoun, John Peter Collett, Michael Aaron Dennis, Gunnar Eriksson, Paul Forman, Bernt Hagtvet, John Heilbron, Karl Hufbauer, Thomas Keiserfeld, Julia Lindqvist, Svante Lindqvist, Anders Lundgren, Everett Mendelsohn, Torsten Nybom, Bryan Pfaffenberger, Erik Rudeng, Simon Schaffer, Francis Sejersted, Hans Skoie, Sverker Sörlin, Spencer Weart, Sven Widmalm, Björn Wittroch, Urban Wråkberg, and colleagues and students at the University of California, San Diego, Department of History and Science Studies Program (1989–1995).

For their assistance, as well as for permission to use and quote from their archival collections, I would like to thank the following: Åbo Academy Library

(Turku, Finland), American Philosophical Society (Philadelphia), Churchill College Archives (Cambridge), Deutsches Museum (Munich), Library of Congress, Lund University Library, Mittag-Leffler Institute Library and Archive (Djursholm, Sweden), Niels Bohr Library of the American Institute of Physics (College Park, Maryland), Niels Bohr Institute Archives (Copenhagen), National Archives (Denmark), National Archives (Finland), National Archives (Sweden), Regional Archive in Lund, Royal Institute Library (London), Rockefeller Archive Center (North Tarrytown, New York), Royal Society Library (London), University of Oslo Library, and Uppsala University Library.

At W. H. Freeman and Company, a very friendly, competent, and professional staff helped ease the process of creating a book. I would especially like to thank my editor, Erika Goldman, for her efforts, patience, and wise counsel. Before she took over, conversations with my first editors, Elizabeth Knoll and Jonathan Cobb, helped me define my goals and launch the writing. In the final stage of transforming the manuscript into a printed book, project editor Mary Louise Byrd's firm hand expertly piloted *The Politics of Excellence* to completion, on schedule. I am most grateful.

Finally, a few persons must be singled out for special mention. Dr. Elisabeth Crawford and the late Professor Sten Lindroth first invited me to research in the Nobel archives and participate in the 1981 Nobel Symposium on "Science, Technology, and Society in the Age of Alfred Nobel." I had finished my doctorate at Johns Hopkins University a few years earlier and was a postdoctoral research fellow in Norway at the time. Their confidence in my abilities gave me courage to embark on this project. For several years, Dr. Crawford (CNRS, Paris) was my closest colleague; she worked on the history of the first fifteen years of awarding the science prizes, I began with the post-1915 period. We shared source materials, ideas, results, and gossip; we criticized and improved each other's work. Her book on the beginnings of the Nobel institution remains the most important monograph to date on the subject; it is a starting point for further studies. Intensive collaboration over several years, no matter how fruitful, often tires out. Although we have worked independently and with little contact for well over a decade, I would like to acknowledge her pivotal role in the early phases of my work and express my appreciation.

My friend George Hesselberg, journalist with the *Wisconsin State Journal,* deserves special thanks for reading an early draft of the manuscript and for pointing out where my prose drifted from the clear and elegant toward academic turgidity.

My adviser when I was a graduate student, Professor Robert H. Kargon (Johns Hopkins University), once again was able to guide me with much insight, wisdom, and knowledge. His comments on an early draft were of tremendous value; his kind words of encouragement were of greater value than he probably realizes.

A very special thanks is offered to Professor Roy MacLeod (University of Sydney) for his close reading of an earlier draft of the manuscript. His enthusiasm and suggestions helped significantly. He took time to edit many chapters and to provide a first line of defense for helping filter out Scandinavian contamination in my English. Whatever value this book has, it would have been much less without the generous help from these three friends.

Although many have helped in small and large ways, in the end, of course, the book remains the responsibility of the author. The opinions expressed in this book are my own; they do not necessarily reflect the position of any of the organizations that have supported my research or of the colleagues who have contributed. Translations from Danish, French, German, Norwegian, and Swedish are my own; when in doubt I secured second opinions from colleagues and friends. Where some ambiguity might still remain, I include the original quotation in an endnote.

Last and certainly not least, for the past few years I have not been able to give many colleagues, friends, and family members the attention they expect and deserve. I pray they can understand and forgive. Although no book is more important than these bonds, I confess to my weakness. Whether motivated by a sense of duty or by vanity and self-indulgence, I, like so many authors, allowed myself to be consumed by my work. To my wife, Trine, who had to share me with this book, I offer my deepest thanks for her support, patience, and love.

Voksenkollen, Norway
July 2001

# Introduction

## Legendary Excellence

*T*he Nobel medallion is etched with human frailties. Both those who select winners and those who receive the Nobel Prize are, well, mortal. And yet precisely because we live in a society awash with hype and commercialism, many people seek comfort in admiring institutions and people who embody nobler sentiments and higher purposes. The Nobel prizes, covering the fields of physics, chemistry, medicine, literature, and peace, are widely accepted as one such institution. Genius and excellence are terms that frequently attend these prizes, especially for those in the sciences. Exclusivity and a matchless level of accomplishment define their status. The lavish, almost mythical award ceremony in Stockholm is frequently seen as the pinnacle of taste—elegance, decorum, and refinement in an age of democratic vulgarity. Most people reject the slightest suggestion that the prize might fail to represent an unambiguous recognition of peerless brilliance. Cynics assume that such awards represent nothing more than calculated self-interest, perpetrated by conspiracy. Neither position is accurate, although resonance of both can be found in the first hundred years of Nobel prizes.

Can excellence involve politics? What an odd formulation. Surely excellence, especially in the natural sciences, must be clear-cut and recognizable. Subjective opinion, taste, and politics might influence the Nobel prizes for literature and peace, but is not science—robed in claims of neutrality and objectivity—sufficiently clear-cut for selection of the best to be beyond such things? When pushed, we might all concede that prizes, by definition, are political, are a form of governing marked as much by interests and intrigues as by insightful judgment. But in the case of the Nobel Prize we have largely been willing to share a fiction of legendary excellence, chosen as it were through a faultless process untouched by predilection or bias.

### THE PRIZE AND ITS HISTORY

Before turning to these questions and the substance of this book, let us examine briefly some facts and the perspectives that underlie this history.

First, a few nuts and bolts that will serve us throughout the book: Nominations for the prize can only come from specially designated individuals. These fall into two categories: those with a permanent right to nominate and those selected annually on an ad hoc basis. Those permanently entitled to nominate are Swedish and foreign members of the Royal Swedish Academy of Sciences, members of the Nobel Committees for Physics and Chemistry (who are not necessarily members of the Academy), previous winners of a Nobel Prize in physics or chemistry, and the professors of physics and chemistry at universities that existed in 1900 in Sweden and other Nordic countries (Denmark, Finland, and Norway). In addition, the committees each year can request that the Academy invite chair holders in physics and chemistry at six or more foreign universities, and an unspecified number of scientists individually, to submit nominations.

Nominations must be received by February 1 each year. The respective five-member committees evaluate the nominations, prepare a report discussing the relative merits of the candidates, and make a proposal to the Academy, usually by early autumn. The Academy's physicists and chemists then vote to approve the proposals or to make their own recommendations. Finally, the full Academy votes. In the first half century of giving out prizes, the individual committees voted in September and the Academy in mid-November; more recently, these dates have advanced by a month.

The prize in physiology/medicine operates basically the same way. The Caroline Institute of Medicine and its Nobel committee, also in Stockholm, are responsible for awarding this prize. Enough indications exist to suggest that those involved with giving out the prize in medicine faced many of the same enormous challenges as those entrusted with the prizes in physics and chemistry. Here, too, the composition of the committee certainly proved critical; judgment, predilection, and interests necessarily entered into its work. Unambiguous, impartial criteria for selecting among several deserving candidates were not at hand. The situation faced by the Caroline Institute in 1950, one characterized as extremely confused, testifies to the enormity of difficulties. It might serve emblematically as a reminder that all prize committees face difficult choices: After four rounds of preliminary voting, no decision could be reached. Three primary alternatives emerged, but going into the decisive meeting, the outcome was still uncertain. When urging a colleague to come to the meeting, one committee member compared the situation that year with recent dramatic events in the British Parliament: If somebody were to catch cold, a completely different outcome would be reached on the prize.

Swedish scientists have been decisive in selecting recipients—and in shaping the process. Their position on the Nobel committees has virtually guaranteed this course of events. An understanding of committee members' scientific orientation and priorities, as well as their professional training, contacts, and ambitions, is vital to comprehending how the awards are given. What specialties within physics and chemistry should receive consideration and what specific works should be rewarded have been influenced by what committee members have considered to be important.

Underpinning the entire enterprise and, of course, the starting point for this history is Alfred Nobel's testament. When made public in 1897, Nobel's last will was a puzzle. When the Swedish government approved bylaws in 1900, these legal guidelines for managing Nobel's fortune and operating the prize institution did not resolve all the original ambiguities; others, less apparent, became visible with time. Part of any serious history of awarding Nobel prizes must also include the history of the interpretation of the bylaws and reconciliation of differences of opinion.

What were committee members actually *doing* when they evaluated and assessed who and what achievements deserved a Nobel Prize? Could they strip themselves of all concerns and prejudices to sit in impartial judgment? The answers to these questions have ranged from "surely not" to "perhaps," and even to a qualified "yes." To appreciate their actions requires insight deeper than simply correlating committee members' scientific tastes and intellectual preferences with candidates' accomplishments. To gain a sense of their possible interests, motives, and biases—and why these could play a role in the process—it will be necessary to broaden the canvass. The world of science is not only one of ideas and personalities; it also entails a socially organized activity. Some perspective on science and on Sweden might provide a crude compass to help navigate the historical landscapes traversed in the book's narrative.

## SCIENCE AS AN INSTITUTION

Brilliant ideas, and a thirst to know, do matter in the advance of science. But science does not flourish simply because individuals wish to comprehend the mysteries of nature. As enticing and satisfying as the cult of individual genius might be for many people, science entails a community, or rather many communities. To admit this fact does not diminish the importance of individual accomplishment. First, rather than speaking of a single entity called "science," we need to bring into sharper focus the many subdivisions that mark bounded territories of pedagogy and research.

The history of the sciences during the past two hundred years reveals a complicated process by which broad realms of natural phenomena have been carved up into fields for specialization. From broad general areas of study related to animate and inanimate nature, there have arisen disciplines such as physics, chemistry, zoology, and geology. By 1900, many such disciplines were already divided into subdisciplines as the enormous amount of knowledge and the need for mastering sophisticated research techniques prompted narrower specialization. Sometimes communication among specialties provided new means to solve old problems; other times specialists locked themselves in their own small duchies claiming xenophobic self-sufficiency. The changing geography of scientific disciplines and their subdivisions did not simply spring up and grow "naturally"; the disciplines' boundaries, standards, and priorities arose from processes, often involving differences of opinion and conflict.

Moreover, a discipline implies the act of disciplining. A productive dynamic field of science cannot get on without means to organize resources, certify knowledge, facilitate communication, recruit new practitioners, and secure publics and markets. How is it decided what are the most important problems facing a field of science? Which research methods give trustworthy results? What types of accomplishment are most praiseworthy? These and many other similar questions that often seem predefined in the day-to-day work of most scientists nevertheless do not have fixed answers for all places and all times. They draw our attention to the fact that authority and power also are constituent to the advance of science. Competition for resources, for setting agendas, for making discoveries, and for reward is, of course, a prominent component of modern science.

During the past century, the researcher seeking to build a reputation, create schools of disciples, and influence the direction of scientific development has increasingly had to function in spirit, if not in action, like an entrepreneur. Using insight and instinct, she or he might define programs for research to satisfy both curiosity and strategic needs for gaining funding and prestige. Patrons are sought; scientific and societal constituencies are primed to make use of the research and hire disciples. Science in the twentieth century involved ever-larger institutions and absorbed ever-larger amounts of money. Modern science, like an army, advances on its belly. Sometimes intended, sometimes unintended, the Nobel Prize provided nourishment for the select few in the form of prestige, authority, and cash.

Juxtaposing the harsh world of cutthroat competition and even intrigues with the sublime image of the Nobel prizes may provoke disbelief, if not ire. Does the history of the prizes bring together these two extremes in scientific culture? Or, perhaps, we need to appreciate the manifold behavior, values, and attitudes within a scientific community at different times. Although scientists are compelled by methods of research to be rational, they are often driven by passions. To seek truth is not the only impulse in their daily work. As one committee member has noted, for better or for worse scientists are not always the most perfect of God's creatures; pure pettiness also enters into their judgments and decisions. In fact, arrogance and modesty, obstinacy and compromise, selfishness and magnanimity are all woven into the history of the prize. Just what has been at stake for those awarding prizes; what did they intend for the prizes? What was actually being celebrated through the prize?

The history of the prize is a history of using the prize. From the moment Alfred Nobel's testament became public, some of those who would likely have an influence on shaping its interpretation and implementation saw strategic opportunities for advancing a number of goals. The subsequent history reveals a succession of uses for the prize as those involved with nominating and evaluating gradually learned to exploit its value as a resource for cultural and scientific agendas. Increasingly, these purposes often served narrow careerist and professional interests. And yet, Nobel linked his prizes with " conferring the greatest benefit on mankind." What did he mean by that phrase, and how did others interpret it?

## VISITING FOREIGN PLACES: AN INTRODUCTION

A history of the awarding of Nobel prizes must take the reader through unfamiliar territory. Those who implemented Nobel's testament and those who gave out the prize were Swedes; to appreciate their actions and thoughts we need to understand their immediate surroundings. For most, Sweden, and especially the Swedish past, is relatively unknown. Although essential bits and pieces of broader history will be introduced as needed, a few brief, broad strokes at the start will help ease us into the narrative.

About 1900, Sweden, on the far northern periphery of Europe, was a relatively minor nation. It had only 5 million inhabitants, as compared with 40 million French or 35 million Britons, and its capital, Stockholm, facing east to the Baltic Sea, had a population of about 400,000. In the past, the Scandinavian country had been a major power. In the seventeenth and early eighteenth centuries, the Baltic was a Swedish sea. Led by warrior kings, Sweden conquered parts of what are now Germany, Poland, Baltic nations, and Russia. Finland belonged to Sweden, and it remained so even after King Charles XII's disastrous losses in the early 1700s. A century later, Sweden lost Finland to Russia, but it was allowed to force Norway into a union under the Swedish crown. Never fully accepted by most Norwegians, the marriage was all the more strained by a series of hard times in both nations. Massive emigration to America etched itself deeply into both nation's psyches. Finally, the union began to unravel during the last two decades of the century. Norwegians agitated for greater self-governing in internal and foreign affairs and, increasingly, for complete autonomy. Crises in the mid-1890s prompted fears that the union could be held together only through Swedish military intervention. Intense diplomatic efforts brought a peaceful end to the union in 1905.

Around 1900, other significant changes were buffeting Swedish society. In some respects, Sweden's hesitant entry into the modern age shared broad features with many other European nations. Agrarian and/or aristocratic ruling classes continued to dominate prior to World War I. Although experiencing a rapid and late industrialization, heavy industry constituted a relatively small sector of the economy. In most nations during the decades prior to the war, neither bourgeois democracy nor the increasingly disruptive labor movement had consolidated into dominating forces. For the traditional ruling classes and royal families, there was a real uncertainty whether the new century along with its new social order was to be rampant industrial capitalist or radical socialist, neither of which held much appeal. Tentative, shifting alliances and efforts to pacify and neutralize disruptive social groups kept national instability to a minimum. The period witnessed an almost explosive growth of national ceremonial traditions across Europe—ranging from royal pageantry to monuments, songs, and fetes—created to instill identity and loyalty to fatherland and social class. Sweden, as we shall see, was no different.

Of course, each nation developed and tackled problems of change differently. Beginning in the 1870s and accelerating by 1900, Swedish society was propelled toward a new era by a relatively late, but intensive, industrialization and urbanization. Characterized by the last full bloom of conservative, royalist nationalism, this period witnessed also the awakening of political and cultural liberalism. As in other European nations, wealthy leaders of industry and commerce gained admission into good society, on an individual basis as always; but as a class the urban bourgeoisie had to overcome the social stigma of parvenu and strove to increase its influence in politics. The accomplishments of this group were clearly driving Sweden into a new age, one that held promise of national rejuvenation. This period is frequently referred to as the Oscarian era, after King Oscar II, who reigned from 1872 to 1907: its cultures, concerns, and values were constitutive of how Nobel's testament was received and given form as a Swedish institution.

Sweden lagged behind many European nations in democratic reforms. In 1900, only 6 percent of the Swedish population had the right to vote. The so-called First Chamber of the Swedish Parliament (Riksdagen) was dominated by political conservatives from aristocratic and other families with long experience of royal state service. The Second Chamber, in principle more democratic, was dominated by prosperous landowning farmers who had received voting rights in the mid-nineteenth century. By the 1890s, a Liberal Party, based largely on urban interests, ranging from industrial and finance capitalists to professionals and intellectuals, demanded the vote for all men, free trade, and reduced military expenditures. Another political party, the Social Democrats, representing the rapidly growing numbers of industrial workers, aimed at more radical political change, including an explicitly antimonarchist solidarity with labor movements in other nations. Contemporary observers noted the accelerating pace of growth in urban, industrial-based society and also commented on the growing divisions in society.

Although the two parliamentary chambers were closed to all but a tiny segment of society, the rapidly expanding mass media of daily newspapers, weeklies, and other periodical literature provided a rich forum for national debate. Both the number of newspapers and the sales grew significantly during the 1880s and 1890s. The media helped launch the Nobel legacy and secured its success at home and abroad.

Politically, Sweden avoided entering into alliances with foreign powers, but culturally and intellectually at this time the country was tied to Germany. Schoolchildren learned German, and academics cultivated relations with German universities. Swedish academic culture was thoroughly Germanic. Academic departments (called institutes) were ruled autocratically by a single professor whose authority to define and control his discipline was indisputable. Both the established official state culture and newer literary and artistic movements thrived in a two-way traffic between the two nations. The opening in 1909 of the train ferry connecting Trelleborg in southern Sweden with Sassnitz in Prussia reduced the already short travel time between Stockholm and Berlin to but a day. Affinity entailed more than direct contact. In the growing European obsession over race, as various Darwinian-inspired theories during the late 1800s entered into the mental architecture of educated peoples, Swedes and Germans—Nordic and Teutonic folk—shared a common biological heritage.

## Institutional Anchorings

During the first half century of the Nobel prizes in physics and chemistry, three institutions played a prominent role: the Royal Swedish Academy of Sciences, which formally awarded the prizes and selected the committee members; Uppsala University, where most of the Academy's physicists and chemists were educated and where many of the committee members held professorships; and the Stockholm Högskola (later University), which was consciously established in opposition to Uppsala's traditions and where significant numbers of committee members held chairs. We will need to appreciate some specifics of the cultures and institutional peculiarities of these institutions. These are not merely background props on the stage of history but constituent of how the Nobel drama unfolded.

The Royal Swedish Academy of Sciences was founded in 1739 for the purpose of bettering Sweden. Its original six founders included men of science and politicians who sought to secure Sweden's future prosperity through practical science. In emulating the Royal Society in London, they obtained a royal charter that bestowed prestige but little funding. They chose as their emblem an elderly man planting a tree, above which the words "For Posterity" float as a creed of purpose. Although the Academy was at first devoted to practical Swedish concerns, spreading rational methods for farming and commerce, it included several internationally prominent researchers such as Anders Celsius, Carl Linnaeus, and Carl Scheele. After a period of relative decline, the Academy entered a new period of strength when in 1818 it elected as permanent secretary the strong-willed and internationally prominent chemist Jacob Berzelius. He intensified contacts with important foreign centers of science and reorganized the Academy's internal structure to reflect the emerging disciplinary classification of science at universities.

But the Academy, just like other national academies, differed from universities. The latter, with roots in medieval times, existed first and foremost for training civil servants: priests, lawyers, doctors, and teachers. Research only gradually entered into university duty during the 1800s and only in the 1900s was original research widely codified as an integral component of professorial obligations. The origins of the Royal Society, in 1660, and the Paris Academy of Sciences, in 1666, are complex, but they, and the Royal Swedish Academy of Sciences, aimed at serving the state through practical scientific research and bringing renown to king and country through the achievements of their members.

At the start of the Nobel enterprise, the practical and the descriptive sciences still held sway in the Academy. Its one hundred honorific members were elected into nine sections. Physics was allowed six members; zoology and botany, sixteen. During the second half of the nineteenth century, the Academy's greatest accomplishments were linked to service to the nation and heroic accomplishment. The Academy, for example, was responsible for organizing Sweden's prestigious polar research, including Adolf Erik Nordenskölds voyage through the Northeast Passage (1878–1880), which brought international glory to Sweden. Neither physics nor chemistry held particular positions of prestige; the Academy's Museum of Natural History was the most venerable part of the Academy's activities. The Academy's prominence also derived from its many services, which included introducing a uniform system of weights and measures, standardizing time and time signals, and organizing a weather bureau.

In a society much structured by hierarchies, deeply concerned with medals, orders, and other paraphernalia of snobbery, membership in the Academy brought scientific and social reward. A sense of clubbiness characterized the Academy. Friendships and rivalries often led to a divided institution, but authority provided a powerful glue. Formally constituted authority, whether in the form of a professorship, election to high Academy office, or seniority, generally determined whose voice was heard. One young scientist complained early in his career about the Academy's overreliance on established authority, titles, and positions rather than on expertise won and renewed through active research. As could be expected, he was more than happy to exercise his own authority heavy-handedly once he secured a high position. The Academy recruited members from academic institutions having very different identities; sometimes the Academy's own culture could not subdue conflicts and antagonisms.

Uppsala University, founded in 1477, was Sweden's major institution of higher learning. Along with Uppsala Cathedral, the university dominated the life of this small town, an hour's train ride north of Stockholm. Then, as now, the distance between the two with respect to style and tone seemed far greater. As in provincial German university towns, students and professors created their own society, focusing inward on the life of learning, with its own sense of privilege and superiority. Uppsala's faculties of law, medicine, and theology trained civil servants. The philosophical faculty (school of arts and sciences) functioned primarily to buttress professional training in the other faculties, as well as to train secondary school teachers. Some professors did research along with their obligations to teaching and examining. But, as in most European and American universities, pedagogy took precedence. Beginning around 1880, many professors desired to have research considered part of their normal workload and to obtain assistance to do research. Progress toward these goals came slowly.

Uppsala professors tended to consider themselves national authorities in their given disciplines. In an academic culture often characterized as more Germanic than the Germans, there were few science professors; departments were ruled autocratically by the senior professor. Uppsala professors tended to define themselves at the top of the academic social hierarchy. One foreign scholar, who had opportunity to observe Swedish academic culture, noted that Uppsala resembled comparable provincial European universities in more ways than one: "Life is as good as stagnant in them.... The professors lead an isolated almost medieval existence that contributes little to the development of new and fruitful ideas. Thereby a kind of camaraderie necessarily arises ... the one professor dragging along the next one with him." One chemist compared Uppsala professors with popes: demanding and expecting respect, but afraid to move beyond the narrow boundaries of their competence for fear of appearing fallible. Over the entrance to the university's main auditorium a tenet of faith was carved in 1887: "To think free is great, to think right is greater." Perhaps this sentiment was appropriate for a university concerned with training professional civil servants to be loyal to king, state, and church, but it also symbolized the attitude of many Uppsala professors. A tendency to "right" thinking characterized many Nobel committee members from Uppsala, and in policing physics and chemistry their attitudes directly affected the Nobel prizes.

The Stockholm Högskola, which in more recent times has become Stockholm University, was from the start a very different type of institution. Privately financed and established in 1878, the Högskola represented the efforts of the capital's commercial and cultural leaders to create an alternative to the considerably older universities of Uppsala and Lund. To their critics, these institutions, especially Uppsala, were mired in regulations and traditions, concerned with producing professional civil servants rather than responding to the cultural and practical concerns of the growing urban middle classes. Many looked with dismay at Uppsala's conservative political, theological, and intellectual atmosphere. All too often, students had no greater ambition than to eke their way through regimented curricula, pass their state-regulated examinations, and enter civil service. Along with colorful academic traditions, student life was tainted by alcohol and rowdiness.

The Högskola was modeled after the Collège de France and the Royal Institution in London, where intellectual curiosity and not the chase after degrees would set the tone. The school rejected Latin in official university titles and eschewed cumbersome age-old academic traditions. Professors would be free to choose their lecture subjects, research would be given greater prominence, and contact with the city's commercial and cultural elites would be encouraged. Progressive social groups considered natural science—its methods, findings, and spirit—to be a weapon against theological and social-political reactionary dogmas; it would be an ally in the modernization of society. Many of the first appointments went to scientists. Unfortunately, internal conflicts, which were widely publicized in the media, blossomed by the 1890s; the dissent, in turn, discouraged patrons. In 1900, the tiny and impoverished Högskola had no building of its own. It rented office space and dreamed of expanding its small staff and minimal laboratory facilities.

International orientation and freedom from tradition were deeply ingrained in the Högskola from the start. Mathematician Gösta Mittag-Leffler, one of the founding professors, and more socially conservative than many of his colleagues, recruited from abroad the highly talented mathematician Sonja Kowalevski, who thereby became the first woman professor in Europe. By contrast, the Royal Swedish Academy of Sciences refused to accept her as a member, claiming that its bylaws specified only "men" as members. Similarly, the Högskola recruited a number of dynamic young Norwegian scientists, which would have been unthinkable at Uppsala. Without state examinations and traditions dictating disciplinary orientation, the new institution fostered intellectual freedom that, in turn, enabled the introduction of new scientific specialties and innovative research methods. Still, personal animosities and internal conflict became part of the Högskola's life in its first several decades.

Swedish scientific life at the turn of the century entailed conflict. A rivalry existed between professors at Uppsala and those at Stockholm. This competition did not necessarily have to produce friction. Although Swedish culture today is often characterized as one that strives toward peaceful consensus among differing groups, whether true or not, in the past, pettiness and hostility among academics certainly stymied collective action. Fault lines of principled and petty difference crisscrossed the Swedish academic landscape, capable of generating tremors of conflict—and even, on occasion, a touch of havoc. The Academy—and, more significantly, the Nobel committees—sat astride this complex seismic zone of personal and intellectual differences.

# Part I

PERMANENT BATTLES WILL SURELY
BE WAGED FOR EVERY PRIZE

Launching the Nobel Enterprise,
1897–1914

# ONE

## The Stupidest Use of a Bequest That I Can Imagine!
### Securing Alfred Nobel's Vision

*W*hen Alfred Bernhard Nobel, the fabulously wealthy inventor of dynamite, died on December 10, 1896, he left behind a riddle as perplexing as his own secretive life. From the time he invented a safe means for using nitroglycerine, and subsequently built an immense industrial empire based on this and other inventions, he captivated the growing European mass media. A loner and a brooder, he puzzled his contemporaries and those who later attempted to understand him. Described as "a retiring, considerate person, who detested all forms of publicity," he never quite succeeded in attaining the privacy he sought or in transforming his financial exploits into personal gratification. Even before his death, journalists and others speculated about the fate of his vast fortune. Early in 1897, a Swedish newspaper revealed the contents of Nobel's testament.

### NOBEL'S LEGACY

Short and succinct, Nobel's testament evoked a collective gasp. Nobody quite knew what to make of it. Soon came strong reactions. Confusion and determination, self-interest and selfless idealism determined the fate of Nobel's vision. The institution that emerged around the Nobel prizes carries the mark of its difficult birth, shaped as much by its implementation as by the famous testament. From the start, interested parties attempted to recruit Nobel's legacy for parochial interests, beginning a history of tension between declared exalted ideals and unstated private agendas. A minidrama in its own right, the story of saving Nobel's testament provides a fitting prologue to the ensuing history of the prize.

The pertinent paragraphs of this testament are worth reading; these are the starting point for this history and ultimately the end point for judgment.

> The whole of my remaining realizable estate shall be dealt with in the following way:
>
> The capital shall be invested by my executors in safe securities and shall constitute a fund, the interest on which shall be annually distributed in the form of prizes to those who, during the preceding year, shall have conferred the greatest benefit on mankind. The said interest shall be divided into five equal parts, which shall be apportioned as follows: one part to the person who shall

have made the most important discovery or invention within the field of physics; one part to the person who shall have made the most important chemical discovery or improvement; one part to the person who shall have made the most important discovery within the domain of physiology or medicine; one part to the person who shall have produced in the field of literature the most outstanding work of an idealistic tendency; and one part to the person who shall have done the most or the best work for fraternity among nations, for the abolition or reduction of standing armies and for the holding and promotion of peace congresses.

The prizes for physics and chemistry shall be awarded by the Swedish Academy of Sciences; that for physiological or medical works by the Caroline Institute in Stockholm; that for literature by the Academy in Stockholm; and that for champions of peace by a committee of five persons to be elected by the Norwegian Storting [Parliament]. It is my express wish that in awarding the prizes no consideration whatever shall be given to the nationality of the candidates, so that the most worthy shall receive the prize, whether he be a Scandinavian or not.

Did Nobel harbor a subtle sense of humor? The testament posed massive legal and interpretive problems. Virtually every sentence bewildered. How, even if able to stand on its own, might this skeleton be transformed into a living being? First, none of the institutions named to distribute prizes had been consulted. Why take on an assignment fraught with obvious difficulties? Second, powerful members of the Nobel family, living in Sweden, Russia, and Germany, could scarcely believe the unfavorable terms of the will, which left them with but a token inheritance. Third, it was not clear which country actually had jurisdiction in Nobel's will. Nobel might have been Swedish by birth, but he had held no formal Swedish citizenship since leaving the country as a child. He lived and controlled his extensive empire largely from homes in Paris and San Remo, on the Italian Riviera. The testament was written and signed in Paris in November 1895. Still, Nobel chose to draft it in Swedish and to have it witnessed by Swedes. Some legal systems could discount the entire testament based on its formal inadequacies. Perhaps the biggest surprise of all, and the wisest of Nobel's actions, was his choice of the two persons he named as executors. They also had not been forewarned.

## Implementing Nobel's Testament

The task of breathing life into the testament fell primarily to a young engineer, Ragnar Sohlman, Nobel's personal assistant since 1893. The other executor, Rudolf Lilljeqvist, was also an engineer, but fifteen years' senior to Sohlman, who had no prior personal association with Nobel. Lilljeqvist was just establishing a Nobel-supported electrochemical plant in a remote area of Sweden, so Sohlman took up the primary challenge. Sohlman had worked with Nobel for years. He was intensely loyal, although he admitted that he never quite understood his employer's inner world of feelings. Still, his devotion, energy, and intelligence proved essential.

The two executors recruited Carl Lindhagen as legal adviser for the Nobel estate. Lindhagen was beginning a highly successful political and legal career. A broad-minded, at times radical member of the Swedish Liberal Party, Lindhagen was just elected a member of Parliament and served as deputy justice in a regional court of appeals. Still, he found time to act as executor and legal council. Equally important as his judicial and political acumen was his familiarity with Stockholm science. His father helped found the Högskola; he replaced his father as its administrative secretary for a decade. This connection turned out to be fortuitous, as the executors' ability to save the testament depended in part on their success in wooing scientists with promises of prestige and institutional enrichment.

With patience, perseverance, and a bit of luck, the trio began in early 1897 to pilot the testament through legal challenges and conflicting claims. Before considering how to implement Nobel's wishes, they first had to obtain legal recognition for the testament, to defuse family plans for judicial countermeasures, and to convince the relevant institutions to accept their assignments. Failure to achieve any one task would mean the testament had little chance.

Sohlman quickly sought legal recognition of the will in Sweden. Moreover, before foreign courts could intervene on behalf of the family, and before Swedish courts pronounced the document to be valid, Sohlman began liquidating Nobel's assets abroad and brought the funds to Sweden. His early clandestine gamble proved successful. In May 1897, the Swedish attorney general declared that the resolution of Nobel's estate was a matter of national interest. Although the state would not play a direct role in negotiations, it would accept judicial responsibility for the statutes; these would have to be submitted and approved by the king-in-council.

Nobel's family certainly had reason to demur. In an earlier testament, which Nobel had declared void, the family was to have received almost three times as much inheritance as they did in the final will. Moreover, in contrast to the earlier will, this one excluded the Nobel family from any role as executor. Of course the best course for the embittered Nobel relatives would have been to have the testament invalidated. They could also have encouraged the institutions named to award prizes to simply decline the honor. Much to the executors' dismay, and the family's delight, the Swedish institutions expressed less than enthusiasm for Nobel's plans.

### The Peril of "Humbug"

News of Nobel's testament left members of the Royal Swedish Academy of Sciences perplexed and dubious. Among those who would be directly involved was Oskar Widman, an organic chemist at Uppsala University and member of the Academy's Chemistry Section. Reading preliminary newspaper accounts, Widman tried to sort out his puzzlement in a letter to a Finnish colleague. Widman expressed concern that to distribute prizes based on such conditions would lead to considerable "humbug." With so much money involved and so little precedence for a truly international prize, he feared for posturing, artifice, and favoritism. How could works during "the preceding year" be evaluated for their significance? Truly innovative work required time to be appreciated. Would not the lure of so much money, he agonized, induce supporters to exaggerate the significance or validity of proposed works? Widman despaired at defining works that "have conferred the greatest benefit on mankind." That phrase, he noted, could scarcely hold up juridically. His immediate inclination, and that of Academy members with whom he spoke, was to decline the task.

Other Swedish scientists echoed Widman's sentiments. The young, ambitious, and internationally prominent physical chemist Svante Arrhenius feared "permanent battles will surely be waged for every prize distribution." His colleague at the Stockholm Högskola, chemist and oceanographer Otto Pettersson, questioned the entire project. The idea of using Nobel's legacy for establishing a competition among scientific works for prizes was "the stupidest use of a bequest that I can imagine! To seek reward for their work is not attractive for scientists." But, most important, everyone appreciated that the vague testament did not bother with legal and institutional niceties and offered little clue as to how possible internally conflicting features of Nobel's text might be reconciled. How were candidates to be selected; how were they to be evaluated? What did Alfred Nobel actually mean by the odd formulations? Amid Nobel's private papers, perhaps, some clues could be found casting light on the benefactor's intentions; if not, the institutions named to award prizes would find themselves "in a great mess." But no "Rosebud" would be found within the Nobelian "Xanadu," neither in Paris nor in San Remo.

When the executors approached the Royal Swedish Academy of Sciences, their appeal foundered. The Academy declined to cooperate. Only after the Nobel family and the Swedish courts accepted the testament could the question of naming formal representatives arise. Sohlman contemplated direct negotiations with family representatives. But Lilljeqvist intervened, and he urged Sohlman and Lindhagen to remain firm. Gradually, they gained an upper hand. Court rulings and media support buoyed their spirits. By 1898, they had decided how to proceed.

## TO HELP DREAMERS OR SCHEMERS?

Even though the Academy of Sciences voted not to negotiate formally with the executors, Lindhagen and Sohlman nevertheless assembled an informal committee to consider how Nobel's vision might be realized. The two other Swedish institutions named to distribute prizes revealed the winning strategy. Those who saw personal advantage were willing to cooperate.

In the Swedish Academy, a faction led by its president, historian Hans Forssell, rejected the task of awarding a literary prize. They doubted whether the eighteen members of the Academy were capable of appraising literature from around the world or even from all the European nations. Tastes and styles differed immensely; the world of culture was in a period of great ferment. Naysayers feared the Academy's members would be exposed to "unpleasantness, pressure and slander." Moreover, serving as an international tribunal might limit the Academy's ability to perform its national obligations.

Not so, thundered the person who had most to gain by accepting Nobel's charge, Carl David af Wirsén, the Academy's powerful and strong-willed permanent secretary. Af Wirsén had long been a self-appointed defender of conservative state-sanctioned culture against the corroding effects of social realism and other budding modernist heresies. Whether as a poet in his own right or as a champion of king, church, and nation, af Wirsén vigorously defended the established order of things. To keep the Academy a bastion of idealistic neoclassical culture, he would promote traditional values of truth, beauty, and goodness. This was also a chance to oppose such popular radical writers as August Strindberg and Henrik Ibsen. Sensing an opportunity to increase his authority, he asserted that if the challenge of awarding prizes was rejected, the Academy would be criticized for turning its back on a great responsibility, one that could propel the Academy to new heights of influence in world literature. Af Wirsén managed to rally members to approve preliminary talks with the executors.

Initial responses from the Caroline Institute also revealed that the testament might float as long as institutional dividends were clearly recognized. Spokesmen for the institute were disposed to work with the executors, but they insisted on maintaining a degree of independence. They wanted to interpret the testament in ways to satisfy their needs, which had more to do with invigorating local research than with distributing international prizes. For starters, a well-endowed chancellery and library for each prize-awarding institution would be necessary. Beyond that, the Caroline Institute wanted latitude and autonomy in allocating its Nobel funds. Without this, Rector Axel Key made clear to Sohlman, he was willing, in principle, to settle privately with the Nobel family in an arrangement enriching both family and Stockholm's research institutions at the expense of the prizes. Key then agreed to provide formal representation once negotiations on the bylaws were begun.

## Aiding Stockholm Science

In part, the desire for compensation was part of tradition. In learned academies, members who served on committees for distributing prizes frequently received some token remuneration. But it seemed hard to escape the thought that an endowment as splendid as Alfred Nobel's must provide fitting recompense for those involved. One distinguished member of the Royal Swedish Academy of Sciences, Adolf Erik Nordensköld, remarked privately that each participant of the juries should receive reimbursement equal to a professor's annual salary. Although hefty compensation for participating in the selection of winners certainly hummed in the fantasies of many Swedish scientists, dreams and schemes for transforming Nobel's testament into a vehicle for advancing local science were set in motion by other impulses.

Rumors had circulated in Sweden that Nobel intended to enrich Stockholm-based institutions. The alleged provisions of an earlier testament raised expectations of a massive benefaction. Before the will was made public, Högskola member Svante Arrhenius had already anticipated "very large sums [of money] for scientific purposes." Nobel had earlier proposed donating 5 percent of his estate each to the Caroline Institute, Stockholm Hospital, and the Stockholm Högskola, as well as 1 percent to the Austrian Peace League. After the family was to receive a relatively generous allocation, all remaining funds were to be entrusted to the Royal Swedish Academy of Sciences for annual prizes. The 5 percent allotments to each of the Stockholm institutions would have created fabulous endowments.

But Nobel changed his mind. Some believed that Nobel removed the beneficence as a reaction against the highly visible and contentious "Feud in the Högskola." Possibly Nobel did not include a prize for mathematics because he understood that Gösta Mittag-Leffler, the Högskola and Academy's leading representative for this field, instigated many of the intrigues and spread malicious gossip in the media. For members of the Högskola, the lure of joining the executors to unpack Nobel's testament entailed the lure of funneling money into local research.

Lindhagen certainly had no problem with this. Sohlman, like Lindhagen, was a Stockholm patriot and able to appreciate the needs of civic culture. Together with

Arrhenius, Pettersson, and other Stockholm scientists, they devised a plan to make Stockholm a leading international center for science. They envisaged a massive Nobel-funded research institute, with model laboratories for physical, chemical, and biomedical sciences. New libraries and reading rooms would be the beneficence of all the prize juries. Moreover the grand research institute would play a critical role in realizing Nobel's prizes.

Nobel gave no indication how prospective winners were to be proposed, judged, or chosen. By bringing together local and foreign scholars, able to keep up with the frontiers of research, the institute presumably could advise the Academies and the Caroline Institute. Preliminary discussions considered the possibility of having twelve full-time researchers at the professorial level as well as invited prominent foreign scientists. In contrast, university laboratories at the time were scarcely designed for research—pedagogy being paramount—and rarely staffed by more than a professor or two, an assistant, and a technician. Budgets for research were either nonexistent or woefully inadequate. Neither universities nor governments considered the regular support of research as their responsibility. In other countries, plans for new large-scale research institutes, funded generally by private patrons, were underway. Still, the Stockholm plan aimed at creating one of the world's major scientific institutions, if not the grandest of them all—and it would carry Nobel's name.

In this plan, the prizes would be retained, but their size would be reduced from the fantastic to the prudent. Once much of Nobel's estate was used to create and maintain the Nobel Institute, remaining funds could generate prizes on the level of generous research grants. Scientists who were awarded prizes would remain at the Nobel Institute to bring their work to fruition. Nobel did not want prizes given to persons already well established and finished with their main investigations, but to "help dreamers, who find it difficult to get on in life. Dreamers such as possess the gift of poetry, but are unknown to the many, or are misunderstood by them, meditative young research workers who are on the very threshold of a great discovery in physics, chemistry, or medicine, but lack the means to achieve it."

This plan to enrich Stockholm science while retaining the essence of Nobel's testament could conceivably have succeeded. The scope of science was then not as daunting as it was later to become; laboratory experimentation was still at a modest scale, requiring relatively affordable equipment and technical assistance. The plan, moreover, would pacify critics who believed that Nobel's fortune ought to benefit Swedish culture rather than be dissipated on foreigners. And most important, a Nobel Institute would be a major boon for Stockholm. Creating senior research positions would solve at a stroke economic difficulties that hindered local institutions of higher learning and research. Open doors among Stockholm institutions would create an interlocking network. Professors savored the prospect. After his initial skepticism, Arrhenius began rehearsing a life of research leisure, freed from teaching and administration as a leader of a laboratory at the proposed Nobel Institute.

During the winter of 1898, Sohlman achieved a decisive breakthrough. Timing proved crucial. Some members of the Nobel family signed a legal complaint against the will, demanding the right to manage Alfred's estate as well as to negotiate directly with the relevant institutions for awarding prizes. Sohlman invited Emmanuel Nobel, head of the Russian branch of the family, to join his meetings with scientists.

From the start Emmanuel had expressed sympathy, in principle, for Uncle Alfred's vision. But having managed the vast Nobel oil-based interests in Baku, he feared loss of control over Nobel Brothers Naptha Company. He balked at openly supporting the executors; Sohlman offered a compromise. Emmanuel Nobel embraced a proposal that allowed him to buy stock options at a favorable rate and, in return, endorsed the idea of a grand research installation, perched on a prominent hill at the northern edge of Stockholm, carrying Alfred's name. Subsequently, Emmanuel declared his intention to work with the executors. He discouraged the other branches of the Nobel family from contesting the will; at the same time, he encouraged Sohlman to remember his obligation to Alfred Nobel: to act according to the Russian expression, *Dushe Prikashshik*—as a spokesman for the soul. As a result, at last the family began negotiating with the executors. By the time the civil suit reached the court, the family had already agreed to accept a significant increase in its share of the estate and to be represented, through Emmanuel, in the proceedings. The case was closed. It seemed, just for a while, that Nobel's dream, in a version acceptable to all parties, was within reach.

Self-interest, once given expression, is rarely contained. The executors envisaged a single set of statutes for all the institutions entrusted with awarding prizes and a single massive Nobel Institute. But the Caroline Institute demanded autonomy, insisting on its own Nobel research institute, and finally the medical representatives rejected the plan. The medical representatives also called for separate statutes for each prize-awarding institution. Consequently, the executors were stopped in their tracks.

## PULLING THE PIECES TOGETHER

Along with representatives from each institution, the executors were forced to return to the original vision of grandiose prizes as the surest means of honoring Nobel's intentions. From the start of negotiations, the executors had seen the need to begin asking questions that the testament had left unresolved. How would candidates for prizes be selected? Who should propose candidates? Who should evaluate the candidates' work? What did Nobel mean by *physics* and *chemistry* or by *physiology* and *medicine?* Might these terms be better defined in the bylaws? And what had Nobel intended by the "greatest benefit on mankind"? No unique set of bylaws and procedures followed inevitably from the testament; local circumstances and personalities would shape the outcome.

As early as the winter of 1897, Sohlman and Lindhagen had begun sketching answers for the negotiators. Some of their proposals received immediate acceptance. To be considered for a prize, possible candidates must have relevant work published; two equally deserving separate works could share a prize. Prizes would be awarded only to living persons, unless a candidate's work was nominated prior to his death. The irksome question of what standard should be used was dispatched by the phrase, "to be of such outstanding importance as is manifestly intended by the will." Other issues required several rounds of negotiations.

### The Nominators and the Evaluators

Who should be entitled to propose works to be considered for prizes and who should evaluate them? Early in the negotiations, Sohlman and Lindhagen proposed that one jury, composed of Swedish and foreign scientists, should both propose and evaluate candidates. They had in mind scientists employed at the proposed Nobel Institute, but once that plan had been abandoned, they considered alternatives. To forestall self-nominations and to try to control who should participate, they suggested certain fixed categories of scientists be given the right to nominate candidates. For the physics and chemistry prizes, the categories revealed a Swedish orientation, contrary to Nobel's insistence on international openness. These categories included all members of the Royal Swedish Academy of Sciences, regardless of disciplinary section and regardless of nationality, whether Swedish or foreign, and professors of science at the Swedish universities and colleges in Lund, Stockholm, and Uppsala. In this narrowly defined group, recipients of Nobel prizes were also included.

The executors and representatives from the prize-awarding institutions had no precedent for a truly international prize. Cultures and regulations common to Swedish academic life inevitably influenced their thinking. Swedish procedure for appointing a professor required that a university-named commission recruit three or four so-called competent experts [*sakkunniga*], mainly from other universities, preferably foreign, to submit detailed evaluations of the candidates. In principle, outside experts stood apart from local intrigues and parochial interests; in reality, the system was open to charges of manipulation and humbug. One of the negotiators for the Royal Swedish Academy of Sciences, its own professor of physics Bernhard Hasselberg, proposed such a comparable mechanism for the prizes. Committees would invite high-ranking foreign scientists to identify and propose worthy candidates. But could there be objective and impartial foreigners?

A few years earlier, Mittag-Leffler and Hasselberg attempted to find foreign experts to block the appointment of Arrhenius as professor of physics at the Högskola. Mittag-Leffler canvassed several European scientists hoping to find testimony against Arrhenius and for an alternative candidate. Although that campaign failed, Arrhenius appreciated how the choice of foreign experts could easily degenerate into the legitimization of entrenched interests and scientific orientations. Not surprisingly, he reacted angrily to Hasselberg's proposal. He warned that it would sustain domination of the "the old guard who have neither the time nor the inclination to be concerned with recent developments." Wouldn't that exclude new ideas, precisely the kind that Nobel wished to assist? Other negotiators feared the limitations of local scientists and proposed a considerable expansion of foreign participation.

In the end, the negotiators agreed to enlarge the number of categories to include professors of relevant fields not only at Swedish universities but at all Nordic universities and colleges in existence in 1900. In addition, each committee could invite, on an ad hoc basis, professors in the relevant fields of science from six or more universities around the world. During the first decades this would generally lead to one German, French, Italian, British, and American university each year. In addition at least one university was invited on a rotating basis from other nations, including Russia, Spain, Poland, and Japan. Furthermore, the negotiators agreed that committees could invite any number of individuals known for their broad knowledge and impartiality. Thus, scientists not connected with invited universities could also be included, such as those working at national research laboratories or less prominent academic institutions.

The *sakkunniga* method of selecting professors seems also to have influenced the prize procedure. The prize-awarding institutions would each name a committee of three to five members. This committee would evaluate the candidates proposed by nominators, which would then be the basis for the award-granting institution to decide. As an expert panel, the committee need not be members of the Academy, or even Swedes. The actual authority of committees was left undefined; they, like the academic panels, were to provide recommendations. But just how much weight would a committee's evaluation carry? Would the Royal Swedish Academy of Sciences open the committee to highly capable foreigners? How should committee members themselves react to the nominations? Might someone such as Hasselberg, who subsequently served twenty years on the Nobel committee for physics, tend to respect nominators outside his own circle of "high-ranking" acquaintances?

## Gains for Swedish Research

Finally, the executors and the negotiators from the prize-awarding institutions also soothed the craving to enhance local science. Although a single, monumental Nobel Institute was not to be, they still wanted Nobel funds to improve research facilities. A sizable chunk of the Nobel estate was set aside to provide a fund for each committee to erect its own institute. Although not as yet sufficient for the purpose, the money was expected to grow through interest on the funds so that institutes could be in reach within a decade or two. The formal rationale for these institutes was to assist the respective committees in assessing the worthiness of works proposed for prizes. The organization and scope of the institutes remained open; over the next three decades their resolution proved contentious and even influenced decisions related to the prize.

Additional provisions for directing funds to local research entered the statutes. Negotiators tackled two issues. First, they did not know what to expect. Would they be flooded with potential candidates, or would the pickings be slim? They agreed that although a formal search should occur annually, the prize would be awarded at least once every five years. But what if no qualified candidate were found in a given year?

One suggestion called for the money from a withheld prize to be put into a fund connected with the relevant committee; interest from the fund could provide grants for purposes related to the aim of the prize institution. For the science committees, the main purpose would be funding for research. But recognizing the potential for misuse of this option, the negotiators agreed to include safeguards. If nobody was found worthy, the prize could be reserved until the following year. If, first at that time, there was still no qualified candidate, only then could the prize be withheld and the money put into the committee's "special fund." Recall that in an age before research councils, the prospect of even a modest fund for research proved alluring. But, just as in the case in establishing Nobel institutes, unpleasantness and accusations of foul play later arose in connection with implementing this bylaw.

Finally, the negotiators considered paying those involved with evaluating candidates. In the initial plan for a single, well-endowed Nobel Institute, full-time research staff, on professorial salary, was to judge. Now, the negotiators conceived of committees composed of academics holding full-time positions at established institutions. Compensation for the three to five members on each committee was to be an annual stipend equal to one-third of a professor's salary. The head of the respective Nobel Institute would receive a full professorial salary, but he would not be allowed to hold other jobs and would be an automatic member of the committee.

Even these comparatively modest proposals for enlisting Nobel's legacy elated Stockholm scientists. Arrhenius bubbled over with delight over the prospects for himself and his local milieu. He and colleagues at the Högskola saw Stockholm on the verge of becoming an international center for research. Physicists whom the Stockholm Högskola tried to recruit were told privately that they could expect to be elected to the committee; thereby they would receive a handsome additional salary and, more importantly, a share of the prestige and resources available to committee members. The prospect of heading a Nobel Institute even encouraged one Norwegian nationalist to consider jumping ship to Sweden. The combined promise of internationally prominent laboratories, as well as fabulously endowed prizes, raised hopes that Swedish scientists would be received abroad with greater prestige and authority.

In 1899, the Swedish negotiators and prize-awarding institutions finally agreed on a set of bylaws. Some issues were left unresolved. For example, definitions of physics, chemistry, and physiology/medicine failed when the Caroline Institute again demanded the right to decide such questions internally. In response, the representatives from the Royal Swedish Academy of Sciences withdrew their proposals for defining the scope of physics and chemistry. They agreed informally upon a broad definition, which later proved ambiguous and nonbinding. The executors did manage to forge agreement on some common features, and they allowed the different prize-awarding institutions to propose supplemental bylaws so long as these did not conflict with the shared provisions.

Sohlman, Lindhagen, and Lilljeqvist submitted the bylaws to the king-in-council, which gave its approval on June 29, 1900. Sohlman, who had spearheaded negotiations from the start, achieved results seemingly at first beyond his grasp. His enormous efforts prompted more than one person to wonder whether the prizes should be called Nobel-Sohlman prizes. Without the young engineer's unwavering efforts to be the "spokesman for Nobel's soul," the testament would never have been transformed into an institution, based on legal regulations and supported by an initial fund of 30 million Swedish crowns. Sohlman helped create the Nobel Foundation, which became the legal trustee for Nobel's wealth. After the negotiations, Sohlman eventually assumed the position of the Nobel Foundation's executive director. As the foundation plays no role in choosing winners (it manages the estate and organizes ceremonies), Sohlman played no overt part in the subsequent history of awarding prizes.

## THE LEGACY OF NEGOTIATION

If the creation of the bylaws proved trying, implementing them was no less so. Conflicting interests influenced the negotiations, yielding ambiguities and internal inconsistencies. This built-in plasticity was not necessarily harmful. In lending themselves to some degree of interpretive maneuverability, the bylaws allowed committees flexibility for renewing the prize institution to meet unanticipated challenges—but only to a point. The bylaws failed to set clear, objective standards by which a committee could propose candidates or for what kind of achievement. Interpreting the bylaws yielded customs and traditions shaped by key personalities and their concerns.

Before turning to the start of the prizes, it will be useful to review some of the bylaws that provide basic coordinates for the subsequent history.

To be eligible for consideration, a candidate must be proposed by a nominator from one of six categories. Those with a permanent right to submit candidates annually are as follows:

1. All members of the Royal Swedish Academy of Sciences, both Swedish and foreign members and regardless of their scientific disciplines
2. Members of the Nobel Committees for Physics and Chemistry (membership in the Academy was not necessary)
3. Prior winners of the Nobel Prize in physics and chemistry
4. Permanent and acting professors in physics and chemistry at Swedish and other Nordic universities and technical colleges existing in 1900 (universities in Uppsala, Lund, Oslo, Copenhagen, and Helsinki; the Caroline Institute; the Royal Institute of Technology in Stockholm; and the Stockholm Högskola)

Those nominators invited on an ad hoc basis each year included:

5. Holders of professorial chairs in physics and chemistry in at least six universities, or comparable institutions of higher learning, to ensure an appropriate representation over different countries and their seats of learning
6. Other scientists from whom the Academy of Sciences might see fit to invite proposals

Nominators did not quite know how to play the role set out for them. Were they expected to propose candidates from their own countries or scientific specialties? How should they interpret the odd bylaws that arrived with the invitation to nominate? Some turned to committee members for advice, and still others worked together with committee members to advance particular candidates. With time, many took clues from the Academy's past decisions. During the first years, approximately three hundred invitations to nominate were sent out annually for each prize. Given that most Swedish and foreign members of the Academy had little contact with either physics or chemistry and therefore only rarely nominated, the mere twenty to "vote" in chemistry and twenty-nine in physics the first year would imply a degree of uncertainty as to what was to be expected. Committee members themselves frequently proposed nominators, so the actual number of outside voices heard was modest. The total number tended to vary year to year, largely ranging between twenty and thirty during the first fifteen years.

The committees consist of as many as five and as few as three members. One member is always the head of the relevant Nobel Institute, and the other members are elected by the Academy, based on proposals from its Physics and Chemistry Sections. Terms of membership are normally four years, reelection being permissible and normally granted for those who choose to return. Denial of reelection entails a major loss of face. The Academy elects for one year at a time a member of each committee to serve as its chairman.

Each September, the committees send out invitations to eligible persons and institutions to make proposals for the following year's prizes. These nominations must be received by February 1. The committees then evaluate the nominations during the spring and summer. In September they meet to discuss the report and work out, if possible, disagreements concerning their opinions and proposals. Reports submitted to the Academy no later than the end of September present assessment of the candidates' merits and recommend action with respect to the prize. Not later than the end of October the respective Academy sections submit their comments and observations. And, finally, the full Academy must make a final decision by the middle of November. Prizes are awarded on the anniversary of Nobel's death, December 10. Deliberations are kept secret; no protest or appeal is permitted.

The interplay among the constituent participants in decision making makes the history of the prize more than simply a history of prizewinners. Awarding prizes entailed how those involved with nominating, evaluating, and choosing understood their roles. These roles evolved over time in response to experience and to changes in the nature of science and society.

## WHO'S IN CHARGE?

Committee members first had to learn the intended and unintended meanings of the bylaws and how to interpret them. Nothing dictated the relative authority of nominators, committees, or the Academy. Conceivably, the committees could be relatively passive. Because their function was defined as advisory in making recommendations to the Academy, they largely could have followed the international nominators' opinions, perhaps nodding to clearly mandated candidates or to those candidates who received most votes for a year or more. Committees also could have assumed a secondary, advisory role to the Academy. The Academy was formally empowered to select winners in physics and chemistry; its own sections for these sciences could assume a dominant role in recommending winners. Nothing in the bylaws specifies that committee members should be members of the Academy. Conceivably, a committee with but one or two Academy members could be supplemented with several foreign scientists, whose recommendations would be considered in relation to the Academy's own expertise.

Those whom the Academy chose in 1900 to implement Alfred Nobel's will brought to the task an assortment of scientific, cultural, and political baggage. It turned largely to its own members to fill the committees and exclusively to scientists in the Uppsala and Stockholm area. To the physics committee, the Academy appointed a strong Uppsala representation: the professors of physics Knut Ångström and his retired predecessor Robert Thalén; professor of meteorology Hugo Hildebrand Hildebrandsson; and the Academy's professor of physics who was an Uppsala disciple, Bernhard Hasselberg. It also appointed Arrhenius from the Stockholm Högskola. To the chemistry committee, the Uppsala professors of inorganic and organic chemistry P. T. Cleve and Oskar Widman; the professor of inorganic chemistry at the Stockholm Högskola, Otto Pettersson; the professor of agricultural chemistry at the Academy of Agriculture, Henrik Söderbaum; and the professor of technical chemistry at the Royal College of Technology, Peter Klason. Most of these names, and those that soon followed, are little known in the history of science. In subsequent chapters, we will become better acquainted with most of them (a list of committee members can be found in Appendix B).

Each committee was an assembly of five members with varying measures of scientific know-how, personal interests, and collective concerns. In the early years, committee and Academy members worked together in establishing the prize as a respectable institution while they also learned to create practical routines and customs. Wisdom and experience, cunning and alliances enabled them to establish and renew the prize institution, as well as to advance specific agendas. The first fifteen years, ending with the disruption brought on by World War I, can be seen as a period in which those involved tested their way forward. Each year's proceedings added a new layer of experience and precedent. A look at some representative episodes will allow us to gain insight into the developing institution of awarding prizes.

# TWO

## Coming Apart at the Seams

### Desperately Seeking Consensus in Chemistry

From the start, awarding Nobel prizes proved a difficult task. Like any new complex machinery, the institutional mechanisms had strengths and weaknesses. The first years before World War I can be considered a period in which the Nobel committees learned to use their statutory machinery, at times groping to find stable routines, at times fine-tuning committee procedures and reacting to unexpected exigency. The Swedish scientists who evaluated candidates had a demanding job. Nominators hardly ever provided clearly mandated candidates; committees frequently had little choice but to draw upon their own expertise and judgment. At the same time, members of the committees and the Royal Swedish Academy of Sciences came to believe that the prizes were, in fact, theirs to dispose.

Launching the prizes was difficult enough in the abstract, but in practice the work never quite floated free above the personal, social, and subjective. Awarding prizes emerged indeed as very human activity. Imperfect judgments and biased interests were woven into the fabric of this history just as much as the desire to rise above parochial terrain and to strive toward disinterested impartiality. Committee members' scientific outlooks and intellectual skills did not operate in a vacuum: Feuds and alliances within the Swedish science community at times erupted onto the proceedings in the committees and Academy. At times, factions within the Academy led rebellions against committee priorities. In learning how to achieve equilibrium among disparate interests, committees at times selected and rejected candidates in a manner more dependent upon the local concerns of the moment than upon an appreciation of their pure merit.

Swedish chemists put their imprint on the prize. The science itself and the peculiarities of committee routine precluded the prize in chemistry from embodying an idealized international consensus of what might be considered the most significant contributions. Chemistry emerged from the nineteenth century triumphant. Its intellectual, institutional, and social gains were staggering. Rather than being thought of as a craft with little right to a place in a proper university, academic chemistry became well established, both as a service subject to medicine and pharmacy and as a science in its own right. Chemistry always had connections with the practical. From its roots in alchemy, mineralogy, and medicine, chemistry's growing importance for munitions, manufacture, and agriculture gave increasingly greater impetus to its support during the 1800s.

In the last quarter of the nineteenth century, British and then especially German organic chemists made great strides in developing means to analyze carbon-based substances and to synthesize these in the laboratory. What was considered a near miracle of science, artificial dyestuffs, came out of the laboratory. Fancifully colored textiles that had been available only to the wealthy were soon mass-produced by gigantic factories that belatedly transformed Germany into an industrial power. In Germany and elsewhere close, if not intimate, relations between academic organic chemists and industrial firms yielded dividends to all. Industries' appetite for organic chemists stimulated growth in enrollments and a flow of funds to chemistry departments. But, by 1900, a tension had arisen between chemists' self-image as academic seekers of truth and their role as spark plugs in rapid economic development. Still, in Germany, and in most nations, organic chemists held enormous prestige and power both in universities and in society. But chemistry was a family of specialties.

Nineteenth-century chemistry itself became fragmented into relatively autonomous subdisciplines. Some chemists still hoped for a unification of these subdisciplines into a "general" chemistry; others at least hoped to master for themselves the whole domain of chemistry. But, as the new century began, such goals were increasingly beyond reach. Chemistry was not so much a unified scientific continent as an archipelago of organic, inorganic, analytic, physiological, physical, industrial, and soon also biochemistry subdisciplines. Complicating matters further, marked differences in national orientations evolved with respect to hierarchies of prestige, priorities, and methods; these diverged further during the twentieth century. Except in the very first years of awarding the prize, this situation yielded an annual jigsaw puzzle of nominations. Nominators rarely provided clearly mandated candidates or consensual standards for what should be taken as the level of achievement meriting a prize. Committee members had to rely upon themselves.

Except for a handful of internationally recognized prominent chemists, who could bring widespread approval as well as prestige to the new prize, the committee had a serious problem in agreeing upon eligibility for reward. Compounding these difficulties, most of the great living chemists had achieved their major discoveries decades earlier. Some of the grand old men who had developed a reputation based on many years of first-rate research had not made a dramatic discovery. The elucidation, over a decade or two, of the complex structure of a family of organic materials did not have the same dramatic impact of, say, the sudden discovery of new phenomena such as the sensational recent news of X-rays and radioactivity. Consequently, the committee faced interpretive tasks related to the bylaws. How should recency of achievement be defined? How might a balance be struck between the formal requirement to reward a specific discovery and the desire to celebrate individuals whose reputations were based on a career of accomplishment?

With or without mandates from nominators, the chemists on the committee made their own decisions. Where there was a will, there was a way: When a committee majority very much wanted to give an individual a prize, the interpretation of the bylaws could suddenly become flexible. Each annual deliberation added a layer of precedent for using bylaws and setting standards; the annual reports evaluating candidates provided rhetorical resources that could subsequently be used for blocking or advancing candidates. In spite of its best efforts to avoid disunity and controversy, the committee frequently steered right into gridlock. To some extent, the problem was part of the process of finding worthy candidates: No individual or group could lay a strong claim. The committee seemed at times to hobble along, arriving at prizewinners as much through exhaustion as by conviction.

## "STRENGTH THROUGH UNITY"

The chemistry committee was not marked by any particular institutional or disciplinary-oriented bias. Most members appreciated that endless debate and dissent would enfeeble the committee's authority with respect to the Academy. The committee did not passively receive and then assess nominations. It became a forum in which its members advanced their own favorite candidates. Most committee members sent in their own proposals; some asked colleagues who were eligible to nominate particular candidates. Consequently, committee members assumed roles as advocates for candidates; they became partisan supporters. But most important of all was the desire to agree among themselves.

In 1900, Oskar Widman wanted to see Emil Fischer, one of the great German organic chemists, receive the first prize. Fischer had a number of significant achievements not too far back in time and was still an active researcher. Widman did not want to send in a nomination himself, thinking that it might restrict his ability to argue the case "impartially" to the rest of the committee. Not knowing what to expect from the nominators, he asked his closest colleague Edward Hjelt, who, as a professor at a Nordic university (Helsinki) and foreign member of the Royal Swedish Academy of Sciences, was entitled to permanent nominating rights. Both Widman and Hjelt had been trained in Germany; they helped institutionalize the newer German organic chemistry in their respective countries. Widman also asked this Finnish colleague to suggest names of others who might be willing to nominate their candidate; he asked Hjelt, too, to offer an opinion as to which of Fischer's many achievements might be most deserving.

Hjelt gladly accepted the task. He and Widman had congruent scientific interests, resulting from their common "scientific upbringing" in Germany. Hjelt considered Fischer to be "a chemist by the grace of God," who had used his unique talents to solve pure chemical problems to a greater degree than any other living chemist. Hjelt was in awe of Fischer's explication of the chemistry of sugars. But, just as others did who received invitations to nominate, Hjelt wondered how to interpret the bylaws concerning eligibility. How strictly would the Academy follow the sentence concerning discoveries or improvements made "during the past year" or whose significance had become apparent recently? If these bylaws did not preclude Fischer, Hjelt would be eager to propose him. Widman urged Hjelt to submit a nomination. Soon Widman understood that most committee members were also submitting nominations, so he sent in his own proposal for Fischer.

Widman's candidate stood in the shadow of the Dutch chemist J. H. van't Hoff, who had provided seminal contributions to chemical kinetics and thermodynamics as well as to the structural chemistry of organic substances. In absolute numbers of nominations, van't Hoff appeared to be the strongest candidate. Of the twenty people who sent proposals, eleven recommended van't Hoff, including three committee members; four nominators suggested Fischer. Widman knew that as long as he had no objection to van't Hoff, who he admitted made an excellent first prizewinner, he should vote for him. Widman felt that it would not be worth debating other criteria, such as weighing the relative significance of the respective scientific accomplishments. Consensus was the most important goal.

At this earliest stage of the nominating process, the committee seemed to prefer counting nominations. After votes were tallied, perhaps a mandate for a particular candidate would become clear. And perhaps it would build a strong case for how this and all future nominating processes should proceed. Would the international community be instrumental in selecting candidates, or would the committee's preferences and opinions hold sway? Which process would result in a more orderly, equitable, and fair consensus?

When asking Hjelt to nominate Fischer again for the 1902 prize, Widman said that he intended to write a detailed argument stating why and for what accomplishment Fischer should be rewarded. Knowing that Hjelt was busy at the time, Widman urged him nevertheless to send a simple, short proposal for Fischer's elaboration of the chemistry of sugar. Widman would take care of the details himself; but it was important for his making the case for Fischer that nominations came from as many different nations and categories of nominator as possible.

In 1902, Fischer received five nominations, as did Arrhenius and Marcelin Berthelot, the elderly senior organic chemist of France. In championing Fischer, Widman could point to the number of nominations. Yet, only one of the nominations came from outside Widman's own circle of acquaintances. Nevertheless, Fischer received the 1902 Nobel Prize in chemistry. Surely few, if anyone, objected; Fischer was, after all, a master of organic analysis and synthesis.

It might be argued that Fischer and some of the other early recipients were inevitable winners. Still, even during the few instances when candidates did receive massive support, swamping all others, the decision to award prizes came as much through active committee management as from outside influence. Widman had taken the initiative for Fischer. The 1904 prize to William Ramsay for his discovery of the inert so-called noble gases was probably as close to inevitable as any decision. The Englishman received twenty-three of thirty-five nominations. These came from leading chemists in many different nations, but they also included nominations from four committee members. Ramsay was the choice of the committee with or without the unprecedented outpouring of support.

More revealing was the case of Svante Arrhenius. In fact, his difficult rise to prominence after a crushing start to his career left deep scars in Arrhenius and the Swedish chemistry and physics milieu. These remained ghosts that haunted Nobel deliberations for over two decades. As a student in Uppsala in the early 1880s, Arrhenius chose as his doctoral research topic the properties of electrical conducting solutions. On the borderline between physics and chemistry, the topic had little appeal to the Uppsala professors in these subjects. Robert Thalén ordered Arrhenius out of the physics laboratory. P. T. Cleve simply didn't know what to make of such work. Allowed to work in the Academy's physics department, Arrhenius collected data to try to make sense of what happens when certain chemical substances dissolve in a liquid. His dissertation contained much that was new and suggestive, but the work was given a sufficiently poor grade to preclude

Arrhenius's plans for an academic career. When a few years later he read work by van't Hoff on osmotic pressure, he took his work another step and laid the foundations for a theory of electrolytic dissociation. To individuals trying to introduce greater physical insight into chemistry—to try explaining the processes behind chemical reactions—Arrhenius's work was a major contribution. Both van't Hoff and the German Wilhelm Ostwald were trying to gain adherence for this new specialty of physical chemistry, but principled resistance from other branches of chemistry made it difficult to gain institutional toeholds. Ostwald traveled to Sweden upon reading Arrhenius's work and invited him to Leipzig to work in his laboratory.

Using his new-found international reputation, and supported by several Stockholm scientists, especially chemist Otto Pettersson, who detested Uppsala's style of science, Arrhenius was named to a junior position in physics in 1891 at the Högskola. After a vicious fight within the Högskola, Arrhenius became professor of physics in 1895. This was not as odd as it may seem from today's perspective; physical chemistry had no clear disciplinary affiliation. Physicists such as Max Planck and Friedrich Kohlrausch—a leader in German experimental physics—worked on electrolytic solutions; and an aspiring physical chemist such as Walther Nernst became a professor of physics. All the while Arrhenius's reputation grew abroad; in stark contrast to those who opposed him at home, Arrhenius attracted increasing numbers of young researchers to Sweden. He had become, along with Ostwald and van't Hoff, a cornerstone in the budding movement to create physical chemistry. They saw themselves as apostles for a new science; but opposition remained strong in spite of a string of apparent intellectual successes. Some of Arrhenius's earlier opponents such as chemist Cleve admitted to having been rash in their preliminary judgment. His friends and supporters idolized him, not the least as a symbol of vindication over Uppsala narrow-mindedness.

Although van't Hoff's strength as a candidate in 1901 arose from contributions in several chemical specialties, physical chemists saw it as a victory for their field. In fact, Cleve and Pettersson had proposed dividing the prize between van't Hoff and Arrhenius. Having clear ambitions to get a prize, Arrhenius let it be known to his closest colleagues that he most coveted a prize in physics: That would add salt to the wound thereby inflicted on Uppsala. But it became clear that no amount of posturing, campaigning, and intriguing would break the physics committee's solid opposition to Arrhenius's nomination in that field. In 1903, Arrhenius dominated the list of candidates for the chemistry prize, but this reflected a local drive to give him a prize. In fact, the vast majority of nominations for him came from either Nordic colleagues who knew and supported him or from fellow physical chemists such as van't Hoff. Fischer joined this roster, surely having been tipped off by his hosts when he came to Stockholm in 1902 to receive his prize that they hoped to see their local star, Arrhenius, crowned with one also.

Committee activism for a candidate rather than an unorganized, compelling mandate from abroad proved decisive in who was to receive the prize. The issue is not whether Arrhenius deserved a prize or not; rather, it is a question of a committee's role in setting agendas and selecting prizewinners—for better or for worse. If the committee's own voice was lost in the flood of nominations for highly popular candidates such as Ramsay, it soon had to make itself heard. Mandates from abroad, whether organized or spontaneous, became rare; uncertainty on how to proceed grew.

Committee member Henrik Söderbaum proclaimed, "Strength through unity." But a culture of compromise in chemistry was not easy to maintain. Söderbaum made this remark to Widman in 1905 just as the pattern of consensual practice was about to end. Temptation egged Widman to break ranks, all too successfully. The peaceful years of attaining agreement were over. Slippery use of the bylaws and growing fragmentation in the international vote soon opened a long era of dispute.

## DEALING WITH THE AGED

In 1901, many of the giants of nineteenth-century chemistry were still alive. But was it possible to award them prizes? Many chemists felt that if the Nobel Prize was to be the greatest of science prizes, then shouldn't the grand old men who had transformed chemistry into an intellectually and industrially prominent enterprise receive recognition? For Fischer, the honor of being selected was diminished by the fact that Marcelin Berthelot and Adolf von Baeyer, the grand old men of French and German organic chemistry, had not first received prizes. It seemed fitting to Fischer and others that prizes should be given first to their own mentors. Similarly, he appreciated that Dmitry Mendeleyev and Stanislao Cannizzaro—two giants who had helped erect some of the more general foundational principles for the science—were still alive. This may well have been a reason why so few nominators sent proposals during the first years; they were waiting and watching for cues from the committee on how the bylaws were to be interpreted.

Committee members tried to decode the bylaws for those who inquired directly while trying to agree on interpretations themselves. Colleagues in physics and chemistry sent letters asking for clarification. One physicist admitted to being confused by the bylaws; he requested a guide for the perplexed. How literally should "the past year" be interpreted? The additional flexibility gained by allowing discoveries whose significance had been recognized "recently" translated into additional uncertainty. "Recently," according to the confused physicist, was "very difficult to define." He assumed, incorrectly, that the clause on "benefit on mankind" meant that pure research would not be eligible for a prize. Similarly, Widman had to explain the intentions of the bylaws to his ally Hjelt; and so forth. Fischer took up a number of questions in person when he came to collect his 1902 prize; he then compared notes with others who had access to the committee.

At first, the committee could dodge the question of age. Illustrious seniors, such as Berthelot, were often nominated for a lifetime of achievement and not for a particular discovery; therefore, the committee had no choice but to reject these proposals on formal grounds. Nominators looked for hints as to how the committees and the Academy were implementing the bylaws. Rumors and guesswork often set the tone; one close French observer of the prizes assumed that the committees worked with a restriction of ten years on the age of a discovery's acceptance. When in Stockholm to receive his prize, Fischer indicated his desire to see von Baeyer and Berthelot receive prizes as well. Widman tried to dampen his fervor: the bylaws prohibited awards for achievements over twenty years old. This message, which represented a tacit rule of thumb rather than formal regulation, was soon widely diffused, especially among German chemists. In reality, there were no formal guidelines.

The first prizes in chemistry provided clues for confused nominators. Van't Hoff, Fischer, Arrhenius, and Ramsay all had produced significant discoveries within the past fifteen years. Clearly a rigid interpretation of "the past year" and "recently" was not being applied. The age of these discoveries might not have been what Nobel had in mind, but in contrast to discoveries made twenty or even forty years earlier they seemed to indicate where the boundaries lay for defining "recent." Or did they?

## Breaking Ranks

Widman, like Fischer, adored von Baeyer; maybe, just maybe, he could find a way to celebrate their former teacher. Suddenly in 1905 von Baeyer received an impressive ten nominations. At the same time, Henri Moissan received massive international support. The Frenchman enjoyed over half of the record forty-one nominations that were sent that year for his extreme high-temperature electric furnace, in which diamonds were produced, and for his isolation of fluorine, a challenge that had eluded chemists for a hundred years. But the strong showing for von Baeyer tempted Widman. Rather than following committee precedent, which would have certainly given a prize to Moissan, Widman saw an opportunity to honor the German organic chemistry tradition to which he belonged.

Widman jumped into action. First, he wrote to von Baeyer, requesting an account of his accomplishments, seemingly under the guise of preparing for an eventual obituary. With such details in hand, he claimed that although most of von Baeyer's noteworthy achievements lay far back in time, a number were relatively recent and, equally important, the seventy-year-old Munich professor was still actively engaged in research.

In its earlier evaluations, the committee had agreed to pass over von Baeyer. His great accomplishment—synthesizing indigo—was ancient history. Claims that its true significance had only recently become clear, through massive commercial use, were questionable since chemical industry used a modified von Baeyer process, established by another chemist. But Widman tried to defuse potential opposition: The number of nominations received in 1905 and the repeated appearance of von Baeyer each year since 1902, he insisted, indicated that the international chemical community favored an award. This recognition was especially important in the early years, when the committees and Academy were both highly sensitive to outside reactions. Widman also appreciated that his committee colleagues shared his own affinity with the German chemical community. They could be expected to support a prize to its most celebrated member, regardless of the issue of age.

Widman secured the committee's support, but he feared the Academy might call his bluff and reject the committee's proposal. Said he, "Rebellion is in the air." Nevertheless, Widman had faith that the Academy would be reluctant to reject a consensual committee proposal that presented von Baeyer's candidacy resolutely. It would not hurt, either, that most of the Academy shared the chemists' loyalty to German science and culture. The Academy approved.

## Breaking Consensus

Once von Baeyer received the prize in 1905, some nominators who had refrained from proposing elderly chemists now reconsidered their positions. They could not know, however, that the choice of which grand old chemical master to be honored rested on the tastes and tactics of the Swedish scientists. Fischer took von Baeyer's prize as a cue that the guidelines were not really so cast in iron. Fischer immediately proposed Berthelot for a prize. Others recommended Mendeleyev for his periodic table of elements. Only nineteen nominations arrived for the year 1906. Mendeleyev received four, including one from van't Hoff and one from committee member Otto Pettersson; both had previously proposed the Russian. Moissan, who had earlier amassed impressive support, collected eight nominations.

The committee believed Moissan worthy; several members had nominated him the previous year. But, in 1905, the committee had claimed that as significant as his work was, it did not have the same wide-reaching impact as the work of either von Baeyer or Mendeleyev. Once again, in 1906, after considerable discussion, a majority accepted this position. The committee recommended the elderly Mendeleyev for his periodic system of classifying the chemical elements. In all likelihood, the committee majority assumed that the considerably younger Moissan could be considered in subsequent years. But following Widman's willingness to break the committee habit of seeking consensus, the committee's representative for technical chemistry, Peter Klason, decided to vote with his heart. He refused to support Mendeleyev and instead proposed Moissan, whom he had nominated. He took his plea to the Chemistry Section and then the full Academy, where he received unexpected support.

Just as Widman had assumed the Academy would not oppose a unanimous committee proposal, Klason now believed that breaking consensus might encourage dissenters to take a stand. His strategy succeeded. After the Chemistry Section endorsed Mendeleyev, the committee majority was ambushed in the Academy. Arrhenius jumped into the debate and brought along his coterie of supporters. Arrhenius insisted that Mendeleyev's periodic system was thirty-five years too old. He disagreed with those who claimed that the full appreciation of Mendeleyev's periodic table was considerably more recent than its initial formulation. He downplayed the fact that it was Ramsay's discovery of the noble gases, which constituted a whole new group in the periodic table, that had prompted van't Hoff and Pettersson to nominate Mendeleyev in the first place. Pettersson and others insisted that this newly established group of elements, a capstone for the chemical edifice, proved the validity of the periodic law, which was not as yet fully integrated into chemistry. Earlier discoveries of individual elements predicted by Mendeleyev might well have helped bring attention to his idea, but the discovery of a whole new group confirmed its full significance.

Arrhenius, however, evaded these arguments and kept up the pressure to reject Mendeleyev. In so doing, he incited Mendeleyev's two advocates to make a tactical blunder. Pettersson and his ally, Gustaf Retzius, admitted later that they probably crossed the invisible line of etiquette in the Academy's culture of debate. They allowed their response to turn into a personal attack on Arrhenius. For most members of the Academy, the difference between the two candidates meant little, but poor manners were another matter. The Academy's majority overturned the committee's and the section's recommendation and instead backed the motion for Moissan.

These events showed that a strong-willed, skillful, and popular advocate could in principle mobilize the Academy to support a particular candidate even if this went against the committee's choice. Based on the precedent set for von Baeyer, it seemed the path was clear for Mendeleyev. For two years running, Pettersson and van't Hoff, among others, had proposed Mendeleyev, but these efforts "ran aground through the absolute opposition of Arrhenius, who spoke very nonchalantly about

M[endeleyev]." Recognizing the impossibility of getting around Arrhenius's opposition in the Academy, van't Hoff informed Pettersson that it would be a waste to propose Mendeleyev again in 1907. When Pettersson learned many years later that Mendeleyev had criticized Arrhenius's dissociation theory, he immediately understood the reason why Arrhenius had acted so ungraciously against the nomination and had spoken so flippantly about the Russian's achievements. Pettersson acknowledged that Arrhenius was not capable of ignoring and forgetting a criticism. As Gösta Mittag-Leffler noted in his diary after the Academy meeting, Arrhenius's intervention for Moissan was, more correctly, one *against* Mendeleyev, "whom he accused of the same sort of humbug that more fittingly characterizes himself."

By now the chemistry committee's age of innocence was over. Members were willing to risk breaking with a consensual majority and as a result counter-recommendations could more easily be advanced in the Academy. The problem whether to give prizes to internationally prominent but elderly chemists gradually faded as their number dwindled. In 1907, Mendeleyev and Berthelot both received a handful of nominations as did the even older Cannizzaro. Von Baeyer showed solidarity by advocating a divided prize to the Russian and the Italian. But soon these men died. This was a relief for some members of the Academy. The annual celebration that accompanied the prize very quickly emerged as the social event of the year. Doddering old men dampened the festivities, if they at all could reach Stockholm in the winter.

Nevertheless, the committee did have to tackle the next generation of senior chemists, who were some twenty years younger but who earned their reputations in the 1880s and 1890s. These included Wilhelm Ostwald, Giacomo Ciamician, and Theodor Curtius, all of whom received persistent and respectable international support. But the real problem was the lack of overwhelming outside support for any one candidate. More serious than the question of giving prizes to aged chemists was the question of deciding what to do when confronted with many candidates, none of whom stood above the rest.

## "KILLING ONE ANOTHER'S CANDIDATES"

At first, by and large, the Nobel chemistry committee had it easy. Few candidates were proposed; and only a few of them monopolized the annual nominations. But, beginning in 1907, nominators rarely gave the committee a majority or plurality of proposals for any one candidate. In that year, for example, of the thirty-nine nominations received, the strongest candidates had only four nominations. Most candidates received two or three. In the following years, at best three or four candidates might enjoy a slight numerical advantage. Although committee members saw nominations as reflecting partisan interests (they knew, they participated in the process), they were now forced even against their will to rely on their own judgment. They were both aided and hindered in their continued quest toward consensus by Arrhenius, who, although a member of the physics committee, had begun taking a more active interest in the chemistry committee's affairs.

### Arrhenius Weighs In

In the early 1900s, Arrhenius was the most internationally prominent chemist—and scientist—in Sweden. For the Nobel chemistry committee, he was both an asset and a handicap. On the one hand, his frequent international travels and correspondence with foreign scientists brought the committee precious advice. On the other hand, his reputation gave him power to challenge committee proposals, as the case of Mendeleyev had shown. Even though he was a member neither of the chemistry committee nor of the Academy's Chemistry Section, his support was critical. His influence was either direct or through one of the committee members. In situations in which the committee was at a loss on whom to propose, Arrhenius's opinion found an eager audience.

Although Arrhenius's actions could enrage his colleagues, he also engaged them with well-planned strategies. He worked to have Ernest Rutherford receive the Nobel Prize in chemistry, rather than in physics, for his work on radioactivity, which nearly all nominators had suggested. Arrhenius alone proposed Rutherford for the 1907 chemistry prize and was joined by Widman in proposing him again in 1908. No other candidate that year could readily gather a consensus in the committee; Arrhenius's opinion prevailed. Although Rutherford was shocked by being transformed into a chemist, he was too delighted with the honor, money, and impressive celebrations to worry much about this sleight of hand. Arrhenius then assisted the committee in breaking yet another impasse.

In 1909, Arrhenius turned his attention to Wilhelm Ostwald. The major force in establishing physical chemistry as an academic specialty, Ostwald had produced innumerable publications, including many textbooks, which together with his popular teaching had tremendous influence on chemists around Europe and North America. His laboratory in Leipzig was a magnet for many ambitious young chemists in the last decades of the nineteenth century; his disciples founded important schools of chemistry in several countries. His Nobel candidacy had been rejected repeatedly on the formal grounds either that his contributions were too old or that he was nominated more generally for his career achievements. In 1909, Arrhenius decided to try to have his mentor, and the cofounder of modern physical chemistry, awarded the prize.

Previously, Arrhenius had remained passive with respect to Ostwald's evaluations, until, in 1908, Ostwald recanted anti-atomism. In the 1890s, Ostwald and a few other philosophically inspired scientists questioned the reality of atoms, which they considered hypothetical constructs. Nobody had direct evidence of atoms; perhaps chemistry and physics could base their laws and principles on considerations of transformations of energy. In contrast to his mentor, Arrhenius championed atomism, which seemed implied by his theory of dissociation, and promoted candidates who contributed to the broader acceptance of atoms as real entities. Finally, studies of the seemingly random, perpetual motion of minute particles suspended in solution—so-called Brownian movement—convinced Ostwald that atoms existed. Arrhenius consequently nominated Ostwald for the 1909 chemistry prize. Ostwald was also nominated by van't Hoff and Georg Bredig, two physical chemists with whom Arrhenius had very close contact. Those in the know understood that Ostwald's moment had come.

In Ostwald's favor was the simple fact that the committee for chemistry faced, once again, a baffling assortment of proposals. It received thirty-three nominations divided among seventeen candidates, none of whom stood head and shoulders above the others. The one nominee who, in principle, would have appeared to be strongest, based on solid support from several national communities over several years, was the fourth founder of modern physical chemistry, Walther Nernst. But he and Arrhenius had become enemies. Arrhenius spared nothing to block him then and for many years to come (see Chapter 10). Ostwald might have had only three nominations, but he had inside help.

An extraordinary career in research, teaching, and scientific entrepreneurship mattered little if it couldn't be fit through the eye of the needle as defined by the Nobel bylaws. Arrhenius and Widman discussed how to advance Ostwald's candidacy, which had repeatedly been earlier disallowed. In a letter, Widman spelled out to Arrhenius the initial discussions and strategies in the committee: At the committee's first meeting after the arrival of the 1909 nominations, Widman let his colleagues talk for a couple of hours, "killing one another's candidates." He then pulled out the names of Nernst and Ostwald, but he promptly let the former drop. He next proposed that Ostwald be considered for his recent work on catalysis, as a way of sidelining the formal statutory objections. Widman considered Ostwald's chances excellent since all the other candidates were weak. But, knowing full well how the committee members tended to operate, he speculated that success would depend on making a case while avoiding provocation. So far, so good, except that the committee assigned Widman, who had pushed for Ostwald, to write his special report. Physical chemistry was not Widman's field. But as Arrhenius's career as physical chemist had all but scientifically sprung from Ostwald's work, Widman asked him to prepare the report for him.

As summer approached, Widman awaited Arrhenius's report. He planned to touch it up during the summer and submit it to committee members under his own name. But as the time for his own vacation approached, he discovered that Arrhenius, who was abroad, had changed his travel plans and would be substantially delayed. Rather distressed, Widman demanded the promised report. To make sure that the committee was not provoked, Widman needed time to edit it. Furthermore, he did not have literature to make a detailed study of Ostwald's contributions: " I feel little competent to write about Ostwald," he stated. And so Widman faced the prospect of submitting an inadequate report. If necessary, he was willing to drop Ostwald as a candidate and instead write a report on the French organic chemist Victor Grignard, whose contributions were in a field that Widman mastered.

Once he had convinced himself that this threat would work on Arrhenius, Widman spelled out his strategy to get Ostwald accepted. Widman guided Arrhenius systematically through possible committee objections. Content, rhetoric, and bylaws had to be molded to fit committee members' attitudes and tastes. Arrhenius had suggested that the official justification, which would go on the Nobel diploma and constitute the formal reason for the reward, should mention Ostwald's contributions to general and analytic chemistry as a result of work on chemical reactions and equilibrium, as well as his systematization of these areas of science. Widman responded diplomatically, first by complementing Arrhenius's "excellent formulation." Then, by wondering whether it might be best to remove direct reference to general and analytic chemistry, as these would draw attention to Ostwald's well-known but aged textbooks, he urged another strategy for success. The official justification must stick close to the statutory expressions "discovery or improvement" and not provoke controversy over the question of "recent."

Widman recommended that they emphasize Ostwald's work on rates of chemical reactions and particularly his recent work on catalysis. Recognizing that Ostwald had not discovered catalysis, Widman suggested that Ostwald's work be considered "an improvement." To appeal to committee members, the work's importance for technology and biology should be noted. But, "for political reasons," it would be advantageous not to emphasize "too strongly" its consequence for studying enzyme action, as then the physiological chemist who recently joined, Olof Hammarsten, might find reasons to disagree: "If it is possible," Widman urged, "polemics ought to be avoided." Finally, Widman added that, regardless of the actual wording of the justification, the report on Ostwald must provide a clear narrative of Ostwald's great significance for new perspectives in contemporary chemistry. It was, after all, Ostwald's lifetime contributions—not the official justification—that were the real reason for the prize. Arrhenius accepted Widman's strategy. When Widman received Arrhenius's "excellent report," he used it almost verbatim; the only changes highlighted yet more clearly the catalytic studies. The tactic worked; Ostwald received the 1909 Nobel Prize in chemistry.

The use of catalytic studies was a smoke screen. Yet it helped one of the late nineteenth century's greatest chemists sneak through the maze of statutory regulations. Historians of chemistry have characterized Ostwald's work on catalysis as among the weakest of his many chemical projects: "Since Ostwald had no theory of catalysis, he proposed superficial analogies"; "no convincing answers were supplied by Ostwald and his collaborators" on key questions concerning catalytic reactions. But what he did have was Arrhenius on his side. Candidates who faced formal hurdles from the bylaws and who lacked such inside assistance did not fare as well.

In the years to follow, whenever nominators provided inconclusive lists of candidates, committee members were only too glad to have Arrhenius help them. In 1911, he engineered the prize for Marie Curie for the discovery and isolation of polonium and radium. Arrhenius helped pave openings for Alfred Werner to receive a prize in 1913 and T. W. Richards to receive one in 1914 (awarded in 1915) by providing compelling arguments for these candidates, who had previously been rejected by the committee. But his sword had two sides. By 1914, Arrhenius's adamant opposition to Walther Nernst was proving an embarrassment. Committee members were quite able to form a consensus in support of Nernst, but they could not keep Arrhenius from opposing them in the Academy. Efforts to circumvent Arrhenius failed. Recently elected Chairman Olof Hammarsten was obligated to apologize and all but concede that Arrhenius held the committee in his power: "[The committee is] obliged to engage the greatest expert knowledge and . . . it would be indefensible of us, if we turned to any lesser expertise without first having solicited your [Arrhenius's] assistance." And yet when Arrhenius did not submit a nomination, or did not actively coach the committee, its members tended to plunge into a state of disarray, or pure muddle, as they attempted to arrive at a decision. What they—and the nominators—lacked were clear criteria for a prize in chemistry.

## Contingency and Bumps

Committee members understood the problem they faced: Obvious winners were lacking. Choosing a winner was seemingly getting more difficult for each year. Oskar Widman noted that in the absence of truly notable and eligible chemists, the committee was all but forced to accept a lower level of attainment. But then, problems arise on how to choose from this larger pool. That was the situation in 1910. The only candidate to receive more than a pair of nominations was Nernst, who received seven of the twenty-eight proposals, but Arrhenius effectively disallowed his candidature. On the remaining nominees, the committee could not agree. Industrial chemist Peter Klason and agricultural chemist Henrik Söderbaum had nominated discoverers of processes for creating artificial fertilizer by "fixing" atmospheric nitrogen—either to Adolf Frank or a divided prize to Frank and Heinrich Caro. Again there was deadlock in the committee, as members hacked away at each other's suggestions. After long debate and many trial votes, the committee's "definitive" vote went nowhere. Widman had voted for the French organic chemist Victor Grignard. Klason remained committed to Frank; Söderbaum dropped his support and instead voted to reserve the prize until the following year. Chairman Hammarsten announced that he could not support any of the candidates, and he also voted to reserve the prize. Pettersson, who was not present due to illness, had sent in his vote claiming that he was inclined to support the American T. W. Richards, but he agreed that under the circumstances—in which nobody else was inclined to support his candidate—he would be willing to reserve the prize. Finding no candidate on whom they could agree, the committee majority invoked the statutory right to reserve the prize. Said Widman: "This year everything was coming apart at the seams. The committee couldn't agree on anybody."

When the Academy's Chemistry Section took up the committee's proposal, another round of debate ensued. All five committee members also belonged to the section, together with five additional Swedish chemists. When the section first met, it fell into stalemate, as its members debated the pros and cons of awarding Frank or Grignard or reserving the prize. One member tried to find a way out by recommending that the committee reexamine Nernst's worthiness; a new meeting would have to be called. At their new meeting, Arrhenius hammered in the point that Nernst simply could not receive a prize. Again, names of all the possible candidates were batted about. As the situation continued to deteriorate, Widman grabbed a possible solution: "At the last moment in the Section, I threw in [Otto] Wallach's candidature and succeeded in getting a majority for him there as well as in the Academy. . . . In the committee I had also spoken of possibly awarding W[allach] but then met solid resistance."

Wallach was known for his prodigious efforts over many years to explicate the chemistry of terpenes and similar alicyclic compounds. But he had been passed over, because it was thought a prize to Wallach would signal a marked decline in standards. Any comparison with von Baeyer, in the same style of organic chemistry, would highlight the difference. Despite Wallach's impressive work on terpenes, the final synthesis was achieved by von Baeyer. These arguments swayed the committee; nobody had voted for Wallach. But when the proceedings moved to the Chemistry Section, Wallach's candidacy suddenly succeeded. Widman was himself not completely convinced that Wallach should receive a prize. But once he reintroduced Wallach's name into the debate, Chemistry Section member Åke Ekstrand joined in, and he found a means to gain a consensus around Wallach. Ekstrand examined again the nominating letter by the Finnish chemist Ossian Aschan. Of course, the committee had previously seen this letter, but now the members became more charitable in their reading. A highly accomplished chemist, Aschan was personally known to many of them. This familiarity helped break the stalemate; Widman, Ekstrand, and colleagues allowed themselves to embrace Aschan's proposal for Wallach in an effort to close ranks around one candidate. Certainly many of Wallach's colleagues were pleased on his behalf, but many in the world of chemistry were left nonplussed. And Nernst was still out in the cold.

In learning to use the statutory machinery, the chemistry committee learned to reach agreement, even if it produced awkward results. Not able to rely on clear mandates from abroad, they tried their best. Widman, for example, avoided direct conflicts with Hammarsten on the new star of German organic chemistry, Richard Willstätter. Inquiries had filtered in, asking why he had not received a prize. Widman replied in confidence, noting the differences between his culture of organic chemistry and that of Hammarsten, adding: "As you know, the physiological chemists poke around much too much with details and don't easily see 'the forest for just the trees'; sooner or late Willstätter will receive a prize." Widman waited patiently, and finally in 1915 got his chance to write a special report on Willstätter, which won approval.

Sometimes the practice of consensus at all costs yielded unsatisfactory results. When the committee was confronted with a multitude of candidates, all with comparable support from nominators and all seemingly equal in merit, problems arose. The list for 1912 posed just such a quandary. A sizable number of nominations arrived that year, thirty-four, but this large number failed to produce obvious winners. Widman nominated the two French chemists Grignard and Paul Sabatier. The chemistry prize had not previously been divided; when a divided prize was awarded in physics, it always included two persons who had both contributed toward a particular accomplishment. Not only were these two chemists linked simply by being French, but the achievements for which Widman had nominated them had been accomplished with collaborators. Others had proposed divided prizes for the respective pairs of collaborators, but Widman's nomination was the only time Grignard and Sabatier were brought together to split a prize.

Widman's reasoning for proposing this odd couple concerned his perception of committee priorities. He decided that it might be some years before the committee might again be willing to recognize French organic chemistry. To avoid pushing all too often in that direction, he chose the more prominent member of each deserving research team. The committee went along with the split; even though the result was an eclipsing of two significant collaborators, each could claim a significant share in the prizewinning work. This packaging both insulted the collaborators and bewildered observers. Grignard regretted that the 1912 prize was not awarded jointly to Sabatier and his collaborator J.-B. Senderens; and at some later date, a prize could then be awarded jointly to him and his mentor P. A. Barbier. Senderens wrote a letter of protest to the Academy of Science; others criticized the Academy's choice more openly. Constrained by the mechanics of procedure, Widman had managed to assemble a consensus for awarding a prize to French organic chemistry. But in so doing, he had let the prize be shaped by circumstance and convenience rather than by achievement. For the chemists, the journey forward was not about to get any smoother.

# THREE

## Sympathy for an Area Closely Connected with My Own Specialty

### Bias in Awarding the Physics Prize

*P*hysicists had good reason to be proud as they looked to the new century; a few of them, however, were worried. During the 1800s, physicists brought the disparate phenomena related to heat, light, electricity, and magnetism under the domain of physical laws, and had shown them to be related through an underlying mechanical worldview. Laws of conservation and transformation of energy gave solid anchoring for theoretical and practical inquiry. Toward the end of the century, some physicists even thought that most of their fundamental work was completed. The young Max Planck was told in the early 1880s not to consider a career in physics, as most of the major discoveries were already made. Fine-tuning, pushing the decimal point of exactitude, and extending the realm of phenomena brought into the realm of physics seemed to many the future that awaited.

Surprises came at the end of the century. X-rays, radioactivity, and electrons pointed to new processes and concepts that had not been previously imagined. A number of theoretical irritants also provoked murmuring. In the beautiful edifice of electromagnetism, some explanatory models assumed fields of forces; others were based on arrangements of charged particles. Reconciliation into a unity did not come about as readily as hoped. Most physicists in 1900 believed that all physical phenomena can be traced to mechanical principles—matter in motion. Some theoreticians, however, advocated making electromagnetism the fundamental basis. Within two decades, a radically different theoretical apparatus was inaugurating a new era.

"Potted" histories such as found in physics textbooks can be deceiving. Such landmark years as 1900 for Max Planck's quantum theory and 1905 for Albert Einstein's special theory of relativity tell little of the reception of such work. Historians have shown the complex patterns of how new ideas and methods only gradually entered into the world of physics. Not only did experimental and theoretical physicists react differently to the challenge of radically new ideas, national communities more generally differed in their reception of major innovations. This was still an age of marked national styles in science. Whereas German theoretical physicists in 1910 might have been debating Planck's and Einstein's contributions, many communities of experimental physicists gave little notice.

Before World War I, chairs of physics were generally considered positions in experimental physics. But changes were underway. Advances in electrodynamics and thermodynamics arose at the borders between theory and experiment. A new type of physicist, represented by Max Planck, avoided experimental work altogether. Planck sought to resolve contradictions among different theories and create an intellectually and aesthetically unified physical world picture. His professorship in theoretical physics in Berlin was one of the few at the turn of the century. Other theoretical physicists began to colonize professorships in mechanics and mathematical physics where they focused on the physical meaning of equations rather than, as was traditionally the case, on mathematical analysis. German experimental physicists accepted, sometimes grudgingly, the creation of new positions devoted to theory, but these remained few in number and low in prestige.

Swedish physics, to some extent, resembled physics elsewhere. Experimental physics largely defined the subject; precision measurement was the measure of a physicist's worth. Use of highly sophisticated mathematics in physics was unusual and generally frowned upon. Theories, it was thought, should be simple and directly derivable from experiment. The main centers for Swedish physics remained resolutely oriented toward laboratory measurement. Beginning with the great mid-nineteenth-century accomplishments of Uppsala Professor Anders Ångström in measuring spectral lines of elements and solar radiation, a tradition emerged that prided itself in precision measurement. Naturally, the pursuit of greater exactness was a cornerstone for experimental physics, but for Ångström, and especially for his disciples, ends and means increasingly tended to merge. For them the aim of physics was the quest for increasingly more precise measurements; this was the only way for science to advance. The belief that the progress of knowledge can be attained only at the laboratory bench, referred to by historian of physics Gerald Holton as a philosophy of "experimenticism," was widespread in the world of physics in 1900. But it was one of many orientations, and in most national physics communities other voices and choices provided alternative modes of doing physics. Swedish physics had a much narrower perspective.

The dominance of this style of physics in the committee influenced the choices made. Equally important, these physicists considered it their prerogative to act as gatekeepers for their science: to turn away those who stood for alternative visions for physics. One member of the committee, Arrhenius, was just such a scientist. Scientific prejudice, gatekeeping, and a good dose of personal antagonism left a clear imprint on the early physics prizes.

## EXPERIMENTALIST BIAS IN PHYSICS

In 1907, Albert Abraham Michelson received the seventh Nobel Prize in physics. This event, the first science prize for an American, has long been celebrated as an indication of the emerging intellectual maturity and excellence of American physics, if not American science more generally. Michelson is mostly remembered with colleague Edward W. Morley in connection with the famous results of precision measurements that aimed to detect the earth's movement relative to the ether, the ethereal fluid that was thought to fill all space. Often considered the start of the ether's banishment from physics, these experiments, and their negative results, prompted the reformation of physics, and they are sometimes linked with Albert Einstein's elaboration in 1905 of the special theory of relativity.

Yet Michelson did not receive a Nobel Prize for his work related to the ether-drag effect. In fact, this work was scarcely mentioned in the few nominations received or in the evaluations of his work. Michelson won his prize primarily because committee member Bernhard Hasselberg considered him a model physicist. Michelson mastered precision measurement. His career was devoted to determining, by precise, objective means, basic units of length. Michelson's achievements in metrology—the science of measurement—dovetailed with committee member Bernhard Hasselberg's ideal for physics: the quest to press the limits of precision to further decimal points of exactitude. Such ideas had long been a prominent characteristic of the Uppsala tradition of physics to which Hasselberg and three other members of the Nobel physics committee belonged. To bestow honor on this particular vision of physics, which was shared by a majority of the committee, Hasselberg championed the American for the prize.

Michelson was by no means an inevitable winner. In 1904, he emerged as a candidate after being proposed by only one person, a Harvard University astronomer. Hasselberg nevertheless advocated him for his instrumental innovation—the interferometer—and for its applications to spectroscopy and metrology. As Sweden's representative on the International Bureau for Weights and Measures, and one who, like Michelson, used spectroscopic measurements for metrological studies, Hasselberg prized Michelson's use of a natural constant in determining experimentally the length of the international meter.

In its report for 1904, the committee accepted Hasselberg's views. It praised Michelson's efforts to define standards of measurement for the "rigor and originality" that placed them at the forefront of contemporary physical research. In putting Michelson forward, the committee underscored how his use of wavelengths of light for units had rendered the international standard of length "for all time to come, indestructible." Unlike the official standard meter held in a carefully controlled environment in Paris, a definition and determination of the meter based on wavelengths promised to be reproducible in laboratories around the world.

The committee concluded that Michelson, along with a few others, should be considered for the 1904 prize. But in a candid phrase, it noted that the choice "to a large degree depends upon everyone's individual notion of the one or the other discovery's greater or lesser scientific or practical significance." Lord Rayleigh, who had been heavily supported two years' running for his work on the densities of gases, which resulted in the discovery of argon, received the 1904 prize instead.

Michelson returned as a candidate in 1907. Americans had not submitted nominations for the preceding two years, and no one else put forward his name. Now an American, who was individually invited, helped bring him back into contention: astrophysicist George Ellery Hale. Hasselberg knew Hale and had probably suggested that Hale be invited to submit a nomination. When the proposals for 1907 arrived and were assessed, Hasselberg confided in Hale that he was prepared "to do all in my power to procure the prize for him [Michelson]." But Michelson was by no means a strong candidate. He received only two other nominations that year. Fortunately, these were not from Americans. Still, one of these nominators actually proposed several candidates and Michelson was not among his first choices. The other nominator, Kristian Prytz, proposed Michelson alone. Prytz, a Dane, shared Hasselberg's obsession with precision measurement as the ideal of physics. The two at times discussed strategies for awarding prizes to like-minded physicists. In contrast to

Michelson, other candidates that year received up to six nominations, but no one received a clear mandate.

However, an even more serious obstacle had to be overcome if Hasselberg was to get Michelson a prize. Hasselberg acknowledged privately to Hale that the statutory requirement could pose a problem; he even asked Hale for advice as to which of Michelson's accomplishments might constitute a "discovery." Regardless, even without a clear-cut discovery, Hasselberg intended to have precision measurement celebrated with a prize; his desire outweighed all other considerations. He could try to persuade his colleagues by writing a report on Michelson's accomplishments.

Committee practice entailed that, at its first meeting after the February 1 deadline for receiving nominations, members accept various assignments. Once committee members decided informally which nominees might be particularly strong candidates for a prize, they distributed the tasks among themselves of preparing detailed evaluations. In addition, one or more members accepted responsibility for writing a general report. Often the chairman drafted this primary document, which offers a cursory evaluation of all the nominees, including a short account of their achievements and, if they are eliminated from contention, the reasons for elimination. When the special reports on the few selected candidates were finished, excerpts were incorporated into the general report; the full texts were affixed as appendixes. The general report, which committee members called among themselves *kapprocken*, closed with a statement giving the rationale for the committee's choice. This short, formal justification for why an individual was being awarded a prize—the *motivation*—came at the very end. All five members then signed the report; if any members dissented, their counterproposals were included in the appendixes.

In his report, Hasselberg described in detail Michelson's investigations with the interferometer. He declared that perfection of this instrument could readily be considered worthy of a prize alone, whether or not it resulted in a discovery of great significance. Precision measurement in itself, he claimed, constituted a precondition for discovery: "The history of the exact sciences shows that almost all major discoveries . . . are ultimately due to increased precision in measurement and that every advance in this direction in itself *can* contain the seed of new discoveries."

A committee majority seeking to strictly interpret the statutes might well have discounted this rhetorical maneuver. But Hasselberg hardly risked losing face; he knew that a majority of the committee, including the chairman Knut Ångström, shared his belief in precision measurement as the foremost means for progress in physics. Hasselberg did fear, however, that Arrhenius might defect from the majority and lead a counterproposal. Arrhenius opposed the Uppsala school's vision of physics, and advocated instead a much greater role for theory. He and Hasselberg had been locked in bitter personal and professional dispute for well over a decade. But much to Hasselberg's surprise, Arrhenius did not frustrate his campaign. The committee and the Academy approved Hasselberg's recommendation that these investigations were rightly considered of "fundamental and epoch-making significance for all of precision physics." The 1907 Nobel Prize in physics went to Michelson's "optical precision instruments and the spectroscopic and metrological investigations carried out with their aid."

Michelson's now-famous and at the time well-known ether-drag experiment received scarcely any mention. His efforts to measure the effect of the earth's movement through a stationary ethereal fluid received only passing comment. At the end of Hasselberg's report, it was mentioned because "it still possesses a significant historical interest in that Michelson first constructed the interferometer for precisely this purpose and with that made possible all the varied precision investigations about which we have attempted here to provide at least a preliminary introduction." For the committee, as well as for many experimental physicists, the negative results of the Michelson-Morley experiment proved more an awkward embarrassment than an inducement for rethinking the foundations of physics. While some physicists appreciated the importance of this warning that the nineteenth-century physical world picture needed adjustment, few were willing to discard the ether. But little of the ferment in European theoretical physics then filtered into Uppsala or even into Stockholm.

## The Stakes: Gatekeepers

Hasselberg clearly wanted to give a prize for work that exemplified his own vision of physics. He acknowledged that his high regard for Michelson was "in some way an opinion of sympathy for an area closely connected with my own specialty. . . . I cannot but prefer works of *high precision*." Awarding a prize to Michelson would enable Hasselberg and his like-minded Uppsala colleagues to argue that precision measurement "is the very root, the essential condition, of our penetration deeper into the laws of physics—our only way to new discoveries." Here was opportunity to celebrate and assert this conception of physics.

Although physics was at the time mostly an experimental discipline, experiment itself was not only about extending results to further decimal points of exactness. But just what were the proper subject matter, methods, and goals of physics? What skills and talents should a professor of physics possess? What types of achievements deserved reward and recognition? These questions enter into the core of any scientific discipline; the answers are rarely obvious or singular. In the case of Swedish physics, such issues colored deliberations on who should receive the prize. Within Sweden, the Uppsala school dominated physics. Anders Ångström began in the mid-nineteenth century a tradition of physics that focused on precision measurement and experiment. Those who followed, such as Robert Thalén, Anders's son Knut Ångström, and Hasselberg, tended to act as gatekeepers for emerging Swedish physics, trying to set standards, boundaries, and goals. Prior academic experiences generally determined how committee members approached their task with respect to the prize.

Their role as gatekeepers was clearly evident a few years earlier when Hasselberg and Knut Ångström attempted to prevent Johannes Rydberg from being appointed professor of physics at Lund University, in southern Sweden. Rydberg had achieved recognition abroad. His studies aimed to elucidate the structure of matter and probe the periodic table of elements through analyzing the spectral lines that are unique to each element. Toward this goal, he discovered laws relating the wavelengths of spectral lines and a fundamental constant that make his name familiar to all students of physics and chemistry. Although the full significance of these findings became clear only after the development of atomic physics some years later, Rydberg nevertheless won the admiration of many foreign physicists. In 1898, Rydberg applied, along with five others, for the chair at Lund.

Following the Swedish tradition, the university named three referees to judge and rank the candidates. Hasselberg and Knut Ångström joined the Danish professor of physics at Copenhagen, Christian Christiansen, as the evaluating committee. Christiansen had no trouble in ranking Rydberg the top candidate. But the two Swedes strongly opposed Rydberg, as he had not performed his own spectral analyses; instead, he used others' results for his studies. Those latter results were characterized by the Swedes as "theoretical speculations" concerning the spectra of elements. Christiansen expressed surprise when he learned that the two Swedes placed Rydberg in third place behind two candidates with fewer accomplishments. Rydberg sought letters of support from prominent foreign physicists. Several leading German scientists wrote enthusiastically in support of Rydberg's results and methods. They extolled his ability to evaluate laboratory observations, his "wit" in being able to find similarities in data that others had overlooked. Hasselberg and Ångström showed little interest in "wit." They insisted that conclusions drawn from others' measurements and experiments must be considered less important than conclusions drawn from one's own experiments. The rigor of a physicist's performed precision should be the measure of worthiness.

Although Rydberg did become a professor many years later, and although his reputation grew abroad—including election as a foreign member of the prestigious Royal Society of London—the Uppsala-dominated Physics Section of the Royal Swedish Academy of Sciences never elected him a member.

Hasselberg and his Uppsala colleagues could not always weed out undesired growth in the garden of Swedish physics. The Stockholm Högskola represented an institutionalization of physics quite different from Uppsala. Hasselberg joined in the 1890s to keep Arrhenius from being appointed professor of physics. Arrhenius worked not only in physical chemistry but also in other areas of physics such as "cosmical physics," which at the time was a growing field that encompassed astrophysics and geophysics: the macrophysics of earth, sea, air, and heavens. But for Hasselberg and for Uppsala physicists, Arrhenius was too speculative. He frequently made bold theories and then experimented and collected data to support his claims. When Arrhenius's friends helped outmaneuver the efforts to keep him from being named professor, the Uppsala physicists then tried to keep him from being elected to the Academy's Physics Section.

On the other side, Arrhenius and his supporters made no secret of their loathing of Uppsala. When he was passed over in favor of Hasselberg for the Academy's professorship of physics in 1888, Arrhenius noted that had he also devoted fifteen years to collecting precision observations of diverse astrophysical and geophysical phenomena and then simply published them, with no analysis, in large volumes that nobody would ever bother to consult, then perhaps he might have had a better chance at being appointed. For that was the kind of effort in physics, he remarked, that brought prestige in Sweden.

In spite of efforts to thwart him, Arrhenius became professor of physics at the Stockholm Högskola in 1895 and a member of the Academy's Physics Section in 1901. Once Arrhenius, aided by his supporters, was successful in getting the Nobel Prize in chemistry in 1903, he never hesitated to deprecate Uppsala physics.

At various stages of this academic guerrilla warfare, Arrhenius tried to exclude Hasselberg from the committee and succeeded in having him removed as chairman. He proclaimed that the lack of scientific insight and depth in the committee—particularly among Thalén, Hasselberg, and Hildebrandsson—compromised its authority, as the Academy found it difficult to rely upon the committee for sound evaluations.

Arrhenius and his local "clique" in Stockholm noted that those who claimed the high ground, by proclaiming allegiance to precision, often missed the fact that the actual practice of science was more complex than just accurate observation. They knew that Hasselberg's intellectual limitations had produced at least one embarrassing fiasco. While leading an expedition in 1901 to subarctic Lapland to study a total solar eclipse, Hasselberg had chosen the coast for setting up instruments and an observation post. Virtually everybody acquainted with the local climate had recommended he use the inland mountains to avoid the frequent coastal fog. Other Swedish scientists remained with their lesser equipment in the mountains and obtained worthwhile data. But in spite of generous support from the Academy and an array of instruments, "Hasselberg's results are = 0," wrote Otto Pettersson.

Hasselberg and his colleagues had to keep vigilance in defense of their authority and their concept of physics. In 1904, Hasselberg learned that Arrhenius had been invited to represent Sweden in a session on astrophysics at an international scientific congress held in America. Angrily, he wrote an emotional six-page letter to astrophysicist George Ellery Hale, who had sent the invitation. Hasselberg condemned Arrhenius's perfidy and lack of understanding. He bemoaned that Arrhenius's recent textbook on cosmic and astrophysical subjects was hailed as brilliant—by Arrhenius's clique of supporters. Although Arrhenius's works did enjoy vast popularity, when it came to these sciences, Hasselberg insisted, Arrhenius was "indeed utterly ignorant." He proceeded to accuse Arrhenius and his colleagues of not understanding the need to "prove by *exact* experiments and measures [sic] the more or less daring hypotheses which they make." Instead, Hasselberg claimed, they declare hypotheses that are not "absolute impossibles [sic] to be accepted as good money" and thereby elevate them to theories. In turn, he stated, they then use these allegedly solid theories to derive consequences, which they display as scientific truths: "And the author [is then] considered a scientific hero. Such a hero is Arrhenius. Perhaps such a sloppy scientific method can be used in physico-chemistry but it won't do in astronomy and astrophysics."

On the issue of Sweden's representation on an international commission for solar physics, Hasselberg claimed he and Ångström would be willing to serve, but only if Arrhenius were excluded. "A *truly* scientific cooperation with him is indeed almost impossible," he concluded.

Hale was taken by surprise. Although well aware that Arrhenius was not an astrophysicist, Hale had become acquainted with him at Mt. Wilson Observatory and had learned to respect his insights. More surprising was the length and depth of Hasselberg's feelings. Hasselberg's willingness to wave Swedish "dirty laundry" revealed the extent to which personality and scientific outlook had become intertwined.

Hasselberg and the Uppsala physicists did whatever else they could to keep Arrhenius's influence to a minimum. They successfully blocked efforts to propose him for a Nobel Prize in physics, and they thwarted his plans for establishing and leading a Nobel Institute for both physics and chemistry. Arrhenius had to settle in 1904 for a narrower Department of Physical Chemistry of the Nobel Institute. They repeatedly blocked the election to the Nobel physics committee of Högskola physicist Vilhelm Bjerknes, whom Hasselberg called "Svante's henchman [*hejduk*]." Having already "more than enough of that d[amned] högskola brat [Arrhenius] in the Academy [*gudinog af det f.b. högskola ynglet i Akademia*]," Hasselberg resolutely opposed allowing another one into the committee.

Amid this rumpus, Hasselberg seized upon Michelson even though he was neither strongly nominated nor able to meet the statutory requirements regarding discovery. But Michelson exemplified precision; a prize to him was also a prize for the Uppsala school of physics. "Rigor" [*skärpa*] of experimental measurement was to be reaffirmed as a guiding value in physics; it is why Michelson was "placed first" in the competition, for the rigor of his work "places it on the highest level of contemporary scientific research." The very few nominations for Michelson might well imply that this belief was not universally shared.

### "Afraid of Theory"

Bias in the committee benefited Michelson. Some observers, then and now, have tried to explain the overwhelming dominance of prizes given to experimental discoveries by assuming the bylaws excluded theoretical research; this was not the case. To understand the imbalance, insight into the committee members provides a guide. Although their predominant experimental alignment shared general features of physics in most other nations, the extreme narrowness and personal rivalries produced a lopsided list of prizewinning achievements.

In a note of protest sent to the Academy in 1911, newly elected committee member Vilhelm Carlheim-Gyllensköld, from the Stockholm Högskola, contrasted the high standing of mathematical and theoretical physics in the scientific world at large with the minimal representation it received through the Nobel prizes. Apart from the shared prize for H. A. Lorentz in 1902 and to some extent the one for J. J. Thomson, in 1906, said Carlheim-Gyllensköld, "the Nobel prizes have up to now been restricted to experimental physicists." He stressed that the neglect of mathematical and theoretical physics was not due to an absence of nominations because, since the beginning of the prizes, prominent representatives had, in fact, been proposed. These included Ludwig Boltzmann, Oliver Heaviside, Lord Kelvin, Max Planck, Henri Poincaré, J. H. Poynting, and Wilhelm Wien. In most cases, they were nominated by physicists with impeccable experimental credentials, including Henri Becquerel, Philipp Lenard, Wilhelm Röntgen, J. J. Thomson, and Pieter Zeeman—themselves all Nobel laureates. "These numerous votes," Carlheim-Gyllensköld implored, "merit attention."

The record was worse than it might appear. The great Dutch theoretical physicist Hendrik Lorentz was given a prize in 1902 only when a fellow countryman and precision experimentalist, Pieter Zeeman, was included for a divided prize. Lorentz's many pioneering contributions to the underlying theoretical foundations of physics were passed over; his award was for a theoretical explanation of Zeeman's important discovery that spectral lines are broadened in a magnetic field: "in recognition of the extraordinary service they rendered by their researches into the influence of magnetism upon radiation phenomena." A proposal in 1908 to reward Max Planck's work on blackbody radiation received stiff resistance by those who believed it "unjust" to award a prize to a theoretician without dividing the honor with an experimentalist. Like Arrhenius, Carlheim-Gyllensköld despaired at the Uppsala school's narrow conception of physics. When visiting J. J. Thomson's laboratory in Cambridge, England, he caustically noted in a letter how wonderful it was to see a laboratory where scientists worked according "to an orderly plan and not for the sake of the instrument—as in certain other places."

Compounding the experimentalist bias within the physics committee was a comparable bias more generally in the Academy. Mathematician Gösta Mittag-Leffler, to his diary, observed this general tendency on several occasions: "The experimentalists [in the Academy's Physics Section] are especially afraid of theoretical physics and they naturally have a majority of the Academy with them. Mathematics of course lies completely beyond the majority's horizon." The prestige of "descriptive" science in the Academy coupled with its hostility toward abstract theory offered little hope to those who wanted to circumvent the committee in advocating a prize for theoretical research. Making matters even worse, those who in principle could have been expected to join forces to advance the cause of theory were stymied by personal differences. The Academy was sharply divided. Hasselberg's and Arrhenius's differences were but the tip of a larger iceberg of personal and intellectual animosity.

## "VEXATION AND GRIEF": A CULTURE OF CONFLICT

Principled differences regarding the proper aims, methods, and subject matter of physics were only part of the problem. Personal antagonisms exacerbated tensions among differing schools of thought. Sometimes the proceedings advanced peacefully; other times detonations resulted in surprising decisions based not so much on commitment to the winners as on a desire to avoid further confrontation. Swedish academic culture was rife with conflict. Behind the proud and solid facade of the Academy, a culture of fiery quarreling festered. Nobel deliberations gave fuel to the flames.

Committee member Hugo Hildebrand Hildebrandsson wrote to his daughter about the quarreling in the Academy: "Tomorrow I'm going in to the Academy of Sciences, where conflicts and intrigues are waiting as usual; I fear more than [the] usual. Often I'm tempted to relinquish the 2000 crowns for being on the Nobel committee just to live in peace. And Ångström has become sick from vexation and grief [förargelse och ledsnad]." Hildebrandsson needed the extra income from the Academy to make ends meet, but some years later, once his ally and protector Ångström died, he finally resigned. Indeed, Ångström's sudden death in 1910 was widely understood as a casualty of battles, first in the 1880s over professional recognition, then over efforts to build a new physics institute in Uppsala, finally, in the puerile cross fire on the Nobel physics committee.

Another observer, Vilhelm Bjerknes, whose candidacy for committee membership became entangled in conflict, expressed privately his relief at not getting elected so that he could avoid getting soiled by "Nobel corruption." Private accounts portrayed these depressing intrigues: Mittag-Leffler expressed despair at how leading personalities in the Academy could be two-faced—declaring their support for an issue when talking to one group of members, and then turning around to vote the opposite.

Other features of Academy culture tended to aggravate the situation. Some members gravitated to popular positions, fearing to find themselves standing outside the majority. Such members rarely voted out of conviction or integrity; it was much better to "howl with the pack," as Carlheim-Gyllensköld noted. Others, and perhaps a majority, constituted a "sluggish" population of "elderly men with one foot in the coffin and the other in bed." In this respect, antagonists needed to mobilize factions whose interest in an outcome might have more to do with loyalties and values far removed from the scientific issues at hand.

Personal conflict influenced deliberations related to the prizes. The professors at the Stockholm Högskola who had divided into factions in the 1890s remained at odds for another quarter century. By the 1900s, the personal feud between Arrhenius and Mittag-Leffler reached absurd levels. Mittag-Leffler's generally conservative social and cultural politics, and genteel life-style, contrasted with Arrhenius's more liberal views and petty bourgeois habits. But both men could be vindictive. Professional success never quite translated into personal equanimity. Although Mittag-Leffler, a mathematician, was not a member of the physics committee, he commanded considerable authority in the Academy. Moreover, he made a practice of cultivating loyal followers, making liberal loans, and playing the role of patrician leader. When launching one of his many "campaigns," he expected support from his allies. Arrhenius also had his followers, who tended to side with him in Academy and Nobel politics. Neither controlled the Academy, but their standing and their ability to count on supporters made them formidable figures in the early years of the Nobel Prize.

### Feuding Over Candidates

Although Mittag-Leffler tried to "open the door for theory," as when he endeavored to have Lorentz awarded an undivided prize, he was willing to keep the door closed if that denied a victory for Arrhenius. When, for example, Arrhenius convinced the committee to back a divided prize in 1908 to Max Planck and Wilhelm Wien for blackbody radiation, Mittag-Leffler helped disseminate within the Academy the news that Planck's work entailed "hypothetical molecules of energy" (the quantum), which were not mentioned in the evaluative report. He watched with glee as the baffled Academy turned down the committee's proposal. In the subsequent disarray, the Academy decided to award the prize to Gabriel Lippmann for his work on color photography. This last-minute decision certainly pleased the Parisian scientific establishment, which had advanced Lippmann as its favorite son for several years. But the decision to give him the prize arose more from the contingency of Academy debate rather than from any form of international mandate.

Not surprisingly, when Mittag-Leffler considered advancing his friend Henri Poincaré for a prize, he had to overcome both the Academy's experimental bias and Arrhenius's vengefulness. Poincaré was one of the world's leading mathematical physicists; he was probably the most eminent scientist living in France. Mittag-Leffler engaged in a massive and complex campaign to overcome the opposition in Stockholm. He sent letters around the world of physics, urging individuals to recommend Poincaré for the prize. The result was an avalanche of nominations, not only from a broad section of French science but also from prominent experimental and theoretical physicists such as former laureates H. A. Lorentz, Albert Abraham

Michelson, and Pieter Zeeman. True, it was difficult to identify which specific achievement should qualify Poincaré for the physics prize. No matter, for many of the reluctant experimentalists on the committee, mathematical analysis was hardly a path toward true discovery and founding of solid fact. It would be difficult to convince those committee members that any part of Poincaré's work merited a prize.

Mittag-Leffler called in his debts. He turned to Hasselberg for support. In a strange twist, Hasselberg, who might have otherwise disdained mathematical physics, found a way to argue in favor of Poincaré. First he rejected what he considered the committee majority's mistake of conceiving mathematical physics to be merely mathematics. He then reasoned that, just as Michelson's interferometer was allowed to be considered a tool for creating discoveries through new levels of precision measurement, so Poincaré's methods of partial differential equations should be understood as an "investigative instrument." He tried to reassure Mittag-Leffler that if Committee Chairman Ångström could be convinced to back Poincaré, then the two other Uppsala-based committee members, Hildebrandsson and Gustaf Granqvist (who replaced Thalén), would certainly follow, as was often the case. Arrhenius would, of course, oppose Poincaré.

But events took a twist. Knut Ångström received a single nomination in recognition for his work on measuring the solar constant. His Uppsala colleagues backed him for the prize in spite of the fact that he had received only one nomination. Recognizing that his best hopes for blocking Poincaré rested in supporting an alternative, Arrhenius joined in to advance Ångström. Arrhenius cared little for Ångström, but he would prefer to support him if that meant sabotaging Mittag-Leffler's campaign. Hasselberg's hopes of convincing Granqvist and Hildebrandsson to support Poincaré were all but gone. Complicating matters further, when the committee began to deliberate, Ångström unexpectedly died. Although the bylaws permitted the awarding of a prize to a person who died after having been nominated, the committee doubted whether the Academy wished to set a precedent. Still, based on the one nomination, a committee majority (Granqvist, Hildebrandsson, and Arrhenius) recommended Ångström, but only on the condition that the Academy be willing to make a posthumous award. When it became clear that there was little enthusiasm for doing this, Arrhenius was prepared.

Arrhenius, indeed, did all in his power to thwart Mittag-Leffler's campaign. He accepted the task of writing the general evaluative report for 1910, a rarity for him. This would allow him to set the framework for discussion. He also wrote a special report. Arrhenius resurrected from the dustbin of previously rejected candidates the elderly J. D. van der Waals (who had been nominated that year solely by the American chemist T. W. Richards) for his work on ideal gases. Van der Waals had been evaluated in earlier years as being in principle worthy of a prize, but he had been rejected because his major accomplishments were far too old, dating back three decades. Arguing that the recent work of Heike Kammerlingh Onnes in liquefying helium was guided by van der Waals's equation, Arrhenius maintained that the older work was only now fully confirmed. In actuality, van der Waals's equations had long been accepted. Arrhenius adapted the strategy previously used by others to advance Mendeleyev's candidacy, which he himself had vehemently opposed. Although the liquefaction of helium was a further dramatic confirmation of the richness of van der Waals's work, Arrhenius's sudden reversal was inconsistent with committee tradition for interpreting "recent."

In the general report, Arrhenius portrayed Poincaré's achievements as primarily mathematical and, when turning to contributions more closely related to physics, argued that none in particular could be singled out as distinctive. True, Hasselberg and Carlheim-Gyllensköld countered these claims in the committee and when the discussion came to the Physics Section and then the Academy. Knowing that neither Granqvist nor Hildebrandsson looked positively on Poincaré and mathematical physics, Arrhenius was ready with his special report on van der Waals as an alternative. In the end, he successfully mobilized the Academy to back the elderly Dutchman.

To place this decision in some perspective, let us recall that Poincaré that year received the most nominations—thirty-four—ever registered for a single candidate. He had previously enjoyed notable support, even before Mittag-Leffler's campaign. Other nominees also received respectable support, dwarfing the one nomination each for Ångström and van der Waals. But Arrhenius's resolute commitment to thwart the grand plan of his longtime foe, coupled with his growing skill in swaying the Academy, forced Mittag-Leffler to swallow defeat. Arrhenius's determination to block Poincaré was probably reinforced by the memory that the Frenchman had conspired in Mittag-Leffler's failed effort to keep Arrhenius from becoming professor of physics in Stockholm.

In the French-speaking scientific world, Poincaré was the supreme authority in mathematics, celestial mechanics, and theoretical physics. A second major effort to reward Poincaré failed in 1911; he died the following year. Certainly, the failure of this well-publicized campaign prompted many nominators to assume that the prize could only be given for experimental discoveries. The prize to van der Waals was no less surprising; everyone had assumed that his accomplishments were far too old.

French scientists were livid, and not the least with Mittag-Leffler, who had all but promised a prize to their hero. French scientists began asking what went wrong. Was it simply that the bylaws allowed only experimental discoveries to be rewarded? In 1911, a new committee member from Uppsala, Allvar Gullstrand, arrogantly brushed aside indignant French scientists by saying that Poincaré's work was exclusively mathematics, which, of course, was nonsense. One dismayed French scientist, Henri Deslandres, complained that Poincaré was being judged as if he were a soldier who had not fired his rifle, whereas his role was that of a general who had influenced the whole direction of modern physics. Enough information filtered back to allow the French to appreciate more fully the mechanics of the prize. Deslandres could well guess the problem: "It is the same as here in Paris, there are intrigues. . . . The candidates naturally are many. And it is the same here as it is in Sweden: some individuals use their influence to intervene in the decisions and thereby become popular. The Nobel Prize is turning into a means for governing. It is, as always, the human comedy [*Le prix Nobel devient un moyen de gouvernement. C'est, comme toujours, la comedie humaine*]."

The Parisians learned that Arrhenius had outmaneuvered Mittag-Leffler and that Arrhenius had great authority in the Swedish Academy of Sciences. Arrhenius capitalized on this. Traveling to Paris soon afterward, he tried soothing bruised French scientists while improving his reputation with the liberal wing of the highly politicized French scientific world. Arrhenius devised a plan to get a prize in chemistry for Marie Curie. After she had been narrowly defeated, in a bitter election, to the Paris Academy of Sciences, Arrhenius and the supportive president of the

Academy, Gaston Darboux, both nominated her for the 1911 chemistry prize for her studies of radium. Nobody had considered nominating her prior to Arrhenius. She had already shared, with her husband, Pierre, and Henri Becquerel, the 1903 physics prize for their work on radioactivity. But Arrhenius had little trouble in convincing the chemistry committee to back his new proposal. He was then elected a foreign member of the Paris Academy of Sciences; his revenge on Mittag-Leffler was now complete.

## Feuding Over Committee Membership

Conflict between Mittag-Leffler and Arrhenius had all but ensured that the Uppsala physicists could keep representation of mathematical/theoretical physics to a minimum. Mittag-Leffler hoped to see it improved, but opposing Arrhenius was more important. When Arrhenius advanced the candidacy of Bjerknes in 1902, 1903, and again in 1904, Mittag-Leffler should have supported the candidacy of the Högskola's professor of mechanics and mathematical physics, who was at the time the only professor in Sweden working in this area. But Mittag-Leffler, who had recruited Bjerknes to the Högskola in 1892, had hoped that Bjerknes would play the proper role as one of *his* henchmen. Instead, Bjerknes sided with Arrhenius and Otto Pettersson in the mid-1890s Högskola feuds. Mittag-Leffler shared Hasselberg's assessement that Bjerknes would only strengthen Arrhenius's position in the committee. This overshadowed virtually all other considerations. Bjerknes may not have been as sophisticated mathematically as Mittag-Leffler preferred, but he certainly offered more scientific insight than the Uppsala school's candidate, Gustaf Granqvist. Moreover, Bjerknes achieved distinction some twelve years earlier while working under Heinrich Hertz on questions related to the discovery of electromagnetic waves. He had also studied with leading French mathematical physicists, including Poincaré. His broad network of contacts with the theoretical physics community included Lorentz, Planck, and Ludwig Boltzmann. By contrast, Granqvist, who was not yet a professor, was largely cut from the same cloth as others in the Uppsala school. His research on electrotechnical problems had not achieved any distinctive results. Nevertheless, Mittag-Leffler did not rally together his factions in the Academy to support Bjerknes; he could not transcend his dislike for Arrhenius, who was singing Bjerknes's praises.

So Granqvist was elected. Arrhenius was enraged—exactly as Mittag-Leffler and Hasselberg had hoped. To foreign observers, the Academy had selected a scientifically narrow and weak technician rather than an internationally respected researcher. Whether Bjerknes's presence on the committee would have made a difference to the prize giving is impossible to gauge. But had he been elected, candidates in theoretical physics would have had a much-needed advocate. Bjerknes would have added expertise, otherwise lacking, for example, to evaluate Boltzmann's pioneering work in kinetic theory, statistical mechanics, and thermodynamics. Boltzmann had been nominated and passed over several times, before the brilliant Austrian committed suicide in 1906.

Bringing Granqvist onto the committee had other consequences. Granqvist's early impact on the committee ensured that applied physics received a prize. Granqvist was primarily responsible for the 1909 prize, awarded jointly, based on his own nomination and evaluation, to Guglielmo Marconi and Ferdinand Braun, for their contributions to wireless telegraphy. Both were minor candidates, or so the number and source of nominations indicated. Marconi could scarcely be considered a serious candidate, as his work on wireless telegraphy had been patented and was bringing him considerable profits. Others who contributed to the physics of wireless telegraphy made the choice of Braun problematic. But, rightly or wrongly, Granqvist's view that Marconi and Braun should divide the prize proved decisive. Members of the Academy, especially the Engineering Section, who desired to interpret Nobel's phrase related to conferring "the greatest benefit on mankind" as meaning practical applications, naturally embraced the proposal. Others were less pleased. Mittag-Leffler noted in disgust his own belief that Marconi was "not much more than a swindler." But to see that, Mittag-Leffler observed, "requires more intelligence than can be ascribed [*tillmäta*] the gentlemen in the Academy of Sciences." However, by burying Bjerknes's candidature, Mittag-Leffler brought Granqvist onto the committee, and that all but guaranteed the 1909 prize for Marconi and Braun.

These early deliberations suggest how difficult it is to grasp the annual decisions. Each year a range of personal and professional considerations, not all of them equally significant, affected the outcome. Receiving a prize most certainly did not involve an automatic reward based on a scientific masterstroke. Who sat in the committee was clearly critical, as were the relations among committee members and with their colleagues in the Academy. Although conflict and rivalry were never far from the surface, the committee still strove, when possible, for consensus. Even when agreement was reached in the committee, the Academy could assert itself differently. Committee members might have had some stake in the outcome as related to their own scientific concerns, but what did the prizes mean to the Academy and its one hundred members representing a broad spectrum of disciplines?

# FOUR

## Each Nobel Prize Can Be Likened to a Swedish Flag

### Excellence as the Academy's Prerogative

*C*ommittees propose, the Academy disposes. To understand the history of the Nobel prizes, primary attention must be given to the committees. They operate the heavy machinery for awarding Nobel prizes, provide the expert evaluation of candidates, assess scientific merit, and, based on interpretations of the bylaws, propose how the prizes should be distributed. But, in the end, the Royal Swedish Academy of Sciences determines the winner. The Academy votes on the committees' proposals—first, the respective disciplinary sections, and then the full membership. The Academy might seem to be passive, simply approving committee recommendations. But it could spring into action when its sense of decorum and institutional values seem threatened. These could have little to do with the weighty technical details of committee special reports or with the balancing act of diverse committee interests achieved by a proposal for the prize.

The Academy represents the public face of Swedish science, a point of intersection between the private world of the individual researcher and broader social and cultural realities beyond. Here, in principle, the best representatives of science in the land receive recognition through membership. Through action, word, and symbol its collective voice embodies and expresses the high ideals of the life of science. Respect for traditional values and formal authority, loyalty to king and nation are woven into the fabric of the institution. Who actually owns the prize; who decides how it shall be given and displayed? The answer, of course, is the Academy. Some further illustration will help clarify how its methods of institutional prerogative influenced how prizes have been awarded and how the prizes could serve its broader national goals.

### "SINK THE STANDARD ... SO FAR THAT EVEN ONE OF US ... CAN GET THE PRIZE"

When, at the start of the Nobel enterprise, Oskar Widman attempted to enroll Edward Hjelt as an active nominator, the Finnish chemist was accommodating. But he added a cautious note: Although the Royal Swedish Academy of Sciences was in a truly wonderful position in its charge to identify and thus encourage science through "more than princely prizes," the task of hitting upon those most deserving of the honor would not always be easy. Hjelt added, "Certainly it can be hoped that personal interests will not come to play all too much a role in these questions." That this formal and cultivated professor stopped nominating after 1904, in spite of his colleague's appeals, might well have been a discrete comment upon the early maneuverings and intrigues surrounding the prizes. Personal interests did enter into deliberations, but so did institutional interests. The latter were generally more subtle than professional or personal agendas, but they also could be blunt and decisive.

Self-interest, tantrums, and a dose of provincial arrogance certainly entered into the early years of the Nobel Prize. This we have already seen. Hjelt was not alone in wondering whether some of his Swedish colleagues were growing with the task, using it to cultivate greater knowledge, wisdom, and morality; or whether they were willing to pull this elevated enterprise down into the muck of academic incivilities. Rather than behaving with the loftiest and purest intentions, some participants proved disappointing in their actions. Maybe such charges were harsh, even priggish, given the practical challenges involved in operating the new statutory machinery. In keeping divergent interests from tearing apart the entire enterprise, the Academy was at times a moderating forum. Like the graphite rods in a nuclear reactor, the Academy's large and somewhat indifferent membership could absorb the highly energetic outbursts from its Nobel juries. But sometimes not even the Academy could prevent meltdown.

## A Culture of Deference

A culture of quarreling did not preclude a culture of civility. Arrhenius's occasional role as provocateur has already been noted. In 1907, he lashed out in the Academy against a member of the chemistry committee. The committee proposed awarding the prize to Eduard Buchner for his astounding experiments that conclusively established cell-free fermentation; that is, processes that were previously achieved only as the result of living yeast cells were achieved artificially with a cell-free juice physically extracted from yeast. Buchner achieved a result that had eluded many before him, including Louis Pasteur. His results demolished what was left of the popular scientific belief in some unknown vitalistic life force, present only in animate matter, that was held to be responsible for fermentation.

As significant as this result was, Buchner's experiments actually entailed little brilliant chemical thinking. The procedure to extract the juice from yeast cells was mechanical rather than chemical; Buchner learned the technique while working with his brother, a bacteriologist, and the latter's colleague. So, although the experimental results were significant, the actual procedure could hardly claim to be on chemistry's cutting edge, nor Buchner's own innovation. Indeed, Adolf von Baeyer confessed that this former pupil did not measure up to the standards of other past students, such as Emil Fischer and Richard Willstätter, but the discovery "will bring him fame, even though he has no chemical talent."

Buchner received few nominations. But one of the two who nominated him, Hans von Euler-Chelpin, was a young energetic chemist at the Stockholm Högskola. He had recently begun working on the chemistry of enzymes; and as he was to do regularly twenty years later when he became a member of the Nobel committee, he vigorously pushed to reward achievements closely related to his own research. In the case of Buchner, he pressed upon committee members, who were not so well acquainted with this growing specialization in chemistry, why they ought to consider

Buchner. It fell to Olof Hammarsten to prepare a special report. Although unsure about Buchner, Hammarsten convinced himself and persuaded his committee colleagues to recognize the significance of cell-free fermentation. The task of convincing the Academy turned into a verbal brawl.

Arrhenius had at first agreed informally with Hammarsten that Buchner should receive a prize. Subsequently a conflict arose between the two on completely different matters. When Arrhenius understood that Hammarsten was preparing the special report on Buchner, he "resaddled" and wrote a private letter to Hammarsten listing objections. The attack came as a surprise. Always civil, the elderly professor reacted with restraint. He continued as usual to address his younger colleague with formal but friendly collegiality, opening with "Highly valued friend [Högt värderade vän]." After rebutting as best he could Arrhenius's points of contention, he graciously closed the letter with a touch of humility as a means for saving face all around. While conceding that of course it was indeed hard to compare achievements in very different branches of chemistry, and while willing to admit that perhaps he had not appreciated all sides of the issue, Hammarsten offered to continue discussing their disagreement. Ending with a polite appreciation for Arrhenius's feedback, he offered "a warm thanks from your old friend, Olof Hammarsten."

When the matter reached the Academy, Arrhenius criticized Buchner's candidacy and Hammarsten's report. He produced a series of reasons why the proposal should be rejected. To many in the audience, the faults seemed perfectly plausible. But then Hammarsten replied. In a cool-headed and systematic response, he presented facts that indicated that Arrhenius had been irresponsible in his attack, using false quotations and having "thoroughgoing carelessness in all his statements."

After receiving this pummeling, Arrhenius called on his allies, mixing science with emotion. Sensing an unbecoming debate about to get worse, the Academy's permanent secretary introduced a motion to reserve the prize until the following year. But this effort to discharge the tension generated yet more shocks. The bylaws insist that reserving a prize can only occur when no worthy candidate can be found, not just to end an unpleasant fracas. When the committee proposal was finally brought to a vote, Arrhenius was roundly defeated. And indeed, according to those in attendance, the vote seemed to have been aimed more against Arrhenius than for Buchner. Mittag-Leffler confessed privately that Buchner and his work did not seem of sufficient significance for a chemistry prize. But he had to oppose Arrhenius, once the latter had made the question into a personal issue. "Had Arrhenius performed honestly and with correct information, then surely also this opinion would have triumphed." For the Academy, the issue touched upon a critical point: Arrhenius overstepped the boundaries of acceptable behavior. Although Arrhenius was probably the more knowledgeable, Hammarsten was the more senior. Propriety and pettiness faced each other—this time the former won. Consequently, Buchner received a prize, worthy or not.

## "One of Us"

Nothing revealed the Academy's "ownership" of the prize more compellingly than its preference for local heroes. In this respect, the continued belief that Alfred Nobel's legacy belonged, first and foremost, to the Swedish scientific community surfaced repeatedly. Of course, committees also tried to advance one of their own for a prize. Following the debacle in 1908, when the Academy was frightened away from awarding Planck and Wien, and instead awarded the physics prize to Lippmann for color photography, many began to wonder whether the time had come for keeping the prize at home. By continuing to make awards even when no truly significant recent discoveries had been made, thought Mittag-Leffler, the Academy would "sink the standard and finally we will come so far that even *Unser einer* [one of us], as some committee members had put it, can get the prize." The physicists soon proposed Knut Ångström based on slender support, but the Academy was more concerned to avoid awarding a prize to a dead candidate. The following year, the physics committee and the Academy were ready to award Uppsala ophthalmologist Allvar Gullstrand a prize for his work on the optics of the eye, but the Caroline Institute beat them to the finish by giving him a medical prize. The only nominations for Gullstrand came from two members of the Academy, neither a physicist. Substantial international support for Poincaré and Planck was ignored. More telling were the times when the Academy insisted on celebrating one of its own, even against the wishes of the Nobel committees. The Academy's sense of values and priorities did not always correspond with those of its committees.

This was the case in the Academy's 1912 rebellion. Against the physics committee's evaluation, the Academy gave a prize to a person who was judged not suitable. The Academy membership included substantial numbers of engineers and applied scientists. Many of these had long been annoyed by the dominant philosophy for giving prizes. They felt that Nobel's legacy obliged the Academy to reward individuals who shared problems with practical implications. At a meeting of the Swedish Inventors' Association [Svenska uppfinnareföreningen] in December 1911, the chairman related a comment allegedly made by Thomas Edison: that should he be awarded a Nobel Prize, he would refuse to accept it as a protest against the way the prize had previously been awarded. Swedish inventors and engineers were annoyed that their own favorite-son candidates, Fredrik Kjellin and Gustaf de Laval, had not been given chemistry prizes. The chair proposed to protest to the Academy, backed by internationally leading associations of inventors. Actually, he and his colleagues soon found another way to register their disapproval.

A leading member of the Academy's Technology Section, Erik Johan Ljungberg, nominated in 1912 the Swedish inventor Gustaf Dalén for his automated lighthouse. This one nomination was the only one received for Dalén. Unlike the previously nominated Swedish inventors, Dalén's work entailed the use of well-known scientific principles; considerable ingenuity and technical insight were invoked but not science-based technical research. The committee understood this, and while applauding the invention, it quickly passed over the candidate. In the Academy, however, the engineers argued against the committee, which had proposed awarding the prize to Kammerling Onnes for the liquefaction of helium. Moreover, Dalén had recently been blinded while working with acetylene; his closest colleagues were determined to celebrate him and the Swedish engineering tradition to which he belonged. It was of little importance for many in the Academy that, in addition to Kammerling Onnes, Max Planck, among others, received substantial international support. The engineers led a rebellion against the committee and the Physics Section. They mobilized a majority of members, thirty-seven to twenty-eight, who accepted the desirability of acknowledging local technical ingenuity while paying tribute to a respected member. This was not the last time the Academy ignored its

Nobel committees and respective disciplinary sections. It revealed a continued tendency of the Academy to consider the prize its possession. Further insight into what the prize meant to the Academy can be gleaned from a feature of the Nobel story of which founder Alfred Nobel never dreamed.

## "THE ROYAL SWEDISH ROYALTY" AND THE "LAND OF SMÖRGÅRDSBORD"

Awarding prizes was never merely a matter of identifying deserving recipients. Once the institutions charged with selecting winners made their annual decisions, the public spectacle began. The wise men who salvaged the testament did not stop at drawing up statutes; they also concocted a vision of grandeur. Modest and retiring, Alfred Nobel never foresaw the hallmark of his legacy: an opulent ceremony and banquet. Representatives from the prize-awarding institutions and the Nobel Foundation developed elaborate rituals that ensured a successful image. The annual observance served as a public stage for celebrating science, culture, and Sweden, or to be more precise, particular images, ideologies, and values associated with these.

At the ceremonies, traditional officialdom—represented by the royal family and conservative political and cultural elites—joined ranks with select members of the urban bourgeoisie. Although Ragnar Sohlman and Carl Lindhagen had strong roots in liberal Stockholm society, they allied the new foundation and its public ceremonies with the conservative power establishment. Members of nobility in high state service dominated the Nobel Foundation. It was the king-in-council that appointed its members, including the chairman of the board of directors, named at the last minute, the recent conservative Prime Minister Gustaf Boström. Many of the respected members of government and civil service also belonged to the Academy, in the catchall section for "general learning."

The ornate prize ceremony provided opportunities for older elites to align themselves with, and perhaps guide the moral development of, the progressive forces of science, medicine, and technology as well as to share in the ennobling practice of "idealistic" literature. For their part, the scientific and medical establishments gained greater political legitimacy and social prestige by association with royalty and the pillars of traditional power. Together they could stage a ceremony that confirmed at home and abroad Sweden's prominent place among civilized nations. And, by bringing honor to the nation, the ceremony justified the elite's continued right to power and authority.

As with other contemporary manifestations of the then oft-discussed peaceful competition among nations, such as world's fairs, this Olympics of culture attracted immense media attention. Consequently, the celebrations became public events, elements of cultural politics. But why was it important to create an ornate festivity in connection with awarding prizes?

Although a comprehensive analysis of the origins, development, and meaning of the Nobel ceremony deserves a volume in its own right, the ritual is part of the prizes itself. At first glance, the description of the award ceremony and accompanying banquet given in the official annual publication, *Les Prix Nobel,* appears innocuous. For the first decades, the full proceedings were laid bare: welcoming speeches, presentation of the laureates, the musical interludes, reproduction of the Nobel diplomas, and narratives of the subsequent banquet, including guests, and numerous toasts. But what was going on when decisions were taken to establish the form and function of the ceremony?

## The Cultural Politics of National Identity

During the 1890s, debate sharpened over Swedish national identity. New forms of patriotism emerged, along with new symbols and rituals for expressing national sentiment. Just as in other Western nations, new nationalist traditions began to be constructed and older ones overhauled. Patriotic organizations sprung up, usually royalist in coloring. Holidays were devoted to nationalistic celebration; public discussion buzzed over the choice of an official national anthem, while frequent and novel uses transformed the national flag into a prominent element of the local scenery. The king's own fondness for displays of loyalty encouraged him to travel around the country, making frequent ceremonial appearances, of which the media never ceased to have its fill—the effect being to reinforce royalist sympathy as a part of nationalism and patriotism.

Artists, writers, and journalists joined in as the simmering efforts to forge national identity occasionally boiled over. As always, nationalist enterprises proved contentious. Radical labor's adoption of the International Workers Movement's May Day and red flag prompted government and conservative countermeasures to offer national alternatives. For their part, socialists and the radical bourgeoisie attacked the growing "quack patriotism." After one of the first efforts to create a flag-festooned national celebration at Skansen, the Stockholm amusement park, in 1894, the Social Democratic newspaper expressed anger that the blue-yellow national colors were used to draw attention away from "so much injustice, oppression, and humbug."

Debates reveal changes in patriotic sentiment and national identity. Sweden had for most of the nineteenth century looked to the past for its identity. Memories of warrior kings and past heroic national sacrifice were expressed in patriotic poems, songs, ceremonial occasions, and schoolbooks. Sweden had been a major European power in the seventeenth century; in its efforts to control the Baltic Sea, Sweden conquered Prussian, Polish, and Russian territory. Legendary kings such as Gustaf Vasa and Charles XII, who led loyal troops into battle for the glory of a united Sweden, long provided the substance of national identity. By the 1890s, critics of various political shades bemoaned what they saw as a sense of national anemia: A lack of vitality and confidence seemed to be national characteristics. The massive midcentury emigration to America also weighed on the national consciousness. Although Norway also witnessed the loss of a significant portion of its population to the promise of a better life in the New World, Swedish critics observed a spirit of optimism in Norwegian cultural life: a culture of youth looking toward the future, contrasted with a tired Swedish culture. In opposition to the officially sanctioned culture, a number of countermovements arose in the 1880s and thereafter. The social realism of the 1880s and the romanticized aestheticism of the 1890s sought new ways of expressing nationalist sentiment and national traits.

Books, articles, and public lectures on the "Swedish national character [*Svenska lynne*]" commanded enormous attention; public demand was insatiable. The fluidity and insecurity of national identity gained a biological dimension as various Darwinian-inspired philosophies and theories of race became widely diffused, just as in the rest of Europe. National traits, social hierarchies, and the sociology of everyday life were widely understood as biologically based phenomena, rooted in race and an inheritable "germ plasm." Increasingly, the Swedish nation was understood to belong to the Germanic race, allegedly the crown of the human species. But where did Sweden actually stand in the hierarchy of civilized nations?

Whatever attitudes might have characterized Swedish culture at the turn of the century, state institutions still embraced a politically conservative philosophy infused with Christian morality, classical aestheticism, and a vision of individual and society as subservient to an idealized state, as embodied by the king and royal bureaucracy. Naturally, antagonisms flourished. At one extreme of the cultural spectrum, Carl David af Wirsén mobilized the Swedish Academy in a crusade against the new literary and artistic movements of the 1880s and 1890s. At the other extreme, cultural rebels, such as August Strindberg, spurned by the Academy and official culture, turned increasingly for support among progressive forces. When the state universities that trained future members of the government began to change and even to house antiestablishment student movements, such as the Uppsala *Verdandi*, King Oscar II participated in academic ceremonies, where he lent his presence to validate official culture. Although members of the conservative academic and cultural establishment might have recoiled from the new industrial society and its cultural manifestations, many understood the need for creating national unity, or at least the illusion of one.

King Oscar II seemed to appreciate the need to make gestures to the new; indeed, he increasingly embraced the new sciences while hoping that older traditional values could still be retained. On one level, he had long accepted his role as patron of Swedish science, as official protector of the Royal Swedish Academy of Sciences. Where the nation's glory was at issue, he provided direct financial support and bestowed honors personally. In particular, he supported Sweden's rise as a major polar-exploring nation. In a European culture extremely conscious of the peaceful competition among nations and the relative ranking of each people's level of civilization, polar exploration gave Sweden an opportunity to prove its fitness. A polar expedition entailed a wide range of challenges requiring a spectrum of skills. Likened to a war, in requiring the heroic qualities of courage, endurance, foresight, and trustworthiness, an expedition also tested a nation's abilities in technology, science, finance, seamanship, and organization.

When Adolf Erik Nordenskiöld and his ship, the *Vega*, returned to Stockholm in 1880 after having navigated the northeast polar passage, they were met by unprecedented jubilation. Throngs of admirers lined the harbor, fireworks erupted in dazzling displays, and a cortege of ships escorted the returning national heroes to the Royal Palace, where ennobling and other honors awaited. Nordenskiöld's achievement revealed to all that science could bring prestige to the nation and helped bring into the public's consciousness the image of "scientist as hero." The scientist, devoted to the pursuit of knowledge, with little interest in personal gain, had long been understood as a dignified calling. In France, Louis Pasteur had achieved heroic status; now the Swedish man of science could wear the laurels of the hero as well.

As the new century approached, the king endorsed the progress of material culture through many public appearances. What was at stake was the hope of mobilizing the nation in the struggle for survival—against nature and competing civilized nations. He and many other national leaders appreciated that belief in progress might yet hold the nation together as a society, a progress that promised peace among the nations and better living conditions at home. But privately, the king had strong reservations about a materialistic worldview, the social consequences of natural science, and especially the refusal of many artists, writers, and scientists to embrace the basic values of faith, love, hope, and humility.

## Monarchy and Modernism on Stage

In 1897, these many strands of national culture came together in three extraordinary events: Sweden celebrated a Royal Jubilee, Stockholm hosted the Scandinavian Art and Industry Exhibition, and the reading public digested the astounding news of Alfred Nobel's testament. Just like the sumptuous manifestations surrounding Queen Victoria's Diamond Jubilee, the events in Sweden marking the twenty-fifth anniversary of Oscar II's reign entailed a "personal propaganda campaign" by the king. His many appearances, as well as mass-produced images, memorabilia, and what we now would call media opportunities, helped shape a proud national consciousness in which the royal court and patriotism shared a common focus. The king was no longer portrayed as a military general leading the nation in battle. Now he assumed the role as a chief, mobilizing the nation for victory in the peaceful battles for higher levels of civilization. The fields of honor were no longer bloodstained battlefields but splendidly constructed fairgrounds, where an Eiffel Tower or a Crystal Palace proclaimed the ascendancy of a new culture.

International exhibitions and world's fairs emerged during the second half of the nineteenth century, captivating huge audiences, through both direct attendance and media coverage. Nations displayed their technologies, crafts, and other demonstrations of cultural attainment, and they competed for prizes. Exhibitions, it was widely believed, spurred progress as a consequence of peaceful competition among nations, civilization being the ultimate winner. Underpinning the extraordinary interest in these exhibitions was a belief in the existence of a universal hierarchy of civilization, according to which the multiplicity of cultures could be ranked in order of material, spiritual, and moral development. Fairs and exhibitions judged a nation's worth, in which technology and science held special significance. A nation's capacity to contribute to the advance of technology and science was understood as a component of the European ideology of fitness. These fields clearly set off the white races from the colored races. For most educated Europeans, science and technology demonstrated the superiority of the Christian West over the colonial peoples under their rule.

Although the Stockholm Exhibition of 1897 was a minor affair as compared to such events as the Paris and Chicago world's fairs of 1889 and 1893, it was nevertheless an unprecedented national experience. Largely planned and paid for by industrial and private capital, the exhibition also received official and royal sanction. As described in detail by cultural historian Anders Ekström, the opening entailed a gala display of royalty and traditional monarchist symbolism, including singing the conservative (if not reactionary) royalist song, "From the Depths of the Swedish Heart [*Ur svenska hjärtans djup*]." An amalgam of old and new values was to provide a future-oriented nationalism. The entire well-designed and well-executed exhibition was planned to give the impression of national harmony and national pride. Established power elites and newly emerging ones could find a common grammar, sanctioned by the king, to express a new Swedish nationalism. On one level, the display of industrial goods, regional crafts, and cultural artifacts that competed with the best of the foreign instilled pride in Sweden. The past need not be the exclusive source of patriotic inspiration. A newly emerging, dynamic industrial society could also inspire national self-confidence. Moreover, the carefully orchestrated images of national unity and harmony, which greeted the flood of visitors from around the country and were reported daily in newspapers, reinforced the message of progressive industrial urban culture, bringing the country to a higher level of civilization. Of course, the exhibition masked the fact that this same industrial society was being forged on a foundation of increasing social conflict and political activism.

The exhibition brought to a focus national self-confidence and patriotic fervor. As one high-ranking official commented, "We Swedes have awoken this year to a sense of our own self worth, of our own strengths, of our own resources . . . and many hearts are warmed by the thought: This is Swedish." Sweden was now more frequently noted as being vigorous and, together with Germany, a virile Teutonic nation. One Uppsala professor wrote that only in Scandinavia could pure Germanic blood still be found: Swedes were the aristocrats of the Germanic peoples. Sweden could be, with its natural resources and healthy blood, Europe's leader. Just as King Gustaf II Adolph had in the past attempted to create a great Swedish empire, so must Swedes now attempt this goal again, but through national work, general education, and thoroughgoing reforms.

The conservative newspaper *Svenska Dagbladet* proclaimed that the exhibition served "as a playing field [*spelplats*] for 'contemporary international creative work,' for industry itself is cosmopolitan." The significance of promoting international competition of those aspects of national cultures that could be compared and judged was clear. "Here the entire nation takes an examination in front of the world . . . displaying its own general level of attainment with respect to material culture." The paper continued: "The Swedish people stood as one before the world's gaze; but there was no need to fear." This notion of the nation testing itself through competition, and thereby attaining further evolutionary heights, ran through European thought. It provided an obvious bridge from the international exhibition to the new international prizes. The National Archivist Emil Hildebrand remarked that the newly liberated Swedish sense of self-worth must submit itself to additional challenges. In the past, warfare strengthened a healthy nation. Today, the repeated experience of peaceful international competition, requiring a united, motivated, and proud people, served the same purpose. Just as in wars past, in international competition the Swedish people's industriousness and diligence would bring honor; moreover, the nation had already assumed a leading role in engineering and natural science.

After the first news of Nobel's testament, on January 2, 1897, Swedish newspapers erupted with debate and discussion of his puzzling legacy. Some papers hailed Nobel's action as the greatest gift to humanity ever made: "Thanks to the important duties entrusted to them, the Swedish institutions in question will be placed in the front rank of similar bodies all over the world as stimulators and promoters of inventions, science, and literature." Although some critics felt that more of the estate should go to Swedish institutions rather than be squandered away on foreigners, most were grateful for Nobel's initiative. Liberal papers tended to follow the line that greater contact with the newest leading intellectual developments around the world would advance progressive thought and politics at home. Said one: "The unprecedented record which the Nobel Prizes will constitute of the advancement of science and liberal ideas and their struggle for recognition will stimulate us as few others have done and will force our minds to adjust themselves to the new things that time will bring forth." Sohlman and Lindhagen, themselves liberals, used like-minded papers as a podium to educate the public.

Conservative papers first attacked the testament. They joined the king in not welcoming Nobel's proposed prizes, fearing Norwegian mischief with respect to the peace prize. During the 1890s, populist Norwegian agitation for greater autonomy, especially in foreign affairs, created a crisis in the political union; the prize could become a pawn in this conflict as the Norwegian Storting (Parliament) was to decide on winners. Conservatives also feared a corrupting influence within the prize-awarding institutions. But gradually other tones entered the conservative media.

The Nobel prizes would be the Olympics of science and culture. The Swedish prizes were cast in this light because the ancient Olympic Games, which were revived the year before, in 1896, were on everyone's mind. They riveted the world's attention, as they provided a way for the newly nationalist-minded countries to compete in the world arena. Racial competition, too, entered the picture, as the individual nations, each seemingly embodying its own national traits, were pitted against one another, much like the Greeks and barbarians competed in ancient times. Like the Olympic Committee with the Games, the institution of the Nobel prizes would host a competition—but involving science and culture. And Sweden would sit in judgment.

What could underscore the importance of the annual Olympics of science and literature than a prize with a money value? The 200,000 Swedish crowns attached to the first Nobel prizes represented twenty times the salary of a professor and a hundred times the annual wage of a factory worker. Prizes had long been a feature of European arts and sciences, but nothing compared with this astronomical sum.

Swedish impartiality would guarantee the integrity of this new playing field, on which the nations would compete. Swedish popular opinion embraced the prizes with extraordinary enthusiasm, seeing a mutually reinforcing reflection of national character and prize institution. Indeed, impartiality was seen as having a biological basis, males of northern European Protestant background being most perfected. When Stockholm hosted the athletic Olympic Games in 1912, local newspapers welcomed the athletes to the home of the intellectual Olympics and reassured them of the national spirit of impartiality.

## The Nobel Ceremony as Oscarian Display

The festivities of 1897 connected with the Royal Jubilee might be understood as a model for the Nobel celebration. The Stockholm event and hoopla surrounding the event underscored the importance of display and festivity for conveying political messages. This was an era characterized by the upper classes' predilection for pomp and ceremony, banquets, and honorific decorations. Swedish officialdom had a marked disposition for ostentatious ceremonies marked by cheering, making speeches, exchanging decorations, and clinking glasses. The Oscarian era in Sweden at the turn of the century incorporated la belle époque and fin-de-siècle mentalité. Swedish upper-class custom welcomed the royal presence, the toasting, and the flags and other national symbols. And as it turned out, this same passion for pomp and ceremony shaped the Nobel prizes' public launching.

Orchestrated extravagance guaranteed the success of the fledgling Nobel Prize at home and abroad. The Nobel ceremony embodied much of the same cultural politics as the Stockholm Exhibition, but it was also a response to the events of 1897. Lindhagen and Sohlman appreciated that Nobel's testament required cultural alliances. The problematic circumstances of the institution and initial lack of royal support had to be overcome. Inspired surely by the Stockholm Exhibition of 1897, and drawing upon prevailing national practices, the prize-awarding institutions created a comparable forum of display.

The media quickly spread the Nobel idea of cultivation at its finest. Sweden's elites could celebrate themselves and convey a message of their legitimate right to national leadership. As Academy member Hildebrand noted: "It is from above and from the few that the higher light comes that shall penetrate the masses. It is this minority with genuine cultivation, which keenly knows, which deepest appreciates, that which is the truly national and the worth of all that which makes us cherish our fatherland."

The royal family was invited to grace by their attendance the first prize celebration. The king did not come. Officially it was said that he was called away on other matters, but unofficially the king was uneasy with the Nobel legacy. He also disliked the fact that so much money was being directed to non-Swedes. Nevertheless, the crown prince was on hand to give the prizes to the winners. The following year, Oscar II was there—toasted and cheered—and began the tradition whereby the monarch gives the prize diploma to each laureate. The success of the first year—the harmonious blending of royalist, national tradition with science and culture—clearly changed his mind. Civic leaders and the industrial bourgeois were on hand, as well as the diplomatic corps. Head of the Swedish Academy and Nobel literature committee, David af Wirsén highlighted the ceremony with a poem stressing the noble Swedish role in bestowing the prizes. In essence: "Neither desired nor sought, the solemn task resting heavily on Swedish shoulders . . . to scrutinize the world, who has probed deepest in nature, brought forth the finest in medicine, which beautiful poem in different lands, shall year for year receive a prize from Swedish hands." To an extent that may not have been anticipated, the banquet ceremony was a showcase for Sweden. To those anxious that Sweden was giving too much to foreigners, the ceremony showed that gifts favor the giver.

Returning the toasts offered in their honor, the prizewinners and foreign ambassadors saluted all things Swedish. Typically, at the 1904 ceremony Hasselberg delivered in Latin a long toast to Lord Rayleigh, praising British science and culture. In turn, the honored guest responded with a speech "full of humor" noting that of all the countries with which he was acquainted none could match Sweden in its relative numbers of eminent and distinguished men. Which, as noted in reports of the ceremony, prompted "vigorous applauds." In 1905, the German ambassador added in his toast that "the prize carries not only the name of Nobel but that of Sweden, through the entire universe [est porté dans tout l'univers]." Perhaps the punch toddy was especially strong that year.

As the ceremony was reported in local and foreign newspapers, never before was Sweden so highly praised and rarely were the Swedish people shown that their cultural leaders—allied with monarchy and state officialdom—could bring such honor and attention to the nation. Although only a tiny group could attend the annual gala, the population could vicariously participate through the media, and they could share the pride of being Swedish.

In some respects the Nobel ceremony was also a cautious response to the 1897 exhibition. Socially, Swedish industrialists had not yet fully "arrived." They were little represented at first among the invited. Older wealth and older power kept entrepreneurs at arm's length. Moreover, the 1897 exhibition was well worth remembering. After four months of messages extolling Sweden's elevated civilization, the exhibition closed in an orgy of plundering. On the night after closing, crowds of working-class and middle-class looters picked apart all the memorabilia they could carry. Silence descended, as the exhibition grounds were stripped bare. This final coda to the well-orchestrated tribute to national pride suggested that much more "light from above" was needed. Industrial capitalism alone could not civilize the people.

## Nobel Day as a National Holiday

Members of the Nobel establishment sought to implant the prizes into the core of national identity. Nobel Day, December 10, the anniversary of Nobel's death and the testament's official birth, became a day for displaying the flag. National music and patriotic songs framed the ceremony and banquet. But the goal was to make Nobel Day a national holiday. Early December, however, is rarely pleasant in Stockholm; fog, blustery winds, freezing rain, and snow frequently greet foreigners lured north. And, at best, there are only a few hours of daylight at that time of year, in pre-electrically lit Stockholm. But would the world come? Elderly and less enthusiastic prizewinners just stayed home. Transport by rail and steamer in winter was difficult. In 1905, it seemed none of the laureates would come to Stockholm in December. Arrhenius despaired that the banquet would end up a parody of Swedish hospitality. Indeed, would the king be willing to take part in "such a comedy"? In the opening speech to the 1909 Nobel ceremony, Count Fredrik Wachtmeister, the Nobel Foundation's new chairman of the board, confessed that the date was less than a festive occasion for distinguished foreigners. Moreover, the time of year limited attempts to celebrate the events of Nobel Day as "great patriotic ceremonies for glorifying intellectual culture."

Committees took up the issue of moving the date. Some preferred June 6, the day that conservative groups proposed as a national Flag Day. Others suggested the summer solstice, which since pagan times was a night for festivity. The Royal Swedish Academy of Sciences voted for June 6, but a lack of consensus from the other institutions deferred action until 1909 when, the institutions still unable to agree among themselves, the king rejected the idea. Patriotic-inspired campaigns to move the celebrations to June continued for the next decade, but Nobel Day stayed in December.

For the ceremony itself, there were plans to build a grandiose Nobel Palace that would dominate the Stockholm landscape. Serious second thoughts, coupled with the building of a sumptuous city hall, brought a halt to the palace. No matter, even in dark and dismal December, Nobel Day quickly became a national ritual. When in

1908 Count Karl Mörner proposed to eliminate the costly ceremony, to send the prize money to the winners, and to direct the resulting savings toward research, he received hardly a notice.

Still, in the early years the ceremonial apparatus creaked, often because of the fixation on national concerns. Speakers and rituals were chosen based on local Swedish concerns with little thought of foreign guests. Thus, Mittag-Leffler's ascerbic diary notes: "Nobel Day [1904]. Celebrated just as bureaucratically as usual. Vulgar, somewhat simple-minded speeches from Swedes [*Tarfliga, delvis enfaldiga tal från svensk sida*]. They don't understand that at such occasions Sweden should allow itself to be represented by the best talent and not of royal ties [*blåa band*] or their kind." Displays of provincialism greeted early visitors. For example, in 1903 the physics laureate Henri Becquerel (the Curies did not attend) was the guest of honor at a meeting of the Academy the day before receiving his prize. There, Mittag-Leffler noted, "Aurivillius held an idiotic lecture on his experiments with bees. Jakob Eriksson held forth for almost an hour boringly and school-like on a botanical theme. All in Swedish for Becquerel's edification. Never has the Academy sunk so low as now." Even the highly polished main event was tedious: "Long boring poorly delivered speeches, which moreover impossible to follow in their entirety as a result of the poor acoustics in the hall." Still, prizewinners who attended came away in awe. If they were not already worshiped, they certainly came away feeling very special. Both their hosts and the media—local and foreign—assured them of this. As Rutherford noted, he and his wife "had the time of our lives." But such gaffes as Prime Minister Boström's Nobel dinner on the day following the official ceremony, to which the distinguished foreign guests had not been invited, revealed all too clearly the Swedish purpose for setting in motion the festivities.

## The Bourgeoisie in Paradise

Although guests had to wear double-woolen undergarments on the way to the banquet, they nevertheless transformed the solemn occasion into the social event of the year. Social conventions and seating arrangements became obsessions that continue today. In this bourgeois dreamworld, the wife of a provincial Uppsala professor could find herself seated close, if not next, to members of the royal family. An elderly professor could loosen up sufficiently, after some drinks, to exchange witticisms with "charming French ladies." A prominent manufacturer could share in an elevated atmosphere of aristocratic taste and charm. And, of course, snobbery and envy were never quite checked in the cloakroom. Scenes from Mittag-Leffler's patrician perspective cluster in our gaze: "[A member of a prominent baker's family had] fingers full of diamond rings and little intelligence"; "[I] sat next to a Norwegian conservative little goose, wife of engineer Sohlmann [sic]. Such are the so-called liberals. More attached to custom than anyone else." Mittag-Leffler might well have had a chance to chat with the crown prince, but he found it difficult to compete for the attention of the crown princess, surrounded by social climbers trying to impress her with family trees. The king at times hosted a separate dinner for the winners and for select representatives from the committees and scientific institutions. On one such occasion, Mittag-Leffler recalled, "the Queen complimented me. I spoke at length with the King." And, of course, the "yellow journalists will be yapping" whenever something went amiss.

Social tensions were inevitable. Some members of the Academy used their tickets to bring loyal house servants to the banquet. Arrhenius himself asked the permanent secretary of the Academy for permission to bring his "serving girls," who had made it possible for him to concentrate on research. Knowing the way the media reacted to the presence earlier of young unmarried women of lower-class background, Arrhenius assured the secretary that his intended guests were all married to respectable men. Scandal had to be avoided at all costs. When in 1911 Madame Curie was accused of having had an affair with a married colleague, members of the Academy tried to prevent her from attending the ceremony. Such a woman could not be allowed in the royal presence! To her credit, the unassuming Curie refused to buckle under, and with help from some loyal Academy members, she collected her prize.

Occasional smudges on the conservative tapestry did not obscure the importance of Nobel Day as a significant social event. Each year, an increasing number of high government officials, members of the diplomatic corps, and prominent members of society attended. Although guests and journalists had few ways of assessing the true meaning of the prizes, bourgeois calculation was easy to make. The prizes were international, and the pomp and circumstance certainly outshone anything else. But there was also the money. When Arrhenius received his prize in 1903, his toast at the banquet made the equation explicit, prompting Mittag-Leffler's comment: "He (according to himself) has received the highest scientific award our age can offer (The power of the word over thought. Highest = largest materially)." Even before the list of winners could proclaim a super-elite in literature, science, and medicine, the prize had attained a prominent reputation from the very start.

Years later, more sober and modest people who cared little about social snobbery, such as Olof Hammarsten, chairman of the chemistry committee, observed that the prize was not necessarily the most important way for signifying overall excellence. In a world already dominated by prizes, popular sensational discoveries could eclipse careers marked by consistent first-rate research and pedagogy. Still, in a European culture obsessed with competition, the Nobel Prize was just too good to be true. The public willingly believed that the competition took its course on a level playing field. Impartial objective judgment over the best in science and literature was assumed from the start. When so much money and so much prestige were involved, only the most noble sentiments would enter into selecting winners. The ceremony tended to confirm this belief.

But such grand ceremony also struck many as typical of what was wrong with Swedish culture. The poet Werner von Heidenstam expressed his disdain for this aspect of bourgeois life. In a letter to Strindberg he called Sweden "The Land of Smörgårsbord," where patriotism seemed best expressed on the menu: love for the nation displayed at banquet tables. The obsession with the royal presence, and the degree to which the court promoted itself, prompted another poet to coin the expression "the Royal Swedish Royalty." Although Stockholm scientists with international reputations, such as Mittag-Leffler and Arrhenius, arranged private receptions for prizewinners and their guests, they were, nevertheless, dismayed by the abundance of food and toasting. The many public and private events were an onerous pre-Christmas burden for those not in full party form.

Speeches and toasts focused more on generalities than details. Praises dominated; few understood the specific merits of the scientific achievements. In 1910, when the winners were overwhelmingly from Germany, members of the Academy and of the German diplomatic corps toasted the deep-felt brotherhood of the Teutonic race. The elderly Dutchman, J. D. van der Waals, broke the brotherly embraces by toasting in French. The formal speech presenting each Nobel winner contained a popular portrayal of the work, often a highly idealized history of the discovery, emphasizing national traditions. That the winners were the best in their fields was taken for granted. Disagreements within the committees and prize-awarding institutions were whisked under the carpet; by the time the ceremonies began, all that mattered was the spirit of harmony and unity, advancing the lofty goals of Alfred Nobel and "Civilization." The giving of the prize by the king held symbolic meanings. Disagreement was dissolved by royal authority. "Go to the King" was an old Swedish proverb as a means of settling disputes. The king, as symbol, delivered the just reward.

The Nobel ceremonies reminded everybody of the virtue of peaceful competition. Chairman of the Nobel Foundation Count Fredrik Wachtmeister, national chancellor of the Swedish universities, in his speech opening the 1908 prizes, said that Nobel's intention was to support those whose work "could contribute to make mankind happier through possession of greater welfare and greater general education." Competition among civilized nations was the key to a better world: "Highminded and noble struggle would make cultivation's fruits available for and within reach of all humanity." This theme repeated itself again and again. Even when world events cast shadows, Wachtmeister repeatedly tried to hold this ideal: "We must nevertheless never relinquish the firm belief that all development [*utveckling*] goes forward, not backwards, led by Him who steers the destiny of people." But how often could he repeat himself? After several years, he asked to eliminate the welcoming speech. He felt all the eyes fixed on him, all expressions asking the same question: "What will he say this time?" Wachtmeister realized, perhaps more than most, that the rhetoric and posturing, the glitter and pomp, had already become mere facade; the light shining from this fete of civilization could not penetrate the gathering darkness.

# PART II

## HAS THE SWEDISH ACADEMY OF SCIENCES... SEEN NOTHING, HEARD NOTHING, AND UNDERSTOOD NOTHING?

### World War I, Biased Neutrality, and the End of a Nobel Dream

# FIVE

## Should the Nobel Prize Be Awarded in Wartime?

### The Cultural Politics of Neutrality, 1914–1915

*W*ars rarely sweep away old worlds. Of course, nobody can deny that the unprecedented horrors of the world war of 1914–1918 precipitated drastic changes. But in many respects the war brought into focus and heightened tendencies already present, conditions conveniently ignored, and assumptions blithely unexamined. In the worlds of science and culture, nationalist passions long kept in check by internationalist posturing erupted into open hostility. When needed, scientists were willing to work for national purposes and, if necessary, forgo transnational cooperation. In neutral Sweden, members of the Nobel establishment were poorly prepared to meet the challenges brought about by the breakdown of international relations. The war revealed the fragility, if not the contradictions, in the values and beliefs that had given meaning to the prizes. Some committee members scrambled to define new roles for their institution. Others passionately tried to shore up the remains of the devastated European world order. Neutrality proved no option. Impartiality was highly political. Both at home and abroad, charges of bias tarnished the Nobel prizes.

The Nobel establishment received ample warning of the impending crisis, but it failed to prepare. At every year's Nobel ceremonies, Count Fredrik Wachtmeister invited his audience to examine the Nobel idea. At the lavish fete, he espoused the creed that peaceful competition propelled civilization. Of course, discretion was always necessary at such gala affairs. Yet, time and again Wachtmeister called attention to disturbing events in the world beyond the glitter and pomp. The Russo-Japanese War of 1904 and the Swedish general strike of 1909 threw a shadow over the prospects of an era of harmony. Repeatedly, Wachtmeister concluded that faith in God and trust in Nobel's dream would bring peace and prosperity. But, in 1912, following news of the vicious barbarism of the Balkan wars, Wachtmeister could not remain silent. It might well be, he hoped, that the bloodshed by these warring races would result ultimately in a healthier civilization. But Sweden could not remain idle.

Bringing politics overtly into the Nobel festivities, Wachtmeister discreetly endorsed the controversial call for increased Swedish investment in armaments and defense. Conservatives and monarchist groups demanded military buildup. Just as the conservative German academic elite had endorsed Germany's campaign for a massive naval buildup, so the Swedish cultural establishment embraced a call for national unity behind its royal navy and army. Although the growing feud over military expenditures forced the resignation of Sweden's liberal government, nobody seemed to consider what to do with the Nobel Prize should there be an international conflict.

As most of Europe went to war, August 1914 found the Nobel establishment preoccupied, once again, with the question of moving its annual festivities to June. A chorus of local newspapers urged change. How could Nobel laureates be expected to think well of Sweden after the usual December fog, slush, and gloomy darkness? By coming in June, they might see that Sweden is indeed a Nordic paradise that everybody will long to visit again. As thousands of men began marching to the battlefields, the Nobel committees continued business as usual. Sweden declared itself neutral; the Nobel committees declared themselves impartial. Committees pushed forward with deliberations on the prizes for 1914 during the summer and early fall. Everybody believed that the soldiers would be home by Christmas. Committee members looked forward to the annual prize ceremony, now postponed until June 1915. They all might wait for the squall to blow over. Soon, however, it became clear how little they—like everybody else at home and abroad—had prepared for catastrophe.

Most German academics, with whom Swedes kept close contact, were joyful at the outbreak of war. Now the German nation could unite in an idealistic cause behind the kaiser's government rather than wallow in petty factionalism. Said one German economist: "The first victory we won . . . was the victory over ourselves. . . . Our own ego with its personal interests was dissolved in the great historic being of the nation. The fatherland calls! . . . When we celebrate this war on a future day of remembrance, that day will be the feast of the mobilization, the feast of the second of August. . . . That is when our new spirit was born . . . the new German state! The ideas of 1914!" Perhaps by next year, Germany might be a bit larger, Russia a bit weaker, and, more importantly, a new European cultural-social rebirth an established fact. For the Swedish upper classes, these were not unpleasant thoughts.

Sweden had long maintained a policy of nonalignment. During the Napoleonic wars it lost Finland to its traditional enemy, Russia. During the 1800s Sweden's political and cultural ties with Germany became the most important of international relations. A German victory could reduce Russian influence in the Baltic Sea. Although some conservative groups wanted to support Germany openly, the moderate-conservative government of Hjalmar Hammarskjöld chose to keep Sweden out of the conflict, hoping that the recent Hague conventions on neutrality would guarantee security. But neutrals also have allies. It became clear that Sweden was to lead a policy of friendly, accommodating neutrality toward Germany and a stricter formal neutrality toward the Triple Entente nations of Great Britain, France, and Russia. The upper classes were largely pro-German and firmly expected a German victory. The leader of the Social Democrats, Hjalmar Branting, was pro-Entente, but some factions in his party, while opposed to the kaiser's militarism, expressed solidarity with the large and powerful German socialist movement. While Sweden was trying to get its bearings, and while the Nobel committees were preparing for important September meetings, the newspaper *Aftonbladet* asked a question that nobody had broached in public: Should the Nobel prizes be awarded during wartime?

The massive international attention given to the Nobel prizes created a dilemma. Would choosing winners now create problems for the prize institutions and for Sweden? Impartiality had been Sweden's claim to moral superiority. The Nobels, it was claimed, embodied internationalism. In Sweden and abroad, belief in unbiased judgment benefited all. Nations could boast that their winners were recognized without prejudice for the excellence of their work. Although nobody in Sweden doubted that the committees would remain unbiased during the war, would the announcement of winners be seen—mistakenly, of course—as a Swedish endorsement for one or another of the warring alliances? Even the best intentions could be misinterpreted.

And yet, without question, and without denying it, most of the Swedish elite felt special bonds with Germany. Just as Swedish politicians had to transform neutrality from a paper concept defined by international conventions into a working practice, the Nobel committees had to reexamine the meaning and implications of their self-proclaimed impartiality.

## 1914: EGOTISTICAL NEUTRALITY?

When he read the question asked by *Aftonbladet*, Olof Hammarsten turned anxiously to Christopher Aurivillius, the permanent secretary of the Royal Swedish Academy of Sciences. As chairman of the Nobel Committee for Chemistry, and a man of principle, Hammarsten could not deny that naming prizewinners at this time could lead to "great unpleasantness and difficulties." He feared that decisions might well be taken as expressions of sympathy for a certain nation, and "perhaps even misconstrued as a departure from the impartiality we are obligated to observe." But, he noted, it would be out of the question simply to withdraw the 1914 prizes. In this respect he—and the others—recognized that the bylaws had no provisions for withholding prizes because of war or other calamities. Only one reason for withholding a prize was stipulated: the lack of a candidate deemed worthy of the honor. Hammarsten knew that the chemistry committee had fixed its attention on two very strong candidates, both of whom, the American T. W. Richards and the German Richard Willstätter, had earlier been declared worthy. No legal ground existed for withholding the 1914 chemistry prize from them. But Hammarsten saw another way. Could the institutions petition the king-in-council to postpone awarding for one year? He did not see a problem in making two awards in 1915; Richards and Willstätter would certainly again be candidates. Admitting his lack of competence in such official matters, Hammarsten asked Aurivillius how to proceed.

To his close friend and influential member of the Academy, Gustaf Retzius, Hammarsten repeated the basic questions early in September. He even confessed a more general anxiety about the prizes, not just for this year but for the future. Sweden's political situation could change within two months. Diplomatic and economic pressures were putting Swedish neutrality under stress. Hammarsten recognized the political sensitivity of the prizes, but was too shaken by the grim news of the war to take an initiative; wiser and worldlier men must act.

### Business as Usual

Despite the imperative of resolving the issue of awarding a prize for 1914, the only decisions taken in September were the usual committee recommendations for allocating the prizes. The chemists agonized about whether to award the prize to

T. W. Richards or to Richard Willstätter. The American had been passed over earlier; his work on precise determinations of the atomic weights of elements had been deemed as not sufficiently significant for the prize. But a number of Swedish chemists and physicists nominated Richards for the 1914 prize, as did several foreign nominators. In 1913, Svante Arrhenius and the 1904 laureate Sir William Ramsay gave long and powerful accounts supporting Richards. Arrhenius was in close contact with the American, socially and scientifically; he was determined to persuade the chemistry committee to reverse its earlier decision. Ramsay, with whom Arrhenius was in contact, was keyed into the process. By 1914, Arrhenius and Ramsay had convinced organic chemist Oskar Widman to withdraw his earlier reservations about Richards; Widman grudgingly backed the proposal.

In the meantime, Willstätter had also collected a relatively large number of nominations, for his beautiful studies of the chemistry of chlorophyll, which were models of painstaking organic chemical analyses. Widman hoped to broaden the committee's appreciation of Willstätter's work and looked to replace Hammarsten, who had not fully understood its significance. Although the committee was highly German-oriented, it appreciated the point made by a Czech chemist that the time had come to award an American a Nobel chemistry prize and that this American should be Richards. To divide a prize between Richards and Willstätter would not be fair; each deserved a full prize. Finally, because Richards had been waiting longer, the committee proposed him for the 1914 prize. Members understood that Willstätter was in line for the next prize; indeed, a postponement of the 1914 prize, such as Hammarsten envisaged, could produce the happy consequence of crowning both of them at once, one from a fellow neutral nation and the other from Germany.

Meanwhile, the physics committee was still working through the revolutionary challenges facing classical physics. Numerous and persistent nominations of Max Planck, for his discovery of the quantum of energy in 1900, had been coming for several years. But the committee members, like all the king's men of the rhyme, still hoped that the Humpty-Dumpty of electromagnetic theory could somehow be put together again—and without the seemingly absurd notion that energy is emitted and absorbed in discrete "atoms of energy." By 1914, the Academy's avoidance of Planck had become embarrassing. Awarding the 1911 prize to Wilhelm Wien for his work on the laws of heat radiation without also awarding Planck prompted flurries of speculation. Although pleased for himself, Wien considered the decision to have been odd, if not awkward. The committee had seemingly ignored that it was through efforts to modify shortcomings in Wien's law that Planck had first introduced the quantum concept. A gathering of leading physicists in Brussels in 1911, the first Solvay Conference, had focused on the challenge of quantum theory. Puzzles remained, but quantum physics in some form had most definitely come to stay. But once again the committee would find a way to ignore the bubbling cauldron of new ideas in theoretical physics, this time by the distraction of a brilliant new line of experimental research.

The announcement in 1912 by Max von Laue of producing a diffraction pattern when sending X-rays through a crystal answered the controversial question of whether Wilhelm Röntgen's phenomenon entailed a form of electromagnetic wave or a stream of charged particles. In favoring the former, the experiment produced a new tool for studying the molecular structure of crystals. In the rush of follow-up research by physicists working with X-rays or crystals, a series of discoveries followed. The work of father and son W. H. and W. L. Bragg in England underscored the importance of von Laue's discovery for exploring the arrangement of atoms in crystals.

Although the committee had been positioning George Ellery Hale and Henri Deslandres for their respective contributions in solar physics, von Laue's discovery appealed strongly to the committee's majority. And indeed, here was a discovery that seemed to fit Nobel's original aim: recent, significant, and achieved by a young researcher. In contrast to the twelve nominations for Planck, von Laue was nominated only by the aging organic chemist Adolf von Baeyer and by Emil Warburg for a divided prize with W. H. Bragg. In his special report on these developments, dated July 3, 1914, Allvar Gullstrand concluded from a study of the published scientific work that W. H. Bragg could not be awarded without also recognizing W. L. Bragg's contribution; therefore, Warburg's nomination could not be used as a basis for a prize. Von Laue alone, therefore, should receive a prize for what Gullstrand considered, without reservation, work of the greatest benefit to mankind. At its September meeting, the committee voted to give the 1914 Nobel Prize in physics to von Laue for his discovery of the diffraction of X-rays in crystals. When the question of postponement arose, the physicists, just like the chemists, understood that this delay would not to be too painful. Several of the key researchers in the newly emerging field of X-ray crystallography could together be feted with the two prizes.

The two committees sent their reports, as usual, to the Academy. For all the chatter immediately following *Aftonbladet*'s provocation, the reports and committee protocol reveal no formal response. By October, all seemed quiet after the initial flurry of responses. But while the Academy's Physics and Chemistry Sections began to digest these documents, in preparation for their usual vote in late October, the cultural firmament, so to speak, caved in. Suddenly, European scientists, writers, artists, and professors of all persuasions joined in the political and military turmoil engulfing the Continent. European culture seemed intent on burrowing headfirst into the muck of divisiveness and hatred. Nobody was quite prepared for such a rapid collapse of the internationalist spirit of the sciences, precisely what the Nobel enterprise necessarily presupposed.

## A Crack in the Internationalist Spirit of Science

Scientists had, of course, noted the growing instability in Europe. But they hoped that reason and goodwill might prevail, at least in the world of learning. When war came, commentators expected a quick resolution, so the first shock to internationalism provoked annoyance rather than despair. When war engulfed most of Europe in August and September 1914, a number of scientists were left stranded in enemy countries. Some had traveled to participate in multinational collaborations, such as the Germans who went to Russia in preparation for a solar eclipse. Some British researchers were working in Vienna and Berlin; Germans assisted in Rutherford's Manchester laboratory. In some cases, they were permitted to return home, others were interned, and still others were permitted to work in local academic laboratories. News of such tremors spread through the networks of correspondence. Although deplorable, these were understood as temporary disturbances. But soon scientific relations would be knocked off the seismic scale of sensibilities.

At the start of the war, the German military high command quickly implemented what was called the Schlieffen Plan, which aimed for a rapid victory over France. To achieve a quick knockout of Paris, and thus France, and thereby avoid a war on two fronts, Germany immediately invaded neutral Belgium on its way to France. This outrage, along with reports of German atrocities against Belgian civilians and vandalism against cultural monuments, prompted heated propaganda campaigns in the Entente nations directed toward their own citizens as well as toward the neutral nations of Scandinavia, Holland, Spain, and America. At first, reactions in academic circles tended to distinguish between the much-respected German *Kultur* and the much-hated Prussian-dominated military state, which was held responsible for the barbarism.

However, in October 1914, ninety-three leading German academics and artists signed a "Proclamation to the Civilized World." Here they declared support for the war machine—and solidarity with all Germans—for the nation's "just" goal of self-preservation. They denied categorically any German wrongdoing. Moreover, they declared that their culture was an organic part of imperial German society; the two nourished each other. To attack German militarism was to attack German culture. Without the military, so the thinking (or lack of it) went, German culture would have disappeared from the face of the earth. Indeed, the signers proclaimed German culture itself had long ago learned to defend the nation from repeated attack: "The German army and the German people are one and the same thing." Seventy million Germans stood united irrespective of education, class, and party. All assertions to the contrary were lies.

The roster of signatories included most of Germany's most prominent men of learning and culture. Leading scientists—such as Adolf von Baeyer, Emil von Behring, Paul Ehrlich, Emil Fischer, Felix Klein, Philipp Lenard, Walter Nernst, Max Planck, Wilhelm Röntgen, Wilhelm Wien, and Richard Willstätter—endorsed the manifesto. Said the proclamation: "To the east the ground is soaked with the blood of women and children slaughtered by Russian hordes and to the west dumdum bullets rip apart the chests of our soldiers. Those who have the least right to act as defenders of European civilization are those who join with Russians and Serbs to offer the world that disgraceful spectacle of fostering Mongols and Blacks as peoples of the white race [*at hidse Mongoler og Negre paa folk af den hvide Race*]." Although unable to answer all their enemies' attacks, these prominent men of culture and the arts called out to those who knew them, those cultured individuals with whom they previously had shared the task of guarding humankind's greatest good: "Believe us! Believe that we will drive this battle to the end as a civilized people, for whom the heritage of a Goethe, of a Beethoven, of a Kant is no less holy than hearth, home, and soil."

The proclamation appeared in translation in newspapers around Europe during the first week of October; copies were sent to individuals, organizations, and institutions. Rarely had such blatant, misplaced patriotism appeared in science. Rarely could it provoke so much antipathy. Perhaps the largely conservative, nationalist German academic community had been too bloated with its own self-importance, too intoxicated by the joy of the outbreak of a war, and too certain of moral and military superiority to gauge the impact of its words.

The manifesto perplexed some, enraged others. French and British savants immediately responded, often with overkill, as they began to appreciate that their former colleagues had no intention of distancing themselves from the German military tactics. Sir William Ramsay led the attack in the British science journal *Nature*: "German ideals are infinitely far removed from the conception of the true man of science; and the methods by which they propose to secure what they regard as the good of humanity are, to all right-thinking men, repugnant." In calling for the need to have the entire German nation "bled-white," Ramsay asserted that this course of events would certainly not retard the progress of science:

> The greatest advances in scientific thought have not been made by members of the German race. . . . So far as we can see at the present, the restriction of the Teutons will relieve the world from a deluge of mediocrity. Much of their previous reputation has been due to Hebrews resident among them; and we may safely trust that race to persist in vitality and intellectual activity.

Thus began a series of charges against not just the morals of the German scientific community but also its allegations of superiority. The editor of *Nature* argued that, during the past generation, German scientific originality had declined in spite of growing productivity because of the heavy hand of military despotism. Even if not ready to begin publicly deprecating the quality of German science itself, one hundred twenty British scholars, including more than thirty scientists, offered their German colleagues a clear response: "We grieve profoundly that, under the baleful influence of a military system and its lawless dreams of conquest, she whom we once honoured now stands revealed as the common enemy of Europe and of all peoples which respect the Law of Nations." Belgian and French response was heated. The Paris Academy of Sciences threatened to expel from its ranks those German members who had signed.

Just to make sure that the original German message was not mistaken, rank-and-file German professors and secondary school teachers circulated a new manifesto in newspapers two weeks later, proclaiming their inseparability from the kaiser, his government, and the military. They dismissed as preposterous the charges of atrocities. How could the boys whom they had endowed with the cultivation of classic culture and scientific spirit burn libraries or shoot civilians? Professors from other countries, in turn, asked how their former German colleagues could sink morally so low, so fast. Indeed, Ramsay speculated that the effects of syphilis must be at the bottom of this German delirium.

For the Swedish academicians, who had begun to reflect on whether Nobel prizes could be awarded without showing partiality, the stakes had changed. Although the escalation of charges and countercharges had just begun, the "Manifesto of the Ninety-Three" (as the proclamation was soon called) and the first well-publicized responses dispelled the illusion that the world of learning stood united in its supranational ideals. The manifesto declared, simply and directly, that all of Germany stood shoulder to shoulder with its military and Prussian leaders. Closer to home, Wilhelm Ostwald, the 1909 winner of the Nobel Prize for chemistry, came to Stockholm as a messenger from the German Cultural Association. Defending the invasion of Belgium, he noted that Germany had hoped Sweden would join as an ally. But, as the situation developed, he declared, it might be just as helpful if sympathetic Swedes could help spread abroad the "truth" about Germany. Ostwald appealed for support based on the superiority of German culture, as evidenced in the number of Nobel prizes awarded to Germans. And this superior culture, he averred, had arisen as a product of the Germanic race's genius for organization. Although earlier known for his internationalist opinions, Ostwald now embraced German expansionist goals: The time had come for Europe to embark upon a higher level of organization and culture, under the guiding hand of Germany. Did culture—and the Nobel prizes—really float above politics?

### Prudence or Cowardliness?

These sentiments stunned the guardians of the Nobel Prize. Those involved with the literary prize reacted first. On October 16, the Swedish Academy asked the Nobel Foundation to call a meeting with representatives from the three prize-awarding institutions to decide joint action. The new head of the Nobel literature committee, Harald Hjärne, was clearly appalled by the sudden "overpowering patriotic fervour" that prompted scientists, writers, and artists "to lose their balance . . . and publicly stoop to virulent attacks. . . . The Academy must regard it as an especially heavy responsibility not to provide, by its choice of new prize-winners, a possible occasion for new outbursts of the prevailing bitterness, which . . . may be directed against our whole country."

Count Mörner, head of the Caroline Institute and a member of the Academy's Chemistry Section, saw matters differently. He wrote to Prime Minister Hammarskjöld before the meeting commenced. Mörner did not like the idea of withholding prizes because of the war. He was not convinced that the warring nations would take the prizes to be a sign of Swedish partiality. And as far as he was concerned, it would be better to make manifest that Swedes, in their remote corner of the world, "are carrying out our peaceful international goals." But he had been informed by both the Swedish foreign minister and the minister for church and education, that they opposed awarding prizes. He conceded that if the government was intent on avoiding further strain on Sweden's neutrality, then two alternatives were possible: First, suppress the awards and direct the prize money into the special funds connected with each committee; second, request the king-in-council to give permission to postpone the prizes to the following year. Still hoping that Hammarskjöld might be swayed to allow the prizes to be distributed, he revealed in a postscript that at least two, and maybe three, of the likely candidates would be from neutral nations, even though no political consideration motivated these choices.

On October 20, representatives from the three Swedish prize-awarding institutions met and agreed to petition the government for two things: a one-year postponement in selecting winners and a June 1, 1916, date for the ceremony for the 1914 and 1915 prizes. Before they received a response, the Chemistry and Physics Sections met and decided in favor of Richards and von Laue. The Swedish Academy's Nobel literature committee proposed the Swiss writer Carl Spitteler, but also recommended in principle that no prize be awarded that might further inflame "nationalistic antagonisms." The Royal Swedish Academy of Sciences proceeded to schedule a mid-November meeting to decide on its prizes; rumor had it that the Academy of Sciences was against postponement.

Nobody seemed to know how the government would respond. Finally, in early November, the government approved a postponement until 1915. Even though by this time it had become clear that Germany would not put pressure on Sweden to declare an alliance, and the Entente nations were not particularly interested in

Sweden, Hammarskjöld wished to keep the appearance of complete neutrality. An expert in international law, he had been involved with the Hague conventions of 1907 on neutral nations' rights and obligations. Of course, Belgium's fate on the first days of the war underscored that international treaties were not always so readily translated into practice.

The decision to suspend the prizes sparked a round of media comment. In the discussion that followed, culture proved to be very political indeed. It soon became clear that definitions of neutrality and impartiality came packaged in political wrappings. Virtually all the major Swedish newspapers regretted the decision to postpone. The liberal *Dagens Nyheter* reluctantly accepted the action, conceding that no matter how well justified the prize decisions might be, they could lead to "the most unreasonable and hateful censure and to considerable discomfort for Sweden and the prize-awarding institutions." The paper was willing to entertain the view that there was no better time to bring Alfred Nobel's high ideals to the attention of the world. But, overshadowing all other considerations, those who give the prizes must avoid having their impartiality called into question.

Conservative media favored awarding the prizes regardless of the circumstances; they also were decidedly pro-German. In fact, the conservative media were largely critical of Hammarskjöld's government, whose professional state bureaucrats were more pragmatic than they were doctrinal conservative, which did not please the political party's more openly pro-German factions. These papers downplayed the possible harm to Sweden's political position if the awards were distributed that year, and instead took the moral high ground. If the prizes were awarded according to Nobel's spirit and bylaws, they argued, then Sweden would be morally invincible and need pay no attention to unwarranted complaints. Even at this early stage, the ground rules suggested that "true" neutrality and impartiality would entail efforts on behalf of imperial Germany. To pause for reflection or to avoid the appearance of being pro-German were branded as "politically motivated" actions.

The influential *Svenska Dagbladet* came out against suspending the prizes, claiming that it was better to risk criticism instead of choosing "disgrace [*smälek*]" by not giving prizes to deserving individuals. Protecting neutrality in this manner might allow Sweden to avoid unpleasantness, but in no way would it bring the nation esteem. A few days later, the paper renewed and sharpened its attack. From the very first, it found the idea of postponing "very inappropriate." The formal decision by the government was "completely wrong." While mourning the fact that science and culture were being pulled into politics by the warring nations, Sweden should and could remain unaffected. By not awarding prizes, however, Swedish institutions had capitulated. They had endorsed the doctrine that culture cannot be kept separate from politics. Sarcastically, the editors commented that if the Swedish academies and government had nothing worse to worry about than the fear of foreign accusations, then all would be quite well. In a final blast, the paper charged that by avoiding unpleasantness, Sweden had now expanded the war zone—not to the North Sea, but to the territory of intellectual endeavor. By having acted in a cowardly fashion, Sweden had failed in its duty to guard these frontiers, and it would receive neither thanks nor respect in return.

Interestingly, the socialists joined in the attack. Strongly antinationalistic, they saw in the "universal republic of science" a reflection, if not a model, for their own international brotherhood of workers. In petitioning the government for a postponement, the Swedish academies had capitulated in the face of nationalism. But as the war continued and the politicization of culture deepened, socialists found this position untenable. It became clear that the campaign of "neutrality" and "objectivity" had been captured by pro-German and antidemocratic interests.

For those who favored close and open ties to Germany, the government's actions seemed to represent a clumsy, legalistic neutrality based on the terms of the Hague conventions. Although, Hammarskjöld enjoyed broad support for his efforts to keep Sweden out of the conflict, his critics on the right charged him with timidity. For them, avoiding war should not imply lack of political backbone. The government's neutral policy seemed to entail a series of muddled responses, seeking convenience and advantage for Sweden rather than exposing principles or respecting shared racial and cultural heritage. *Svenska Dagbladet* resolved to keep the Nobel prizes before its readership. Just before Nobel Day, December 10, 1914, the paper printed highlights from an interview with the Swedish establishment's national poet and now head of the Swedish Academy, Werner von Heidenstam. The article opened by asking whether the traditional courageous Swedish spirit of past eras was still alive—and then followed with the simple statement that equated the cancellation of the Nobel prizes with cowardice.

Von Heidenstam claimed that Sweden's action with respect to the Nobel prizes evidenced a decline in national character. It exposed a loss of proto-Germanic strength and patience. Before the flame of true Swedish nature was completely extinguished, he pronounced, new life must be encouraged. Just as German professors and artists had welcomed the war as a means to wash away the pollution of materialism, individualism, and political division, von Heidenstam claimed that Sweden, too, had to find its place in the fight for survival among nations. In so doing, it might be possible to forge national purpose through the cultivation of the Germanic virtues of duty, character, and patience. In this respect, the "most foolish and absolutely the most cowardly" thing the Swedes could do was to postpone the awarding of the Nobel prizes. According to him, Sweden had chosen to ignore its self-respect as a civilized nation; it had chosen to abrogate its cultural obligations. To have given out the prizes, to have refused to capitulate, would have resulted in a neutrality that could have been respected rather than scorned.

Underlying these charges was a broader message for social politics: Neutrality should not mean national passivity and lack of resolve. Neutrality must engage the nation, bind the nation through a willingness not to acquiesce with other nations. Sweden must not be afraid to state its sympathies for one or another power; otherwise, neutrality was merely cowardliness. How else could Sweden partake in the glorious spiritual renewal being brought about by the war? The same birth of a new spiritual era, the same unleashing of new character-building forces, and the same driving out of cynicism and decadence that were destined to generate a new and victorious Germany must also bring about cultural renewal in Sweden. He added that although socialism might seem to be gaining, in the end socialism was doomed to defeat; a new age of conservatism was dawning, just as German intellectuals were proclaiming.

For von Heidenstam and most members of the Swedish elite, continued close ties with Germany and renewed efforts to create a fellowship of Germanic peoples based on Germanic values were very important. But, unlike Germany, Sweden had a duty to administer the Nobel prizes—and this duty entailed an expression of national cultural politics. For them, the question remained: Would Sweden live up to its true heroic nature or wallow in indecisive cowardliness?

## 1915: TAKING A STAND

Insistence on the nonpolitical nature of elite science and culture remained a guiding light for the Nobel establishment. But October's winds of antagonism grew to gale and then storm force in 1915. Was this the start of a long-term change in cultural climate, or was it just a short-lived freak of nature?

### Internationalism in Science

The myth that the sciences were never at war had long proved popular, especially among scientists. Images such as that of the English chemist Humphry Davy receiving honors from the Paris Academy of Sciences in the midst of the Napoleonic wars enabled many nineteenth-century commentators to distinguish between the nasty partiality of politics and the tranquil rationality of science. It was fashionable to point to science as a model for nations to follow: networks of rational, objective experts working together across national borders interested only in the advance of knowledge, uninfluenced by passion. Wise men, if there were any, understood the gossamer quality of the prevailing prewar internationalist rhetoric. True, by the end of the nineteenth century the world of science and learning had become increasingly supranational. International scholarly associations started and the number and frequency of congresses had increased. National learned societies exchanged publications. Scholars visited and corresponded. National academies coordinated activities through an International Association of Academies. International commissions sprang up in several sciences such as meteorology and astronomy, for which exchanges and coordination of data were necessary.

Internationalism and service to high-sounding idealistic goals were, in part, self-serving ideologies, inflated and displayed, manipulated and discarded as circumstances required. Some people clung to the ideal, even when waves of nationalism seemed to undermine the most fundamental of all beliefs, that is, the validity of scientific findings across national boundaries.

The rapid growth in institutional science during the last decades of the nineteenth century paralleled the equally rapid growth in science's importance for government. Scientists were increasingly integrated into nationalist politics and agendas. In the decades before 1914, belief grew in science as a resource for the nation. The German case was most persuasive. A Prussian victory over France in 1870 and the subsequent establishment of a German Reich under Prussian leadership underscored the importance of science as a real and symbolic factor of national strength. Under imperial Germany, scientists achieved unprecedented prestige, as well as state and corporate support for research. Having convinced themselves and their leaders of the importance of research for the nation, German scientists accepted their roles in the rapidly industrializing and militarizing unified Reich.

Scientific research as a resource for national strength and national pride created increasingly tighter dependency between scholar and patron. Similar processes, often in response to German developments, were under way in Britain, France, and, to a lesser degree, in other Western nations. Scientists, especially academic physicists and chemists, learned to balance a number of hats. To their academic colleagues they pleaded their purity in the search for knowledge. To politicians and industrialists they proudly pointed to their ability to contribute to economic and military competitiveness. Bringing honor to their nations and proclaiming patriotic fervor went hand in hand with calling attention to their service in the temple of international learning. Reputation and recognition abroad always helped secure greater prestige at home.

Much of the organizational infrastructure for international science actually arose from efforts to strengthen national activities. Scholars knew that requests for funds at home had greater hope for success when pitched in terms of national honor and of a civilized nation's duty to contribute to international cultural cooperation—and competition. Equally, scientists juggled a number of loyalties and identities: to nation, to institution, to discipline, to transnational colleagues. But while leaders of national scientific communities could modulate the rhetoric of nationalism and internationalism, the war stripped bare the veneer of internationalist civility to reveal in far too many cases a brutish xenophobia festering beneath.

The escalation of hostilities raised the question of whether anything resembling the prewar community of international cooperation could ever be reestablished. Scientists in the Allied nations believed that they could not. And for many, such a negative outcome was to be desired.

Years of resentment against German academic arrogance now turned into attacks against German research. Many Americans and Europeans concurred with the statement that the war "would pull down from its pedestal and shatter for ever the notion of the German super-man in science, literature, art, of ingenuity created by German self-assertion, and supported by the effusive adulation of a few professors of our own, proud of a smattering of second-rate Teutonic learning." The ever-growing number of *Jahrbücher, Zeitschriften, Zentralblätter,* and numerous other forms of academic journals and monograph series that were the symbol of German scholarly dominance were now openly called into question. Germans, it was said, obsessed over minutiae of detail, generally losing sight of the point of the scholarly work while weighing the reader down in ponderous language. English scholarly writing possessed a good-natured and discursive tone, while French writing, as seen from the English-speaking world, offered a "form which is sometimes a little vague but always suggestive and elegant."

There were suspicions that much of the apparatus for international cooperation were actually instruments of German cultural domination. The editor of the Italian science magazine *Scientia* reminded the English that just as German industry had practiced a policy of dumping its manufactured products on world markets, the same aim lay behind the practice of "scientific dumping." According to the editor, "[These have been] increasing in number and volume, have gradually monopolised the whole of the scientific production of the world by gathering widely, and even demanding, the collaboration of learned men of all countries. Thus were apparently built up international scientific organs, but are in reality German instruments of control and monopoly of science."

Charges of intellectual dishonesty by German researchers grew in intensity. Allegedly, Germans claimed precedence for the achievements of others. They used international congresses to insist on the priority of discovery and the right for setting nomenclature and research agendas.

Some German scholars reacted vehemently. The winner of the 1905 Nobel Prize for physics and then-emerging hypernationalist Philipp Lenard charged that scientists in the Entente nations often neglected to credit German scientists: "Away with humility before the graves of Shakespeare, Newton, and Faraday! Indeed, the present-day Englishmen are not of the same type: the souls of these great men have moved on to other countries." The 1911 prizewinner Wilhelm Wien composed a proclamation calling for a German response to the hostile English attacks. In this *Aufforderung,* Wien, joined by fifteen German physicists, called for a boycott of English physics: no publishing in English journals except to respond to personal attacks, no citing of English literature except when absolutely necessary, and no writing in any language other than German. Lenard demanded that the Nobel committees in physics and chemistry stop sending invitations to nominate in French—the only language used other than Swedish. The committees agreed to send him and others who recoiled at the sight of an accent égu an invitation written only in Swedish.

But there were some signs of moderation, perhaps with a day of reckoning in mind. Max Planck, who was attempting to serve as moral pathfinder, did not sign Wien's call for action. But when a countermanifesto calling for internationalism and a renunciation of nationalist passions circulated in Berlin, virtually nobody joined its authors, Albert Einstein and a prominent physiologist, in signing. Planck regretted the decline in the internationalist spirit, and he therefore regretted having signed the Manifesto of the Ninety-Three. He did not publicly repudiate its content, but through private channels he urged his colleagues in the Allied nations to appreciate his belief that "there are domains of intellectual and moral life that lie beyond the struggles of nations, and that honorable cooperation in the cultivation of these international cultural values and, not less, personal respect for citizens of enemy states are indeed compatible with ardent love and energetic work for one's own country." Soon after the manifesto, Planck was communicating with Arrhenius to minimize damage to the international spirit that they both so dearly admired. Neither of these idealistic moderates succeeded.

## Defining an Honorable Neutrality

By the first months of 1915, the war, especially on the western front, stumbled into stalemate. By the first months of 1915, Swedish commentators began to accept that a return to the prewar order of things was unlikely. *Svenska Dagbladet* continued its efforts to steer Swedish cultural politics under a banner of honor. Waving the flag of impartiality, the paper's editors continued to campaign for Sweden as a self-proclaimed neutral arbiter in the civilized world. But Sweden and its important Nobel-awarding institutions simply could not ignore the growth of hatred and distrust.

In March, as news circulated of the charges and countercharges among European savants, the paper sent a questionnaire to selected foreign scientists, humanists, and artists, asking answers to two questions: What will be the consequence of the war for future international cultural cooperation? And what measures will be necessary to overcome the conflicts that have arisen because of the war? Beginning in May, *Svenska Dagbladet* published the responses regularly under banner headlines.

It then published sixty-three of them in a book that received considerable attention. No clear pattern emerged, yet the collage contained much of value, especially for the Nobel establishment. In an effort to summarize, an unnamed editor noted two approaches. One school of thought—largely German—maintained that political animosity should not be transferred to intellectual culture. The other—largely French—asserted that any future resumption of international cooperation could only come about by excluding Germans. English respondents divided between the two. In reality, neither the choice of respondents nor the analysis of their replies could be claimed to be impartial.

Germans were largely portrayed as willing to forgive and to turn the clock back to pre-1914. German respondents assumed a position of noblesse oblige: As members of the leading cultural nation, they readily acknowledged French and English contributions to civilization, and they were willing to continue cooperation. They were even amenable to taking the first step, but they would respond indignantly should they be rebuffed. But these German responses, and the Swedish editor's annotations, were written as if the Manifesto of the Ninety-Three did not exist. Historian Hans Delbrück asserted that it was foreign scientists who made nationality an issue in culture. Others pointed out that Germans were always able to appreciate Shakespeare, Rembrandt, and Michelangelo; foreigners who rejected Luther, Dürer, or Goethe because they were German were unreasonable. It was part of German cultivation [*Bildung*] to appreciate other cultures.

Some Germans addressed their Swedish readers directly. Berlin professor Hugo Convents reminded the Swedes of the closeness between their two peoples. He recalled the remarks of Swedish historian and member of the Nobel Foundation Oscar Montelius, who said that the ferry connection between Sassnitz, in northern Prussia, and Trelleborg, in southern Sweden, went back to the Bronze Age. Convents hoped that the war would renew this age-old relationship and bring closer scientific, artistic, and technological cooperation between the two nations. He called upon the scientists from the neutral nations to take the lead in reestablishing international relations. Sweden, in particular, should grasp the challenge as part of its important and challenging role as the custodian of Alfred Nobel's testament. Delbrück also addressed his Swedish audience, claiming that he would like to see the Swedes join in the defense of Germany against "Cossack domination," but he accepted that they and other small nations desired to remain neutral. But these countries, and especially Sweden, could help out later by mediating the resumption of international cultural cooperation.

The German poet Richard Dehmel reminded the Swedes that, actually, nobody could appreciate the French and Belgium cultural monuments, which were destroyed in the war, better than the cultivated German soldiers. Clearly aware of the upper-class readership of *Svenska Dagbladet,* he put the real blame on the "hurrah patriotism" of the vulgar middle class and parvenu in all nations. Dehmel reassured friends in Sweden that Germany was still essentially a nation of "poets and thinkers, dreamers and brooders." Germany's goal was a peaceful one: to spread its advanced culture around the world. Said Dehmel: "We are fighting for mankind, for the sake of divine peace."

Still, even among the measured, restrained German responses, clear bitterness occasionally boiled over, especially among the scientists. Nobel laureate Wilhelm Wien wondered whether it paid to seek personal association with a people "who have so little ability and good will to understand other people" as the English. He shared the reaction of many of his colleagues: Precisely because they had so respected the English, and had worked to create true bonds of brotherhood, that English attacks on Germany caused such deep bitterness. Toward their French scientific colleagues, the German respondent showed little antipathy. But because the Paris Academy of Sciences had voted to expel its German members, it would undoubtedly be impossible for a long time to come to reestablish personal relations. More generally, Wien called for greater emphasis on cultivating *national* science and forgoing international congresses. This need not be a disadvantage; each people should focus on strengthening its own cultural strengths while learning to respect those of others. The Swedish people, through the Nobel prizes, could then proclaim, impartially, which accomplishment deserved acclamation and show that cultural work belongs to humankind and must be respected across national boundaries.

In summarizing the French and English responses, the editor made no mention of what aroused these emotional outbursts in the first place. *Svenska Dagbladet* had loaded the dice. For example, the editor noted that among those expressing anti-German sentiments, the two 1912 Nobel Prize winners of chemistry, Paul Sabatier and Victor Grignard, had asserted that the chasm between the two nations could never be filled; not even public penance would enable a restoration of solidarity. These comments were presented simply as an example of French intransigence. Yet the full texts of their replies gave ample reason. Sabatier claimed that French scientists had close friends among German colleagues, which was why the appearance of their names among the ninety-three signers of the manifesto caused such pain. The burning of Louvain University, the bombardment of the cathedrals, the shelling and other horrors against women, children, and priests were what stood between German and French scientists. Even more painful was the assertion that German culture was supreme. Grignard observed that when men who should be pursuing the search for truth, beauty, and goodness not only attempted to justify crimes but also made an apologia for them, human comprehension was sorely taxed.

French respondents believed that some form of international cooperation would ultimately be restored. But this cooperation could at best only be on an impersonal, formal basis, such as in the organized exchanges of publications. Sociologist Emile Durkheim, for one, found the thought of shaking hands with a German sufficient to send a shock through his being. Personal relations were out of the question, as the Germans, he declared, have "excluded themselves from civilization and have shown us how distant they are from its spirit and ideal." French writers and painters were now "vaccinated" against German influence. French scientists still intended to acknowledge German accomplishments with the intellectual honesty that always allowed the French people to recognize foreigners' discoveries. Several writers warned the neutral nations to understand that the conflict between Germany and France went deeper than political or national hatred. Germany had violated civilization. Only by renouncing militarism and by cleansing themselves of "Prussian predatory savageness" could German men of learning hope to regain a human soul. If neutrals attempted to reestablish international relations without appreciating these feelings, their efforts would be doomed.

Even more disturbing were the admonishments to those who awarded the Nobel prizes. Historian Frédéric Masson asserted that the Nobel prizes for literature were being awarded preferentially to Germans, "to the indignation of all

French writers." Grignard challenged the Nobel committees to consider that not even an impartial elector has the right to bestow an international prize to a researcher who has not "a sufficiently high-minded spirit to enable him . . . to allow the voice of justice and truth to be heard." Sabatier warned that in spite of their neutrality, Swedes must appreciate that nominations could not now be considered impartial. German scientists could only see the Frenchman in a French researcher, and vice versa. The prize should not be conferred. He made oblique references to what were seen as conspicuous actions of "certain Swedish scientists" on behalf of Germany while turning their backs on past relations with France; this debased the idea of the prizes as expressions of impartial judgment.

Finally, English hard-liners insisted that German science and culture no longer held attraction. Literary historian Edmund Gosse said that he would not mind if the cleavage dividing French and English culture from that of the "Teutonic spirit" continued to deepen. H. G. Wells suggested that the imperial German state had systematically corrupted its schools and universities while making its teachers and professors slaves. Sir William Ramsay proposed that Germans and Austrians be ostracized indefinitely from future international congresses; for generations to come, treating them as equals would be impossible. Acknowledgment of significant work should always be forthcoming regardless of national origin, but (in a clear message to his Swedish audience) he said that any effort to renew former relations would fail. Scientific congresses that included Germans and Austrians would be boycotted.

For academic internationalists, such as Arrhenius, whose career depended upon international congresses and plying international networks, the future looked bleak. Calls were being made to end these practices. As one Italian historian remarked, with the destruction of tolerance, international cooperation would suffer. But science and literature could progress even when their practitioners hated one another. Congresses will be more difficult—"but the serious researcher should regret that less than hotel owners." The 1913 Nobel Prize winner in medicine, Charles Richet, was even less charitable. He declared categorically that the ideals that sustained the Nobel prizes were gone: "Our highly prized European civilization has ended in an unprecedented massacre: noble Belgium is burnt, plundered, martyred." International intellectual cooperation was finished. But this was not a problem. Neither scientific discoveries nor aesthetic creations required international cooperation. René Descartes created analytic geometry alone; Antoine Lavoisier discovered oxygen without any scientific congress. Individuals, not groups, create. For many elitists, such as Richet, the war confirmed that the individual person was powerful and able to bequeath great and beautiful things, but as a collective organism, humankind was completely unable to achieve even the most paltry intellectual effort.

No member of the Swedish establishment could read these responses without being depressed. For the conservative's cultural program, the message was clear: Swedish neutrality should defend Germans from being cast as barbarians. Sweden must take a leading role in reestablishing harmonious relations in Europe. Here was a niche in international cultural affairs for an impartial and morally upright Sweden. And just as the Swedish Foreign Ministry had secretly allowed some of its embassies to let their German counterparts exchange telegrams with Berlin, quiet, small actions that did not awaken distrust could serve as gestures of continued friendship among the Germanic peoples. Was this the way forward for the Academy?

## To Award or Not to Award, Which Is Neutrality?

Nobel committee members could not avoid questioning the significance of the prizes at a time when the civilized world seemed to be headed to collective suicide. Newspapers carried details of barbarism: hundreds of thousands of soldiers mowed down by machine guns, gas, and submarine warfare. One poet described the details vividly:

> *The wheels lurched over sprawled dead*
> *But pained them not, though their bones crunched,*
> *Their shut mouths made no moan,*
> *They lie there huddled, friend and foeman,*
> *Man born of man, and born of woman,*
> *And shells go crying over them*
> *From night till night and now.*

How should committee members respond to this reality, much grimmer than the name-calling of academics? The shock of this war, coming in such contrast to expectations of an era of peace and harmony, opened depths of despair. In the prevailing eugenic thinking of the day, shared by both sides of the political spectrum, European civilization faced catastrophe. The best genetic material was rotting on the fields of Flanders. By mid-1915, nobody dared to predict how long the carnage would continue.

Although long delays and censorship created some difficulties, an extraordinary communication network among scientists continued throughout the war. Of all the committee members, none had so many and so widespread contacts with foreign researchers as Arrhenius. His widespread network allowed him to serve as a conduit for news and views from warring and neutral nations. As one of the most prominent scientists in Sweden, Arrhenius did what he could to mediate and steer. Not all members of the Academy shared his liberal political preferences, but most shared his anguish at the loss of international community.

Arrhenius was distraught. He foresaw that, as the war continued, Europe would be devastated, America would make profits. Europe would sink back in standards of material culture by fifty years. In 1915, he seemed to be hoping for a negotiated peace, a new Christian alliance, by which nations might sacrifice small pieces of territory to Germany, which in turn would remove their debts and protect them. Inability to achieve such a settlement could be blamed on England's opportunism as well as Russia's willingness to use its size—coupled with an inferior level of civilization—to keep fighting. There might even come a permanent mobilization, possibly leading to more wars. Still, after the devastation, perhaps the peoples of Europe would no longer support militarism and maybe even embrace peaceful coexistence.

Trying to keep his affection for German culture while abhorring German militarism, Arrhenius appreciated the difficulties facing the future of international scientific relations. When Arrhenius signed a protest statement against German war methods, Ernest Rutherford wrote to express his approval, understanding how difficult this act must have been for the Swedish scientst:

> I feel, however, that if the neutral world makes no sign of protest, the Germans will continue more and more to use promiscuous murder of non-combatants as

their chief method of warfare.... I am expecting even worse things in the future. As far as I can interpret English opinion, it seems to me that whatever may be the result of the war, all social and scientific intercourse with Germany will be practically stopped for this generation.

Rutherford wondered what was being done about the Nobel prizes "in these troublous times." No clear answer could be given. Arrhenius was saddened over the blatant use of the prizes in propaganda, as when his old friend Ostwald equated the large number of Nobel prizes with Germany's cultural superiority. Arrhenius ruminated over the dubious giving of prizes in such a world; maybe the whole institution should be overhauled. Many smaller grants would advance the progress of culture better than the one huge prize.

Still, the machinery of processing and evaluating for the annual decision carried on. During the spring and summer of 1915, the committees continued their usual work. Invitations to nominate had been sent out the previous fall; committees had met in the winter to act on the nominations that had arrived by February 1. Although the number of nominators who responded was greatly reduced in physics—from thirty-eight to sixteen—responding chemistry nominators increased from twenty-eight to thirty, partly because of increases from the Academy and committee. As usual, the committees divided up assignments for special and general reports.

The outcome was all but given. In chemistry, the nominators gave T. W. Richards and Richard Willstätter clear superior support, confirming the committee's earlier enthusiasm. Ramsay noted in his new proposal for Richards: "It would, I think, be a graceful act if the prize for 1915 were to be given to an illustrious citizen of that great country, and under present circumstances it might be especially appropriate to take that consideration into account.... I am certain that this award would be acclaimed by all chemists of whatever nationality." In physics, the same candidates who in 1914 had largely divided most of the nominations returned to dominate the list: von Laue, Hale, Planck, and W. H. Bragg. Now, however, W. L. Bragg entered the list to divide a prize with his father based on a lone nomination. Although some committee members had hoped to see candidates in astrophysics and cosmical physics awarded, the prior evaluations of von Laue and the Braggs seemed to point to how the prizes would be distributed, if they were to be awarded at all.

Although the scenario was set, one nomination directed toward both committees from Arrhenius might have given pause for reflection. Arrhenius proposed awarding a prize to young Henry Moseley, either alone or together with collaborators for the extraordinarily beautiful and significant results gained through studying X-ray spectra of elements. These suggested a physical explanation for the periodic table. But the committees simply noted this achievement as promising and waited to assess its full significance. The chemists saw Moseley's work as being primarily significant for "theoretical chemistry," which in the context of that committee was not necessarily a compliment. By autumn, both committees lost their chance to reward Moseley. News came of his death in one of the many disastrous British military actions, this time at Gallipoli.

The loss to science of Henry Moseley, certainly the most promising physicist of his generation, coming so soon after he had unraveled the meaning of the periodic table of the elements, shocked researchers in all nations. Leading British researchers used his death to pressure the British government to keep scientific talent off the battlefield and directed in other ways toward the war effort. Although the Germans and the French had begun using some scientific talent to aid national military needs other than as combatants, it was clear that whole generations of university students from all the warring nations were being slaughtered.

Moseley's death did not pass without comment in Sweden. *Svenska Dagbladet* continued its campaign against withholding prizes by leaking the news that one committee had almost agreed to give a prize to "a leading English researcher," when news arrived of his death. Certainly, the paper asserted, potential prizewinners have already been killed; any further delay would deprive deserving scientists and writers their rightful reward. Perhaps Moseley was more valuable as a symbol; the paper made no effort to retrieve the Englishman's claims. The bylaws permitted awarding a deceased person who had been nominated prior to death. Might Moseley, as heroic Oxford prizewinner, killed in a stance against waves of Turks, attract too much sympathy for Britain?

News of the tragedy deepened Arrhenius's gloom. But, in contrast to *Svenska Dagbladet,* he concluded that giving out prizes would be a mistake while the war continued and while emotions were so frayed. He shared Planck's belief that it would be best in general not to provoke further antagonism; best to keep a low profile. Although he had not voted to reserve the prize when the physics committee met in September, he proposed at an Academy meeting in October that the prizes be withheld until after the war. But an ambush awaited him. The Academy's president and permanent secretary, as well as a number of prominent scientists that included members of the Nobel Committee for Chemistry, insisted that the prizes should be awarded that year. After considerable discussion, Arrhenius modified his proposal to request a postponement only until 1916, rather than for the indefinite duration of the war. After still more discussion, those few Academy members present split the vote, giving Arrhenius a slight majority (twenty-one to nineteen). Consequently, the Academy petitioned the government for permission to postpone. But Academy President Eriksson, Permanent Secretary Aurivillius, and chemistry committee members Söderbaum and Ekstrand, among others, formally went on record in opposition to the vote.

Nothing was very secret about these deliberations. Newspapers reported and invited debate. Again, conservative newspapers repeated the need for Sweden to take a bold stand by awarding prizes. *Svenska Dagbladet* emphasized that the vote in the Academy revealed that almost half were in favor of awarding prizes. Some claimed that the jagged edges of hatred were now being smoothed down; the prizes would not provoke strong feelings. Just maybe the response from the warring nations would be respect and appreciation. In any case, at least Sweden would have done its duty by raising a voice "for something larger and more elevated than hatred and rivalry between nations" and thereby evidencing "the cultural solidarity within the entire civilized world."

Admitting to the difficulty of getting universal acclaim for prize decisions even in the best of times, those in favor of postponing argued that any claim of impartiality would be seriously threatened. Advocates of postponement also warned that plans, such as those of the literature committee, to restrict prizes to neutral nations would equally hurt their reputation for impartiality. Instead, they urged delay.

Subsequently, the payoff might be an opportunity to host a multinational Nobel celebration once the war ended. A truly untarnished Sweden could then begin the work of bringing the hostile factions together again.

Before the government took up the petition, the Caroline Institute announced that it was awarding its reserved 1914 prize to Robert Bárány, a young Austrian medical researcher, then being held prisoner in Russia. Immediately, the Nobel Foundation then voted not to support the Academy's petition. The government, which after the harangue from *Svenska Dagbladet* and other conservative papers, was no longer interested in postponement. The danger of being drawn into the war seemed to have past. It quickly rejected the proposal, giving as a formal reason that once one of the prize-awarding institutions had announced an award, it would be awkward to allow another to postpone. Given that Prime Minister Hammarskjöld was a member of the Academy and would be well aware of the likely prizewinners, he had little trouble in rejecting postponement. The committees were proposing what turned out to be a politically satisfactory group of prizewinners for the physical sciences: one for neutral America, one for Britain (split between father and son), and two for Germany.

Without a government-sanctioned postponement, the Royal Swedish Academy of Sciences had to confront at its mid-November meeting whether to make awards or not. Formally, it had little choice. The committees had evaluated candidates and recommended winners, and it would be a violation of the bylaws to withhold the prizes. At least this was the view of those who opposed a postponement. Alternatively, an argument could be made that only nominations that had arrived by February 1914 should be accepted as untainted by the subsequent corrosive effect of the war. Nominations for the 1915 prize were too politicized to be useful. The physicists made it clear that if only one prize was to be awarded, then it must go to von Laue. Whether alone or together with others, at least one German would be celebrated, and if the 1915 prizes were to be awarded, then yet another German, Willstätter, would also be feted.

Unlike the October meeting, when only forty members of the Academy attended, in November 1915 over eighty were assembled. Arrhenius held that the best way to ensure a renewal of international contacts after the war was to keep the level of hysteria to a minimum. Every act of restraint could help. Actions such as those of the German Chemical Society, which had decided not to act on a motion to expel Ramsay until after the war, offered hope. Arrhenius was placing bets on the belief that once the war ended "all the idiocies" being tossed about would be forgotten and simply "written off." Therefore, he urged the Academy, don't award prizes during the conflict. Just before the Royal Swedish Academy of Sciences met, a rumor spread that the Swedish Academy was about to award the Nobel Prize in literature to the pacifist Swiss-based ex-patriot French author Romain Rolland. The rumor prompted howls of indignation from the French. Frothing with anger, French newspapers attacked the Swedish "jerry-lovers [*bochophiler*]" who were awarding the prize for Rolland's propaganda rather than for the ex-patriot's literary achievements. Arrhenius did not need further proof of the volatile international situation; here was a glaring illustration.

The Royal Swedish Academy of Sciences decided to hold two meetings to consider the prizes for 1914 and 1915. At the first one, Mittag-Leffler quickly entered the fray on the same side as Arrhenius. Perhaps, just like Arrhenius's, Mittag-Leffler's strong German sympathies were balanced by strong friendships in other nations, especially with France. Both men were unusually well connected internationally; both appreciated the need to start preparing for a healing process. The debate dragged on well into the night. The prime minister took the floor. He reminded his fellow members of the Academy of the statutory imperative to award the prizes; he reassured those assembled that no political reason existed for withholding the prizes. Hammarskjöld was well aware that the committees had recommended a constellation of candidates who could well cast Sweden and the Nobel legacy in a favorable light while still managing to offer a solid hand of friendship to Germany.

After much discussion, the Academy decided to use an old tradition for sensitive votes—blank ballots—as a means of not voting for a proposal. According to Mittag-Leffler, a combination of factors led to the final result: Hammarskjöld's intervention, the ability of those in favor of awarding to claim that von Laue's work was truly deserving, and the many friends of Germany who firmly desired ". . . to be able to give the prize to a German." The final vote was sixty-five for von Laue, with eighteen blank ballots. Once this decision was made, the choice of Richards for the chemistry prize followed quickly, and the meeting ended at eleven o'clock in the evening. The next day the Academy voted in a like manner for the 1915 prizes, with only eight to ten blank ballots. The Braggs got the physics prize; Willstätter, the one for chemistry.

Arrhenius was soon able to breathe easier. He tried to convince himself that the international media were less interested in which nation, and to whom, the science prizes were awarded as compared with the more emotion-laden prizes for peace and literature. In fact, the Swedish Academy, deeply divided, decided that rather than award a prize to Rolland, it would postpone awarding prizes for literature until 1916. The Academy did not want to appear to be trying to sanction Rolland, whose literary worth was strongly criticized in France.

The distribution of science prizes among Germans, Englishmen, and an American—and all of indisputable merit—brought comments of praise and appreciation. Ramsay was delighted that his support had helped Richards. He even noted a ray of optimism: "I hear from neutral friends that the German 'intellectuals' are now beginning to recognise that they have been deceived, & that German military methods are detestable. It is time they did; for I can't believe that many of our old friends can have acted as they did except through ignorance."

Was this decision a fairy-tale ray of sunshine in the midst of European horrors? Richards thanked Arrhenius for his congratulations, and he expressed gratitude to the Academy. For a moment, his own worries about the war and the future of scientific relations melted in the glow of prestige and money. Richards acknowledged that the prize would help with his work and would remove the worries that a professor with limited funds and a dependent family must inevitably have. Most importantly, he related to Arrhenius the great interest generated by the prize: "It would be impossible to exaggerate my pleasure in this delightful event. . . . I venture to think that it has had a happy political effect also, for every American who knows of it and appreciates what it means is immediately thereby drawn towards Sweden, and the unity of the neutral nations in this terrible war is especially to be desired." Richards, like Arrhenius, only heard praise of the prize distribution. "It was particularly a happy thing that von Laue and the Braggs should receive prizes together; Willstätter also is everywhere regarded as entirely worthy." Arrhenius's many contacts in England, France, and Germany reported universal approval of the choices. Von Laue used Arrhenius to extend congratulations to the Braggs, and he was moved that these highly talented researchers wanted to extend the same to him. But Richards's letter, among others, revealed that Arrhenius would have to accept that even the science prizes were well noted abroad.

The Academy managed to avoid controversy this time, thanks in part to chance. It might even seem that the conservative pro-German establishment had been right all along. Sweden could continue to play its role as the impartial arbiter of international culture without losing esteem, and it could even lend a hand in advancing the cause of its beleaguered colleagues in Germany. The war was far from over. Nobody, even in the despairing air of late 1915, conceived how much longer the war would last and how much worse a catastrophe was still waiting for Europe. The luck of the Swedes would also soon run out.

# SIX

## While the Sores Are Still Dripping Blood!

### Nobel Passions: Defending *Kultur*, 1916–1919

*T*he prize decisions in 1915 provided a respite, a taste of vintage cultural optimism from years past. However satisfying it was to award the prizes without incurring the wrath of foreigners and domestic critics, members of the Nobel establishment sensed that the world of European culture would never be the same. Celebrating peaceful competition among nations rang hollow. And yet many of these men longed to maintain the order of things past. How might the Nobel prizes find purpose amid cultural and political turmoil? To what extent did those involved appreciate the sea change underway—or accept the need to alter ideals, values, and beliefs? Individuals reacted to the situation as it unfolded, fumbling to make sense of the challenges and threats sweeping them into a new era.

### STAKING OUT A NOBEL ROLE

In 1916, the question of reserving and withholding prizes again loomed large. The end of war was no closer. Reports and rumors from both sides predicted another year of war, maybe two years, even more. Now that the government had indicated a reluctance to intervene, the prize-awarding institutions would have to either play the roulette wheel by venturing to award prizes again or begin the process of reserving their prizes.

Looming just as large as this procedural question was one of principle. Committees were required to evaluate nominations and propose winners without taking into account nationality. How should this directive be weighed against falsely declaring that no candidate could be found worthy of a prize, in the name of promoting international cooperation?

The Nobel establishment began to discuss this problem as soon as the 1915 prize decisions were made. More generally, they tried to tackle the broader question of how to avoid the strain of awarding prizes annually. Such feelings had been mounting over the years. The war brought them into sharper focus. Those involved were odd bedfellows: Svante Arrhenius was joined by Harald Hjärne, chair of the Nobel literature committee; Sven Hedin, the very popular and aggressively pro-German explorer; and Oscar Montelius, Germanophile and member of the Nobel Foundation's board of directors. They agreed that the bylaws regulating the prizes needed to be changed. Regardless of their individual agendas and preexisting concerns, they all understood that Sweden had to take a leading role in reestablishing international cultural cooperation once the war ended. But to do so would require creative uses of Alfred Nobel's legacy. Maintaining the prestige of the prizes and generating funds for research were central concerns.

Swedish scientists who earlier had hoped to erect monumental Nobel-funded laboratories saw a new opportunity to revive the issue. Earlier discussion had emphasized the goal of bringing international scientists to these grand installations. Now, in the name of reestablishing the broken links of scientific collaboration, Nobel institutes could serve as attractive, neutral meeting grounds. Alleviating Swedish needs for better research facilities could be combined with reviving European culture.

Count Karl Mörner took the initiative by telling the press that he hoped funds could be raised for establishing a preliminary modest Nobel Institute for medical research. Although the funds originally set aside for this purpose in 1900 had accrued interest, further injections of cash would be necessary. For members of the Nobel establishment, two problems could now be addressed simultaneously. If prizes were first reserved and then withheld because of a declared lack of qualified candidates, then that money could be directed toward research. Of course, a change in the bylaws might allow the prize-awarding institutions the option of bestowing prizes less frequently, perhaps only once every five years, without having to squirm about declaring annually a lack of worthy candidates. But without such a change, for which the Ministry of Justice seemed to have little sympathy, committees would have to learn to make novel use of the bylaws' ambiguities.

Count Mörner preferred a solution that would ensure funds for research while avoiding ethical improprieties. He recalled the negative reactions in Germany and Austria following the medicine prize to Robert Bárány. Senior professors in Vienna marked their disapproval by refusing to promote the young instructor [*privat-dosent*] to an associate professor [*extra-ordinarius*], claiming that his work was too insignificant. Some even charged that young Bárány plagiarized his prizewinning achievements related to the physiology and pathology of the inner ear from their own research projects. Whether motivated by anti-Semitism, envy, or a combination, the hostility shown Bárány prompted Swedes to create a position for him on the Uppsala medical faculty. It also prompted Mörner and others to claim that to avoid a repetition of such scandalous charges against a Nobel laureate, they must raise the standards for receiving a prize, at least for the duration of the war. If Bárány could not win universal acclaim, it was argued, then the committee had better search for higher, uncontestable levels of accomplishment. It did so by insisting that only a specific, very recent discovery of great significance would be considered prizeworthy. In this way, even though awards previously had been given out annually, and even though several potential candidates had been under consideration for a prize, the medical committee could now state in its evaluation reports that no candidates could be identified as being eligible for a prize.

Some ideas are too good to be held secret. The Nobel Committees for Physics and Chemistry learned quickly from their medical colleagues how to withhold prizes without losing sleep. Drafts of evaluation reports on candidates for the 1916 Nobel

Prize for physics show the process at work. During the summer, committee members agreed that several candidates were clearly worthy of a prize. Astrophysicists George Ellery Hale and Henri Deslandres, Max Planck, and Johannes Stark were all found deserving. Stark, a vocal German nationalist, was nominated for his experimental discovery of an electric field's ability to split spectral lines, which in turn offered experimental opportunities to study atomic processes. Stark appealed to the committee majority's experimentalist tastes.

The committee proposed Stark for the prize; the other candidates were declared worthy but would have to wait until a later date. These conclusions were in the report discussed on September 13. By the next meeting, however, on September 21, the text had been altered significantly. The committee members had, in the meantime, agreed to follow the medical precedent and raise the Nobel standard. Its earlier draft was now sanitized. All references to candidates being worthy of a prize were crossed over. Penciled in were statements that no matter how important these discoveries might be, they required further monitoring to assess their full significance. Since each annual report entered into the canon of established evaluation, committee members had to make sure they did not create future obstacles for these candidates. The rhetorical device, that a given work, although of great importance, had to be watched further, emerged as a means to keep candidates in a holding pattern until their turn arrived. Finally, the report ended with a newly sketched conclusion saying that no candidate had been found deserving; therefore, the prize would be reserved until 1917.

The final draft, cleanly retyped, signed by the committee members, and sent to the Academy, gave no hint that any candidate had ever been deemed worthy of a prize. The same conclusion came from the chemistry committee. The Academy approved. The first step toward raising additional funds had been taken under the cover of raised standards. Perhaps just as important, Arrhenius had managed to convince his physics colleagues that by not feting a vocal German nationalist today, they could help all physics tomorrow. Although the official protocol maintained the illusion that only scientific criteria had entered into the proceedings, the newspapers reported leaked inside information that one or another candidate would have received a prize had the *war* not prompted reserving the prizes.

The Swedish Academy, the institution responsible for awarding the Nobel Prize in literature, chose a different strategy. Members of its Nobel committee were less interested in saving money for research. The literary Nobel Institute required only modest funds: a library with reading room and a budget for books. For committee members, the question of withholding prizes was less a matter of economics as of continuing to play at being a beacon of impartial judgment. Under its culturally and politically reactionary leader, C. D. af Wirsén, the committee and the Academy tried to shore up neoclassical idealism, at home and abroad, against the corrosive perils of modernism, naturalism, realism, and national romanticism—impartially, of course. Although under Hjärne's leadership the committee began to change aesthetic criteria, the war prompted clear political measures.

The literature committee decided that the prize should be awarded to writers from neutral nations, preferably from the smaller and often ignored nations, as long as the war continued. It paid no attention to the fact that this strategy formally conflicted with the bylaws, which insist that nationality play no role. Regardless, the committee was soon reminded that neutrality was contentious and ambiguous. In 1916, it arrived at a proposal to award the reserved 1915 prize to Benito Perez Galdós and the 1916 prize to two Danes, Karl Gjellerup and Jakob Knudsen. But the full Swedish Academy, aware that a prize to Romain Rolland would enrage the French and actually please some Germans (Rolland was highly critical of France), discounted the committee's recommendation. Instead, it awarded Rolland the reserved 1915 prize. Although some committee members challenged his literary merits, his idealistic pacifism attracted broad support.

Rolland's flowery, visionary response in 1915 to *Svenska Dagbladet*'s questionnaire surely appealed to the Academy's majority. As was reproduced in the newspaper and in the compendium of replies, Rolland waxed poetic on the fate of humanity, which takes precedence over all the fatherlands. He repeatedly had hit the right notes: Nothing can prevent the resumption of intellectual ties between the warring nations. To deny this would be to commit suicide. The idea of mankind's brotherhood will explode into the open, once the muzzle of civilian and military dictatorship is removed for all time. Holding his convictions dearly, Rolland was not the least worried about the future of European society. He declared that he shared the suffering of the millions of innocents killed. This moment of war was in reality the baptism in blood of a new, invigorated Europe. Rolland's bombast seems to have appealed to Academy members, sitting at their desks in a neutral country. Although Rolland's pacifism was not popular among German men of culture, he nevertheless enjoyed German popularity as an uncompromising critic of France.

For the 1916 prize, the Swedish Academy also rejected its committee's recommendation and instead turned to its own leader, Werner von Heidenstam. His recent publicized attack on Swedish cowardliness with respect to neutrality made it clear that he stood for an active supportive policy toward Germany and traditional German culture. Interestingly, Carl Spitteler, whom the Academy originally had proposed for the 1914 prize, had now fallen completely from grace. The Swiss poet had incurred the wrath of Germans and Austrians because of his "incomprehensible position" against Germany. Although his aesthetics could be questioned, the committee and Academy eventually did award him a prize once the war, and his political liability, ended.

## 1917: WAITING FOR SOMETHING TO HAPPEN

With the autumn of 1917, a new round of decisions loomed. Europe had sunk yet deeper into despondency. Nobody dared guess when the war might end. A ray of hope came when the Americans entered the war in April, but how many months or even years would be required before they might tip the scale? Perhaps, as growing numbers of Germans urged, a compromise peace could be negotiated. Regardless of hopes for an end, the European reality in 1917 was one of spiraling economic and political collapse. Food shortages were rampant; famine appeared in eastern Europe. There were new massacres on the battlefields to gain, temporarily, a few miles of territory. Meaningless military offenses trampled the spirit of European civilization further into mud and rot. Unrestricted German U-boat attacks on commercial shipping, including ships from neutral nations, brought emotions to a boil, but it did not break the Triple Entente's will to fight. In April 1917 alone, torpedoes sank more than a million tons of British and neutral shipping. But by the autumn, convoys, submarine detection devices, and means to decipher secret codes enabled the Entente to blunt the edge of German submarine warfare.

The collapse of the Russian war effort and the uncertain future of the Russian Empire raised new questions for Swedish neutrality. Germany was ready to intervene in Finland to drive the Russians out. Swedish assistance, openly or passively, could result in that country's regaining the Åland islands in the Baltic Sea as well as removal of the traditional Russian menace from the east. Heavily weighing against this opportunism was the fact that Sweden had become dependent on trade with Britain and America, especially for grain and food imports. In 1917, domestic concerns overshadowed international diplomacy.

Blockades, unrestricted submarine warfare, and shortages of basic commodities from abroad contributed to social and political instability in Sweden as elsewhere. Great wealth was being created among merchants, speculators, large farm owners, and shipping magnates. Union-organized workers in industry were forcing higher wages from the owners who were themselves reaping large profits. But as food prices skyrocketed, those on fixed incomes, such as teachers, academics, government employees, small farmers, and non-factory workers, experienced a sharp loss in buying power. The unusually harsh winter of 1916–1917 created misery for much of the population as exorbitant prices for coal produced freezing homes and workplaces. Strikes, protests, and civil disobedience marched onto the national landscape. Demands for greater democratic rights, including universal suffrage for all men (and even for women) and a truly parliamentary system of government, gained considerable momentum.

Hammarskjöld's moderately conservative government had so far kept Sweden out of the war, but he stood for the old Swedish tradition of rule by elite experts rather than by elected party representatives. The decline in social stability discredited the government and those who advocated resistance to parliamentary rule. In 1917, Hjalmar Branting, the leader of the socialists, emerged as the most popular politician in Sweden. But when Hammarskjöld stepped aside, the king turned to the head of the Liberal Party, Uppsala history professor Nils Edén, to build a left-of-center coalition with the socialists. This government was decidedly more friendly toward the Entente powers than its predecessor. Through rationing, Edén's government tried to alleviate what were becoming the worst living conditions since the 1860s. Growing social unrest took on an even darker coloring when news of the Bolshevik revolution arrived. Equally unsettling, bloody civil war between reds and whites broke out in neighboring Finland after the end of Russian imperial rule. Could Swedish, German, and French Social Democrats turn into Bolsheviks? Would whatever was left of the prewar world and its values be swept away by revolution?

## Arrhenius Mobilizes

Arrhenius followed and judged the changing political, military, and economic fortunes of the war; he found it difficult to do much else. Others in the Academy, less agile or less willing to confront the world, became lethargic. The Nobel establishment had other reasons to be despondent.

Foreign scientists and academics not only were engaging in propaganda and politics but were also lending their skills to the war effort. Not only was internationalism in ruins, the nineteenth-century ideal of creative individualism was shattered.

Amid the growth of democracy, anonymity, and mass production, scientists and artists alike claimed that, unlike factory laborers, they did not—and could not—produce quality goods through organized, administered teams working toward set goals. And yet just such cooperative teams were emerging in all the combative nations, using their research skills to solve war-created problems. Unprecedented funding from state and industry served to fertilize teams of academic researchers working on war-related projects. World War I has often been called the "chemists war" because of the widespread use of this science to create "ersatz" materials, produce munitions, and spawn deadly gas weapons. But, in actuality, most sciences were mobilized to resolve myriad concerns—psychology, geology, physics, and meteorology—all playing conspicuous roles.

Arrhenius mourned the decline of the Germanic race, as many of Germany's best students were sent to the front. Others, such as Vilhelm Carlheim-Gyllensköld, mourned as well the death of science as an innocent activity pursued by men of independent means. Since nation-states had regressed to barbarism, he mused over the possibility of an international cooperation based on a joining of "rich, private people in the countries where scientists exist who are working on things other than explosives and poison gases or forecasting weather for the Zeppelins."

News of foreign, and especially German, colleagues continued to filter into the Swedish scientific community. Stories were told of professors losing sons and students to shrapnel and field-hospital disease. Confessions came from professors who now appreciated that the war was not a valiant cause but instead an evil rotting away the fabric of the social order. Although Arrhenius grudgingly conceded that socialist rule might actually be desirable, to eradicate militarism and profiteering, many of his colleagues began to dig in, to resist as best they might the rising forces of democracy in politics and culture.

Not much enthusiasm could be mustered for the annual Nobel rituals in 1917. Committees and the Academy mechanically proclaimed that no deserving candidates could be found. Hugo Hildebrandsson reported that the meeting of the Academy's Physics Section lasted but fifteen minutes; for once, there was no "squabbling," as everyone was in agreement. The money from the 1916 prizes went to the respective committees' special funds; the 1917 prizes were reserved until the following year. The Swedish Academy continued its policy of finding authors from neutral lands—but only those embracing aesthetic and political ideals to its liking.

## 1918: PREPARING FOR A POSTWAR ERA

By spring 1918, the war seemed to be approaching a conclusion. Germany had made peace with the Bolshevik government, and it could now concentrate its armies on the western front. New technologies such as tanks, bomb-carrying airplanes, and heavy, long-distance cannons—"Big Berthas"—foreshadowed a break in the stalemate. Toward this goal, German General Ludendorff launched a major offensive against the French and British front. Now, too, American troops and equipment were entering into the equation. Deepening fears in most countries that social discontent was accelerating into revolution gave Germany incentive to consider a political settlement. President Wilson announced his Fourteen Points as a guideline for achieving a peace without revenge.

The political climate in science had also changed. Allied scientists began discussing plans for new international cooperative arrangements, but few spoke for a rapid reconciliation. British, French, Belgian, and American scientists began organizing new associations for international cooperation. Learned men from the Central Powers would be excluded from them for some indefinite period, while neutral nations would be permitted to join only if they accepted the boycott. Swedish coziness with German colleagues could easily become a point of contention with the most aggressive advocates of boycott.

Such was the general background when the Nobel committees began assessing the nominations that had trickled into Stockholm in late 1917 and early 1918. Pride, principles, and pragmatism conflicted as committee members scrambled to situate themselves in the new power structure emerging in international science. The physics and chemistry committees reacted differently to the changing national and international political climate. In part, this difference arose from the fact that Arrhenius—still a key factor—was a member of the physics committee, where he could influence his colleagues directly, but only an unofficial, informal adviser to the chemistry committee.

## Arrhenius Takes Charge

Arrhenius assumed that the German defeat at the Marne would result in a negotiated peace by spring 1919. His postwar plans included a range of scenarios to buoy his own career, to revive international cooperation, and to nudge the Swedish state into recognizing the strategic importance of research. Arrhenius had constructed his professional identity in terms of international contacts. His frequent jovial presence at meetings and his never-tiring correspondence gave him an unusually strong anchoring. Moreover, he had accepted the role of a public scientist. His frequent interviews in media as well as his immensely popular lectures and books brought him stature not only as an interpreter of science but also as an advocate of science in modern society.

Arrhenius believed in science. He also believed that rational, nondogmatic thought would defeat reactionary politics, injustice, and other social ills. His strategy for the Nobel prizes, as peace loomed on the horizon, was guided by three principles. First, Sweden must invigorate local research facilities. Both for economic and social recovery, as well as for creating a neutral meeting ground for foreign researchers, Sweden must increase its investment in science. Second, German research, the fountain of inspiration for international science, must not be allowed to dry up. Third, Sweden must make strong positive links with Allied scientific communities, both to avoid isolation and to better assist the reintegration of German cultural workers. The authority and resources of the Nobel Prize would be crucial on all levels. Arrhenius appreciated that Sweden and its Nobel establishment must act with alacrity to anticipate the postwar reconstruction of Europe. Only by being perceived as truly neutral could they fulfill this mission. But as political turmoil increased, Arrhenius and his colleagues seemed to resemble men on a raft swept along by a flooded stream and desperately trying, each in his own way, to maneuver to safety.

At the 1918 Royal Swedish Academy of Sciences' Commemoration Day [*Högtidsdag*], Arrhenius reached out to the scientific community, politicians, and general public for support. In his lecture "Science and War," Arrhenius drew attention to the new role of research for national self-sufficiency. The collaboration among scientific research institutions, industry, and state in Germany had prevented the nation from accepting an early defeat once the war entered stalemate. In America, even while still neutral, new organizations and collaborations based on a mobilization of scientists energized the local industries for national self-sufficiency. The lesson was clear: Arrhenius urged a greater investment in pure scientific research, which held the key for future economic competition among nations.

Just as in Germany, where the Prussian Academy of Sciences had taken a leading role in creating new institutions for research, Arrhenius hoped that the Royal Swedish Academy of Sciences might do the same. The various Kaiser Wilhelm Institutes for the Advancement of Science (renamed the Max Planck Institutes after 1945), in Germany, established to promote pure research on topics of economic and national strategic importance, could find their Swedish equivalents in well-endowed, partially state-subsidized Nobel institutes.

Having underscored the need to emulate German models in organizing and funding research, Arrhenius noted the importance of breaking Swedish academics' monogamous dependence upon German science. They must turn west and look to Britain and America, even France, for inspiration and collaboration. Moreover, if they intended to play advocate for their German colleagues, they must avoid appearing one-sidedly friendly toward them.

Arrhenius therefore began cultivating institutional contacts with the English-speaking world. He worked to help establish the Swedish-America Foundation for promoting cultural and educational exchanges. He urged using some of the Nobel committee funds to join the campaign of Allied scientists to pay homage to the recently deceased Sir William Ramsay by contributing to a plan for an internationally oriented institute in his honor. An even clearer message of friendship could be sent with a Nobel Prize.

## A Prize to Please the British

In the efforts to recruit state assistance and to turn his countrymen westward, Arrhenius worked in tandem with Carlheim-Gyllensköld. The latter, a self-styled aristocratic bohemian who loathed authority, held no warm feelings for German academic culture based on rigid hierarchies and loyalty to established ideas. He had chosen earlier in his career to work in J. J. Thomson's Cambridge laboratory and communicated sparingly with Germans. Carlheim-Gyllensköld might have at times been a gadfly to Arrhenius—and even more so for other committee members—but now he was a worthy ally.

In 1918, a respectable number of nominators (twenty-seven) returned the invitation to submit candidates (twenty-four) (several nominators, as often the case, proposed more than one candidate; some proposed splitting the prize). As usual, Planck received solid backing from German and Austrian nominators. The other well-supported candidate was Albert Einstein, who had begun to receive recognition in 1917 for his work on relativity theory. Einstein was one of the few Berlin professors who had worked openly and actively against German nationalist and militarist propaganda. He was nominated in 1917 by the French Pierre Weiss, the neutral Dutch Arthur Haas, and the ever sober-minded, highly respected elderly German experimental physicist Emil Warburg. Now, in 1918, a number of leading Germans added their votes for Einstein, some suggesting a division between Einstein and the venerable leader of international theoretical physics, H. A. Lorentz, for their collective contributions to relativity. In principle, this division could win much approval, at least politically. Both Einstein in Germany and Lorentz in neutral Holland had worked from the time of the Manifesto of the Ninety-Three to redress the wounds caused by the charges and countercharges. They embraced and exemplified the spirit of internationalist devotion to science as a supranational activity. But their science was totally unacceptable and incomprehensible to the Nobel Committee for Physics. They had no chance for a prize.

In 1918, Arrhenius and Carlheim-Gyllensköld managed to steer the committee away from Johannes Stark, the candidate whom it had earlier declared to be first in line. Stark had received one nomination, from his fellow archnationalist Philipp Lenard, who shared Stark's antipathy for theoretical physics. Although one nomination might seem a slender mandate, the committee, of course, set its own agendas. Stark, who appealed to several committee members, would subsequently receive the prize based again on the lone nomination of Lenard. Moreover, a single nomination was all that had to come for the candidate who did get the prize in 1918.

One lone nomination arrived from Britain. Arrhenius appreciated that the nominator, Ernest Rutherford, was by now Britain's leading physicist. Unlike some of his colleagues, Rutherford had not publicly engaged in rabid anti-German propaganda. In private he even revealed what for Arrhenius must have sounded like a relative degree of open-mindedness. Therefore, Rutherford's nomination of his not very well known countryman, C. G. Barkla, received immediate attention. Barkla had not previously been nominated. Rutherford alone proposed him. Barkla was at best a problematic candidate.

Barkla was actually burnt out. He had earlier engaged in a heated and at times personal dispute with W. H. Bragg on the nature of X-rays, which in the early 1900s was one of the major divisive questions in physics. Barkla preferred a wave interpretation over the corpuscular theory. Although he had correctly insisted that X-rays consisted of transverse waves like ordinary light, his less acute talent for theoretical reasoning resulted in his having been outmaneuvered during several years of quarreling. But, most important, Barkla studied the nature of secondary radiation emitted by metals exposed to X-rays. He discovered that these yielded a specific characteristic of each irradiated element. He found two series of such radiations, dubbed K and L. These proved to possess increasing penetrating power as the atomic weight of the element increased. The emission of this characteristic radiation therefore seemed related to the nature of the atom.

In 1918, however, Barkla was no longer in the midst of the rapidly growing enterprise of X-ray spectroscopy, which was nurturing the emerging specialty of atomic physics. His discovery of the K and L radiation had been of great value, one of the basic building blocks for all subsequent work. But it was the efforts of others, particularly Henry Moseley, that transformed this discovery into a significant aid for exploring the atom and thereby established the importance of Barkla's find. Barkla was soon unable to participate in the exploitation of these insights. Historian of physics Paul Forman has shown that by 1916 Barkla's sense of scientific judgment had eroded to the point that he had lost touch with newer findings and relied almost exclusively on his own personal insights and interpretations. When he gave the prestigious Bakerian Lecture that year, Barkla dogmatically rejected well-received theory and promulgated unfounded speculations as fact.

Rutherford gave no clue why he nominated Barkla. His letter of nomination was very succinct, offering no explanation or rationale for his proposal. Was Barkla to be a stand-in for Moseley whose work built on the characteristic K and L spectra of elements? But Barkla's studies could scarcely be said to entail a comparable level of brilliance, nor had Barkla followed the path that Moseley's work had opened. Had Barkla perhaps suffered a nervous breakdown that was responsible for his insistence that he was being unjustly ignored and for his increasing rigidity in accepting new findings? Did Rutherford hope to bring prestige to British physics and remind the world of Moseley's pivotal investigations?

Whatever the case, Arrhenius and Carlheim-Gyllensköld knew that Barkla was Rutherford's candidate and, therefore, Britain's candidate. That, in essence, was what mattered. After the defeat of Ludendorff's armies at the Marne during the summer, it was only a question of time before a peace treaty would come. Germany had no chance of winning the war. Carlheim-Gyllensköld took responsibility for the special report on Barkla.

Maybe the report was a mere formality to a foregone conclusion. Carlheim-Gyllensköld submitted a weak evaluation. Possibly the recent British science journals had not found their way to Sweden, or perhaps it was best not to probe too closely into Barkla's recent work. In contrast to the usual practice of reviewing a candidate's scientific genre, providing a bibliography, Carlheim-Gyllensköld's short report omitted much of the usual apparatus for seeking completeness and for assuring the committee and Academy of the candidate's overall worthiness. Barkla's studies received little attention other than the basic results of his pioneering studies; nothing is mentioned of his work and views after 1911. The largest part of the report focused on the work of others who subsequently made important discoveries, at times dramatic ones, using Barkla's results. Investigations by Moseley, by Manne Siegbahn (a young Swedish experimental physicist in Lund), and by the Munich theoretical physicist Arnold Sommerfeld all revealed how, in the hands of others, Barkla's discovery had become significant; Moseley and Sommerfeld's work occupied half the report.

What Carlheim-Gyllensköld did not include of systematic review of the candidate's publications he made up by rhetorical flourish. Moseley's work was shown to rest squarely on Barkla's. Moseley, he reminded his prospective readers in the committee and the Academy, had been considered worthy of a Nobel Prize, but he was killed before he was proposed. More significantly, for the audience being addressed, Barkla and his achievements were identified as the result of the style of science preferred by the Academy's physicists: "With simple experimental means and without any heavy scholarly apparatus [*vidlyftig lärdomsapparat*] [he] managed to find a new and unpredicted phenomenon of fundamental importance for our notion of the atom's constitution."

The committee had surely agreed in advance, or during the meeting of September 4, 1918, when, after "considerable discussion," they were persuaded by Arrhenius that Barkla, and only Barkla, should receive a physics prize that year. A prize for Stark, whether scientifically warranted or not, would only confirm anti-German scientists that Sweden could not be trusted in an international organization for cultural collaboration. While the new international organization was still fluid, and while a ray of optimism remained that maybe the neutrals could as yet work with Allied moderates to broker Central Powers representation, it was best to avoid provocation.

The committee, the Physics Section, and finally the full Academy all agreed to award the reserved 1917 prize to Barkla and to reserve the 1918 prize for the following year. Rarely had a prize been awarded on such slender documentation. Neither Rutherford's letter of nomination nor Carlheim-Gyllensköld's report met the standards of thoroughness that had become accepted practice in the committee. It would seem that the desire to make some symbolic action to honor Moseley, and to acknowledge British physics at the moment of Germany's collapse, had resulted in a lowering of standards. Not that Barkla's discovery was insignificant, but the committee's tradition of seeking proof of a candidate's continued high level of research after having made an earlier discovery was not followed.

This point was brought home when Barkla gave his Nobel lecture in 1920 and promptly made some gaffes. He insisted that his studies revealed the continuous nature of energy absorption in phenomena that already had been almost universally accepted as being ruled by discontinuous quantum processes. He proposed the existence of a J series; yet such softer penetrating radiation had been ruled out by developments in atomic theory, especially by Niels Bohr's work. But Barkla's strong stance against quantum theory, although an anachronism in most quarters of physics, was certainly sweet music to many of the Academy's physicists who were still resisting capitulation to the new world of quantum and relativity theories. Arrhenius might not have endorsed Barkla's opinions, but he achieved a victory in trying to steer the Academy's foreign relations. Barkla paid off politically. At the 1920 celebration, Barkla was the only laureate from the Allied nations to attend. He extolled German science and revealed the spirit of forgiveness and internationalism that Arrhenius had so strongly advocated. But by the summer of 1920, this gesture, in many respects, was already too late. The chemists and the rest of the Academy would ensure that it was.

## Defiant Chemists

The years 1917 and 1918 were considerably more turbulent in the chemistry committee, where different sensibilities prevailed. Aging committee member Peter Klason was determined to have his way. Although a member of the committee from the start, he had rarely been able to gain a consensus for his favorite candidates. Trained as an organic chemist in Lund and various German universities in the 1870s and 1880s, he turned to industrial chemistry, especially after having been named to a professorship in this specialty at the Royal Technical College in Stockholm. His autobiography, written on the occasion of his sixtieth birthday in 1908, reveals a learned man who studied Latin and liked to pepper his texts with quotations from classic and modern literature. He confessed that it went against his nature to be quarrelsome, but, regrettably, he frequently had to fight for his beliefs. In fact, doggedness characterized his battles at the Royal Technical College and on the Nobel Committee for Chemistry.

By 1918, Klason had been a thorn in the side of the Nobel committee for over a decade. Before the war, he often dissented from the committee's majority recommendations. He pushed candidates in industrial chemistry, but his singular lack of success might well have had as much to do with his inability to choose candidates wisely as with principled opposition of his colleagues. Poor health, including bouts of agoraphobia, might well have contributed to his difficulty in persuading his colleagues. The chemists had adopted a loose principle of trying to satisfy each of the major subfields of the discipline, at least those represented by members of the committee. Klason maintained that industrial chemistry, having been the source of Alfred Nobel's fortune, deserved greater respect. As the war drew to an end, he was intent on giving a prize to a new hero of industrial chemistry. His candidate could draw on committee sentiment for German chemistry.

Klason had long looked to acknowledge pioneering work that dealt with world food shortages. When a British researcher announced in 1898 that the supplies of Chilean guano, on which the major nations depended for nitrogen-rich fertilizer, would probably run out by the 1930s, the Malthusian threat prompted much discussion. The solution for staving off massive starvation, it was argued, had to come from the laboratory. Nitrogen was the key; it was available in abundance in air, but for over a hundred years the problem of "fixing nitrogen" (by which it is combined with hydrogen and oxygen to create useful compounds) confounded researchers. The basic chemistry was not terribly difficult; the problem was that the conditions required enormous energies or pressures, well out of reach of known techniques.

In the 1900s, the Norwegian physicist Kristian Birkeland and engineer-entrepreneur Sam Eyde developed a process for making fertilizer that made use of abundant, cheap electric power from waterfalls. The start of Norsk Hydro, and its industrial society in the hinterland of Telemark, seemed to offer a real solution to the fertilizer problem, except that few other nations could copy the Norwegian process because of the lack of comparable cheap electric power. When Klason repeatedly tried to advance a prize for Berkeland and Eyde or for Adolph Frank and Heinrich Caro for a competing German process, his efforts were rebuffed by committee majorities, which claimed that neither of these processes offered a definitive solution to the problem. It may also be that some of them were hoping for a Swedish breakthrough in finding a successful process. Complicating the matter yet further were Klason's poor tactics. He shifted position, one year backing the Norwegians, then the Germans, and then only one of the partners from each team. Before the outbreak of the war, another method was developed that became the universal solution: the so-called Haber-Bosch process. Klason was determined that Fritz Haber, the academic half of the team that created this process, must be the next prizewinner. Haber's hands, however, were bloodied.

The Berlin physical chemist Fritz Haber was a talented researcher and German patriot. By the end of the war, probably no scientific name was as well known. Along with assistants, Haber had patented a process in 1909 that fixed nitrates from atmospheric elements, which achieved results using modest electric power. The chemical equation was critical, as were the laboratory trials, but these did not translate into an economical process. The chemist Carl Bosch, working for the mighty corporation BASF, led the research program that scaled up Haber's findings to commercial production. Although the results were widely publicized, details of the Haber-Bosch process were kept secret. Before the war, Germany began to build plants in an attempt to become self-sufficient in nitrates for agricultural fertilizers and explosives. In 1912, Haber was named head of the new Kaiser Wilhelm Institute for Physical

Chemistry. At the outbreak of war, he shared in the exhilaration of national unity and purpose, and he spearheaded efforts to bring the importance of chemistry to the attention of Germany's military leadership.

As feared, the British blockade restricted Germany's imports and forced the country to rely upon the Haber-Bosch process. New and larger factories were built, which enabled Germany to remain in the war. A man of wide cultivation, Haber sought to prove himself both as chemist and as patriot. As the war in the west collapsed into stalemate, Haber sought other means by which chemistry could benefit Germany. His colleague Walther Nernst and others had suggested the use of gas weapons; Haber stepped in to direct the program. In their first use of gas, above Ypres in April 1915, the Germans released several thousand canisters of chlorine. A several-mile-long cloud wafted across French lines, spreading horror and chaos; the hoped-for massive retreat immediately followed. But the German commanders were not prepared to take advantage of what could have been a decisive breakthrough. Soon, chemists on both sides were at work devising protective gas masks and new, more potent gases. Haber's laboratories spewed nightmare weapons that dragged the already unbelievable levels of horror and misery of the war to further depths. By 1918, almost half of all German artillery shells fired on the western front were filled with gas. The Allies soon responded in kind. These were the weapons that inspired soldier-poets to etch into civilization's memory the barbarism of modern science in the service of war:

> *Gas! Gas! Quick, boys!—An ecstasy of fumbling,*
> *Fitting the clumsy helmets just in time;*
> *But someone still was yelling out and stumbling*
> *And flound'ring like a man in fire or lime . . .*
> *Dim, through the misty panes and thick green light,*
> *As under a green sea, I saw him drowning.*
> *In all my dreams, before my helpless sight,*
> *He plunges at me, guttering, choking, drowning . . .*

In 1916, Klason had championed Haber's candidacy for the work on nitrogen fixation. Although one nomination called for a division between Haber and Bosch, Klason backed the two nominations for Haber alone. But committee chairman Hammarsten pointed out that Klason's report on Haber lacked information. The extreme secrecy surrounding the sensitive industrial process held back details of its actual working. More information would be needed to assess whether Haber should receive the prize. Finally, Hammarsten noted that it would not be consistent with Sweden's neutrality to award a prize for a process that helped Germany to wage war. Over Klason's protests, the committee majority voted to reserve the 1916 prize.

Haber returned to the list of candidates in 1918, a single nomination from the Munich chemist Wilhelm Prandtl. Few nominators had responded that year: a sprinkling of eleven nominations for eight candidates. Clearly, many people agreed with nominators who believed that no prize should be awarded in chemistry, which had become so thoroughly identified with military operations. Furthermore, chemistry had few recent major discoveries, or at least few that eligible nominators considered to be significant. Indeed, what did this list represent? Could such a scanty number of nominations be taken as an accurate reading of the international chemical communities? Perhaps it revealed the silent shame chemists felt about activities that Alfred Nobel had hoped to see ended. Why strive to make an award?

Klason provided a first report during the spring, concluding that Haber deserved a prize. He conceded that Haber's process (according to Prandtl's lone nomination) had provided Germany with munitions. But, since the Americans had also used the Haber process for their ammunition production, he argued, it could be said that both sides were benefiting. A prize to Haber should not, therefore, be seen as favoring Germany. And finally, no other innovation in chemistry had so greatly benefited agriculture. Haber's process should be understood as truly "a benefit on mankind."

Prior to its critical meeting in September 1918, the committee seemed to mirror the recent situation on the western front: disarray and counterattacks. In March, German commander Ludendorff had thrown into offensive all his troops, and failed. The Allies counterattacked. But the American generals had not learned from French and British strategy; they sent their fresh troops to attack at the German strong points, resulting in a further 100,000 deaths added to the millions already killed. Klason also, once again, threw into his battle all he had, but would that be sufficient? In a preliminary vote during the summer, three members of the committee, Klason, Ekstrand, and Söderbaum, backed Haber; but Hammarsten and Widman opposed the motion on the formal grounds that critical information on the process was still lacking. Klason was asked to provide more information in a new report. He assaulted his colleagues with massive undigested material ranging from statistics of fertilizer production to whatever else he could find that seemed relevant. Privately, Hammarsten informed Widman that in spite of all his supplemental information, Klason still had not answered the questions raised in 1916. Widman had to agree, but, given his sympathy for German chemistry, he asked fellow member Söderbaum to coax further facts from Klason that might be useful in making a case for Haber.

The committee met on September 2 in Klason's home. Here he provided further information. But Hammarsten, concerned with committee formalities, countered Klason's popular but messy campaign with a four-page reality check. He had no choice but to point out that, just as in 1916, the committee could not honestly claim to have at hand sufficient information to make a decision. Unlike the physics committee, which had chosen to ignore the inadequacies of the report on Barkla, the chemists were evaluating a highly controversial candidate. Moreover, Hammarsten was a chairman of unusual integrity. Should Haber receive the prize alone, or should he share the prize with collaborators Robert Le Ronsignal, who held the patent rights with Haber, and Carl Bosch? Hammarsten reminded the committee that it should never award someone at the cost of another who might also be worthy of sharing in the award. He also did not accept the argument that a prize for Haber was not politically controversial since the process helped both sides (indeed, it had not; the Americans could not build a Haber-Bosch factory until the 1920s when they paid for BASF patent rights). Finally, while it might be true that the Haber method would aid agriculture to a greater extent than any previous chemical discovery, the record so far indicated that it served primarily to extend the horrors of war.

"For reasons of expediency," Hammarsten recommended against a prize to Haber in 1918. Yes, he conceded, the committee must evaluate candidates strictly on scientific merits. By contrast, the full Academy could take up broader questions of values. Hammarsten addressed his fellow committee members "as a member of the Academy"; he urged them to reject Haber because of the current political conditions. His appeal failed; the other four members voted to propose Haber for the 1918 Nobel Prize in chemistry.

> *If in some smothering dreams you too could pace*
> *Behind the wagon that we flung him in,*
> *And watch the white eyes writhing in his face,*
> *His hanging face, like a devil's sick of sin;*
> *If you could hear, at every jolt, the blood*
> *Come gurgling from the froth-corrupted lungs . . .*

In late October, the German army and imperial German government gave up. Germany had accepted President Wilson's conditions for an armistice. Or had they? On the day the Academy's Chemistry Section met to vote on the committee's proposal, the kaiser dismissed General Ludendorff. Rumors were swirling like desert dust as to whether the kaiser and his leading admirals, eager to unleash their U-boats again, would actually accept an armistice.

At this meeting, Klason replied to Hammarsten's criticisms, adding the point that Germany, in order to assist agriculture, had begun building Haber factories before the war. It would be a mistake, therefore, to consider these factories to have been motivated by military use. Further stretching credulity, Klason added that the prolongation of the war actually was the consequence not of the Haber process but of Wilhelm Ostwald's process for converting the ammoniac into saltpeter, which in turn was used for manufacturing explosives. Hammarsten again voted to place the 1917 reserved prize's money into the special fund and to reserve the 1918 prize. Otto Pettersson had sent in his vote: no prize for Haber since the ammoniac process was still held secret. He maintained that only work which was open for full examination should be considered; anything less violated the spirit of the Nobel legacy. The final vote was seven for Haber, two against, and one member not present.

On November 12, 1918, the day after the sudden signing of the armistice, there was a meeting of the full Academy. On the issue of Haber's candidacy, the Academy calmly followed tradition: respect for authority. The Academy elects the chairman of each Nobel committee, who is entrusted to uphold impartiality and respect for the bylaws. And Hammarsten had gone on record. His voice had to be respected more than the collective call for Haber from the committee and Chemistry Section. Those who knew Hammarsten knew also that he had only contempt for the English, and yet, formal principles kept him from giving voice to his real sympathies—for now. The Academy voted not to make an award in chemistry in 1918, but not without a certain amount of quarreling.

Although some political acumen might be ascribed to the Academy in taking this decision, the events in 1919 would indicate that the essential issue here was the institution's insistence on respecting formal authority. The vote might have dismayed the most ardent German supporters, but it gave some cause for optimism to Arrhenius and like-minded members, who hoped to strike a balance between the warring nations. Just possibly, the Academy could position itself to help bring the hostile factions together in a resurrected transnational scientific community.

## 1919: THE END OF THE BEGINNING

At the 1911 Nobel festivities, Allvar Gullstrand had compared the prize to a Swedish flag, displayed in such a way that the entire world can gaze upon it. Who—and for what achievements—should be found worthy of being draped with the blue and yellow colors of the Swedish flag? In 1919, Europe struggled to raise itself from the mud, blood, and misery of the war only to totter without direction, tripping and stumbling on its way into a new era of uncertainty. In 1919, the Royal Swedish Academy of Sciences and the other institutions responsible for awarding Nobel prizes signaled to the bludgeoned remains of the civilized world. Contrary to the intent, these signals marked the end of the original Nobel vision. They were to bring shame to Alfred Nobel's legacy. They revealed the topsy-turvy politicization of culture, the inability of that culture's moral compass to navigate the tragedy of war or to stake out a sure path into postwar realities.

Beneath the rubble of war and social upheaval, a few shards of a shattered dream about peaceful competition glittered amid other failed ideologies, but they attracted few collectors. In November 1919, the Academy announced three winners of Nobel prizes, all German. Much had changed in the course of a year.

### Postwar Aftershocks

In 1919, bloodshed continued in Europe, as revolution and counterrevolution swept Germany, Hungary, and Finland, not to mention Soviet Russia. Food and material shortages, continued inflation, and high unemployment added to the landscapes of ruin. President Woodrow Wilson's dream of a peace to prevent further war was swept aside by revenge-seeking Allied political leaders and by the newly elected isolationist Republican-dominated Congress in the United States. Economic dislocation resulted in considerable hardship for fixed-income state officials, including academics. Major advances by democratic social movements rubbed salt into these wounds. Factory workers and stevedores were earning as much as junior-level professors and secondary school teachers.

Sweden shared in many of these broader European political and socioeconomic changes. Universal male suffrage was accepted; full parliamentary rule and a socialist government were ready to bloom. Members of Swedish society, just like their counterparts around Europe and North America, were disgusted by the rise of a new gaudy materialistic commercial middle class that sprang up amid wartime profiteering. Mammon seemed to be the only true victor. The cultural leaders of postwar Sweden were dismayed, anxious, and enraged. With the culture of the country largely compromised, and declining economic status bringing discomfort and social slippage, even liberals such as Arrhenius were angered by the image of factory workers earning as much as professors and dissipating their wages in orgies of materialistic spending. Science students turned to engineering and careers in industry; academic careers were not attractive. No political party offered hope for culture and learning. Belief that men of science and culture would be able to heal the wounds of war proved overly optimistic. Even the idea of the neutrals taking a lead in the healing process met hard resistance. They could do little to prevent the exclusion of German academics from international cooperation. Like tufts of fog rolling in from a cold, indifferent sea, despair and bewilderment enveloped members of the Nobel establishment.

Swedish scientists turned first to their German colleagues. Surrounded by economic ruin, shamed by humiliating defeat, buffeted by revolution, and cast as international pariahs, German academics needed encouragement. Declarations of support were forthcoming. For many Swedes, there was no alternative; their loyalties never swayed. Gullstrand stated that Swedish scientists had no intention of turning their backs on German culture. Respect for German science had not diminished; he, for one, had no plans to publish in any other language. German professors received invitations to visit Sweden and give lectures; a conference in Lund invited physicists from neutral and Central Powers nations to meet. The Nobel Committee for Physics gave a grant to Arnold Sommerfeld in Munich from its special fund, one of the very first grants, and the only one ever given to a non-Swedish resident foreigner.

German scientists appreciated the Swedish efforts. Those in the know understood the importance of internationally prominent committee members for defending German culture. Stark, for example, as dean of arts and sciences at Greifswald University, arranged in autumn 1919 an honorary doctorate for Arrhenius. Stark called Arrhenius's work "genial." More importantly, Arrhenius was a true son of the Swedish "*Volk*," who through ancestry, religion, and cultural and political history had embraced the German "*Volk*" to their hearts.

Arrhenius could not fathom the bitterness that had infested the world. Sitting in isolated Stockholm, nurturing internationalist fantasies, he could not fully appreciate the tragedy. He received stern warnings from abroad that Sweden would risk exclusion from international cooperation if it attempted to bring Germans to meetings. Although some Entente scientists feared that a constellation of scientists from neutral nations and the former Central Powers nations could form their own world of learning, excluding the Allied nations, extreme factions led by French and Belgian scientists made attempts at compromise impossible. Even colleagues whom Arrhenius greatly respected refused to forgive and forget. T. W. Richards and others declared that they could not shake hands with any German who had not indicated sincere remorse over having signed or supported the Manifesto of the Ninety-Three. Apart from Max Planck and Emil Fischer, Richards knew of no one who had expressed regret. Still, Arrhenius kept his sights on rebuilding the international scientific community while also aiding German research. He dreamed of a Nobel ceremony in which, somehow, winners from both sides could embrace and begin a new era of cooperation.

## Defiant Nobel Committees

Reactions to the prizes in 1918 were favorable. But then, again, most scientists, like virtually everyone else, were preoccupied with the end of war and subsequent political turmoil. The Swiss chemist P.-A. Guye applauded the decision not to award a chemistry prize in 1918. He assumed that this action arose from Swedish "hesitation and scruples" and in recognition of Alfred Nobel's desire to advance the cause of peace. The results for 1919 would give him reason to reconsider.

Having acknowledged British physics by the award to Barkla, the physics committee felt freed from making a further gesture to the victors. Barkla was an unexpected bubble of expediency that diverted committee members' attention from other agendas. Earlier, the committee had indicated a desire to reward Johannes Stark. Although Stark continued to be proposed solely by Philipp Lenard, his candidacy was too attractive for some committee members to neglect. Equally difficult to ignore, Max Planck remained a formidable candidate, but his quantum theory, which the committee had branded as "revolutionary" and "threatening," remained for most members as vexatious as ever.

During the war, experimental and theoretical studies had largely secured acceptance of the quantum. Although physicists were bedeviled with problems of reconciling old and new theories of electromagnetic radiation, and trying to digest theories of the atom that relied on quantum processes, most researchers accepted that in some form or other the quantum was an established fact. Many experimental phenomena that could not be explained using classical theories, such as the Stark effect, were elegantly tackled through quantum theory. But not all physicists were willing to accept or comprehend the theory—Stark being one of them. Most of the Academy's physicists also fell into that category.

Most Swedish physicists still hoped that the quantum developments were but a digression, a stopgap measure to keep physics going until some new discovery would direct the flock back to the familiar world of classical physics. While waiting, some were willing to wrestle with the implications of Planck's theories; others just continued grazing on familiar patches of science without paying much attention to the changes around them. By contrast to the Academy's physicists, Carl Wilhelm Oseen, a theoretical physicist in Uppsala who was as yet waiting offstage to make his appearance in the Nobel committee, urged colleagues not to ignore quantum theory. The theory was actually not to his liking, but it could not be ignored. As early as 1913, Oseen disliked that his Swedish colleagues were too eager to dismiss quantum theory; he would prefer that they "with open eyes see how fruitful the quantum hypothesis is as a working hypothesis." In 1918, he published a nontechnical article on the significance of quantum theory for electromagnetic theory, "as a service for my less informed countrymen, not the least for those who are members of the Academy's Physics Section." Some engagement came about in 1919 when a number of foreign visitors gave lectures in Sweden on new developments in physics. The visitors noted the widespread hostility to quantum theory, but they also came away believing that their efforts would contribute to a thaw. Still, the Nobel committee was troubled.

There was no doubt that Planck had to get a prize. Planck had emerged as a major spokesperson for German science. He was the leader of German physics; the respectful and affectionate celebration in 1918 of his sixtieth birthday spoke to his central position. As revolutionary as the quantum theory had become, Planck was of temperament and outlook basically a conservative man, but moderation, balance, and a keen sense of morality allowed him to transcend rigid doctrine and dogma. His efforts to undo the damage caused by the manifesto and to work toward reestablishing the broken links of civility and international cooperation had been well received abroad. For Arrhenius, Planck was a key figure for the rehabilitation of Germany's international scientific reputation and for restoring German science's strength amid revolutionary political change. Planck, as symbol of traditional German academic culture, was highly attractive to the committee. But what could be done with quantum physics, which the committee had resolutely refused in the past to accept?

Wording of the formal justification changed three times in committee; the question was how to have Planck, without embracing quantum theory too passionately.

Arrhenius prepared three consecutive drafts, containing three different texts for a justification to meet objections and gain a consensus. After much discussion, he contrived a formulation that won assent. In essence, it was not so much an embrace of quantum theory as an acknowledgment of the role Planck's concept had played in stimulating research. A thaw? Perhaps, but only a grudging one. Had Planck not emerged as the spokesperson for German science, the committee might well have continued to hold him at arm's length, in spite of a decade of tremendous international support. In contrast, Stark was a shoo-in.

In fact, Bernard Hasselberg and Allvar Gullstrand first proposed that Stark should receive the reserved 1918 prize, ahead of Planck. In their draft of the general report the reasons stood clear: to underscore the priority of experiment over "theoretical speculations." But the chronology of the candidates' work would reverse the order. Although the special report on Stark revealed Gustaf Granqvist's increasing appreciation of recent quantum physics, the report made clear Stark's rejection of these developments, which made him attractive to the committee majority. The lone man in the committee to oppose both German nationalism and experimentalist bias, Carlheim-Gyllensköld, insisted that because of the one lone nomination and the contents of the special report, Stark did not merit a prize. He advocated that one of the two prizes available that year should go to Niels Bohr, for the young Dane's quantum model of the atom. He also probably appreciated the political importance of balancing Planck with a non-German. But the rest of the committee claimed that Bohr's atom conflicted with physical reality, and therefore he could not be considered for a prize. Albert Einstein and Robert Millikan, both well supported, were passed over. In the end, the committee majority allowed chronology rather than prejudice to dictate the order; their political preference impelled the choice of winners. The final recommendation: the reserved 1918 prize to Planck, the 1919 prize to Stark.

In the meantime, the chemistry committee was exercising its prejudices. Provincialism, arrogance, loyalty, or some combination of these may help explain the actions in 1919 of the Nobel Committee for Chemistry. After the Academy's decision in 1918 not to award a chemistry prize, members of the committee regrouped to ensure a victory for Fritz Haber. Klason and Ekstrand sent their own letter proposing the German for the prize. Moreover, Hammarsten was no longer bound by the political situation defined by the war, and he could more freely give voice to his German sympathies. He could help the committee overcome the formal problems that had earlier hindered his support of Haber.

The committee set about to pave the way for a prize to Haber. He was the committee's candidate, not the nominators'. The only other proposal for him had again come from the Munich chemist Wilhelm Prandtl, with whom the committee was in direct communication. The German chemical community retreated for the moment from public contact. No other German chemist submitted a nomination. Physicist Wilhelm Wien sent in a nomination for Emil Fischer for his work on amino acids and the chemistry of egg white. Fischer had already received the 1902 prize, but Wien undoubtedly understood that Fischer was the only German chemist who might be acceptable internationally. Although during the war Fischer had been a key mediator in mobilizing chemistry for the war effort, he had expressed regret for having signed the Manifesto of the Ninety-Three. The committee had received few other nominations.

Hammarsten kept the discussion moving. Klason's report again urged giving the prize alone to Haber for the greatest chemical achievement in living memory, for aiding agriculture, and for stimulating the growth of industrial chemistry: in short, for benefiting mankind. Bosch's share was dismissed as not being prizeworthy. Pointing to the alleged fact that Americans seemed to have managed to build ammonia factories without the aid of Bosch's efforts at BASF, the committee concluded that Haber's work was the breakthrough that deserved the prize. Here the committee, racing to embrace Haber, erred. Even Hammarsten seemed to have forgotten his own strictures. Nothing in the report revealed a grasp of Bosch's critical role in the complex research effort to scale up the process to allow efficient industrial production. Had the committee chosen to wait an additional year or two, it would have gained an opportunity to sort the problem. Committee members would have learned, among other things, that the Americans had failed to build successful factories based on Haber's equation. As it turned out, a strong sentiment of injustice forced the committee to reward Bosch twelve years later.

Klason's eagerness to reward German chemistry overcame any concern that millions of men had just been killed in a war fueled in part by Haber's innovation. In his report, Klason used statistics to show the high agricultural yields made possible by fertilizers manufactured by the Haber process. He then turned to the crucial role of nitric acid: "For all civilized peoples [*kulturfolk*] nitrogen in suitable compounds is necessary for their economy, industry, communication, and military system." True enough. Klason then tried to defuse the contention that the Haber process prolonged the war. The nitrogen-based fertilizer and munitions enabled Germany to survive the illegal British blockade. Klason blamed the blockade for prompting Germany's unrestricted U-boat warfare as well as for prolonging the war; he also attributed to it the fact that all major nations would in the future have to make their own nitric acid. That is, they would have to establish their own factories based on the Haber process. Indeed, it is hard to believe that Klason was not involved in efforts in 1919 to initiate, with German technical assistance, just such a factory in Sweden. No mention of Haber's work on gas warfare is made. It is even possible that Klason's and other committee members' insistence on celebrating Haber alone might have been a desire to defiantly show solidarity with this most hated of all German chemists.

In May 1919, the committee met for a preliminary assessment; all members agreed to support Haber. One member, Söderbaum, expressed a slight reservation, stating that Haber's candidacy would only get stronger if the committee waited an additional year or so for the universal value of the process to be fully acknowledged. But he was willing not to oppose an immediate award; the process had already demonstrated its benefits for agriculture and other economic activities. Haber's candidacy sailed through the committee, which gave unanimous support in its September 1919 vote.

In October, the Academy's Chemistry Section voted its support. The same month, fifty members of the Academy sent a petition to academies and scientific journals abroad, urging Entente scientists not to ostracize German researchers. Reactions came swiftly: Not until German men of learning repented openly for supporting Prussian military atrocities could they be invited back into the circle of civilized peoples. No matter.

The Academy met on November 12 to vote on the prizes. Late at night, the Academy voted to award three prizes to Germans—Haber, Planck, and Stark—

and to reserve one prize. The Academy had made its statement; it hoisted its Swedish flag.

> *Obscene as cancer, bitter as the cud*
> *Of vile, incurable sore on innocent tongues,-*
> *My friend, you would not tell with such high zest*
> *To children ardent for some desperate glory,*
> *The old Lie: Dulce et decorum est*
> *Pro patria mori*

Disgust and dismay soon followed.

## THE PRIZE AND THE ACADEMY SHAMED

Early in the war the conservative establishment in Sweden accused the prize-awarding institutions of not having maintained the ideal that culture remain above politics. Culture could and should remain aloof from the barbarism engulfing Europe. Following the announcement of the 1919 prizes, Hjalmar Branting, the Social Democratic leader who was soon to become prime minister as well as a winner of the Nobel Peace Prize, attacked the Academy. In a newspaper article that was reprinted far and wide in Sweden and beyond, Branting charged that the prize to Haber served only to equate culture with barbarism. By applauding Haber, the Academy and its Nobel prizes became part of the same barbarism. By celebrating Haber and German chemistry, the Academy had turned away from the most elementary sense of tact as a neutral institution. Branting called for a vigorous protest and, in particular, a protest from within Sweden. Would a comparable French or British chemist, who had worked directly for the war effort, have been given a prize at this time "while the sores are still dripping blood? But because Haber is a Prussian professor we find it acceptable!"

Vigorous charges and countercharges followed. The conservative newspapers interviewed Klason and Hammarsten, defending the Academy. Klason repeated the rationalizations he had used in his reports. Further, he claimed that the Academy had voted unanimously for Haber. Hammarsten defended the action, saying that the Academy was required by law to take into account only that which is stipulated by its statutes: objective scientific merit. Of course, that was not what he had said to his own committee in 1918. Hammarsten insisted that no political consideration entered into the decision. No one within memory had made a discovery of such importance for agriculture and nutrition. Söderbaum seconded this statement. Wilhelm Palmær, the committee's secretary, regarded the criticism as "pure nonsense," which was also what Ekstrand told journalists. Indeed, they and the conservative press considered Branting's criticism pure politics. Their responses, as well as those of other committee members, shared a common theme—a defense of the Academy and its basic tenets, which they feared were being undermined.

The debate was about yet bigger issues. Even more significant for the Academy than solidarity with German science was the fear of democracy. Rule by politicians elected by the masses threatened the Swedish tradition of governance by expert, objective evaluation. Conservatives claimed that parliamentary democracy was a British institution and totally alien to Swedish national traditions. A state bureaucracy consisting of nonpolitical experts was the accepted bedrock for governance. The Academy and the state universities were part of this culture of experts. Swedish election reform had now opened the door to rule by democratically elected representatives. Workers and others with little education and with no shared sense of culture could now vote. The Social Democratic Party was certain to win the next election. For conservatives, this meant a repudiation of their most fundamental belief. Not the voice of the many but the voice of impartial expertise yielded sound government, in society and in science. The Nobel enterprise and the Academy were imperiled. Similarly, the conservatives held that although the imperial German government had fallen, Germany's elite culture remained. Democratic revolution threatened. And so, the committee felt, it was important to honor that elite culture by the three prizes of 1919.

For Hammarsten and his colleagues, the only real issue in the media debate about Haber's prize was the fact that the legitimate experts—the committee—had agreed. The decision on Haber had been based, as he and his colleagues saw it, on an impartial evaluation of the facts. Branting and the socialists, in their view, sought to entangle the Academy and the prize in politics. The Academy's chemists had remained loyal to the bylaws and the tradition of impartiality.

But the story was not over yet. Carlheim-Gyllensköld broke the bylaw imposing strict secrecy on Nobel proceedings. He revealed to the press that, contrary to what Klason maintained, the Academy's vote for Haber was not unanimous. A minority had voted against him. Moreover, not everyone had voted exclusively for Germans, and he, in particular had withheld support for Stark. He further accused Klason of making the false statement about unanimity to harm those who were working to improve relations with Entente researchers. He might have been thinking of Arrhenius's efforts to protect Sweden from charges of being a German ally. Confusing matters further, a private account of the Academy's meeting indicated that actually no protest had been registered against Haber.

The controversy reached far and wide. German media and scientists rejoiced at this "impartial" confirmation of the superiority of their science. They played upon the theme of a clear *victory* over Allied culture. Elsewhere, most scientists were too shocked to comment, but editorials in scientific journals and newspapers expressed disbelief. A French newspaper, tipped off with inside information, reported that Stark had scarcely been nominated but had still received a prize. Here and elsewhere the point was made: Haber's work on gas warfare had broken the rules of the Hague Convention. Was it possible that the members of the Academy were unaware of Haber's wartime activities? The paper referred to the massive information available that described the gas warfare. Newsreels of the battlefront filled with images of soldiers ravaged by gas had brought tears in neutral countries such as Spain, but had they in Sweden? The French newspaper asked, "Has then the Academy of Sciences in Sweden, shut behind Nordic mists, seen nothing, heard nothing, and understood nothing?"

Alfred Nobel's vision, like the odd flower pushing through the rubble of bones and rust in the fields of Flanders, was not completely eradicated by the war. But it was an incongruous one amid the realities of postwar Europe. The Nobel establishment had lost its bearings and had yet to ask new questions: What did the prizes represent in this new world? What role might they play in the reconstruction of civilization? Its efforts to shore up the eroded order of things had failed, in part, miserably. Similarly, those who had hoped to create Nobel institutes for fostering international cooperation were forced to accept indefinite postponement. Spiraling inflation and a series of onerous new taxes braked the movement toward using Nobel money for those purposes. The government's only response to the call for emergency aid for academic research was to fund an institute for racial hygiene.

Even Arrhenius, the most internationally linked and widely traveled member of the Academy, let his own single-mindedness blind him to the anger and bitterness in the scientific communities touched by war. His desire that the Academy might hold a Nobel festivity in neutral Stockholm in which scientists from both sides of the conflict would shake hands, make up, and renew their loyalty to internationalism did not result in catharsis. Barkla was the only non-German to come to Stockholm in June 1920 for the first Nobel celebration since 1913. The Braggs, like Richards, had no intention of embracing an unrepentant enemy. Although Arrhenius enjoyed the spectacle of Barkla shaking hands with Germans and praising the quality of German science, the others were conspicuously absent. At this first Nobel ceremony specially arranged to be held in June, it seemed appropriate that the weather was cheerlessly wet and the wind cold. The ceremony returned to December.

Uppsala professor of literature Henrick Schück, the new chairman of the Nobel Foundation and new liberal leader of the Nobel Committee for Literature, made a confession in his opening speech at this first Nobel ceremony after the war. Alfred Nobel's dream of peace and happiness, brought about through progress, was no more. He added that the feeling of pride Swedes had first felt over being entrusted with Nobel's donation and vision has in many quarters been transformed to dejection and shame. How might Nobel's legacy be brought back to life? He had no firm solution.

In the new era just then emerging, developments within the world of science—intellectual and institutional—as well as changes in the broader cultural-political landscape prompted new ways of thinking about and using the prizes. Belief that the peaceful competition among civilized peoples leads to peace and social harmony might have lost its hypnotic power, but much of the rhetoric from the prewar age of illusion lived on. Through this lost dream and its language, the prize was able to provide robes of idealism about humankind and the abundance of science generally to cloak what was eventually to emerge increasingly as the naked quest for furthering narrow professional interests and personal careerist gain.

# PART III

## SMALL POPES IN UPPSALA

Arrogance and Agenda:
The Physics Prize, 1920–1933

# SEVEN

## Einstein Must Never Get a Nobel Prize

### Keeping Physics Safe for Sweden

PREAMBLE: PHYSICS BETWEEN THE WARS

*P*hysics changed dramatically between the two world wars. Intellectual ferment at the frontiers of research, for those who participated, reached heights of exhilaration. Not the least, for those who created a new science of the atom, this was a golden era. Physics as a research enterprise and as a profession beyond its traditional pedagogical tasks increasingly set the tone for the discipline. This was a period of transitions. In 1920, a physicist, with the help of a technical assistant, could still do cutting-edge science by designing, constructing, and operating his own laboratory equipment. An annual budget of a few hundred dollars or the local equivalent for running a laboratory was, for many, as much as could be hoped for. On the eve of World War II, those participating in the new field of experimental nuclear research thought in terms of many thousands, even hundreds of thousands, of dollars for a laboratory filled with massive complex machinery run by teams of technical assistants. Demand for funding always outpaced the few available sources, usually generated through private patronage. The statement in 1939 that a handful of physicists was more important for national defense than a battalion of soldiers pointed to even more radical change just around the corner.

During and immediately following the general bewilderment of World War I, physicists also sensed crisis and confusion in their science. Efforts to make sense of data and laboratory phenomena that neither classical theories nor the new quantum theory could readily explain prompted disillusionment in some and intensive research in others. In the mid- and late 1920s, pessimism turned to excitement. Dramatic new quantum mechanical depictions of the atom forced physicists to confront radically new interpretations of physical reality. On the subatomic level, classical laws of causality—the most solid foundational brick of the science—seemed no longer valid. Events could only be comprehended in terms of probabilities; uncertainty emerged as one of nature's features. Light, which was proven in the 1800s to be a form of electromagnetic wave, was shown also to possess properties of a particle. And the electron now had to be considered not just a particle but also a wave in space. A broad international group of specialists met frequently in conferences and workshops to discuss the many surprising and conflicting theories as well as new experimental findings. All of this followed upon Albert Einstein's special and general theories of relativity, which demanded their own radical perspectives on space, time, mass, and energy.

By the early 1930s, a quantum physics of the outer part of the atom—its electronic shell—was largely in place, even if its interpretation was still open to heated debate. Interest turned to the nucleus—the minute and extremely dense core of an atom—which became increasingly fashionable as a research topic as methods were devised to probe its secrets. In the 1930s, new, previously unimagined particles entered into physicists' vocabulary, discovered through both experiment and theory. The positively charged proton and the much tinier negatively charged electron suddenly had company. Proton-sized particles without charge, electrons with positive charge, and something completely different and inexplicable—mesons—kept physicists enthralled. New instruments and techniques allowed probing the atomic nucleus and achieving the dream of alchemists: transforming the elements and even creating new ones. The gold of the modern alchemists was not yellow but glowed radioactively.

Physics entailed more than the physics of the atom. Physics helped to gain new understanding of the earth's atmosphere and weather change, how stars and galaxies evolve, the nature of chemical reactions, and the behavior of matter in extreme physical conditions, among other important phenomena. Yet many physicists, especially on the Nobel committee, preferred to define their science narrowly and to marginalize specialties that were not to their taste.

Physics not only made great strides intellectually during this period but also grew tremendously as an enterprise. Although postwar economic crises in Europe dampened recruitment to the field and strangled expansion of facilities, an American boom in higher education and industry during the 1920s stimulated the growth of a thriving national physics community. Industrial research and science-based engineering created new markets; students flocked to physics courses as never before. Universities responded by hiring new professors and by providing additional funds for laboratories. To some degree, elements of this pattern occurred in most industrial nations.

Physics as an enterprise also experienced qualitative change. Growth coupled with intellectual ferment accelerated processes of fragmenting the world of physics into semiautonomous enterprises as new subspecialties came into being. Opportunities arose for increasingly narrow research specialization. In many academic institutions, research schools grew in size as individual professors directed larger research facilities staffed by increasing numbers of assistants and technical experts. During the interwar years theoretical physics emerged as a vital part of most national disciplines. New professorships and institutes specifically defined for theory contributed to changing the intellectual and institutional landscape of international physics.

Although the formal boycott of the former Central Powers scientists ended in 1926, informal contacts began much earlier. The young Danish theoretical physicist Niels Bohr established after the war an Institute for Theoretical Physics that provided a neutral meeting ground where researchers from several nations could work together. What Arrhenius and other Swedish scientists had hoped could be achieved through Nobel institutes, Bohr attained in Copenhagen. Aided by private foundations, physicists traveled as never before, attending congresses and visiting foreign universities. By 1930, strong transnational networks of specialized researchers had become commonplace, at least for those scientists working on the research front.

The Nazi rise to power in 1933 destroyed what had been the world's greatest scientific community. German physics had already experienced rough times as economic crises after 1918 impoverished universities and decimated the professional middle classes. Still, many German universities, including Berlin, Göttingen, Hamburg, Leipzig, and Munich, among others, revived as dynamic powerhouses for physics. But this situation changed once the Nazi regime quickly dismissed academic staff who were Jewish, socialist, or dissident. Other academicians and scientists who were not immediately threatened but who possessed great moral conscience voluntarily chose to leave their homeland. And a few anti-Nazi nationalists remained behind to try to ride out the storm and protect academic values. Extreme experimentalists—those who for over a decade had declared that the true German intellect sought direct contact with physical reality and eschewed abstract mathematical theory—set the agenda for Aryan physics. They declared as decadent, and as the alleged product of Jewish charlatanism, relativity and quantum theories; these ideas were not to be taught.

Those scientists with reputations—and, not the least, those with Nobel prizes affixed to their names—found refuge in other nations, especially in the United States. Numerous less prominent scientists and especially younger researchers had to scramble to find safe haven and employment. The Great Depression and considerable anti-Semitism did not make for very many happy endings, but all was better than the fate facing those who could not flee. Another wave of refugees came from Mussolini's Italy when in the late 1930s the fascist regime began implementing racial laws and exercising greater militancy. American physics, which had begun to flourish and produce its own first-rate researchers, benefited greatly from the intellectual riches that fled from tyranny.

Most academic physicists continued to justify the prestige of their science in terms of the morality of disciplined work and self-sacrifice toward uncovering eternal truths of nature. Some physicists still saw their professional identities as being cultured individuals who shared the responsibility of civilizing society. But during this period other tones became audible, tones that after the Second World War dominated. Physics was increasingly heralded as a motor for economic growth and national strength.

Even when trying to transcend parochial concerns, the task of the Nobel Committee for Physics in identifying worthy candidates was made difficult by growing partiality among nominators. The boycott of Central Powers scientists through 1926 and the German counterboycott influenced the choices taken by many nominators for their proposals. Moreover, in European nations the dismal financial support for science and research prompted many nominators to think globally but act locally. That is, it was well understood that a Nobel Prize to a local scientist could yield important dividends not only for the individual and his home institution but also for the national scientific community. In some nominators a desire to support potential local heroes dampened the inclination to nominate purely on perceived merit.

More generally, Nobel fever continued in spite the disaster of World War I. In the hands of media and scientific leadership, a prize provided opportunity to proclaim excellence at home and boast national pride abroad. Prewar belief in progress through the peaceful competition among civilized nations lived on in more vulgar forms of nationalist boosterism. Collective mourning and collective memory by nations that had sacrificed millions of men also required living testimonies of national grandeur. Led by the mass media and interest groups, public opinion responded to Nobel prizes, Olympic medals, and other internationally visible forms of competition as reassurance of national honor, if not superiority. In the United States, the media, regional patriots, and academic entrepreneurs gradually nurtured a local cult of the Nobels as Americans began to bring home increasing numbers of prizes.

But equally significant, and counterbalancing nationalist tendencies, some nominators felt increasing loyalty to their transnational specialist networks, which challenged at times allegiance to their home universities or national scientific community. Just what the nominations represented, how they should be interpreted, and what signals were being sent between committee and nominators made the task all the more difficult for those entrusted with distributing the prizes to find level ground for their delicate balancing act.

By looking at the activities of three pivotal committee members—Allvar Gullstrand, Carl Wilhelm Oseen, and Manne Siegbahn—it will be possible to examine changes in managing and using the prize during this important period. The three, referred to disparagingly as the "small popes in Uppsala"—which was not a compliment in Lutheran Sweden—dominated the committee. Their attitudes, insights, and agendas proved critical to how the lush flora of scientific achievement in this period was assessed for reward with Nobel prizes. The three committee members can be seen as emblematic of the transition in professional physics as narrow professional goals increasingly eclipsed broader cultural concerns.

Gullstrand's sensibilities remained in the world of pre–World War I German and Swedish conservative academic culture: the professor as safekeeper of elite culture, which included natural science, in the name of spreading cultivation in society. Oseen had one, maybe two, feet well planted in that world; indeed, he agonized over the cultural and theological consequences of modern physics. But he also flirted with the growing trend of academic scientists as being more than educators: the professor as a professional researcher devoted to producing knowledge and training disciples. In his professional life, Siegbahn cared little about broader cultural issues and cultural heritage. All that mattered was building bigger research facilities, training and hiring teams of assistants, and catching the waves of new research frontiers that might allow him and his school to achieve renown. Siegbahn thought little of cannibalizing hundred-year-old antique instruments for parts that could be used in experiments. In contrast, and indicative of an alternate culture for physics, Vilhelm Carlheim-Gyllensköld considered preservation and respect for physics' past, as embodied in these instruments, as part of the living culture of the science.

For the Nobel Committee for Physics, the coming of peace held many hazards. Democratic politics and hostile international relations within science were not the only sources of anxiety. In particular, quantum and relativity theories threatened the "commonsense foundations of physics," which the Academy had sought to honor for twenty years. Nobody denied that conflicts between old and new physics, between theory and experiment, pointed to an unsettled situation. At first, only Max Planck was a serious, persistent candidate, but then Niels Bohr and Arnold Sommerfeld, theoreticians who sought to establish models of the atom based on quantum theory, entered the lists. Even more challenging to the committee's sense of physics, Albert Einstein's general theory of relativity, published during World War I, presaged radical changes in conceptions of space, mass, gravitation, and motion. How the committee and Academy reacted to Einstein became a drama in several acts. Provincialism, arrogance, and expert judgment faced off in a controversy that was as much about the nature of physics and its implications for general culture as it was about Einstein's eligibility for a prize.

## EINSTEIN ON THE WORLD STAGE

If it seemed to the physics committee that it had diffused the threat of quantum theory in 1919 by giving a prize to Planck, an even greater disturbance exploded almost simultaneously. On November 6, 1919, at a joint meeting of the Royal Society of London and the Royal Astronomical Society, the results of an expedition attracted world attention. Two groups had traveled to the coasts of Brazil and West Africa in May where during a total eclipse of the sun they attempted to collect data that might help confirm or disprove Albert Einstein's general theory of relativity.

The president of the Royal Society, the retired Cambridge Nobel laureate J. J. Thomson, announced the results of the eclipse expeditions and declared Einstein the winner. The ever-so-minute bending of starlight by the sun's mass was recorded on the photographic plates; the effect was real and not the result of random experimental error. Around the world the news spread of the overthrow of the Newtonian world picture. A Swiss Jew working in Berlin had proposed a theory that now had been confirmed by a British team of astronomers and physicists. The international spirit of the endeavor contributed to the newsworthiness. In fact, the leader of the expedition and the head of the Royal Astronomical Society, Arthur Eddington, a Quaker and an internationalist, might well have pushed the interpretation of the results to be able to use this sensation to defy the boycott of German science, which he opposed.

Almost overnight, it seemed, Einstein was transformed into a celebrity. But the implications of the general theory of relativity, both those that Einstein had proposed, as well as a torrent of popular interpretations, set in motion a range of reactions that nobody could have predicted. In a world still recovering from four years of war, and still trying to make sense of the tremendous political and social upheavals arising from that calamity, news of a revolutionary theory that claimed to overthrow the foundations of physics riveted media attention. Prewar culture had long linked truths about social and human affairs with the truths of nature discovered by scientists. The two-hundred-year-old Newtonian inheritance of a mechanics based on absolute, objective notions of time, space, and mass provided a foundation for belief in corresponding absolute moral and religious values. Now, following Einstein's theory and its confirmation, these assumptions no longer seemed valid. Space was curved; no fixed, prioritized frames of reference existed from which to measure physical reality; and at extremely high velocities, time contracted and mass expanded.

Naturally, in the political and cultural agitation coming on the heels of the war, many thinkers, artists, journalists, and others who saw only ruin and worse in traditional European culture grasped relativity theory. The term *general relativity* was erroneously interpreted to be a declaration of the general relativity of all spheres of human activity. For many who mourned the decline of the old prewar order, confirmation of the theory reinforced their fears of a new relativism shattering former absolutes in ethics, theology, and political thought. Those who looked to build a new and different culture on the rubble of a discredited society embraced relativity as a guide for overhauling the arts, moral values, and social theory. It seems that in this age of anxiety, people transformed Einstein into a prophet of hope—or doom. No easy equation could be found correlating political orientation with reaction to Einstein's theory. And, of course, the media fueled the fascination by the myth that only a tiny handful of persons around the world actually understood the theory.

Einstein himself was a sufficiently unorthodox figure in appearance, manner, and belief that the cult of relativity—and its sworn enemies—gained yet further momentum from the man. An internationalist who refused to back the Manifesto of the Ninety-Three, a socialist who in contrast to most Berlin professors welcomed the initial revolution in 1918 and the subsequent liberal Weimar Republic, a physicist who allegedly did not conduct experiments, and a Swiss Jew living in Germany who supported the growing Zionist movement, he stood apart. His informal appearance, unruly hair, and soulful eyes also spoke for a unique individual who differed from the image of a Prussian professor—or, for that matter, any professor of his day. Einstein, who disliked publicity, was caught in a maelstrom of media attention. He who overthrew Newton from science's throne and whose theory produced "Jazz in the Heavens" and brought Dadaism into rational science, bemoaned that he was a King Midas, but instead of gold, all he touched turned into newsprint. He shrugged off many responses that did not at all reflect his own values, and some, no matter how preposterous, had to be taken seriously.

Some German physicists, especially the ultranationalists who also were firmly experimentalist in outlook, considered Einstein and relativity theory the embodiment of all that had gone wrong. Einstein had refused to support the kaiser and the war effort; he had signed and circulated internationalist manifestos; he embraced pacifism and democracy. Einstein, "the magician" who through his Jewish charlatanism conjured up speculations and tricked the world into believing them (or so the right-wingers thought), mocked all that they and their discipline stood for: disciplined experiment, reluctance to engage in speculative theory, and avoidance of advanced mathematics that distanced the researcher from reality. Both his methods and his conclusions intimidated their notion of physics. Although Einstein's more recent generalized theory of relativity and the British eclipse results precipitated the frenzied attention, his original special theory of relativity from 1905, before the war, had already antagonized steadfast experimentalists.

## Special and General Relativity

Unlike other theoretical physicists such as H. A. Lorentz and Henri Poincaré who were then working on theories of relativity to account for the Michelson-Morely measurements that failed to detect the earth's motion through the universal ether, Einstein approached the problem from another end. He probed the logical basis for mechanics. Beginning with the basic premise that the speed of light had to be constant for all observers and all frames of reference, he elegantly derived a series of postulates on the measurement of mass, time, and length in linearly moving frames of reference—hence the "special" or limited theory of relativity. The young unknown physicist working in a Swiss patent office made a number of startling deductions that followed from his reasoning. It was necessary to reject the belief in an absolute, objective privileged framework for all measurement. He also demonstrated theoretically the equivalence of mass and energy, but that insight was much less explosive at the time than his call for eliminating the ether and absolute space. These remained the basis for physicists' worldview in spite of accumulated paradoxes and conflicting theories.

Einstein's terse 1905 article in the *Annalen der Physik* seemed to critics merely to entail rearranging postulates and dragging the reader through logical twists and turns. Theoretical physics as a semiautonomous subdiscipline was still in the process of coming into being. But for those who devoted themselves to seeking the unification of physical phenomena, the unraveling of contradictions in the foundations of their science, such as Planck, Lorentz, and Poincaré, among others, began early to take Einstein's theory seriously. Indeed, Einstein was working on a broad front of theoretical problems.

In what is often referred to as Einstein's year of miracles, 1905, he published articles on no less than four significant problems, each calling for radical revisions of thinking in physics. After he began attending German science congresses a few years later and began to impress the participants, he gradually won the respect of important physicists. After holding academic positions in Zurich, Prague, and again Zurich, Einstein was recruited by Planck to Berlin in 1913. Certainly much to the surprise and displeasure of some experimental physicists, Einstein was given, all at once, professorship in Berlin, membership in the prestigious Prussian Academy of Sciences, and directorship of a Kaiser Wilhelm Institute for Theoretical Physics. In the years before the war, Einstein's theory was well accepted by many theoretical physicists, but it prompted disbelief, annoyance, and anger among those who were reluctant to abandon a mode of thinking with which they had become accustomed. Ideological and personal prejudice against Einstein and his theoretical physics sharpened after the war.

## Postwar German Reactions

Amid the avalanche of responses to relativity theory after November 1919, those in Germany took the lead in belligerence and intensity. At first, the news from London was met with pleasure. Here, in the shadow of defeat, a clear victory for German science shined through. Coupled with the announcement a week later of the three German Nobel prizes, commentators could remind the populace that German science and culture were still the world's greatest, able to bring honor to the fatherland. For some people, the theory seemed to be a blow against deterministic materialist philosophies, which were being attacked by such popular writers as Ostwald Spengler as being responsible for the decline of civilization.

But, soon enough, reaction set in. Einstein himself came into greater focus, and his theory was seen to be a blow against older conceptions of physics, representing the alleged degeneracy that brought down the prewar order. Nobel laureates Philipp Lenard and Johannes Stark joined with another extreme right-wing nationalist, Paul Weyland, to organize a new science association that would, as they saw it, represent true German researchers, an alternative to the decadent Berlin-based association dominated, in their view, by Jewish scientists. Their main aim was to combat relativity theory and Einstein; in so doing, they would combat democracy and other supposedly un-German movements in politics and culture. Lenard had long opposed the institutionalization of theoretical physics; a good experimental physicist possesses more than enough theoretical insight. There was no need for specialists who sat at a desk and worked with only a blackboard. He defended the ether theory from what he called the discredited hypothesis of relativity. Stark shared these views. The two Nobel laureates placed all their scientific authority in the name of their political and scientific biases. They began to write books and articles denouncing relativity as a hoax and Einstein as a publicity-seeking charlatan.

They introduced a theme that was to gain legitimization under the Nazi regime some thirteen years later. For them, true German physics entailed painstaking experimental work, direct contact with reality, and commonsense theories based on a clear visualization of phenomena. Jewish physics, not rooted in a pure national culture, replaced these qualities with speculation, metaphysics, and overly abstract, mathematical theories that contributed to decadence.

Einstein never sought attention for himself; he despised publicity and the media intervention in his life. Yet, in 1920, the worlds of science, culture, and popular media transformed Einstein and his theory into an unprecedented focus of attention. He defended himself and his science at the first public attack in 1920, but as the campaign of hatred mounted in a climate marked increasingly by reactionary terrorist attacks on popular radicals and democrats, Einstein considered accepting offers to move abroad. His Berlin colleagues, and especially Max Planck and Max von Laue, did what they could to defend him and his work as well as to convince him to remain in Germany. Naturally, many observers wondered whether the Royal Swedish Academy of Sciences would give the 1920 Nobel Prize to Einstein.

## EINSTEIN AND NOBEL PROVINCIALISM

Einstein was no stranger to the Academy's physics committee. He had been nominated in 1910 when Wilhelm Ostwald suggested that the special theory of relativity deserved a prize. In the years just before the war, Einstein was recognized for his many contributions to theoretical physics, but he received only modest support from nominators. No doubt, nominators who might have been favorably disposed toward Einstein were still trying to open the Nobel doors to theoretical physics with Planck. Nevertheless, before and after the war Wilhelm Wien nominated Einstein and H. A. Lorentz for a divided prize for work on relativity. During the war, Einstein's internationalist stance alienated most of his German colleagues; but some French and neutral physicists continued to nominate him. Prior to the Royal Society's announcement on the results of the eclipse expedition, some key German and Austrian physicists again nominated him. The committee responded to these proposals simply by dismissing Einstein, claiming that, at best, his work needed to be followed to see if it possessed significance for physics.

## 1920: A Swedish Surprise

Following the news of the eclipse results in November, nominators began to register the first aftershocks. Although still modest in number, nominations for Einstein dominated the 1920 list. From the committee's perspective, the nomination from Niels Bohr might have been taken as one revolutionary thinker advocating the work of another intellectual radical, but other nominators could scarcely have been more respectable. The elderly Berliner Emil Warburg, who had often won the committee's respect for his nominations, was one. Others were Dutch prizewinners H. A. Lorentz, Heike Kamerlingh Onnes, and the archexperimentalist Pieter Zeeman.

None of these nominations made much impact upon the committee. In its 1920 report, the committee lightly danced past Einstein, making a minimal gesture in his direction. In his special report on the degree to which relativity theory had been confirmed, Arrhenius reviewed the three predictions that Einstein had made based on his theory. The first of the predictions for the general theory had been satisfied, that is, it accounted for the eccentricities in Mercury's orbit around the sun that Newtonian mechanics could not explain. Arrhenius claimed that the results of the British eclipse experiment could not be admitted as evidence, as questions remained on their degree of exactness. Ignoring the British conclusions, Arrhenius followed the skeptics' line that the margin of possible experimental error was larger than the effect that was to have been measured. Finally, he noted the inability of researchers to find a so-called shift toward the red end of the solar spectra, which cast further doubt on the soundness of the theory. The committee agreed with Arrhenius that Einstein's theory could not be considered.

Arrhenius's report was decidedly tilted toward the negative opinions and interpretations in circulation. He duly noted J. J. Thomson's and Arthur Eddington's points of reservation, but he did not include their widely quoted positive remarks. He referred to a number of minor criticisms, as well as those leveled by Lenard and his coterie, to discredit Einstein. Arrhenius called attention to the preparations for the German Natural Science Association's meeting in August, where efforts were being planned to defang the relativity threat to physics and culture. He closed with a quotation from Lenard on the irresponsible nature of Einstein's work. Here, and in the coming years, Arrhenius seems to have tried to assess and follow German opinion on Einstein.

Nobody on the committee seemed to mind this less than comprehensive report, and members took little notice of what the nominators found valuable in Einstein's work. Committee members not only were firm in their categorical rejection of Einstein but were also actually working with quite a different agenda. In fact, from the start it would seem the committee wanted the prize in physics for 1920 to go to Charles-Edouard Guillaume.

Guillaume was the Swiss director of the International Bureau of Weights and Measures, who had been nominated for the Nobel Prize solely by the Genevan physicist C. E. Guye. For over two decades, Guillaume devoted himself to improving the precision and reliability of standardized measurements. He had accidentally discovered a nickel-steel alloy that, unlike virtually all materials, was relatively unaffected by changes in its environment. This discovery prompted Guillaume to devise a form of this alloy, called invar, that resisted changes in volume in response to alterations in temperature and pressure. Once found, this alloy enabled improvements to be made in precision measuring instruments, especially those used for geodetic surveys. Swedes had used such instruments in the Arctic and achieved excellent results. Clock and watch mechanisms that used invar performed with greater precision; and incandescent light bulbs were fitted with filaments based on invar rather than considerably more expensive platinum. Nobody would deny that this work, as well as the other efforts by Guillaume and the staff of the International Bureau of Weights and Measures, was of importance to science and commerce.

But, in 1920, when the world of physics had entered an intellectual adventure of extraordinary proportions, it was remarkable to find Guillaume's accomplishment, based on routine study and modest theoretical finesse, recognized as a beacon of recent achievement. Even those who opposed relativity theory found Guillaume to be a bizarre choice for the prize. As one classical physicist remarked: Whatever may be said about Einstein, he made his mark *on* physics; whereas Guillaume at best was marked *by* physics. Others expressed disbelief. What, then, was going on in the Academy?

The committee and the Academy had other matters with which to concern themselves. Bernard Hasselberg, the Academy's professor of physics and member of the Nobel committee, had been ill for several years and was planning to retire. The committee wished to pay him tribute. It was the custom in Swedish academic life to allow a retiring professor whom colleagues wished to honor a decisive voice in selecting his successor. Hasselberg had been a colleague of long standing and the Academy's own professor for thirty years, so his choice of a nominee would have to be respected.

After helping Albert Abraham Michelson win the prize in 1907, Hasselberg indicated his preference for two pairs of candidates whose work reflected his own. In precision metrology, he favored Guillaume and his colleague at the International Bureau, René Benoît; in astrophysics, George Ellery Hale and Henri Deslandres. After his success with Michelson, Hasselberg discussed with his Danish colleague Kristian Prytz the possibility of getting the next Nobel Prize for the two leaders of the International Bureau. Prytz shared Hasselberg's mania for precision measurement; he had already proposed on several occasions a prize for Guillaume and his colleagues. Although Prytz and others responded to his call for action, Hasselberg could not convince the rest of the committee. Guillaume's and Benoît's work warranted merely a simple notice.

During the next several years, Guillaume and Benoît received an occasional nomination, but never more than one at a time and always by the same two elderly individuals, who themselves, like Hasselberg, were long-standing members of the International Bureau. Hasselberg managed to leave his fingerprints on the 1912 general report in the statement that Guillaume's discovery of a nickel-steel alloy "surely seems to fulfill Alfred Nobel's requirements." Not many others agreed with him. Nobody proposed Guillaume until 1920. During the war years, Hasselberg backed his alternative favorites, the two astrophysicists Hale and Deslandres, who had considerably better chances of winning. But, in 1920, as Hasselberg lay seriously ill, only one of his prior favorites had been nominated, his friend and colleague Guillaume.

Although well respected, Guillaume's name was scarcely on the lips of world physicists. Meager and infrequent nominations underscored this fact, as did

Guillaume's own surprise at receiving the award. For committee members adamant on honoring Hasselberg, it was an advantage that Guillaume was a neutral Swiss citizen, who had worked for most of his life in France, where the International Bureau was located. For Arrhenius, who might well have balked at tipping his hat respectfully toward Hasselberg, his long-time adversary, or to a style of physics for which he had little enthusiasm, the idea of bestowing a prize, in essence, on a central institution for international cooperation was an attractive choice. The Bureau, a pinnacle of the internationalist spirit in science, was suffering severe economic hardship. Moreover, Arrhenius certainly also desired to keep relations in the Nobel Committee for Physics harmonious, especially as he had his own agendas for which he would need support.

Just as in the case of C. G. Barkla, the committee gave Carlheim-Gyllensköld the task of preparing a special report for a candidate on whom it had already tacitly agreed. Again, he prepared a sketchy report. He provided no compelling reason why Guillaume should get a prize. In early September, a bedridden Hasselberg wrote to the committee that he would be very happy to see Guillaume receive the prize. Perhaps fearful for his health, he registered in advance his vote for the committee and Academy meetings. As could be expected, the Academy's vote delighted Hasselberg and Guillaume. It brought applause from French and Swiss commentators, and it bewildered others. Swedish newspapers noted that Guillaume was virtually unknown and had difficulty in explaining why a metallurgist was awarded a physics prize. But, as usual, the media effused—this time over French officialdom's enthrallment with Sweden's wisdom and good taste.

## 1921: Allvar Gullstrand versus Albert Einstein

In 1921, Einstein's place in physics received unambiguous confirmation. Nominations from France, England, Holland, Finland, and Sweden gave him broad international support. Indeed, Einstein was the only candidate proposed by nominators from Allied nations during the boycott of German science that lasted until 1926. Einstein's position in the world of physics was portrayed by some nominators as being that of a giant, the likes of which had not been seen since Newton. From Germany, Einstein received support from the Berlin establishment, including Planck and Warburg, and also from classic experimentalists. For some, it would be difficult to consider other candidates without first seeing Einstein recognized.

Fourteen of thirty-two nominators proposed Einstein. Numerically, this result was noteworthy; a peek at the other nominations reveals that Einstein was alone in commanding a mandate. Other nominators provided scattered votes for favorite sons. Lenard proposed fellow political reactionary and antirelativity companion Ernst Gehrcke. Both men claimed that those aspects of Einstein's work on relativity that were of value had been plagiarized, namely, from Gehrcke.

Many who had the right to nominate in 1921 did not do so. Perhaps confusion, or disappointment, over the Academy's priorities and prejudices stifled interest. Some knew the committee's outlook on scientific matters; others wondered whether legal obstacles prevented awards in theoretical physics. One nominator insisted that Einstein was his candidate, but if it proved impossible to give a prize to a theoretician, then Friedrich Paschen should be considered for his spectral measurements.

Some nominators might have thought that Einstein was the obvious candidate, but they did not share enthusiasm for either the man or his physics; they failed to vote.

When the Nobel Committee for Physics met in late winter 1921 to assess the nominations for the 1921 prize, Swedish Einstein-mania was in full bloom. Einstein and relativity theory were emblematic of the new age of anxiety. Radicals proclaimed relativity as liberation from social, cultural, and intellectual traditions. In Sweden, as elsewhere, academic philosophers and cultural commentators feared that Einstein's theories implied a relativity of values. A few physicists borrowed the tone of the German attacks, but they rarely invoked racial slurs, at least in public. Although a diffuse anti-Semitism was prevalent among Swedish elites, the issue did not have the same emotional charge as in Germany. Of course, the snap, crackle, and pop of media noise need not have had any direct effect on the Nobel committee. But, in actuality, evidence points to the fact that committee members could scarcely conceive of this bushy-haired political and intellectual radical standing at the Nobel ceremony, as the pinnacle of physics, and receiving a prize from their king.

In 1921, Allvar Gullstrand took it upon himself to report on Einstein's contributions to relativity and gravitational theory. Gullstrand, who was one of the most distinguished members of the Academy, simply did not understand Einstein's work. Nevertheless, he resolved that Einstein must not receive the prize.

Gullstrand seemed too impressed with himself to be troubled by happenings outside his own intellectual pond. In 1911, the physics committee tried to give him its prize for his work on physiological optics; but the medical committee gave its prize to him first. Although a medical doctor by training, Gullstrand demanded and received a personal professorship as both ophthalmologist and physicist. In his own path-breaking studies on the optics of the eye and on astigmatism, Gullstrand developed his own intricate methods of analysis. When new advances in mathematics offered tools to simplify analysis, Gullstrand remained unmoved. His command of higher mathematics and theoretical physics was at best limited. Still, he decided that he should be elected to the physics committee as a representative for mathematical physics. At the time of Gullstrand's election to the Nobel Committee for Physics in 1911, Mittag-Leffler shook his head in disbelief. The Uppsala contingent in the Academy ensured Gullstrand's election over the internationally renowned professor of mathematical physics at the Stockholm Högskola, Ivar Fredholm. Neither Gullstrand nor Fredholm worked in what was then emerging as theoretical physics, but Fredholm at least was on the research front with respect to the mathematics involved in the new theoretical physics, and he was able to follow many of the developments. Gullstrand's accomplishments were many—some were outstanding—but he was unable to sit in judgment of physicists working with theoretical research vastly different from his own.

Gullstrand embodied the best and worst of Uppsala's academic culture. As a teacher, he was respected and feared. He demanded as much of himself as of others. He was fond of saying that "a professor whose hands do not shake from exhaustion by the end of the academic year has not performed his duties properly." Gullstrand received all the honors, titles, and royal orders Uppsala University and Sweden could bestow on a professor. In a small, isolated, but locally prestigious academic environment, arrogance, like mold in a damp cellar, tends to thrive. Arrogance prompted him to confront Einstein, to prove to the committee, the Academy, and Swedish society that special and general theories of relativity were of little significance. In the Academy, as elsewhere, titles and authority sometimes, but not always, could be mistaken for expertise.

Gullstrand had no intention of admitting relativity theory into the comprehensible world of physics. The theory's broader implications were troubling: Where is God in the fourth dimension? The theory itself contained numerous errors, according to Gullstrand. But his own published critique of Einstein's account of the irregularities in Mercury's orbit itself was shown to rest on a mistaken understanding. No matter, he rejected this criticism and returned with new objections. While preparing his special report on relativity and gravitational theories for the committee, Gullstrand turned for help to his Uppsala colleague and friend, professor of mechanics and mathematical physics C. W. Oseen. Gullstrand presented bits and pieces of his critique to Oseen, who in turn revealed Gullstrand's erroneous analyses on matters Einsteinian. For a period, Gullstrand found new objections virtually every day and Oseen corrected his friend's misunderstandings every day. Oseen himself harbored strong doubts about the validity of the relativity theories, but he grudgingly was willing to give Einstein a fair hearing to the best of his intellectual powers. Oseen confessed privately the catastrophe for the committee that as the representative for theoretical physics Gullstrand had to evaluate things he did not understand.

Gullstrand's exceptionally long fifty-page special report might appear comprehensive, but it was written from a basic premise: Einstein cannot be right. For his assault, Gullstrand mobilized any negative claim, any article that was published in a scientific journal that provided alternative interpretations to those advanced by Einstein. His own analyses of special and general relativity worked toward defusing those aspects of Einstein's theory that called for an overhaul of the "common sense foundations" of mechanics. According to Gullstrand, that which remained of the theories once Einstein's errors and unproven assertions were eliminated could best be treated successfully by classical mechanics. He rejected the suggestion that the bending of starlight by the sun could be regarded as a critical test for the theory. Even if this effect were successfully demonstrated, its confirmation would not at all mean that Einstein's theory was right. Gullstrand supported traditionalists who maintained that classical ether physics might also explain this phenomenon. Besides, Gullstrand rejected the British results of the eclipse experiment as useless.

Even on its own terms, as an intellectual accomplishment Einstein's theory did not hold "the significance for physics for which an awarding with a Nobel Prize can come into question." Gullstrand concluded that neither special nor general relativity theory, nor Einstein's gravitational theory, alone or in combination, warranted a Nobel Prize.

Gullstrand had no trouble blocking Einstein in the committee. No member approved of relativity theory or had been convinced that the eclipse experiment had yielded proof in its favor. The remaining strong experimentalist core of the committee, Granqvist and Hasselberg, had little sympathy for Einstein's aggressive use of theory. Hasselberg wrote from his sick bed: "It is highly improbable that [Alfred] Nobel considered speculations such as these to be the object of his prizes." Hasselberg and probably most of the other committee members could not reconcile themselves to accepting such work as physics. Einstein did not conduct laboratory experiments; he did not derive his theory from experimental investigation. His manner of revising fundamental assumptions, and of seeking to bring the disparate fields of physics into a unifying theory, seemed to these experimental physicists work of a metaphysician rather than a member of their scientific tribe.

For Swedish critics who most strongly agitated against Einstein and relativity theory, Hasselberg and Granqvist helped keep the Nobel Prize from promoting this "diseased movement." Hasselberg was not alone in expressing concern about the eroded foundations of European civilization; Bolshevism was a danger, poised to eradicate what was left of prewar order. For many who opposed relativity, Einstein and his theory exemplified the erosion of past certainties; an example of Dadaism in science standing in direct opposition to universal notions of truth, beauty, and goodness rooted in Western civilization's classic Greek heritage.

The less rigid experimentalist members of the committee largely acquiesced. Carlheim-Gyllensköld was not openly antagonistic to either Einstein or the theory, but he wanted to have experimental confirmation of general relativity before pushing for a prize. He later declared that in voting with the committee majority, he was not agreeing with all of Gullstrand's report; his vote was not to be taken as a final refusal to consider relativity. He knew full well how rigidly his colleagues viewed newer developments in theoretical physics. His favorite candidate since 1919, Niels Bohr, had been dismissed a bit too breezily for devising a model of the atom that allegedly conflicted with physical reality.

Arrhenius had his antenna up, trying to follow signals from abroad. He was the committee member who followed most closely Einstein's activities. Arrhenius continued to visit Germany as well as other European countries; he also received numerous guests and letters from abroad. Arrhenius wanted to help German science. A prize to Einstein might bolster the self-respect and esteem of German science, but it might also alienate important factions at home and even in Germany. Among other things, Arrhenius was in the midst of trying to improve the financial situation of physics in the Academy and to generate support for pure research. The government had turned down his requests while supporting a new academy for engineering sciences. Would a prize to the outlandish relativity theory lower the esteem of academic physics as well as irritate the intellectually and politically conservative majority of the Academy?

Equally important, what price would have to be paid to oppose Gullstrand, who enjoyed great respect and who was putting all his scientific authority into blocking Einstein? When Gullstrand presented his report to the committee, concluding that neither Einstein's theories of relativity nor their combined significance warranted a prize, nobody dissented. They agreed that no candidate could be identified as worthy and therefore the prize should be reserved. The Academy's Physics Section said the same.

Gullstrand might have won the backing of the Academy's physicists, but he had not proved Einstein wrong. When the full Academy met to vote on the 1921 science prizes, a few solitary members attempted to argue for Einstein and against Gullstrand's evaluation. As usual, no official account of the meeting exists, but a number of sources allow us to piece together the events. The Academy's discussion revealed gaps in Gullstrand's command of physics as well as his prejudice. Might the Academy be willing to award Einstein a prize for his contributions to quantum theory or for his other highly significant contributions to theoretical physics? Some members understood that Einstein was more prominent than the Nobel Prize. Would it not be an embarrassment for the Academy not to recognize the man who became the greatest scientific celebrity since Newton?

Again, the question was, to whom and for what achievements should the prize be awarded? For the conservative majority, the question reduced to: Should the prize, in Gullstrand's words, being a highly visible "Swedish flag," be given to a media-created sensational yet scientifically dubious theory? Just as in 1919, the Academy had no intention of backing away from its own beliefs. Gullstrand remarked privately, "Einstein must never receive a Nobel Prize even if the entire world demands it." A creed of expert evaluation and not democratic opinion placed the Academy's own committee evaluation on a higher plane than journalistic articles, popular books, or even letters of nomination from internationally prominent physicists. Serious attempts to propagate understanding and acceptance of relativity theory, such as the widely read book by Nobel laureate Max von Laue, were dismissed by Gullstrand as "partial." To some observers, the Academy had painted itself into a corner. Debate dragged on late into the night.

If the issue was simply Gullstrand's faulty evaluation, then in principle, the Academy was free to act once this was brought to light. But most members of the Academy had little inclination to give Einstein a Nobel Prize and no desire to slight Gullstrand. It mattered little that for years leading European professors of experimental and theoretical physics had praised Einstein as the greatest living representative of their discipline and had declared his accomplishments in relativity theory to be among the most significant in the history of science. Swedish "expertise" had spoken; the Academy guarded its own authority and its own right to assess and judge. As the clock approached midnight on November 12, 1921, the Academy voted not to award that year a Nobel Prize in physics.

## 1922: Enter Oseen

The Royal Swedish Academy of Sciences had once again made a statement. This obvious rejection of Einstein was troubling. Those invited to nominate candidates for the two available physics prizes to be given in 1922 had to reflect on the meaning of the Academy's decision. Could the Academy be persuaded to recognize Einstein, or would a proposal for him be a wasted nomination? Not sure how to second-guess the Academy's recent decisions, many nominators suggested a choice of alternate candidates or divisions among candidates. Einstein still received overwhelming support, but now Bohr also appeared as a candidate. German-speaking nominators who proposed Bohr for his quantum model of the atom also nominated Arnold Sommerfeld for his subsequent elaboration of the quantum atom, as well as Paschen for his spectral measurements. Those nominators from the former Allied nations who sought to recognize the new atomic theories eschewed the two Germans. One lone nominator offered, as he had in 1921, a quirky justification for an award, based on Einstein's discovery of the law of the photoelectric effect. The nominator, Oseen, chose his words with care. He had a clever strategy that just might defuse the ticking Einsteinian bomb without unleashing its theoretical shock wave; at the same time it might rescue Bohr from committee bias.

The law of the photoelectric effect emerged in connection with one of Einstein's brilliant 1905 papers. In this work, "On a Heuristic Point of View Concerning the Production and Transformation of Light," Einstein argued, to the disbelief of readers, that light behaves at times not as widely believed and verified as continuous waves but as discrete, individual particles. What began as a hypothesis, Einstein developed beautifully into a firm theory within two years. The individual packets of light possessed a specific amount, or quantum, of energy.

Light-quanta (later called *photons*) seemed to defy all the triumphs of nineteenth-century physics that showed light to exhibit all the phenomena of waves, such as interference (when the crests and sloughs of waves cross), diffraction, and polarization. James Clerk Maxwell's electromagnetic theory and Heinrich Hertz's experiments established light as part of a continuous wave of energy propagating at finite speed through an ether fluid. That which appears to be continuous waves of energy, according to Einstein, were actually averages of innumerable individual "atoms" of light, each possessing a discrete amount of energy. The phenomenon that could be explained by this radical suggestion was the so-called photoelectric effect. It was found that light directed at metals could, under certain circumstances, induce the emission of electrons. But the ejection of these charged particles seemed to depend only on the frequency of the light and not the intensity. A wave theory of light in which the energy is distributed over its entire surface could not account for this odd phenomenon, but it could be explained if each frequency actually corresponds to minute particles containing a specific amount of energy. The electron absorbs a discrete quantum of energy upon impact, so when enough energy is contained in a given frequency, it could knock the electron free. Einstein's law of the photoelectric effect relates the maximum energy of an ejected electron in a linear fashion to the frequency of the incoming light.

Very few physicists accepted Einstein's claim for the particulate nature of light. (Ironically, Johannes Stark alone had endorsed it for a while before he turned vehemently against all quantum-based theories.) When in 1916 the American physicist Robert Millikan painstakingly accumulated reliable experimental data with which to disprove this theory, he was shocked to find that Einstein's law was beautifully confirmed. But this did not convince him, or most other physicists, to accept the "bold, if not to say reckless hypothesis" of light-quanta. The law of the photoelectric effect held, but what was the physical reality behind it?

The committee met as usual in late winter to assign reports. Gullstrand accepted the responsibility of updating his evaluation of relativity theory. It seems to have mattered little that his limitations and prejudices had been exposed in the Academy debate. But the committee accepted the need to recruit Oseen. Hasselberg had just died, leaving an empty place on the committee. The protocol for the meeting on May 30, 1922, records a motion to co-opt Oseen as an ad hoc member assigned to preparing reports on Einstein's theory of the photoelectric effect and on Bohr's atomic theory.

The committee met after the summer to assess the evaluations. Once again, Gullstrand axed relativity. He dismissed the vast literature that favored the special theory. Books on the subject tended to be too popular; these did not give the reader a critical examination of the underlying assumptions: "Dogma [*trossatser*] . . . are mistaken for facts." Although the special theory had been long accepted by most leaders of international physics, Gullstrand insisted that "the question of the special (and with that also the general) theory of relativity's justification is a matter of faith [*trossak*]." After reviewing a number of technical points and recent literature, he again asserted that Einstein could not receive a Nobel Prize for either the special or general theory of relativity or for the combined significance of the two.

Oseen unleashed on the committee his vastly superior command of theoretical physics as well as his unusual critical skills. He made no secret of his frustration with the committee and with the Academy Physics Section's dilettantish understanding of theoretical physics. A year earlier, in 1921, Oseen had already nominated Einstein for the discovery of the law of the photoelectric effect. To evaluate this and related nominations on Einstein's treatment of this phenomenon, Arrhenius had prepared a short special report. The report portrayed Einstein's contribution as merely a natural extension of the work of others, at best ambiguously confirmed by experiment. Arrhenius missed the mark, having failed to reveal the radical content of Einstein's contribution or its consequences for quantum theory. Arrhenius concluded that it would be seen as "strange" if this work were awarded a prize while Einstein's more significant accomplishments such as relativity and Brownian molecular movement were passed over.

Oseen had managed to get himself invited to the Physics Section for its vote on the 1921 prize. He tried at that time to persuade its members that Einstein did indeed deserve a prize precisely for the law of the photoelectric effect. All he could pry out of the Physics Section was a statement acknowledging its "great significance" and a decision not to take action until a later time when more certain knowledge could be gained. Oseen thereby prevented the Academy from rejecting the photoelectric effect. As a relatively new member of the Academy, in the Mathematics Section, Oseen learned firsthand the importance of being recognized as a formally constituted expert. Now, in 1922, he was invited to play just such a role.

Oseen knew what he was after. Once he joined the committee, he insisted on a demarcation between his nomination, which specified the discovery of the law of the photoelectric effect, and the formulation used in specifying his committee assignment. The committee protocol was altered after his entry onto the committee: His assignment changed from assessing the theory of the photoelectric effect to assessing the discovery of the law. In the final draft of the general report, Oseen's nomination for the discovery of the law was now listed separately from nominations that specified the theory. As odd and perhaps incomprehensible as this surgical procedure might appear to a physicist today, in the context of physics in 1922, and especially in the context of the Academy, it reflected unusual acumen.

Einstein's work on relativity and gravitation might well have been appreciated by admirers as a brilliant intellectual masterpiece, but Einstein's theory of the photoelectric effect was perhaps even bolder—and called for an even more radical adjustments in physical thought. Even Einstein's admirers, including Max Planck and Niels Bohr, refused to accept the claim that light possessed qualities of discrete particles. When he proposed Einstein for the post in Berlin, Planck felt compelled to excuse the brilliant theoretician's intellectual exuberance with respect to this one particular claim. Bohr, who was pushing the use of quantum theory into atomic domains, balked at accepting this extremely radical proposition. Indeed, in 1922 this aspect of Einstein's theory had won very few adherents. But the actual law relating discrete absorption and emission of energy to the frequency of light had been meticulously proven. Still, Robert Millikan and others took the position that although the law had been proven valid, the actual physical reality underlying the law remained a mystery. In 1922, physicists such as Bohr and others working with quantum theory used Einstein's law, but they employed a variety of linguistic and conceptual alternatives to the light-quantum concept.

Oseen very much wanted Einstein to receive a prize, but not for relativity. Oseen was aware that the prestige of the Nobel prizes had begun to wane, not the least because of the sinking value of the prize money and because of such embarrassing awards as that to Guillaume. A prize to Einstein would, of course, bring fame to the Nobel institution. Equally if not more significant, Oseen wanted his younger friend and colleague Niels Bohr to receive a prize. Bohr's atomic model was based on the idea that electrons spinning around a nucleus jump from one orbit to another when absorbing or emitting discrete amounts of energy, Planck's quantum of energy. Surprisingly, the model accounted for a number of experimental phenomena that could not be explained by existing models of the atom, such as the Stark effect and the distribution of lines in the hydrogen spectrum. For Oseen, Bohr's model and its unexpected successes were "the most beautiful of all the beautiful" in recent theoretical physics.

Oseen understood the Academy's prejudices. Actually, he shared many of these, but he was able to see a bit farther than most. In one brilliant stroke, Oseen saw how he could meet the objections against both Einstein and Bohr. A master strategist entered center stage in the ongoing Nobel drama.

First, by restricting the motivation for the prize to just the law, and not the "theory," Oseen could safeguard the proposal against opponents who might maintain that the existence of Einstein's light-quantum was still disputed. Second, in his special report, he noted that Einstein's various achievements appealed to different types of researchers. Here, he could avoid the problem that had arisen the previous year, when objections had been raised to Einstein's being rewarded for a less prominent achievement than relativity, which some might consider strange. Oseen identified a variety of scientific orientations and outlooks; each one might consider a different Einstein contribution to be the most significant of his many extraordinary accomplishments. Theoretical physicists might be drawn to Einstein's work on quantum theory. And for "the measuring physicist"—the type of physical scientist most admired in the Academy—no work of Einstein's could compete for interest with the discovery of the law of the photoelectric effect.

Appealing to the committee's convictions about the nature of physics, Oseen opened the door for Bohr. The young Dane's atomic models should enjoy high status because (as Bohr sometimes argued) his radiation theory rested on Einstein's law, which had now been shown to be a fundamental natural law. Previously, the committee majority had been willing to acknowledge some of the successes that Bohr's model had won in explaining spectral phenomena, but it would not give a prize for work that stood "in conflict with physical reality." By emphasizing, if not exaggerating, the very close bond between Einstein's thoroughly proven law of nature and Bohr's atom, Oseen masterfully defused the charges of speculative theory. But, in reviewing the recent modifications in Bohr's atomic model, Oseen was able to claim that even if the model were only a first step, it nevertheless served to stimulate research. Just as Stark and Planck were rewarded not for their actual discoveries, per se, but for their role in provoking new lines of research, Bohr deserved recognition.

Oseen respected and liked the young Danish physicist. Oseen had helped Bohr get a chair of theoretical physics at Copenhagen University, but now Bohr was struggling as much with establishing his own institute as he was with extending his atomic model. Clearly, Bohr could use some help. Not only would a Nobel Prize provide much-needed money in a very strained economy, it would also certainly help him attract funds from other sources. Both Bohr and Oseen appreciated the need to develop institutional resources in neutral countries that might serve as meeting places for a divided world of science. And Oseen appreciated that here was a candidate who really did fulfill the spirit of Nobel's testament: He was young, a dreamer with an unfinished project of great significance, and he could use the money—and prestige—to achieve great things.

Oseen's wish to reward Bohr resulted in Einstein's prize. Oseen's agenda directed the committee's preliminary discussions. In the first draft of the general report, Arrhenius and Granqvist singled out from all of Einstein's many theoretical works the discovery of the law of the photoelectric effect and the law of specific heat of solid bodies. According to this draft, discussed at the September 6 meeting, Einstein was to receive the reserved 1921 Nobel Prize for his general contributions to theoretical physics and especially for discovering these two laws. During the meeting, references to the specific heat of solids were crossed over. More significantly, Oseen insisted on yet another change. The original justification for giving the prize referred to these two laws "which he [Einstein] derived from quantum theory, which in turn received a new specially vigorous renewal." This formulation was also sanitized according to Oseen's strategy to read simply, "the discovery of the aforementioned law [of the photoeffect], through which quantum theory received a new specially vigorous renewal." Reference to the law's derivation from quantum theory had been dropped. Although Oseen provided in his special report an analysis of how Einstein derived the law, he strove to emphasize that the prize was to be given for the discovery of a fundamental law of nature. Oseen hammered in the point that this law, thoroughly confirmed by experiment, could stand independent of Einstein's troublesome theoretical assumptions and the many conflicting physical interpretations of the law then circulating among physicists. In one masterful stroke, he disarmed opponents to both Einstein and Bohr.

When the committee came to vote, there was hardly a whisper of dissent. Gullstrand knew that Oseen understood theoretical physics; he could not challenge Oseen's proposals. For his part, Arrhenius was transformed into an enthusiastic supporter. He had met Einstein in Berlin and had witnessed the immense respect and affection accorded him by members of the Berlin scientific establishment. He learned that the once-respected Philipp Lenard and Johannes Stark were now discredited and held in contempt by the mainline leaders of German physics. While monitoring Einstein, Arrhenius appreciated that the Swiss Jewish theorist in Berlin not only stood for, and worked for, internationalism, but could also contribute to its reestablishment. Einstein's many international lecture tours—such as a 1922 lecture series at the Collège de France—offered hope that communication was possible between the two hostile camps. Indeed, for his more open-minded German colleagues, as well as for leaders of the new liberal Weimar Republic, Einstein emerged as the representative of German culture able to break the Allied boycott. Einstein, the pacifist, internationalist, and, above all, brilliant scientist, could uniquely help maintain the traditional high prestige of German science.

## EINSTEIN'S TRIUMPH

When the full Academy voted in November, questions arose about the wisdom of not rewarding Einstein for relativity theory. Predictably, strong sentiments against the theory quickly surfaced. But Oseen steered the vote in favor of his own formulation, which the committee had backed. The Academy finally voted to give the reserved 1921 prize to Einstein for his contributions to theoretical physics and in particular his discovery of the law of the photoelectric effect. It then voted to award the 1922 prize to Bohr.

Rumblings, of course, continued. Reminders came to Arrhenius and Oseen that no mention of relativity would be tolerated on Einstein's diploma or at the ceremony. Censorious shots were fired at the Academy from the extremely conservative *Nya Dagligt Allehanda* by an elderly Swedish physics professor who claimed that Oseen's publications on Einstein showed the extent of decadence in modern science. Without Granqvist and Hasselberg, the professor bemoaned, the Academy could no longer defend itself from sensationalist, scandalous physics. Lenard sent a telegram to the committee condemning the prize. His action soon reached the German and Swedish press: "Nobel Prize Recipient Lenard Protests Against Einstein's Nobel Prize"; "Serious Accusations Against the Swedish Nobel Committee"; "Nobel Prize in the Firing Line."

It was the *Nya Dagligt Allehanda* that broke the news in Sweden of Lenard's attack. According to Lenard, Einstein's work on the photoelectric effect was unoriginal and, where novel, failed to take into account basic knowledge on the nature of light. Lenard considered Einstein to be a publicity-seeking Jew whose work and actions were alien to the spirit of true German science. The editor of the *Nya Dagligt Allehanda* was in a bind. He felt it necessary to express surprise that an internationally renowned scientist such as Lenard would drag racial issues into his attack on Einstein. Still, he guardedly defended Nobel Prize winner Lenard—"a scientific authority on light of the first rank"—and tried to limit the criticism of Einstein more principally to that physicist's publicity seeking. But the dominant conservative paper, *Svenska Dagbladet,* perhaps wanting to defend the Academy and the Nobel institution, condemned Lenard's accusations as reprehensible. An article, probably written by Arrhenius, pointed out that few scientists had followed Lenard into an anti-Einstein camp; he and his band of antirelativity warriors had disgraced themselves through their dirty—and misinformed—campaign. Turning Lenard's own rhetoric against him, the article concluded that in this latest action Lenard had not at all revealed "clear Germanic intellect [*ande*]."

Humility ultimately vanquished arrogance, this time. Einstein might not have received the prize for his contributions to relativity theory, but he was soon able to have these celebrated in spite of the Academy. Einstein learned of the Academy's decision in a telegram that reached him while he was en route to Japan. He had been compelled to leave Germany for a while. Previously, the unswerving support of his Berlin colleagues and his own affection for them, and for Berlin, prompted him to stay. But following the French occupation of the Rhineland in 1922, political unrest aimed at left-wing liberal personalities unnerved Einstein. Among those murdered by right-wing assassins was the Jewish foreign minister and Einstein's acquaintance Walther Rathenau. Threats on Einstein's life came more regularly; Einstein started to take these seriously.

Einstein's successes only fueled reactionary propaganda. His efforts on behalf of progressive movements in postimperial Germany, at a time of economic catastrophe and international humiliation, put him in danger. He assumed a low profile at home and gladly accepted the opportunity to go to Japan for several months. Having received notice of his Nobel Prize while at sea, Einstein could happily avoid the media frenzy in Germany. Finally seeing the end of his immediate economic worries, he could sit at rest on the deck of the steamer, heading for the Far East.

Einstein was relieved to receive the prize. He had been counting on it since 1919 when, after divorcing his wife, he assumed that the prize money would pay his alimony. Subsequently, not only did he not receive a prize, the German economy went into a tailspin, placing him under great financial stress, which affected his health. Given the weakening German mark, being paid the prize money in Swedish crowns would make a difference. He certainly also knew that Arrhenius had helped Planck and Haber avoid the astronomically high postwar taxes by keeping their winnings in a foreign bank. Although he never expressed any special reverence toward the Nobel Prize, Einstein later told a Swedish journalist that the prizes brought about a "social regulating" of age-old injustices: "Through the prizes scientists finally could harvest gains [*skörda vinst*] from their work just like businessmen." And that made him happy.

Einstein undoubtedly found amusing the instructions informing him that when he delivered his Nobel lecture, he must lecture on the topic for which he has been awarded the prize—that is, there would be no mention of relativity. Providentially, his trip to Japan would keep him away from the formal ceremonies of December. He could avoid the stiff formalities and media attention, both of which he loathed. He arranged to receive his prize the following summer. And when he did, he went not to Stockholm and the Academy but to Sweden's second city, Gothenburg, where he addressed the Seventeenth Congress of Nordic Natural Scientists.

Einstein stopped off in Palestine and Spain on the way home. Japan had been a great reprieve for the beleaguered celebrity; he found Japanese simplicity much to his liking. But what was waiting in Sweden? Einstein understood that opposition to relativity theory was strong. In Germany, his life had been threatened; would his enemies seek to disrupt the occasion? In advance of Einstein's visit, the German engineer Paul Weyland visited Sweden. Weyland, representing the ultranationalistic scientific association, claimed to be seeking Swedish partners for this alternative to the Berlin-based association that he and cofounders Stark and Lenard felt was infested with Jews and liberals. Rumor spread that his real reason for visiting was to transplant their extreme anti-Einstein agitation to Sweden in time for the Congress. The question arose: Was he helping to plan an assassination, metaphorical or otherwise? Even the *Nya Dagligt Allehanda* tried to play down the visit, noting that Swedish opinion for and against relativity was not as polarized as in Germany. Sweden had not experienced anything resembling Germany's national trauma. Moreover, as the press made clear, nobody quite knew when Einstein was going to speak. And, whenever Einstein did give his Nobel lecture, the paper added, he most definitely would not speak on relativity, as his prize was not awarded for that.

Einstein arrived late. Wary and worldly, a seasoned celebrity, showman, and genius, Einstein sneaked into the back of the Gothenburg Congress's opening banquet ceremonies. Some say he misunderstood on which day he was expected to arrive; others claimed that the train from Copenhagen was delayed. Or maybe Einstein was simply taking precautions. He was expected to sit on the dais with the king and other dignitaries. But the awkward intruder was soon recognized and whisked, against his will, to a place of honor.

His triumph began. All who came in contact with him fell under his spell. His charm, his humility, his voice, his face—nobody was left untouched. Even when he lectured on the most incomprehensible subjects, reported one journalist, he entertained. His face could be that of a musician; his voice was such that the most convoluted physics sounded like a romantic poem by Heine. More than a thousand people came to the huge auditorium in Gothenburg's famous Lisenberg amusement park. Arrhenius acted as host. Not being a formal Nobel ceremony, Einstein felt free to choose the topic for his lecture, or else he simply did not care. He spoke on recent developments in relativity theory. Carefully and pedagogically, Einstein reviewed earlier studies on special and general relativity, including those of many contributors who together had created the exciting new understanding of physical nature and the cosmos. Sitting in the front row of the audience was one most attentive spectator who let it be known that he very much wanted to learn something of relativity theory: King Gustav V.

# EIGHT

## To Sit on a Nobel Committee Is Like Sitting on Quicksand

### Disciplining and Its Discontents

### TWO PHYSICISTS SEEK CHANGE

*T*he year is 1919: two Swedish physicists, theoretician C. W. Oseen in Uppsala and experimentalist Manne Siegbahn in Lund, both discontented with the state of Swedish physics, both contemplating ways and means of improvement. Fast-forward to 1923: Both men are now in Uppsala, both men on the Nobel Committee for Physics, both determined to use their positions to develop thriving schools of theoretical and experimental physics that might bring renown to Swedish science. Their efforts to invigorate Swedish physics—in their image—and the opposition these actions prompted, influenced the Nobel Prize. To understand the actions of these new committee members, some insight is needed into their broader goals and interests.

#### Oseen: Making Sweden Safe for Theoretical Physics

Oseen was a man of great intellect and broad cultural interests. Although he was by no means an empire builder, he nevertheless understood that money, blended with authority, was the oil necessary for a smooth running of science. Following his success in defusing the Einstein problem, he joined the Nobel committee as a regular member and moved from the Mathematics to the Physics Section. His recent experience primed him to mobilize and focus resources, including the prize institution, for the betterment of Swedish physics. Proud, arrogant, and talented, Oseen was bound to cause a stir.

Early in his career, Oseen wanted to be part of a truly *Swedish* scientific discipline. Brimming with nationalist pride, Oseen had believed that "for us the most important [thing] is to try to create an independent Swedish science, which dares pose and cultivate its own problems whether or not these happen to be fashionable abroad." Oseen was keenly aware of the intellectual weaknesses of Swedish science, but these were not an insurmountable obstacle. During two and half decades he never submitted a paper to a foreign journal. But in 1913, when he learned that the

Royal Swedish Academy of Sciences could not afford to accept all worthy manuscripts submitted for publication, he thought, "Then it is of course ridiculous" to work toward an autonomous national science. Frustrated and ashamed, he refused to identify his nationality on papers published abroad; he could not accept admitting the inferior Swedish "vassal" relation to foreign science.

Such obvious deficiencies prompted Oseen to work for the improvement of Swedish physics. During and immediately after the war, the Academy's economy worsened. Publishing the research of its members proved even more difficult. Moreover, physics had to make room for the "descriptive sciences" that dominated in the Academy. The permanent secretary, the entomologist Christopher Aurivillius, was outright hostile to physical science and failed to understand the urgent calls for rapid publication. Oseen believed that Swedish physics deserved to be better appreciated by nonspecialists at home and by colleagues abroad. When plans emerged at the end of the war to create a national physics society made up of academic scientists, teachers, and serious amateurs, he joined in the discussions. By uniting all the nation's physicists into a single interest group, he thought that greater visibility and even funding for research could eventually be attained. Oseen acted on conviction; he became president of the new society in 1921.

Oseen considered himself a theoretical physicist and wished to see this specialty institutionalized in Sweden. Chairs in mechanics and mathematical physics, such as his own, could be colonized by theoretical physicists. But mathematical physics focused more on mathematics than on physics; these positions required little more than paper and blackboards for research. For Oseen, the worst mistake a theoretical physicist could make was to lose contact with reality, to study only a world on paper. Controlling theory through experiment and deriving theory from the study of laboratory phenomena were the foundational pillars of his creed. He learned that professors of theoretical physics abroad had access to laboratories. Hospitality in his university's physics laboratory was dependent upon the professor of physics. This laboratory had been designed primarily for instructional purposes and could accommodate only so many projects before delicate measurements and fragile experiments were threatened. Gustaf Granqvist squeezed space for doctoral students and also gave Gullstrand a corner for his optical investigations.

Oseen desired to transform his professorship in mechanics and mathematical physics to one in theoretical physics. He had been working on problems in hydrodynamics, the physics of fluid motions. He entered a heated dispute with a leading German professor in this field, Ludwig Prandtl. Oseen worked mathematically; Prandtl had recourse to experiment that gave hard data. Wanting to have recourse to the same, Oseen petitioned Uppsala University for a modest grant for space and equipment. He argued that theoretical physicists often had recourse to experiment; unlike mathematical physicists, who worked with highly idealized models, he wanted contact with reality. He was turned down. Oseen understood that the Nobel committee could, in principle, offer support. Oseen conceived his ultimate goal to be a Nobel Institute for Theoretical Physics.

When Granqvist died, in 1922, Oseen began to upgrade Uppsala physics. Selecting the right person to take over the chair in physics could benefit his own plans. Oseen turned to Manne Siegbahn, a young professor in Lund who already had gained a name abroad. Oseen maintained that Siegbahn was the most qualified physicist in Sweden. Both men were disappointed over the lack of contact between local experimental and theoretical research, and both felt that perhaps they could rectify this chronic problem in Swedish physics if they worked in Uppsala.

## Precision and Money: Siegbahn as Entrepreneur

Siegbahn also had ideas for improving Swedish physics, or, more specifically, on expanding his research activities. He learned about developments in X-ray spectroscopy and in the early years of World War I began to measure spectra. The leaders in this new field of endeavor—such as Maurice de Broglie and the Braggs—were occupied with war-related actitivies; Henry Moseley was killed. Siegbahn didn't advance, however, merely because others fell behind. He had a gift for instrument design, and equally important, a talent for organizing laboratory activity. Capitalizing on a rapid increase in students studying physics, he recruited some of the most talented ones for advanced studies. As student numbers declined during the war, he secured space in the teaching laboratory for doctoral projects. Immediately upon becoming professor in 1919, he began to expand the scope and space of the institute. The quest for more, bigger, and bigger yet characterized Siegbahn's subsequent career.

Siegbahn believed that work in X-ray spectroscopy could play an important role in atomic physics. During the war, his laboratory provided critical data needed by foreign theorists. Taking the lead in making precise measurements of the spectral wavelengths of elements, he and his assistants cornered the market in the field. Unlike spectra derived from visible light, the spectral lines derived from X-rays provided information on the electrons close to the atomic core. He entered into a close collaboration with Arnold Sommerfeld, the energetic professor of theoretical physics in Munich. Sommerfeld was busy expanding the quantum model of the atom; in his leap into mathematical abstractions, he needed Siegbahn's data.

Siegbahn's prestige depended upon the atomic theorists. As long as he could provide critical data for their enterprise, his laboratory could remain prominent. He also needed to understand what the theorists required. Siegbahn hosted a conference in 1919 to which he brought Sommerfeld and others from devastated Germany together with Nordic physicists, including Bohr. The chaotic state of quantum theory was a boon for Siegbahn. By increasing the precision of the measurements and extending these to heavier elements in the periodic table, he could maintain prominence. His technical ingenuity, coupled with the experimental finesse of some of his assistants and the unusual dexterity of his local instrument maker, gave Siegbahn reason to be optimistic, even when other laboratories were beginning to take up the same research. All that Siegbahn required to assure continued success was money.

Increased precision—and with that, Siegbahn's career as physicist and discipline builder—boiled down to an economic question. He confessed to Oseen that a new instrument designed for attaining greater exactness in measuring X-ray spectra would cost about 5,000 crowns—or, simply put, the entire annual research budget for his institute, a situation that posed "difficulties." Oseen urged Siegbahn to move to Uppsala, the best-endowed physics institute in Sweden. But even that institution's resources would not be sufficient for Siegbahn. Siegbahn accepted the offer, but immediately began discussing his needs to obtain extra funding for equipment and technical assistance.

Siegbahn also wanted membership in the Academy and Nobel Committee for Physics. In Lund, he scarcely could follow the Nobel events and deliberations. Seeing that Granqvist's position on the committee wasn't being filled, he reminded Oseen that the committee needed an experimental physicist. Siegbahn got his wish in 1923. But the attempt, led by Arrhenius, to hold the position vacant as a means to save money was a sign that even the Nobel Committee for Physics had economic limits.

Following Siegbahn's move to Uppsala, Oseen expressed an unusual degree of optimism in declaring the start of a new era. Actually, it was a period of frustrating setbacks for Swedish science, and especially physics. Plans during the war years for establishing the Nobel institutes had come to naught. Efforts to engage the government to contribute to the establishment of these laboratories had also failed. The Academy's worsening financial difficulties forced a decision to eliminate the professorial position of Academy physicist, held last by Hasselberg.

Oseen and Siegbahn were not alone in their concern over money for research. Arrhenius also despaired over the worsening financial situation. Arrhenius's failed campaign to raise money for the Nobel institutes transformed into a campaign to save money. In an effort to cut expenses, Arrhenius went so far as to recommend that the various tasks of the Academy's permanent secretary be divided among members, so as to save a salary. Making matters worse, following the imposition of new taxes on the Nobel Foundation's funds, the prize itself was declining in monetary value and, according to some observers, losing some of its appeal. Arrhenius's own Nobel Institute for Physical Chemistry, which traditionally was bustling with local and foreign researchers, now stood almost empty, just when the need to help German, Austrian, and Russian colleagues was greatest. Repairs, improvements, and operational costs required grants from the Nobel committee's funds. In a situation in which the economic situation was "simply terrible" and "the prospects for the Academy's future are . . . quite hopeless," Arrhenius kept his eye on the Nobel pie. Concerted and drastic action was needed to concentrate funds. Arrhenius and the Uppsala committee members all understood that fewer slices meant portions possibly large enough to make a difference. To do this meant changing more than two decades of procedure based on a broad understanding of physics with respect to the prize.

## 1923: ELIMINATING ASTROPHYSICS FROM NOBEL PHYSICS

What did Alfred Nobel mean when he stipulated a prize in the field of physics? What subject matter, methods, and professional orientations can fit under the umbrella of physics? With respect to the prize, the question was never solely academic. When the prizes came into being, physics was a broadly conceived field. At the first International Congress of Physics, which met in Paris in 1900, the organizers used a definition that aimed at inclusiveness. Similarly, when the Academy appointed its first physics committee, everyone assumed that Nobel intended a wide understanding of physics. Meteorologist Hildebrandsson, astrophysicist Hasselberg, and physical chemist as well as cosmical physicist Arrhenius were elected to the first committee. The committee set priorities, but it did not aim to exclude.

Some grumbling was heard over the scope of physics open to the prize. With money and media attention associated with the prizes, to some people it seemed best that competition be restricted to a narrow field of entrants. Ernest Rutherford opposed Gösta Mittag-Leffler's appeal to nominate Henri Poincaré in 1910. Rutherford was little interested in admitting mathematical physics for a prize. That would only increase yet further the competition for prizes, making it more difficult for deserving experimentalists. But other than occasional debate in the Academy on the role of mathematical physics and theory in Nobel matters, little attention was devoted to the question prior to 1923. Until then, nominations for contributions in meteorology, astrophysics, cosmical physics, theoretical physics, and physical chemistry were not declared ineligible for consideration. Some of these proposals received serious consideration. Knut Ångström had been evaluated worthy of the prize in 1910 based on his measurements of the solar constant. Astrophysicists George Ellery Hale and Henri Deslandres entered the ranks of those candidates declared worthy and lined up to wait for their time to come.

What did come, in 1923, was a will to restrict the scope of physics. Oseen and Siegbahn brought a different set of priorities for their discipline. They, together with Arrhenius and Gullstrand, considered the question of focusing resources and prestige more narrowly on fewer specialties to be urgent. They first eliminated astrophysics and the candidacies of Hale and Deslandres in one clean surgical procedure; meteorology, not so easily disposed of, eventually followed once Oseen rhetorically bludgeoned the favored candidate and his supporters.

Awarding or withholding prizes for contributions in specific research specialties could conceivably assist in promoting or hindering the advance of these fields in Sweden. The case of Hale and Deslandres is worth considering, as it highlights the campaign to redraw boundaries and define new priorities; it also reveals how, independent of the nominators, committee members themselves could either propel a candidate toward a prize or, just as readily, eliminate him from consideration.

## Astrophysical Contenders

In the late nineteenth century when George Ellery Hale began his career in Chicago, he sought to bring the methods and theories of physics into astronomy. As a pioneer astrophysicist, he attempted to apply spectroscopy to the study of the physical and chemical characteristics of stars and in particular the sun. He invented an instrument, called a spectroheliograph, that advanced solar research. Among other important discoveries, he found magnetic fields associated with sunspots. This revelation both opened a new appreciation of the sun's nature and provided added incentive to explore solar physics.

A group of American scientists in 1909 first brought Hale to the Nobel committee's attention. Hale next appeared on the list of nominees in 1913. That same year, a lone French nominator proposed as his second choice the French astronomer and meteorologist Henri Deslandres, who independently and subsequent to Hale also had created a spectroheliograph. Nobody, then or later, outside the committee ever considered having Hale and Deslandres share a prize for the same achievement. Undoubtedly, Arrhenius's and especially Hasselberg's own close relations with Deslandres made the two men sympathetic to his assertion of priority. Very few others took Deslandres's claim to heart, at least with respect to nominations for the Nobel Prize. Nevertheless, the committee linked Hale and Deslandres together as equals and noted that their respective work in solar physics deserved to be "followed with the greatest attention, and which—sooner or later and rightly so—ought come to be regarded as deserving of the Nobel physics prize."

In 1914, American nominators again proposed Hale. The committee declared him, along with von Laue and Planck, to be among the most deserving of the prize, but they declined to evaluate his work more thoroughly, as Deslandres had not been nominated that year. To ensure that Hale and Deslandres would be evaluated together once again, Arrhenius proposed the two in 1915. The committee declared them worthy of a Nobel Prize but put them on hold because of the "especially fortunate" circumstance of being able to back the British Braggs whose work and nationality harmonized with the postponed 1914 prize for the German von Laue.

In 1916, Hasselberg sent in a proposal himself for Hale and Deslandres. Hale, alone, also collected additional support from two Dutch Nobel laureates, J. D. van der Waals and Pieter Zeeman, as well as an American. In the first draft, before a decision was taken to withhold the prize because of the war, Hale and Deslandres were declared worthy of a prize. Hale enjoyed further nominations in 1917 and 1922 from Italy, Germany, Spain, and the United States. And finally, in 1923, after the dramatic dispatching of Planck, Einstein, and Bohr, very few nominators responded. Hale was proposed by a German and an American; Carlheim-Gyllensköld picked up Hasselberg's earlier pairing of Hale and Deslandres, proposing them himself.

Gullstrand, who had become committee chairman after Granqvist, reviewed the nominations that arrived for 1923. The chairman, as was the custom, then suggested assignments for special reports. Based on past criteria and prior evaluations, Gullstrand understood that the strongest candidates that year were Hale and Deslandres—but only if the committee still wanted to consider astrophysics part of physics.

When this question had first arisen, in 1900, Hasselberg wrote that the prize could be awarded not only to achievement in "pure Physics properly so-called," but also in "the sciences most closely connected with Physics and for the cultivation of which physical methods are employed." He included physical chemistry and astrophysics: "This I suppose will certainly answer to the proper meaning of the testator. Whereas thus . . . pure Astronomy or celestial dynamics . . . will scarcely be included, any work in Astrophysics of great importance deserves certainly a thorough consideration." No discussion arose earlier whether Hale's and Deslandres's work fell outside the scope of physics. Similarly, the nominations of astrophysical work by W. W. Campbell (1901) and W. S. Adams (1920) or for that matter those for meteorological achievement such as that of Julius von Hann (1913), Vilhelm Bjerknes, and W. N. Shaw (1921), among others, were not disallowed for not being part of physics. But by 1923 Hasselberg and Granqvist were dead; Carlheim-Gyllensköld soon discovered that he stood alone in the committee as a champion for the preexisting broad understanding of physics.

## The Stakes

For Oseen, Siegbahn, and also Arrhenius, the time had come to act. With respect to astrophysics, the committee's past assessment was not to have a future. If the Nobel institution was to benefit Swedish physics, the number of specialties and individuals having access to the funds would have to be limited. They had reason to appreciate astrophysics and the broader cosmical physics as a threat to their own priorities.

Discussions had begun in earnest during the first years of the war on creating the remaining Nobel institutes; at that time, it was assumed that by 1922 sufficient funds would be generated to establish a Nobel Institute for physics and one for chemistry. Although Hasselberg had been disappointed that the Academy's monumental new building had no room for physics, he warmed quickly to a suggestion to house the Academy's collection of instruments in a department of the Nobel Institute for Physics. Hasselberg was already fancying a place for himself in the new institute. He wrote to an American colleague in 1916: "As however the Nobel [Foundation] . . . will in some time erect a great laboratory for general physics it is perhaps possible, that the question can be solved in such a way, that my laboratory, in the form of an astrophysical department of it, can become a special pavilion in connection with the proposed new [Nobel] physical institution." Hasselberg's eagerness to promote Hale and Deslandres as candidates for a Nobel Prize dovetailed with his eagerness to have his own department of the Nobel Institute. Naturally a prize would underline more clearly astrophysics as part of physics with respect to matters Nobel.

Carlheim-Gyllensköld also had his eye on the Nobel Institute. In 1915, he broke ranks with Hasselberg and nominated his own candidates, whose work more closely resembled his own. Carlheim-Gyllensköld nominated and evaluated positively two Norwegian cosmical physicists, Kristian Birkeland and Carl Størmer, who had breathed new life into the age-old riddle of the northern lights (aurora borealis). He hoped for a department of cosmical physics at the Nobel Institute. He also harbored designs on the Academy's instrument collection, which might form the basis for a museum of the history of the exact sciences. Recognizing that the formal rationale for creating the Nobel institutes entailed evaluating candidates' work, he pointed out in his special report on the Norwegian researchers that unfortunately a thorough examination of their results could not be provided because as yet no department of cosmical physics at the Nobel Institute existed.

Whether to call the proposed department astrophysics or cosmical physics was not a problem. Hasselberg backed the plan; a report to the Academy noted that cosmical physics in general had a strong research tradition in Sweden without being formally represented in any university. Be that as it may, the Nobel Committee for Physics repeatedly tabled motions to make a formal recommendation to the Academy. In private, Arrhenius expressed scorn. Hasselberg had managed his job as Academy professor so poorly that the position was now considered expendable. Furthermore, Arrhenius had little faith in Carlheim-Gyllensköld's designs on the Nobel Institute. For his part, Granqvist sensed Hasselberg's efforts to define the department largely in his own image. Perhaps also wanting a comfortable, prestigious research position, Granqvist noted the importance of not allowing the department and its choice of leader to be defined exclusively by astrophysical spectroscopy.

In 1917 and 1918, jockeying for position continued; subcommittees estimated costs for alternative options for a cosmical physics department in the Nobel Institute. But growing inflation as the war dragged on and the ensuing economic crisis forced a suspension of plans for the Nobel Institute.

Cosmical physics jeopardized in other ways the majority's plans for mobilizing the modest Nobel funds for their own causes. The very modest nest egg of money generated from those Nobel prizes withheld during the war—the committee's special fund—was already being decimated. First, the committee had given a gift to Arnold Sommerfeld to assist his institute in Munich. Then Arrhenius had begun making claims for his Nobel Institute, and then cosmical physics joined in. Carlheim-Gyllensköld requested and received two grants for his study of secular variations in the earth's magnetic field, and Granqvist, before dying, received a modest grant for his investigation of solar radiation. Granqvist justified his grant by claiming the need to evaluate the nomination of C. G. Abbott, proposed for work on the solar constant. Even though Abbott was certainly not a major candidate for a Nobel Prize, the principle seemed clear: As long as these specialties remained eligible for consideration for Nobel prizes, committee members could request funds to help evaluate relevant proposals.

## Assassination

Not surprisingly, Gullstrand's message that Hale and Deslandres could well claim the 1923 prize triggered a flurry of action. Oseen informed Gullstrand that he doubted whether their contributions belonged to physics. Oseen claimed that he was willing to yield to Arrhenius's authority on this question. He understood that both Arrhenius and Siegbahn shared his own desire to focus resources. Gullstrand played the role of neutral committee chairman, but actually he maneuvered the situation masterfully against the astrophysicists.

To prevent surprises, especially in the Academy, Gullstrand rigged the game. He convinced Carlheim-Gyllensköld to show his cards first. Gullstrand urged him to prepare a draft of his special report on the astrophysicists' work as quickly as possible; this action would allow the other committee members to begin deliberating before the summer. Gullstrand baited Carlheim-Gyllensköld into forgoing the then-customary use of the summer for such reports. He warned that if the committee could not achieve a clear consensus on a candidate, the Academy would surely vote to award the physics prize to Ernest Rutherford. In reality, this tactic allowed other committee members to prepare counterarguments to the expected positive assessment of Hale and Deslandres.

Carlheim-Gyllensköld pressed forward. He delivered a report by June extolling the astrophysicists' accomplishments. To shore up support, Carlheim-Gyllensköld arranged to hold a lecture on cosmical physics for the full Academy in conjunction with the committee's June meeting. His efforts were of little avail; ambush awaited him.

Once the committee had the draft as a basis for discussion, Arrhenius prepared a devastating counterstatement. He concluded that astrophysics had expanded so rapidly since 1900, when it had been declared eligible for the physics prize, that it now encompassed all of astronomy. With a rhetorical sleight of hand, Arrhenius then declared that astronomy had therefore become astrophysics. But since Alfred Nobel clearly had not intended astronomy to be eligible for his prizes, astrophysics no longer could be eligible. Arrhenius emphasized the importance of withholding a prize to Hale and Deslandres, regardless of the earlier declaration of their worthiness. He warned that a prize to them would set a "precedent" legitimizing astrophysics as being part of physics. Numerous nominations would then arrive, some for candidates such as Arthur Eddington, the leading British astrophysicist, who could not be ignored. But much more problematic, a flood of nominations for contributions to astronomy could be expected. These, in turn, would then have to be evaluated as to whether they belonged to astronomy or to astrophysics, thereby strangling the committee. The way out was to support Arrhenius's contention: Astrophysics had become astronomy. Therefore, Hale and Deslandres had to be eliminated on the grounds that their field fell outside of physics.

Carlheim-Gyllensköld protested vigorously. He denied as ludicrous Arrhenius's equating astrophysics with astronomy. As for the feared deluge of astronomical nominations, Carlheim-Gyllensköld held that the scarcity of nominations for astrophysics to date attested otherwise. Moreover, Hale and Deslandres, he reminded his colleagues, were nominated, after all, by physicists, including winners of the Nobel Prize and by the Academy's own physicist.

Although his ploy was a fiction, Arrhenius knew that as an accomplished researcher in cosmical physics he commanded authority to counter Carlheim-Gyllensköld's response. Moreover, he understood that the committee's majority backed him fully. They edited the final draft of Carlheim-Gyllensköld's special report to remove the concluding laudatory comments in support of dividing the prize between Hale and Deslandres. In Gullstrand's draft for the general report, he left blank the concluding comments for two other candidates, the American Robert Millikan and the German James Franck. It was clear by this time that one of these two would likely be the majority's candidate. Finally, the committee majority diminished Hale's accomplishments by claiming that they had not resulted in solid achievements for an understanding of the sun. They raised, at least for the moment, the standard for a prize: A discovery must not just open new significant insight, it must also offer a definitive solution to past puzzles. In contrast, the work of the proposed winner was declared to have transformed contentious problems once and for all to one of science's solid achievements.

In one of the many ironies than run through the history of the Nobel prizes, the 1923 physics prize went to Hale's colleague at the California Institute of Technology, Robert Millikan. In Hale's efforts to transform Cal Tech into a major center for science and technology, he had recruited Millikan from Chicago. At the start of his career in the 1900s, Millikan, a student of A. A. Michelson, produced lackluster but competent achievements—nothing that would allow him to fulfill his ambition to be a leading physicist. Not being especially adept at theory and not being particularly enthusiastic about the rise of the new revolutionary ideas in physics, he used a keen talent for defining a niche where he might yet make a major mark. By bringing great precision to measurements of controversial physical phenomena, he believed he could play a role of importance in the world of physics. His now famous oil-drop experiments from 1910 not only provided extremely accurate determinations of electronic charge but also claimed to show that electrons were all unique, identically charged particles. Although some doubts were raised about his findings, by the time Millikan emerged as a candidate in 1919, few doubted his claims. His verification of Einstein's law of the photoelectric effect, which actually was intended to disprove quantum theory, nevertheless strengthened his acclaim as a meticulous master of measurement. But was he worthy of a Nobel Prize?

Millikan was by no means a sure winner. His earlier support was always limited to Americans: his patron Hale, chemist T. W. Richards, and naturalist H. F. Osborn of the American Museum of Natural History being among the most loyal nominators. In 1923, Millikan was nominated solely by Edgar Meyer, from Zurich, as a second choice after the German spectroscopist Friedrich Paschen. Competition that year included Ernest Rutherford's candidacy for a physics prize for his work with alpha particles that, among other discoveries, led to the nuclear model of the atom and to the discovery that atoms of lighter elements could be disintegrated by these energetic particles. But what gave Millikan an edge, once astrophysics was banished from the Nobel garden of physics, was Siegbahn's entry onto the committee. The two shared similar scientific temperaments and career strategies. Moreover, Siegbahn, in contrast to his committee colleagues, looked upon the emerging American colossus favorably; he aimed at forging closer ties with American scientific institutions and was already planning to visit California the following year.

So Hale and Millikan, colleagues working closely together in the United States, do not share the same place in history. This is evidenced in the halls of the epicenter of American physics, Cal Tech's Athenaeum. In the dining room is a group portrait of Hale and Millikan, along with chemist A. A. Noyes: the three founders who launched the institution toward international prominence. But in the barroom, where individual portraits of the Nobel laureates are, Milllikan parts company with Hale; Hale is not among the group who are toasted from time to time. Millikan's place there, and not Hale's, can be attributed to internal politics on the Nobel Committee for Physics, not to a clear mandate from the world of physics.

## Mopping Up

After this first successful attack on the traditional broad understanding of physics, the committee majority continued its assault against less favored specialties. Recent successes in elucidating, with the help of physics, the dynamics of atmospheric processes prompted a scattering of nominations for physicists working on these difficult problems. Traditionally, meteorologists occupied a number of the openings in the Academy's Physics Section. After the deaths of two Academy members who were meteorologists, the same constellation of Nobel committee members decided "to act to make sure that we do not elect yet another weak meteorologist." They organized a strategy behind the back of the senior Academy meteorologist, Hildebrandsson, by agreeing on candidates in advance so as to preclude discussion in plenum. Hildebrandsson reacted angrily. Privately he threatened a showdown in the Academy if the committee members did not back a meteorologist the next time.

Part of the motivation for the committee's action was the simple fact that it seemed that most of the local Swedish meteorologists were not particularly strong scientifically. A number of recent bitter quarrels over the nature of cyclones (low-pressure disturbances), coupled with expansive plans for meteorological activities, only added to the committee's wariness. Although Oseen had closed his published 1919 lecture series on theoretical physics by suggesting that the physics of atmospheric changes might become the next frontier after the physics of atomic phenomena, his dismay with his local meteorological colleagues caused him to reconsider. Oseen then claimed that terrestrial physics was by nature intrinsically inferior since it could not achieve exactness through laboratory experiment.

The committee appreciated the need to act prudently and avoid the appearance in the Academy of being hostile to meteorology. Oseen accepted the challenge to keep the prize from going to any physicist who contributed to atmospheric science. A prize would legitimate the field as being part of physics, precisely what they were hoping to undo. Oseen did all he could to dismiss the candidacy of Vilhelm Bjerknes, in the mid-1920s and again in the late 1930s, arguing so emotionally that meteorology cannot be allowed to be considered physics that the Academy was prepared to overturn the committee's clear prejudice.

The precedent of not awarding Hale and Deslandres made it easier for the committee to dismiss subsequent candidates who nevertheless sporadically popped up on the list of nominees. Although no formal statutory changes had been taken, a new tradition for interpreting the statutes had been initiated, at least for the time being. During the next few decades, leading astrophysicists who were nominated were generally summarily dismissed: Hans Bethe, Ira Bowen, Arthur Eddington, Edwin Hubble, Meghnad Saha, H. N. Russell. The committee noted that regardless of how important the astrophysicists' achievements might be for the specialty field of astrophysics, these did not have sufficient significance for the field of physics in general (as the committee defined it) to warrant a Nobel Prize. Notwithstanding the committee's desire to shrivel the domain of physics, the disciplinary boundary line necessarily remained diffuse and the gatekeepers acted as overzealous guardians.

Work by Saha and Bethe, for example, certainly could not justifiably be dismissed as being solely significant for astrophysics divorced from a mainstream of physics. Indeed, Bethe, who was not an astrophysicist, proposed in the late 1930s a brilliant theory for explaining how stars produce energy that drew upon insight from nuclear physics and paid back, with intellectual interest, to that field by providing new insights on nuclear fusion. The committee repeatedly dismissed the enthusiastic proposals that regularly appeared in the 1940s on the grounds that this work was not part of physics. It undoubtedly took both major changes in the committee's composition and relentless pressure from prominent physicists to open the gates in 1967 for Bethe to receive a Nobel Prize in physics. And yet an individual such as Edwin Hubble, who demonstrated in 1930 that the universe was indeed expanding, with galaxies hurtling away from a primordial center with a velocity proportional to their distance, might arguably be said to have contributed more to astrophysics than to "mainline" physics. But that did not stop some physicists from nominating him. More importantly, a definition of physics that could not embrace the significance of such a striking cosmological finding was surely impoverished. But the prize was already becoming something quite different than a celebration of intellectual triumph.

## SAVING MONEY AND SAVING SIEGBAHN

Tightening the boundaries of what could be considered physics was only one response to the unfavorable conditions for research. Committee members also tried to generate more money for research. They desired to withhold the prize as often as possible. By first reserving a prize and then not awarding it the following year, the committee—with the Academy's assent—could assist in building up either the main fund, interest from which provides money for the prizes, or the committee's special fund, interest from which could be used for research grants. Adding money to these funds became a committee obsession.

Fixated with the need to raise funding for research, Arrhenius devoted himself to the task. He spearheaded a crusade to have the government free the Nobel Foundation from the extra postwar taxation. He finally succeeded in 1925, setting in motion the possibility of increasing the funds for proposed institutes and strengthening the value of the prize. He continued to plan expansion of his own institute while other committee members entertained their own dreams. Siegbahn, who was now firmly in place in Uppsala, already had begun searching for more money for new instruments. Recognizing the ease with which grants from the committee's special fund could be obtained, he also concurred with the aim of trying not to find worthy candidates.

When all members could agree on the agenda, the committee could dispatch and dismiss the nominees, assuring the Academy that no candidate could be found whose achievements fulfilled the high standards stipulated by Nobel's testament. Aiding the process, Oseen's hypercritical stance and keen rhetoric skills could find fault with virtually any candidate.

Oseen accepted the role of critic. Although his own work entailed hydrodynamics and the physics of crystal lattices, he sat in judgment of all theoretical physics, and especially atomic physics. He wielded a censorious Samurai sword to dispatch sloppy reasoning, incomplete theories, or logically inconsistent arguments. Just how fairly and consistently did he apply his critical judgment? He set almost impossibly high standards for himself and for others. Any candidate who happened to work on problems with which Oseen had direct interest came in for extra rigorous scrutiny. Being a man of culture and having a foot in the tradition of a professor as bearer of high culture for society, he took to heart the intellectual turmoil and radical changes in theoretical physics. For Oseen, the physical world picture held consequence for broader social, cultural, and theological values.

Oseen was not happy with how physics was developing. His own demand for clear logical consistency drove him to despair over partial and temporary solutions to the deep crises of atomic physics. He longed for a giant to enter the world of physics who might sweep away all the inconsistencies, all the culturally alarming implications, and make physics right. As such, he had his hands full holding the world of physics under control, at least in his evaluations, and maintaining his ability to influence how prestige and recognition should be distributed based on awarding of the Nobel Prize. Oseen saw little reason in celebrating half solutions, temporary scaffoldings, and piecemeal steps toward an unknown future.

Unbridled arrogance coupled with great intellectual powers is not necessarily the best quality for a Nobel committee member. Nobody mastered all of physics or all of theoretical physics, even in the as-yet-modest magnitude of activity before World War II. Oseen's special reports differed from those of other committee members. To a greater degree than his colleagues, Oseen often played prosecuting attorney, judge, and jury. When committee or Academy members disagreed, he countered with scathing counterarguments as well as personal hurt. Of course, Oseen was capable of sheathing his sword and pleading leniency for the sake of causes close to his heart. Oseen's refusal to concede a touch of uncertainty, a hole in his intellectual armor when modesty or even prudence might have dictated otherwise, left deep tracks in the Nobel competition. The paucity of awards to theoretical achievement during Oseen's two-decade reign reflected his sensibilities rather than formal obstacles or scarcity of candidates.

## The Years 1924–1925

After a sprinkling of responses in 1923 for physics, the eligible nominators answered in larger numbers the call for proposals for nominees. Still, the results were modest as compared with those of earlier years. Those nominations that did arrive revealed adherence to the boycott and counterboycott sensibilities that divided scientists from the former Central and Allied powers. Nominators and candidates from the neutral nations remained the only sign of transcending hostilities. In part, intellectual orientation also played some role. The physics communities of these nations differed in orientation and style of research; moreover, in the midst of economic hardship many scientific leaders increasingly appreciated the prize as an important means for opening money chests at home. No clear candidate emerged from the nominations. Germans largely backed James Franck and Friedrich Paschen; French and Dutch were for Jean Perrin (study of Brownian movement and proof of the existence of atoms); English support went to C. T. R. Wilson for his work on cloud chambers and atmospheric electricity; and a mixture of nationalities suggested the Johns Hopkins experimenter R. W. Wood for his discoveries in fluorescence and physical optics. Among those nominating, several former winners of the prize indicated their conviction that deserving candidates existed. Bohr, Einstein, and von Laue were for Franck; Guillaume, Kammerlingh Onnes, Lorentz, and Zeeman for Perrin; Rutherford and J. J. Thomson for Wilson. The Nobel Committee for Physics concluded, however, that no candidate could be found who deserved the 1924 prize.

By 1925, a flurry of activity was beginning to advance atomic physics through its crisis; the situation was still, at best, unsettling. Little of this growing ferment was as yet reflected in the proposals. In many respects, the pattern of nominating resembled the previous year's. Nominators remained modest in number, perhaps reflecting not just indecision over recent claims in physics but also uncertainty as to what the committee considered Nobel quality. The lack of an award in 1924 prompted nominators to wonder what to do; some repeated their choices, others tried new candidates, while a few simply decided it wasn't worth the time and effort to make a proposal. Many of those who did return the invitation to nominate felt strongly that regardless of how the conflicting models of the atom and energy radiation eventually would be resolved, those people who helped launch the early quantum theory nevertheless deserved to be honored with a Nobel Prize. Arnold Sommerfeld, for his efforts to extend Bohr's atom, was the most popular of them, but Paschen, Franck, and Wood once again also received noteworthy support for producing critically important experimental data. Sommerfeld had long been proposed by a handful of nominators, including Planck, but now he managed even to enjoy the support of Millikan, who seemingly defied the boycott.

Actually, Sommerfeld was chanceless: Oseen was dead-set opposed to his nomination. He saw little reason to reward a partial solution, even if it was fruitful for stimulating further work. He also opposed Sommerfeld's style of theorizing. Beginning with the "answer," Sommerfeld then worked backward mathematically to create a model that could account for experimental data. The mathematics might work, but what did it mean physically? Oseen opposed such so-called mathematical formalisms in theoretical physics; he insisted that physical interpretation and visual comprehensibility must remain central to the study of physics.

Unlike the nominators, Oseen opposed rewarding work that had played a significant and even dramatic role in moving physics forward but that no longer was at the absolute cutting edge, such as Sommerfeld's quantum model of the atom. But Oseen also had little stomach for that which was going on at the cutting edge of theoretical physics. He, like many physicists, did not believe that any resolution would be in reach in the foreseeable future. It was claimed that whole generations of physicists would have to work theoretically and experimentally before resolving the quantum dilemma. Perhaps that was why committee members were optimistic that the prize could be withheld fairly regularly; the bylaws demanded that a prize be awarded at least once every five years.

## Siegbahn Gets By with a Little Help from His Friends

Some committee members were willing to reserve and withhold prizes for at least three years. Others were largely in agreement, except for one candidate, whom they felt compelled to aid. Oseen and Gullstrand combated Arrhenius and Carlheim-Gyllensköld over Siegbahn.

Siegbahn was not a very strong candidate. In 1923, when nominated for the first time (by the Russian Orest Khvol'son), he requested that he not be considered. In 1925, Siegbahn returned to the list of nominees. Max von Laue dropped his repeated nomination of Franck, whom he assumed would again have little chance of success. This time, von Laue proposed Siegbahn, adding that such a prize could also be considered a recognition of the late Henry Moseley's pioneering work in the line of research that began with von Laue's own discovery. Another nomination for Siegbahn came from Stanford University's president, David Starr Jordan, who was impressed with his recent Swedish guest. And from Vienna, Stefan Meyer called attention to Siegbahn while also proposing C. T. R. Wilson for his cloud chamber as an alternative. These proposals could not be interpreted as constituting a mandate for a prize; committee traditions and statutory regulations also posed a stiff barrier for anyone wanting to promote Siegbahn.

Nobody could deny that Siegbahn's X-ray spectroscopic measurements were playing a significant role for atomic theory. The problem immediately confronting the committee entailed the question of whether Siegbahn actually had a significant discovery or invention to his credit, which the statutes specify as a condition for receiving a prize. His instrumentation, highly innovative in many respects, drew upon the work of Moseley, the Braggs, and, especially in the preliminary design of the equipment, Maurice de Broglie. Siegbahn perfected that which others had created. Moreover, none of the nominators designated a particular discovery or invention; they merely noted in very general terms Siegbahn's highly precise measurements of X-ray spectra.

As was often the case, evaluation involved more than assessing candidates' merits. Differences in research styles and orientation as well as personality had long contributed to an on-and-off rivalry between physicists at Uppsala University and the Stockholm Högskola. New jolts arose when Siegbahn strongly opposed providing any assistance from the Nobel committee's special fund to Hans Pettersson, a young physicist with Stockholm connections then working in Vienna. Siegbahn refused even to discuss the issue. He had no interest in seeing the meager funds directed to what he considered less serious research than his own. According to one observer, Siegbahn's aggressive response seemed more like wrangling over awarding prizes or professorial appointments rather than a reaction to an application for a few thousand crowns. Arrhenius, and other Stockholm scientists, recognized Siegbahn's actions as symptomatic of the authoritative control that Uppsala physicists—"the small popes in Uppsala, especially Siegbahn"—sought over Swedish physics.

If Arrhenius had personal stakes, so too did Oseen and Gullstrand. They supported Siegbahn's efforts to expand Uppsala physics. Oseen still believed that he and Siegbahn might eventually discuss physics regularly, specifically the mutual dependence between theory and experiment that both had talked of cultivating but still had not gotten around to pursuing. Also, clearly, the Siegbahn school was the most prominent representative of Swedish physics abroad. Most important, they appreciated that Siegbahn's energetic and ambitious plans had just slammed into a brick wall.

Siegbahn's trip to the United States in 1924 was an exhilarating but also disturbing affair. Received with respect and shown great courtesy, he also had his eyes opened. Physics in America had a vitality that took Siegbahn by surprise. American physicists were building bigger laboratories, hiring more research assistants, and forging links with industry, which in turn provided money, equipment, and jobs. He appreciated that his own research school was highly regarded as having been the best for X-ray spectroscopy, but others were now entering the field. Money was a key to dominance. He turned to one of the new splendidly endowed private foundations that had begun funding scientific research, the Rockefeller Foundation's International Education Board (IEB). Stipends for very promising young scholars and grants for centers of excellence were to promote international exchanges among the top echelons of science: "Make the peaks higher" provided the guiding philosophy during the 1920s.

Siegbahn requested $30,000 for instruments and for hosting foreign research fellows. The situation in his laboratory had become so overcrowded that he was considering reducing the number of doctoral projects. Siegbahn's annual laboratory budget came to the equivalent of $6,000. His plans for expanding his research programs to push the measurement of X-ray spectra toward both shorter and longer wavelengths would require equipment purchases of $30,000.

Augustus Trowbridge, the American physicist who directed European funding from the IEB, understood that Siegbahn's recent visit to the United States prompted the application. Trowbridge noticed that, just like so many other recent European visitors, Siegbahn was surprised to find American laboratories to be "enormously better equipped than his own, and in fact most of the European laboratories of physics." Work comparable to Siegbahn's was being pursued in at least two major American physics laboratories, and, thought Trowbridge, "I believe he has come to the conclusion that, in order to continue to do work in his field, in competition so to speak with these American laboratories, he will need something comparable in the way of equipment." If the IEB were to provide this grant, Siegbahn would still have to raise an additional equivalent of $3,000 per year for running the laboratory and for upkeep of the instruments. Grants from the IEB required matching local funds. Even if Siegbahn had exaggerated the expected costs, which was what IEB officials believed, he still clearly felt that to remain competitive in his own research niche he would have to have access to considerable patronage.

In visiting the IEB's headquarters in New York, Siegbahn had impressed the president as being a "young man of pleasing traits and tremendous vitality," but more was required to obtain sizable funding. Trowbridge was initially favorable toward Siegbahn, but later he felt compelled to inform the main office of an about-face. He learned as part of his ongoing survey of European scientists that Siegbahn might not warrant a major grant: "I have an opinion of several of my physical [science] colleagues that Siegbahn's work has the merit of great accuracy and [is] prized for that quality by all. However, he is not regarded as at present doing work of very great originality." In the spring of 1925 the question of funding for Siegbahn was suspended until some indefinite future date.

After having approached the Rockefeller Foundation in 1924, Siegbahn eyed the Nobel committee's special fund as a possible source for some of the matching funds. Here was perhaps one reason contributing to his anger over Pettersson's application for the same funds. There was not very much money available to begin with. Here also was a good reason for Siegbahn to look favorably at reserving the 1924 prize. But in 1925, when the IEB dashed his hopes for American support, his needs became all that much greater. Oseen and Gullstrand appreciated Siegbahn's situation; they understood that a Nobel Prize could provide much-wanted money and increased prestige, which could help attract further assistance.

Although not an experimental physicist, Oseen nevertheless prepared a special report on Siegbahn and his assistants' accomplishments. He concluded that Siegbahn brought X-ray research to a new height. For atomic theory, asserted Oseen, Siegbahn's discoveries were of great significance: "Not a few theories that had been greeted with great acclamation upon being presented, have through Siegbahn's measurements been shown to be in conflict with reality." Moreover, "the picture of atomic structure that we now do have, in many and important respects, is based on these measurements." Arrhenius and Carlheim-Gyllensköld could not readily dispute the importance of the measurements, or whether other investigators shared in the production of data for atomic theory, but they did find the evidence of a "discovery" to be lacking.

For similar cases in the past, the committee declared that regardless of how consequential the research might be, a candidate was not eligible for consideration without a specific discovery or invention. Recognizing that he was treading on thin interpretive ice, Oseen conceded that although he found Siegbahn deserving and although he planned to vote accordingly, he was willing to defer to the Academy's opinion as to whether Siegbahn's work fulfilled the statutory requirement concerning "discovery." The committee split, two for and two against, but in his capacity as chairman, Gullstrand could break the tie by having his vote count as two. The Uppsala faction proposed Siegbahn for the reserved 1924 prize. Carlheim-Gyllensköld and Arrhenius dissented. They proposed withholding the 1924 prize and placing the money in the special fund. All agreed that no other candidate deserved a prize; the committee unanimously agreed to reserve the 1925 prize until 1926.

To try defusing potential opposition when the vote came to the Academy's Physics Section, Oseen and Gullstrand moved to co-opt a medical professor of X-ray radiation, Gösta Forssell, to comment on the issue. He argued that Siegbahn had provided evidence for a new series of spectral lines, the M-lines, which could be considered a discovery. But even if this was the case, this assertion did not alleviate the challenge; none of the nominators had mentioned this particular achievement. Committee practice was to exclude candidates for whom no specific discovery was named by nominators.

But now, suddenly, Oseen appeared not brandishing his critical sword but waving a flower of tolerance. He did an about-face, urging that the time had come for the committee and the Academy to be more liberal when interpreting the bylaws. Gullstrand backed him, saying that the stringent requirements demanded in the past should now be relaxed. A closer reading of the situation reveals that they were trying to soften the criteria only for their colleague Siegbahn. During the very same year, Oseen argued that Vilhelm Bjerknes's discoveries in hydrodynamics and in atmospheric physics, while significant, could not really be considered discoveries according to the restrictive demands placed upon the committee by the statutes. In fact, Oseen subsequently continued to use the statutory requirement for a specific discovery like a blunt instrument to eliminate candidates not to his liking.

Arrhenius and Carlheim-Gyllensköld did not swallow this sudden call for flexibility. Carlheim-Gyllensköld noted that Siegbahn was preparing a new means of measuring spectral wavelengths with new equipment. These measurements might yield a significant discovery. In any case, by waiting it might be possible to assess more certainly Siegbahn's merits. An additional formal objection was raised: Siegbahn had not worked alone; his assistants and coworkers obtained many of the most significant measurements. None of these arguments had any effect. The Physics Section voted six to three for Siegbahn, and the Academy backed the majority. Although Siegbahn received the prize, disgruntlement in the Academy with the committee soon came to the fore.

## A Pope Crowned

No doubt, Siegbahn was a first-rate designer of experiments and instruments. Most certainly, he possessed a keen professional instinct for identifying problems that suited his talents and that could command attention abroad. But, based on the standards the committee was applying to others at this time, did Siegbahn deserve a prize? If Siegbahn was worthy, he surely should have shared the prize with others.

Not to belittle Siegbahn and his students' accomplishments, these nevertheless could not actually fit through the needle hole of statutory requirements that Oseen and others held up to filter out other candidates. The great experimental physicist Friedrich Paschen was regularly dismissed on the grounds that although he had produced critical precision measurements of spectra and other phenomena relevant for atomic physics for over a decade, he could not claim a particular major discovery. Paschen's record of being nominated far exceeded that which Siegbahn enjoyed in number, duration, and reputation of the nominators. Based on the continued persistence, at times very strong, of support for Paschen, even reaching into the 1930s, it is hard to believe that a divided prize that included Paschen would have been received by anything less than applause in most centers of physics. Similarly, a division that included Sommerfeld would also, based on the record of nominations, have been largely welcomed. Without Arnold Sommerfeld's insatiable need for ever more precise X-ray spectroscopic data, Siegbahn would not have engaged so single-mindedly in this field. Just like Paschen, Sommerfeld commanded considerably greater support from nominators than did Siegbahn. The interrelated work of these three physicists marked an important phase in the development of atomic physics, one that, although ultimately inadequate, nevertheless proved a critical step forward. In fact, such an award might have made the blatant favoritism for Siegbahn seem somewhat less apparent.

Even if Siegbahn was not a brilliant physicist, the prize conferred authority. When Siegbahn tried again to get a grant from the Rockefeller Foundation a few years later, the magic of the prize still shined through, in spite of nagging uncertainty. Noted one foundation member, "Siegbahn has, as you know, obtained recently the Nobel Prize, and while in the opinion of physicists generally he does not measure up to the standard of the former recipients of this prize, it is nevertheless a great distinction and an evidence of the man's standing in his profession." Once again, the illusion of anointed Nobel laureates being all but divine was hard to dispel. In a comprehensive survey of European science a few years later, Rockefeller Foundation officers noted, "Siegbahn and C. W. Oseen form a sort of scientific-political clique, with Siegbahn as the dominant member, and this clique has great prestige and authority in Swedish science." Reporting what informants had told them,

> [Siegbahn's work] *is the incarnation of Swedish precision and mechanical genius,* and yet ... granted that spectrum lines could be photographed at all, it was clear that in due time, someone would do a superlative job of measuring them. His actual work, that is, is excellent, but his problem is routine. . . . Siegbahn, moreover, has the limitations which often go with such intense and long continued specialization. He is a poor judge of any works outside his own narrow field and, in fact, is not interested in discussing broader aspects of his own work.

Others also questioned Siegbahn's actual understanding of physics. Possessing an eminent talent for instrument design might well produce valuable results, even without mastering theory or following in detail developments outside one's own little highly specialized duchy, but should such a person be elevated to a national scientific leader and sit in judgment of others?

Far less well known to the Rockefeller network was the fact that this very same clique of Siegbahn and Oseen would also ultimately dominate the Nobel committee for two decades; Siegbahn would sit in judgment even longer. Oseen, arrogant and judgmental; Siegbahn, engrossed with expanding his laboratory and maintaining his reputation: They emerged as the strongmen on the committee, but not without conflict. Undoubtedly, to some members of the Royal Swedish Academy of Sciences, the Uppsala physicists reeked of partiality. Moreover, some found irksome the lightness by which the committee more generally dismissed several well-supported, long-standing nominees such as James Franck, Jean Perrin, and Robert Wood. The committee was about to be waylaid.

## REBELLION

November 11, 1926: A dreary, dull autumn day in Stockholm, but in the Royal Swedish Academy of Sciences a storm of controversy whipped up passions. Several weeks earlier, the committee claimed, once again, that none of the candidates merited a prize. It proposed withholding the previously reserved 1925 prize and placing the money into the committee's special fund. The current 1926 prize would be reserved until the following year. Inadvertently, the committee set the stage for an assault on its authority. The committee had been so preoccupied with trying to save money, as well as with trying to ride out the rising storm of quantum mechanics, that it seems to have been oblivious to broader concerns. It was soon reminded of them.

Carl Benedicks, professor of metallurgy and earlier of physics at the Stockholm Högskola, had been elected in 1924 to the Academy's Physics Section. He held little enthusiasm for quantum physics and even less for the concentration of authority in Uppsala. He experienced dismay and annoyance when, as a member of the Academy, he saw how the committee so readily dismissed candidates who surely deserved more careful consideration. He also observed how the committee applied various dismissive phrases, based on the authority of the bylaws, erratically. Benedicks' first notable action as a member of the Physics Section was to join his Stockholm colleagues in voting against Siegbahn's prize.

Benedicks had long hoped to see a prize awarded to Jean Perrin for his work two decades earlier related to proving the existence of atoms in a study of Brownian movement, the irregular movement of minute particles suspended in a liquid. Although Perrin had consistently received firm support from French and Dutch nominators, the committee had long since closed the book on the Frenchman. During the past few years, the committee had merely responded to repeated nominations for Perrin by claiming that nothing new could change the group's earlier judgment: His research was not of sufficient importance. Moreover, Perrin's last significant work, *Les Atoms,* was published in 1913; hence, according to the committee, his achievements were too far back in time to be considered. Perrin's candidacy appealed to Benedicks for several reasons: to advance a contribution that did not rely on quantum theory; to reward French science, especially "in the brilliant form of Perrin"; to forward a chemistry prize to The(odor) Svedberg, the young Swede who worked on the same problem; and to challenge the committee's authority. Benedicks no doubt wondered how he might advance Perrin's candidacy in light of the committee's solid, long-standing consensual rejection.

Had the committee not been so self-obsessed, it might have noted the wispy high clouds of discontent that heralded the approaching storm of rebellion. Prior heated exchanges between Stockholm and Uppsala factions in the Academy over grants from the special fund, as well as over Siegbahn's candidacy, all pointed to an atmosphere increasingly unstable with dissatisfaction. When, in 1926, the committee proposed not to award any prize, the stage was set. When the Physics Section met on October 23, the first gale warnings were already flying.

Benedicks appreciated what the committee had completely ignored. The year 1926 was the twenty-fifth anniversary of the first awarding of the Nobel prizes. The Academy, Nobel Foundation, and others close to the prize were looking forward to a splendid celebration. Without sufficient prizewinners, and with them their respective nations' diplomatic representatives, it would be difficult to have a grand party. In 1924, only two prizewinners were named, and both stayed away (Einhoven for medicine and Reymund for literature), which effectively put a damper on the ceremony. The festivities for 1925 also disappointed. Siegbahn was the only prizewinner present. For all his dynamic entrepreneurial activity, he proved to be a boring speaker. Again, there was a lackluster celebration. Those in the select social circles invited to the festivities were now looking forward to a resplendent tribute to Alfred Nobel and his legacy—and to celebrate themselves.

At the meeting of the Physics Section, Benedicks charged that the committee once again summarily dismissed strongly nominated candidates with inadequate justification. He proposed that the reserved 1925 prize should be given to Perrin. He sought to find some weakness with which he might produce a counterargument in the committee's rejection of his candidate.

The committee had dismissed not only Perrin but also Robert Wood and Arthur Compton on the grounds that their work did not have sufficient significance within physics. Since all three candidates had been solidly nominated by Nobel laureates, among others, Benedicks assumed that the phrase was simply deployed as a means for dispatching unwanted candidates. Indeed, Compton's discovery had, by 1925, received acclamation after first having been met by skepticism. Compton convinced physicists to accept the notion of light-quantum. A candidate who enjoyed endorsement from a collection of laureates as internationally and intellectually diverse as W. L. Bragg, Albert Einstein, Robert Millikan, and Wilhelm Wien most certainly had contributed momentously to physics. But Compton's discovery advanced quantum physics, which Benedicks found distasteful. Still, the committee's evaluation of Compton reinforced the impression that it was predisposed to withhold prizes.

Benedicks preferred to focus his efforts on Perrin, who again received impressive support from French and Dutch nominators, including laureates Guillaume, Lorentz, and Zeeman. Desperate to find an opening in the solid wall of committee opposition, Benedicks asked for support to Perrin on the basis of an alternative interpretation of the bylaws. Of the two relevant clauses in Nobel's testament—"discoveries or innovations within physics" and "benefit on mankind"—emphasis almost always went to the significance for physics. Benedicks asked, What discoveries or inventions in physics have benefited mankind? His rationale was that definitive proof of the existence of atoms naturally had many scientific, technological, and philosophical consequences. It might have been worth a try, but Benedicks did not come very far. Gullstrand rebutted angrily by insisting that Perrin's work had not at all benefited mankind. In the final vote, Benedicks received support only from the two meteorologists in the section; there was a seven-to-three defeat as the majority supported the committee proposal to place the 1925 prize money into the special fund.

Committee members subsequently learned that Benedicks still planned to advance Perrin's candidacy in the Academy. Having stumbled badly in his initial performance, Benedicks began rehearsing. He wrote to a French nominator of Perrin whom he knew well. He discussed strategy for how to present Perrin's achievements to the Academy so as to rebut the committee's negative scientific judgment.

Faced with Benedicks' mobilization, the committee needed to act. Arrhenius wrote to Oseen, admitting their naivete in not putting forth any candidate. Newspapers had already picked up the story that no winners were proposed for the two physics prizes and wondered whether the lack of winners could affect the celebration. Arrhenius conceded that the committee should recall how indebted it was to the Nobel Foundation; it would be insulting and unfortunate not to provide a winner to be feted at the twenty-fifth anniversary. But, said Arrhenius, "I must thwart most resolutely this [Benedicks'] proposal, primarily for the reason that P[errin] has been proposed many times for the Nobel Prize, but the Academy [based on committee evaluations] rejected these proposals." He underscored the growing age of

Perrin's work as an unequivocal reason for strenuous opposition. By awarding a prize to Perrin, the Academy would be refuting the committee's authority. To defuse the situation, Arrhenius planned to propose at the Academy's meeting that the reserved 1925 prize should go to James Franck and Gustav Hertz for their experiments over a decade earlier on the process by which electrons collide with atoms. Several times earlier, Oseen had evaluated these two as well as Franck alone. Although generally favorable, the special reports never recommended a prize to Franck, whom most of the nominators preferred alone, or to both of them. In 1923, Oseen had first suggested that Franck merited a prize, but he dropped that clause to allow Millikan.

Arrhenius wanted to be sure he would have backing from Oseen and Siegbahn. Whatever differences they might have had in the recent past, they all appreciated that in an assault on the committee's authority they had to rally together. Oseen shot back a letter the following day. He confessed that he was greatly in doubt over his own decision not to propose a division between Franck and Hertz; under the circumstances he now enthusiastically supported their candidacy. Siegbahn also appreciated the need for action; although he would prefer awarding the prize to the American, Compton, he was willing to back Arrhenius's proposal. Gullstrand gave his support. Arrhenius was responsible for communicating with Carlheim-Gyllensköld. They assumed that by offering the prize to both Franck and Hertz, the Academy would be appeased, the threat of rebellion averted.

## Showdown

The protocol for the Academy's meeting on November 11 provides few clues as to what happened. Private correspondence provides glimpses: "Yesterday evening offered much of interest in that corporation called the Royal Academy of Sciences." Benedicks took the floor and reminded the Academy that this year was the twenty-fifth anniversary of awarding the Nobel prizes, "which shall of course be celebrated." Benedicks had no objection to awarding the reserved prize to Franck and Hertz, an action that received full support in the Academy. But he proposed that rather than reserving the 1926 prize, it should be given to Perrin, regardless of the committee's dismissive evaluation. Discussion ensued; disagreement prompted more than one round of voting. It appeared at one point that Perrin would not receive a majority, but in the end Benedicks won against the committee's objections. A previously buried and forgotten candidate had now been resurrected.

Benedicks was not finished challenging committee authority. The chemistry committee and section proposed reserving its 1926 prize. Benedicks took the floor again, this time in favor of The Svedberg, who had been previously passed over. (For further discussion on Svedberg, see Chapter 11.) Again, Benedicks rallied the Academy over the objections of a Nobel committee.

Benedicks' ability to convince the Academy to overturn unanimous proposals shocked committee members. This unprecedented double rebuke and rejection of past principles prompted senior chemistry committee member Oskar Widman to exclaim, "To sit on a Nobel committee is like sitting on quicksand; there is no firm ground to stand on." Oseen was outraged. Such rebellions threatened the Nobel enterprise, which was based on a culture of expert judgment. Was this wholesale rejection of the committee's authority simply an eccentricity, an expression of desire to have a full house for the twenty-fifth-anniversary festivity, including a local Nobel hero to celebrate in Svedberg? Or perhaps members of the Academy finally gave expression to their growing discontent with the committees? Whether deserving or not, Perrin received a prize that would certainly never have come to him without Benedicks' maverick campaign.

Although the committee adamantly rejected Perrin, once the rebellion was over, the illusion of inevitability and singularity of the choice needed to be maintained. First and foremost, secrecy and facade had to be upheld. There would be no sour notes to newspapers; no protests lodged in the media. Anything less than a magnificent reception for all the prizewinners was, of course, unthinkable.

In France, Nobel magic sparked nationalist ardor. Perrin was hailed a hero. Media descriptions waxed poetic: The glory of French science! The heritage of Lavoisier! The triumph of Perrin, the triumph of his atoms! At that time, French science was in a state of depression, both economically and otherwise. Victory over Germany did not mean victory over German science, which, especially in physics, still enjoyed international high esteem, with or without the boycott. Perrin was a prominent leader among liberal and left-wing professors who sought a rejuvenation of French science, including its economic infrastructure, organization, and relations with society. Benedicks' quirky ploy helped transform Perrin into a national scientific leader with authority, who in the 1930s helped bring about a major overhaul of French research.

When Perrin came to Stockholm, the committee feted him. Benedicks, who had considerable personal wealth, now joined the ranks of the old Academy elite in hosting private Nobel dinners connected with *le grand événement*. Perrin and Svedberg joined their patron, along with well-chosen guests, for a jolly good time. The two physicists understood Benedicks' role in their elevation to the ranks of laureate. Svedberg commented that although he would formally receive the prize from the hands of the king, he knew it was actually being given by Benedicks.

Oseen and Siegbahn's new epoch had run into turbulence. In their effort to bring sharpened discipline to the good ship Nobel, they encountered loose cannons. They were not at all ready to abandon their goal of disciplining the committee and Academy to enable the prize to become a strategic resource for Swedish physics. All that was needed would be more tact, intrigue, and even questionable use of authority.

# NINE

## Clamor in the Academy
### Taking Charge of the Committee

*C*arl Wilhelm Oseen tolerated dissent poorly; any challenge to his authority became a highly personal affair. His perfectionist tendencies tired him out, as did his hypercritical attitude to others' as well as his own research. His nerves and health were repeatedly frayed. Oseen devoted tremendous physical and emotional energy to, first, preparing his committee reports and, then, defending them. "You must try to be a little less conscientious," he was told by a colleague. "[T]hat Swedish science shall suffer is of course not the intention of Nobel's donation." His efforts to gain prizes for Einstein and Bohr left him exhausted, and his maneuvering for Siegbahn debilitated him for weeks: "Awarding the Nobel Prize . . . a busy and strenuous time. Now it is over for this time, but the attack . . . has been so severe, that I am now overwrought." None of this gave him cause to reconsider his role as self-appointed sentry for physics and the Nobel Prize.

### COUNTERINSURGENCY

Oseen—formal, rigid, and demanding—faced an impasse. Benedicks' rebellion infuriated him. He subsequently spearheaded a committee effort to reassert power over the prize, but this maneuver prompted yet more bickering. Dissent needed to be brought under control or eliminated. At stake were Oseen's and the committee's authority. In what resembled a chess tournament, Oseen and his opponents set in motion a series of strategies to gain an upper hand; candidates for the prize became pieces in their match.

The events of 1926 put a stop, for the moment, to the committee's efforts to withhold prizes. Whatever sentiment still supported this practice, the committee knew that business as usual could only resume once the threat of rebellion had ended. Was Benedicks now satisfied and willing to accept the committee's authority and expertise? Just as Oseen feared, it soon became clear that Benedicks wanted to champion other agendas: more prizes for practical contributions and for individuals whom he believed had not received fair hearings. Benedicks relished making life miserable for those committee members who radiated self-centeredness. He sought out foreign nominators and recruited them for his campaigns, often suggesting the wording of the proposals.

## Benedicks Strikes Again

Benedicks wasted no time. Even as Jean Perrin was being feted, Benedicks discussed with him strategy on behalf of victims of the committees' heavy-handedness. They agreed that the British creator of the cloud chamber, C. T. R. Wilson, deserved the prize. In the hands of Ernest Rutherford and other Cambridge physicists, Wilson's invention was an instrument that was beginning to open nuclear and particle physics as scientific specialties. Benedicks considered these developments much more to his liking than the emerging quantum mechanics of atomic physics.

Wilson invented this ingenious device in which a piston could evacuate the air from a dust-free chamber to produce a supersaturated fog. Wilson was interested in questions of atmospheric electricity and cloud physics; he found that when an electrically charged subatomic particle passes through the chamber, it ionizes the molecules in its path. Miniscule water droplets then condense on these, thereby making its path through the chamber visible. In the hands of Cambridge scientists, this invention proved to be an important instrument for studying nuclear physics.

All the while that most physicists were focusing on the electronic shells of the atom, a few tried to investigate nuclear questions. The energetic particles emitted from radioactive sources were the only known means for probing the nucleus. Rutherford and his assistants, first in Manchester and then Cambridge, had been studying the so-called alpha particle emitted from radioactive materials. These investigations led him, in 1911, to announce the nuclear model of the atom: the image of a tiny, dense nucleus surrounded by orbiting electrons. Further, the alpha particles were revealed to consist of two positive particles, the nucleus of a helium atom. Often standing in dark cellars for hours on end, squinting into a microscope to count scintillations made when alpha particles hit a phosphorescent screen, the scientists were studying these phenomena with, at best, limited methods. In 1924, making use of a Wilson cloud chamber coupled ingeniously to an automated camera, student assistant Patrick Maynard Stuart Blackett confirmed Rutherford's 1919 dramatic claim of having disintegrated a nitrogen nucleus and thereby transformed an atom of one element into another. Wilson's invention would, in the 1930s, be used to produce a series of startling discoveries that helped open nuclear and particle physics as thriving specialties. But until then, few physicists were involved.

For the moment, in 1927, center stage was clearly dominated by atomic physicists. Although most of the findings were still too new and too hotly debated to register in the nominations, Arthur Holly Compton and his method of studying the collision of light particles with matter continued to receive broad support, now also from leading German physicists. Wilson, as in years past, enjoyed nominations from his British colleagues Rutherford and J. J. Thomson, as well as now also, after Benedicks' hint, from Perrin.

It was no secret that Benedicks sought the 1927 prize for Wilson. To avoid more embarrassing defeats in the Academy, which would further erode its authority, the physics committee tried to accommodate by compromising. In the committee, Oseen prepared a special report on Compton; Siegbahn took Wilson. The accounts of the recent discoveries by Blackett at Cambridge, which both Rutherford and Perrin drew upon to validate the cloud chamber as a highly significant new instrument, helped change the previously negative attitude toward Wilson's candidature. Both Wilson and Compton were found worthy of the prize.

Compton's candidacy was certainly stronger, as reflected in the nominations and the active research fronts in physics. The committee took an unusual step to avoid a new round of aggravation and possible embarrassment. It proposed a division between the two candidates, even though each one had been deemed worthy of a prize and each for work largely independent of the other. This decision proved expedient. To give Wilson a prize alone would once again raise eyebrows abroad whether the committee was willing to consider any of the newer developments in quantum physics. To propose an undivided prize for Compton, would precipitate a new uproar from Benedicks. In fact, when the committee proposed its compromise solution to the Physics Section, Benedicks was ready with a seven-page polemic arguing for an undivided prize to Wilson, but he failed to convince the majority. The others voted for a split.

But in the Academy, the vote was not certain. At one point, a motion to give an undivided prize to Compton seemed to be gaining support, but it lost after a "masterly speech" by Benedicks on behalf of Wilson. The Academy eventually approved a split prize. Benedicks was seen as being responsible for Wilson's part of the prize. The elderly and somewhat aloof Wilson expressed much surprise and joy upon learning of the award. He immediately thanked Benedicks for his part in the sudden reversal of fortune. It seemed that, for the moment, a bit of compromise by the committee could go a long way toward avoiding unpleasantness.

## Stabbed in the Back

Benedicks tried for a hat trick: to make 1928 his third year of influencing the Nobel choices. Oseen knew what was coming. Arrhenius died in 1927; Benedicks informed the Physics Section that Arrhenius's last proposal deserved attention. At stake for Benedicks was a desire for the Academy to pay tribute to Arrhenius. He also wanted to give a prize to a practical contribution that more directly benefited humankind. Arrhenius called for a division between J. A. Fleming, the British inventor who had developed the diode vacuum tube, and the British physicist O. W. Richardson, who had provided an initial understanding of this device two decades earlier with a theory explaining how electrons are emitted from the surface of hot metals. Subsequently, many other inventors and scientists continued to develop the theory and improve these devices, finally resulting, after the war, in widespread radio communications and broadcasting. Benedicks claimed that these candidates deserved renewed attention.

Arrhenius had been impressed by the recent developments in radio, including the advent of Swedish broadcasting. It was Arrhenius who paired Fleming and Richardson in a letter of nomination on the last possible day. He had read others' proposals for each of them and decided to link the two. He felt that significant benefits to humankind brought about by radio broadcasting now made it worth considering the earlier breakthroughs in research and development.

Benedicks acknowledged difficulties facing the nomination. The committee had repeatedly rejected Fleming and Richardson in the past because too many investigators had worked on the same problems, all of whom had made important contributions. It seemed impossible in some cases to sort out the priority of originality. A history of litigation over patent rights and accusations of plagiarism tended to confirm the committee's reluctance to single out one or two contributors in this field. Still, Benedicks insisted that it should be possible to identify a few investigators worthy of claiming the major credit. And precisely those persons named by Arrhenius should be in the main spotlight. For the 1928 prize, Benedicks repeated Arrhenius's proposal.

As the committee turned to the nominations for the 1928 Nobel physics prize, it was clear that no single candidate stood out. Those giants of quantum mechanics such as Werner Heisenberg, Erwin Schrödinger, and Louis de Broglie, who for several years had been astounding the physics community, scarcely registered as yet. Based on nominations, a fairly even match could be seen among a number of candidates, all of whom habitually entered the lists. Among the few who proposed Richardson, Bohr suggested splitting the prize between Richardson and Irving Langmuir, who had considerably improved upon Richardson's original work leading to a comprehensive understanding of the physics and chemistry of the electron tube.

Fleming had no chance. Richardson by no means enjoyed an advantage over other convincingly supported candidates. More importantly, Oseen had decisively and definitively rejected Richardson, as well as Fleming, as recently as 1927.

Now, committee evaluations flip-flopped this judgment. They claimed that Richardson should be rewarded a Nobel Prize, alone. At least he was a well-respected physicist; Fleming, an engineer, did not at all appeal to committee members. Although Langmuir's work at General Electric's industrial research laboratory could arguably be said to have eclipsed that of Richardson, Oseen had little enthusiasm for mixing business with physics. First-rate science, he and other purists in Sweden believed, could only be created in an academic setting. Richardson alone seemed the best compromise. His continued productivity and renown would render a prize to him popular in many circles abroad, as well as defusing any mischief Benedicks might be planning in the Academy. Having confidence in Bohr's opinion, Oseen was determined to backtrack from his earlier categorical rejection of Richardson. He collided with Gullstrand.

Benedicks was pleased; others objected. Committee chairman Gullstrand, supported by Carlheim-Gyllensköld, expressed reservation over the reversal. The new evaluations did not adequately address the prior grounds for rejecting Richardson: the age of the innovations, dating to the beginning of the century, and the priority of just Richardson's contributions over those of others. In advance of the Physics Section's meeting, Gullstrand moved to co-opt the Academy's new permanent secretary and veteran member of the Nobel Committee for Chemistry, Henrik Söderbaum. Obsessed with how a candidate who had previously been declared not worthy could be given a prize, Gullstrand hoped that Söderbaum could bring greater wisdom to the interpretive task at hand.

To counter Gullstrand's charge, Oseen entered into a long discursion on why only now did he understand the true significance of wireless telegraphy. He mentioned the dramatic search during the summer of 1928 following the crash of the airship *Italia* north of Spitsbergen. Oseen's conversion, based allegedly on the use of radio in finding Captain Nobile and some of his crew in the Arctic ice drifts, did not, however, convince Gullstrand, Söderbaum, or Carlheim-Gyllensköld of a sudden new significance of Richardson's discovery. After a heated and long debate, Oseen triumphed. The Physics Section voted eight to three to propose Richardson. Offended, Gullstrand protested by resigning from the committee.

Gullstrand's pride may well have been bruised, but he nevertheless triumphed in the Academy. Here, the authority of title still prevailed. Gullstrand, the chairman of the Nobel Committee for Physics, and Söderbaum, the permanent secretary of the Academy, both claimed that no sufficient reason had been provided to reverse the earlier negative evaluations; the rank and file took notice. Here, Oseen and Benedicks found themselves on the same side—what proved to be the losing side. The Academy voted against its expert majority and instead chose to support its senior notables: It voted to reserve the prize for 1928.

## Controlling the Damage

Gullstrand kept his word about resigning; it was a question of principle and not one of breaking with his old friend. Still, Oseen had to proceed cautiously to prevent yet further damage to the committee. Fearing that Gullstrand's departure could easily be exploited as evidence of "personal differences" in the committee, Oseen asked for his cooperation in staging the resignation. Gullstrand agreed to remain in place a bit longer and formally offer his resignation as a consequence of the committee's need for greater expertise in modern physics.

Oseen became committee chairman. But his allies could offer but shaky support in steering a firm course forward. Gullstrand, still in the Physics Section, could scarcely follow the new atomic physics unfolding. Moreover, his uncompromising rigidity on matters of protocol had begun to overwhelm all other sensibilities. Meanwhile Siegbahn was swamped with teaching, advising, researching, and, of course, raising funds. He had neither time nor interest to follow the paths being cut elsewhere. The new committee member who replaced Arrhenius, Henning Pleijel, was professor of theoretical electronics at the Royal College of Technology in Stockholm. What he lacked in intellectual breadth, he made up for in loyalty to Oseen and Siegbahn.

Gullstrand's replacement, Erik Hulthén, posed no problem for Oseen, but he also offered only limited help. In 1930, Hulthén suddenly found himself not only professor of experimental physics at the Stockholm Högskola but also a member of the Nobel committee. Although his research on so-called band spectra, which enabled study of molecular processes, was of high caliber, Hulthén's expertise was narrowly defined by this work. He barely squeaked past Hans Pettersson for the chair. His task, therefore, was not enviable: having to enter a committee dominated by older men who had already learned how to get the results. Hulthén's immediate recruitment onto the committee so frightened one leading candidate for the Stockholm chair in mathematical physics—a former Oseen student—that he withdrew, remaining in Gothenburg where he was safe from being pressed into committee service.

That left only Carlheim-Gyllensköld to subdue. He and Oseen simply did not mix, either intellectually or socially. Oseen had to pilot the committee with a crew not quite up to the task. Having already experienced mutiny, he knew what had to be done. Some further weeding was needed. Committee tranquility required silencing one last dissident. Without a unified committee to agree unanimously, the

Physics Section would stay ripe for rebellion against a divided committee. Carlheim-Gyllensköld had to go. Oseen was determined to eliminate this last dissonant note or, as an alternative, silence it.

"Clamor in the Academy of Sciences [*skrälla kastrullocken*]!," Otto Pettersson called the subsequent uproar. "Oseen and Siegbahn proposed that Carlheim-Gyllensköld be dismissed from the Physics Committee (71 years!) and reduce the number [of members] to four." Oseen, with Siegbahn's support, proposed that Carlheim-Gyllensköld not be reelected. They claimed he was too old and then proposed reducing the number of committee members until a new professor in mathematical physics was appointed in Stockholm. Oseen had high hopes that one of his disciples would receive the post. But veteran from the very beginning of the Nobel enterprise, Otto Pettersson thundered, "I am appalled by the coterie domination that a little group from Uppsala exerts not only over the Academy but over the entire Nobel Foundation." Pettersson charged that, if anything, a lower age limit might be more appropriate than an upper one: Only those scientists who have already secured a solid reputation at home and abroad should be entrusted with committee evaluations. The mood turned against Oseen and Siegbahn.

Oseen took a last drastic step to remove Carlheim-Gyllensköld. He claimed once again that the committee needed increased expertise in atomic physics to evaluate the huge number of candidates in this turbulent field. He proposed in the Physics Section the unprecedented act of electing a foreigner to the committee: Niels Bohr.

Much to Oseen's dismay, Gullstrand voiced his repugnance for the latest developments in quantum physics. What was needed, Gullstrand insisted, was a committee member with "common sense." He opposed Bohr, who "clearly possesses much more scientific talent than common sense." He then quoted a Danish diplomat who sat through Bohr's immodestly long 1922 Nobel lecture and exclaimed: "The devil take me that Bohr can't manage to explain in two and a half hours why he received his prize." But Gullstrand had even less enthusiasm for Carlheim-Gyllensköld. He suggested flip-flopping some of the committee posts. Henning Pleijel, who was filling the remainder of Arrhenius's term, should be elected instead of Carlheim-Gyllensköld to the four-year renewable position. Since Arrhenius's place on the committee was as director of the Nobel Institute, that position could then be filled on an annual basis, until a new director of the Nobel Institute was appointed, which would likely be Oseen. Soon, too, a new professor of mathematical physics in Stockholm would be in place. Carlheim-Gyllensköld could be squeezed out.

Oseen concurred and brought the matter to a vote. By majority of five to three, the four-year mandate went to Pleijel over Carlheim-Gyllensköld, and then by the same numbers the one-year renewable position went to Bohr over Carlheim-Gyllensköld.

When the full Academy voted, it agreed with the section by twenty to three on the full term going to Pleijel. It then expressed a clear message of institutional values: loyalty to its own Swedish, rather than foreign, members. By an astounding vote of twenty-two to five, it rejected Bohr in favor of Carlheim-Gyllensköld. Other than Oseen, Siegbahn, and their three supporters from the section, the rest of the Academy understood that Carlheim-Gyllensköld had been unnecessarily humiliated. Collegiality outweighed scientific expertise. Maybe Oseen had appreciated that the vote would result in defeat for Bohr; he nevertheless managed to demoralize Carlheim-Gyllensköld into passivity: the committee voted no confidence. Oseen's strategy to discipline the committee seemed to be succeeding.

## Oseen Takes Charge

Oseen achieved control over the physics prize. Perhaps the mortifying setbacks in his campaign egged him on. For the two prizes available in 1929, he intended not to be denied. Benedicks was no longer a problem, at least for now. The two men understood in 1928 that they could agree on a compromise for Richardson. But Richardson was even weaker as a candidate in 1929; in fact, several candidates received significantly stronger support from nominators for achievements that more readily satisfied statutory requirements. But, for Oseen, it was also a matter of pride to redeem last year's blow in the Academy. He had every intention of sticking to his 1928 declaration of Richardson's worthiness and of compelling its acceptance by the Academy.

In 1929, the floodgates opened. Eligible nominators responded in large numbers; quantum mechanics and a host of other accomplishments enjoyed bountiful support. Some of the audacious theoretical and dramatic experimental achievements from earlier in the decade had withstood scrutiny. In particular, a large number of nominators enthusiastically supported Louis de Broglie, Erwin Schrödinger, and C. J. Davisson for contributions revealing wave characteristics of electrons. A number of repeat candidates from the past year also commanded greater support than did Richardson.

Oseen was now chairman. He again pushed for Richardson. He also evaluated the contributions to the wave properties of the electron. Although in 1928 he had criticized de Broglie's manner of theorizing, the results of which were "saved" by Schrödinger, he now advocated an undivided prize for de Broglie.

Other committee members held other opinions; but these were stifled. First and foremost, Carlheim-Gyllensköld's special report on Paul Langevin and Pierre Weiss's work on magnetism, which declared them to merit a prize, was ignored. The evaluation was not even included with the general report. Carlheim-Gyllensköld protested silently; he refused to back the committee proposal and boycotted meetings. Similarly, Hulthén prepared a meticulous evaluation of Robert Wood's contributions that had accumulated persistent nominations over the past years, including support from former laureates Niels Bohr, James Franck, and Wilhelm Wien. Hulthén's report received little attention. Oseen and Siegbahn also managed to dispatch summarily other strongly nominated candidates.

Oseen prevailed in the committee: Richardson was selected for the reserved 1928 prize, de Broglie for the 1929 prize. In the section, Gullstrand objected, but he stood alone. In the Academy, Oseen was now the one able to enjoy the authority of chairman. The Academy voted with Oseen for Richardson and de Broglie.

The prize to Richardson made many people happy. As Bohr and Rutherford both expressed, it was only right that those who had to take the heat in the early days should now be duly recognized. Those who supported Richardson largely favored him for his overall contributions to physics; he was extremely prolific, making many important findings. Although his earlier work had been superseded, it set in motion valuable research. Oseen normally rejected this line of justification for a candidate.

He tended to write scornfully of candidates whose major discoveries had once played an important role only to be left high and dry by new intellectual currents. On Langmuir's crucial further elaboration of Richardson's work on the thermiotic effect, Oseen charged the industrial physicist's manner of researching with "grave ethical problems." Without any further elaboration, he insisted upon rejection. Equally odd in this episode was Oseen's assertion that the rescue of the *Italia* survivors convinced him of radio's significance and with that also Richardson's work. It sticks out like a sore thumb: In over twenty years on the committee, Oseen never hinted at anything vaguely similar as a justification when considering a candidate; indeed, he usually ignored just such direct appeals to benefiting humankind through the practical relevance of a scientific achievement.

Oseen backed Richardson as part of his effort to take charge of the committee. That his respected friend Bohr wanted to see Richardson given a prize made the decision easier, but by no means could Bohr dictate choices, especially not to Oseen. In 1928, the Academy snubbed Oseen and the committee. In 1929, Oseen felt compelled to push this candidate again, even though the questionable conflicts with bylaws remained and other candidates could claim to be equally if not more deserving based on committee principles. Oseen transgressed his own standards to save face. True, Richardson was a first-rate physicist, but just like Wilson, it is doubtful he would have received a prize had a most peculiar combination of personalities not engaged in a private *comédie humaine*.

## OSEEN UNBOUND

Although Oseen attained a dominant position in Swedish physics, he had become increasingly equivocal about how this science was developing. The most recent advances in quantum theory reinforced his growing cultural pessimism. A few years earlier, when for a short period of time it seemed that American astronomers had measured the long-sought-after ether drift, Oseen despaired over how foolish so many physicists had been to accept, even tacitly, Einstein's whimsical special and general relativity theories. He seized these new data as a last chance to avoid embracing Einstein's general theory of relativity and with that a descent into "cultural relativism and pessimism." Soon the measurements proved flawed; in fact, a series of observations placed Einstein's work on an even more solid evidential basis. And then came quantum mechanics in earnest.

Rather than finding, as Oseen had hoped, a means to reconcile the chaotic claims of quantum physics with the classical foundations of the subject, researchers were proposing ever more outlandish theories. Werner Heisenberg—young, brash, and ambitious—proposed in mid-decade that the traditional goal of visualizing atomic processes should be relinquished. Sophisticated mathematical equations able to generate numerical solutions that agreed with observed data would be the way forward in atomic physics. Oseen balked. He complained to Bohr, whose institute was a hothouse for quantum atomic theories. For Oseen, Heisenberg's highly touted solution introduced into their science yet more formalistic mathematics rather than physical insight. This was not to his taste. And then came claims that, on the atomic level, probability rather than determinism reigned supreme.

These and comparable unsettling tendencies reinforced Oseen's desire to withdraw from active engagement with physics. He had, in 1923, indicated a desire to devote but ten more years to physics; other cultural and even artistic concerns awaited him. But he showed no desire to relinquish the committee's tiller.

### Trying to Resist Quantum Mechanics

Beginning in the late 1920s, Heisenberg and Schrödinger started to receive solid support from nominators for their complementary approaches for a new quantum mechanical account of the atom. Some nominators preferred Schrödinger's more visual depiction, with electron orbits conceived as a form of wave mechanics. Older theoretical physicists such as Einstein, Planck, and von Laue preferred this approach to Heisenberg's more drastic step of nonrepresentational models of atomic processes. Moreover, implications arising from Heisenberg's work seemed to overturn physicists' traditional belief in causality. A number of physicists working in close contact with Heisenberg, such as Bohr, Wolfgang Pauli, and Max Born, were opening the door to a subatomic world that differed radically from the physics of larger-scaled phenomena. Still, Heisenberg's and Schrödinger's theories seemed to be working and to be drawing nominations for the two men from leading physicists.

Oseen did what he could to avoid recognizing Schrödinger and Heisenberg. He might have favored Schrödinger's approach, but he understood, just as nominators maintained, that the two should be rewarded together. Regardless, Oseen created hurdles for their candidacy, some outlandish, some peevish. His twisting and turning to find some way of withholding a prize to the new quantum mechanics was not based on impartial evaluations of merit but on a relentless search for identifying grounds for postponement. These delay tactics were rooted in intellectual temperament and practical expediency.

In 1929, Heisenberg and Schrödinger entered in earnest into Nobel competition. Oseen responded by claiming in his evaluation that their theories had not as yet sufficiently matured "from a logical point of view" to permit a systematic depiction. Moreover, he could not declare them eligible for a prize, as their theories had not resulted in any discovery of fundamental importance. That is, Oseen tried to hide behind the bylaws.

Support for the two physicists continued in 1930. Again, some nominators preferred Schrödinger, others Heisenberg, or a division between Heisenberg and Max Born, who shared in creating the theory. But such differing Nobel laureates as Max Planck and Jean Perrin endorsed the prevailing thrust to reward Schrödinger and Heisenberg. To counter Oseen's intransigence, The(odor) Svedberg nominated Heisenberg and emphasized that the theory had predicted and had led researchers to an important discovery, a new form of hydrogen molecule. Oseen first responded sarcastically that perhaps Heisenberg should be considered for a Nobel Prize in chemistry. Although Oseen later conceded that awarding a physics prize for theoretical work resulting in a chemical discovery was not unthinkable, he again refused to endorse the two for a prize. Perhaps the problem was, as several nominators suggested, that it would be unjust to have the two share a prize when each deserved a full prize for his respective achievement. Why should two intellectual giants be forced to accept half-prizes when others might later receive full prizes for work of lesser accomplishment?

Oseen found a convenient detour around the entire problem and in the process also reinforced his power in the committee. The Indian physicist C. V. Raman suddenly emerged as a popular candidate based on his discovery of a new process by which molecules scatter light. Although some nominators, such as Bohr and Franck, still included Wood, whose work actually pioneered this line of research, many former winners of the prize, such as de Broglie, Rutherford, Stark, and Wilson, endorsed the Indian alone. Raman had been campaigning for several years, appealing as much for a prize for his own work as for the importance such an award would have for research in a developing Asian nation. New committee member Hulthén evaluated Wood and Raman; he recommended the two should split a prize. But Oseen ignored Hulthén's conclusion and insisted on awarding Raman alone.

In 1931, the number of nominations for the pioneers in quantum mechanics dropped, but so did the total number of nominations that were received. Did nominators want to waste their votes on candidates to whom the committee seemed adamantly opposed? Again, the world of physics was small; many nominators understood who were sitting in judgment and what biases they held. The highly critical but brilliant theorist Wolfgang Pauli commented at this time that there were no theoretical physicists in Sweden; he scornfully dismissed Oseen. Some nominators no doubt were puzzled and withheld their proposals. But without any deeper reflection, Oseen glibly claimed that the decrease of support for Heisenberg and Schrödinger evidenced a "cooling off" of enthusiasm for their work. He attributed the lack of support to the fact that these theories had not been successfully broadened to include relativistic effects in electron motion: "This problem lies so deep that a completely new thought is necessary for its resolution." Oseen added the reflective pause: "How this new and as yet inconceivable breakthrough will impact these theories only the future can tell." He urged the committee to agree with him that Heisenberg and Schrödinger must wait.

Here, again, Oseen's impossibly high standards prompted him to demand a complete theory, a quantum of success. Either a theory was fully capable of explaining all relevant phenomena or it was not worth recognition. And yet, nobody denied the need to include relativistic effects. But this sought-after goal did not at all diminish the nominators' esteem for Heisenberg's and Schrödinger's momentous intellectual achievements. Oseen and the committee once again forgot that the spirit of Nobel's testament actually was to aid highly significant work-in-progress rather that to place a monument over a finished job. Granted, Oseen and the committee might well have been trying to buy time so that Heisenberg and Schrödinger could each receive a full prize the next year.

Still, in the following year, 1932, nominators began registering impatience. Some even questioned the committee's willingness and ability to evaluate Heisenberg and Schrödinger. Pauli nominated Heisenberg alone. He wondered whether the committee perhaps could not decide between the two approaches. Speculating that this might be the reason for no award to either of the two, he claimed that Heisenberg's was the more original contribution since Schrödinger derived his from de Broglie. His curt voice became almost audible in a letter of nomination seething with annoyance: By this time, Heisenberg had well fulfilled the requisite conditions stipulated by the bylaws and by Alfred Nobel's intentions. Give him a prize! Even Einstein, who nominated only occasionally, decided to take time to send a proposal for the two. He remarked that he personally preferred Schrödinger's formulation, but he conceded, in all modesty, that he might be mistaken. As both theorists were of such importance, he would prefer that a prize not be split between them. Einstein wanted to see Schrödinger awarded first, if only one of them could be rewarded. And Bohr also proposed the two of them. Bohr clearly understood these theories' limitations; he even feared that a whole new physics would ultimately be necessary to comprehend the atomic nucleus. He knew these were not the end point but an important start. In contrast to Oseen, Bohr once again maintained in his proposal that Heisenberg's and Schrödinger's contributions had unexpectedly not only provided a satisfactory perspective on known atomic phenomena but also led to predicting a series of new aspects of atomic phenomena. They have thereby "to the highest degree" enriched our knowledge of this area of natural science. He proposed using the two available prizes to award both of them.

In response to these nominations in 1932, the committee allowed Hulthén to prepare a special report on the relation between quantum mechanics and experimental atomic research; Oseen updated his own earlier reports. Hulthén analyzed the reciprocal relation between theory and experiment; Heisenberg's and Schrödinger's theories had made sense of crucial data as well as stimulated significant experimental and theoretical investigations. While agreeing that further breakthroughs would be necessary to apply quantum mechanics to the innermost electrons nearest the atomic nucleus, the theories' startling successes within a restricted domain had to be appreciated as an epoch-making chapter in atomic physics. Oseen dug in again. He admitted it would be "without doubt" of value to be finished with the question of awarding Heisenberg and Schrödinger. But he was compelled to oppose them yet again based on his own interpretation of the bylaws and on his own evaluation of the theories' weaknesses.

Once again, Oseen stretched and squeezed all he could to find arguments for withholding the prizes. He again appealed to a strict interpretation of "discovery." On the one hand, he demanded that a significant discovery arise from the theories. On the other hand, he pressed into service his own adherence to classical physics, claiming that the bylaws' meaning of "discovery" was the same as the general public's understanding: "significant progress in knowledge of actual reality." Oseen did not think the bylaws had been fulfilled. Some deeper insight into his thinking might be gained by stopping to examine more carefully his actual rhetoric. His use of the expression "actual or factual reality [*faktiska verkligheten*]" to define "discovery" is overdrawn. Why did he feel obliged to emphasize *actual* or *factual* reality? What was wrong with knowledge of nature or simply reality? Oseen, it would appear, could not accept some of the broader philosophical and scientific implications of quantum mechanics. Just as Einstein recoiled at the probabilistic interpretation of subatomic reality that did not follow conventional notions of causality ("God does not play dice"), Oseen pondered the cultural, theological, and scientific ramifications of the theories. But although Einstein might not have approved of Heisenberg's contributions, he acknowledged their worthiness, even if he hoped it was but a stopgap measure of great brilliance. Both Einstein and Oseen might have longed for a future remedy, but Oseen seemed bent on sulking until that day arrived.

Among the nominators who called for a prize to quantum mechanics, sooner rather than later, the three new professors in Stockholm argued persuasively: Oskar Klein, David Enskog, and Hulthén. The first two were newly appointed professors enjoying international reputations in theoretical physics. Klein was a significant contributor to the new atomic physics. Having worked at Bohr's institute for many years, he was "in the loop" of informal communications among the leaders of theoretical physics. He admitted the weaknesses of quantum mechanics, but the challenges that still remained did not detract from the tremendous accomplishment already attained.

David Enskog, at the Royal Technical College, proposed Schrödinger and Heisenberg in a lengthy letter of nomination. He cautiously presented considerable insight into the most recent developments in atomic physics, concluding with a measured reasoning why the two warranted prizes. In some respects, Enskog's voice should have served to remind Oseen that even he could be mistaken in evaluating physics. A decade earlier, Oseen had disparaged Enskog's dissertation, giving him a marginal grade. Having accepted the verdict and the banishment from academia, Enskog received a pleasant surprise when his work was soon discovered abroad and declared to be a brilliant contribution to the theory for gas diffusion. Through the intervention of foreign scientists, Enskog was resurrected as researcher. Enskog was proof that once again Uppsala popes were fallible. Still, Oseen was determined to remain the sole judge in Sweden of "the true, the beautiful, and the good" in theoretical physics. He disregarded their nominations.

Oseen again repeated that a satisfactory atomic physics must account for the effects of relativity; hence, Heisenberg and Schrödinger simply did not make the grade. He urged placing the 1931 prize money into the committee's special fund and reserving the 1932 prize. The committee majority voted with Oseen; Hulthén reserved his vote and instead proposed a division of the 1931 prize between Heisenberg and Schrödinger. Oseen kept discipline in the Physics Section, where only Svedberg defected to Hulthén (in a vote of seven to two). In the Academy, Hulthén undoubtedly could rely on members' appreciating that even in Sweden other expertise than Oseen's existed. Although Klein and Enskog were not members of the Academy, their voices, along with those of so many prominent physicists, gave Academy members reason to reflect on Oseen's intransigence. In the protocol papers belonging to the Academy's permanent secretary, the vote—normally never recorded—was jotted down, revealing considerable defection from the committee majority: forty for withholding and twenty-three for awarding Heisenberg and Schrödinger

Finally, in 1933 Oseen, accepted that the time had come for quantum physics to be acknowledged. A savior was on the horizon. He learned through his gifted student, Ivar Waller, that noteworthy advances toward a relativistic quantum mechanics had been achieved. In contrast to Oseen, Waller attended international conferences and visited important centers of physics research. Waller was all but spoon-feeding a reluctant Oseen the news from Cambridge and from Copenhagen about Paul Dirac's theoretical masterpieces, as well as experimental findings, that provided support for Heisenberg and Schrödinger. Actually, only two nominators added Dirac to their nomination of the other two physicists.

Overwhelmingly, nominators indicated a clear desire to reward the two pioneers before considering any other worker in the field, be it Dirac, Pauli, or Born. At its presummer meeting to discuss the prizes, the committee voted tentatively to award the reserved 1932 prize to Heisenberg and the 1933 prize to Schrödinger. Oseen's preliminary report on Dirac revealed how he evaluated. He asked whether the brilliant British theorizer could be compared with Planck, Einstein, and Bohr, who suddenly seemed to have become the standard for a Nobel Prize, a standard that, say,

Siegbahn and most other winners could not have matched. Oseen, the judge, ruled negatively. But he wondered whether bad timing rather than native ability forced this assessment: Upon entering physics, Dirac had to confront Heisenberg. Dirac had to devote his creativity and energy to resolving contradictions in the German's theory. Noting that most of Dirac's important work had only just been published, Oseen held out faith that sometime in the future this new star in the firmament of physics would achieve something truly great.

By September, Oseen had a change of mind. He now urged dealing Dirac into the division of Nobel spoils being distributed. Oseen had blocked until 1933 prizes for Heisenberg and Schrödinger, both of whom had been strongly nominated for several years and both of whom were declared as being truly worthy of receiving each a full prize. But it was Oseen's judgment that elevated Paul Dirac at the last moment, who had scarcely as yet been nominated, to a prizewinner at this time. Or, rather, it was the chance occurrence that Waller reported on the latest international gathering of the best and brightest in physics, where no member of the committee had been invited. Waller reported that Dirac's odd prediction of the existence of a positively charged electron had been confirmed through two independent experiments. Here, then, finally, to Oseen's relief was a significant "actual fact" that had been discovered resulting from quantum mechanics. A discovery that "transformed one of the most difficult reservations against the new atomic theory to a support for this theory." For the committee meeting in early September, Oseen included Dirac in the same special report as Heisenberg and Schrödinger. He now linked the three as standing head and shoulders above others. In a surprising turnaround, Oseen called for Heisenberg to receive the undivided 1932 prize, emphasizing the discovery of alliotropic hydrogen; both Schrödinger and Dirac ended up sharing the 1933 prize.

Oseen ensured that neither Wolfgang Pauli nor Max Born, both of whom played critical roles in the development of quantum mechanics, would receive a prize, at least as long as he lived. After receiving the good news of his prize, Heisenberg immediately wrote to Born to express dismay that they were not sharing the prize. Oseen knew Born's earlier work on crystal lattices all too well, as he and Waller had worked on the same problem. Pauli, according to Oseen, was past his prime. Although Waller tried to convince him that maybe Pauli's slower frequency of publication just then had more to do with the relative difficulty of the problems he chose to tackle than to his alleged intellectual exhaustion, Oseen decided that Pauli should not share in a prize. Dirac certainly merited the reward, but the manner of distributing prizes to the pioneers of quantum mechanics was quirky and perhaps less than fair.

Oseen continued to call the shots on theoretical physics. For the next decade, while he lived, Oseen could find no candidate worthy of a prize. When Carlheim-Gyllensköld died in 1934, the committee elected neither Klein nor Enskog, both of whom had excellent reputations. Instead, it recruited Axel Lindh, a competent but minor experimental physicist in Uppsala. Oseen preferred to monopolize authority in theory. Siegbahn took the same position with respect to experimental physics. Numerous problems arose in the committee evaluations during the 1930s. Several candidates who were unfairly overlooked or rejected received prizes much later once the committee's composition changed; others were not so lucky.

Although Oseen's critical predilection—and, for that matter, the committee's composition—proved critical for the distribution of prizes in the 1930s, the annual outcome was also skewed by growing bias in the nomination process. Some, such as Millikan, emerged as overbearing boosters for candidates from their home institution. Moreover, some tried to choose their candidates in accordance to perceived "signals" from Sweden. The Great Depression, the Nazi regime, and the start of international tensions prior to the outbreak of war all disrupted patterns of nomination. Some of the odd omissions, delays, and winners during this period can be understood through this double focus of the committee's internal dynamics and of the politics of nominating.

By the mid-1930s as revolutionary discoveries in nuclear and particle physics multiplied, Oseen increasingly turned to historical and philosophical studies in his attempts to make sense of the recent intellectual upheavals. At the same time, Siegbahn turned to raising enormous funds to retool and join the nuclear revolution, not as an active researcher but as institutional entrepreneur. In actuality, the sheer overload of excellent science that was beginning to register on the list of nominations proved too much for the beleaguered committee. Before turning to these developments and the start of the age of so-called Big Science, it will be useful to examine the parallel history of the chemistry committee.

# PART IV

## DON'T SHOOT THE PIANO PLAYER, HE'S DOING THE BEST HE CAN

The Quest for a Nobel Standard in Chemistry, 1920–1940

# TEN

## It Can Happen That Pure Pettiness Enters

### Out of Touch, Out of Depth:
### A Bewildered Chemistry Committee

*T*he Nobel Committee for Chemistry had seen problems before World War I. Consensus was always difficult to reach. But following the war and Fritz Haber's prize in 1919, the committee staggered along even more indecisively. The committee's procedures could not cope with the vast changes underway in chemistry. National differences sharpened. Germany had been the international leader, but now chemists from the Allied nations shunned their former colleagues. German academic chemists responded by largely ignoring developments abroad. Still, during the next two decades, international chemistry grew tremendously, in part sparked by wartime experience. More than ever, chemistry was understood as a key for economic and social progress.

New specialties flourished; new experimental techniques and sophisticated theories transformed existing fields. For observers in Stockholm, the increasing wealth of research in specialized communities contributed to a chaotic pattern in nominations. Compounding this problem, the net of nominators was woefully inadequate; many significant developments scarcely made a showing. Only once in these twenty years did any one nominee enjoy a clear-cut consensus from nominators; on occasion, the number of candidates almost equaled the number of nominations. Decisions necessarily had to come from within the committee.

Committee members necessarily asked basic questions: How should "worthiness" in chemistry be defined? Indeed, how should chemistry be understood as a discipline? What role should advanced physical theory play in chemistry? Many chemists took pride in craft skills in the laboratory; physics had no place in their science. Should greater emphasis be given to rewarding conceptual or instrumental innovation? Was biochemistry part of chemistry "proper," or did it belong to biomedicine? The members of the chemistry committee, like chemists everywhere, had difficulty in agreeing on answers. Consequently, the committee created Nobel laureates who were characterized by its own priorities and peculiarities. "Excellence" in chemistry, as defined by the Nobel Prize, was therefore necessarily skewed, if not actually flawed.

For much of the 1920s, the Nobel Committee for Chemistry consisted of elderly chemists who had retired from active teaching. In 1924, the average age of the committee was seventy-two years—ranging from Henrik Söderbaum at sixty-two to Olof Hammarsten at eighty-two. These were alert and concerned scholars, but they had little opportunity to engage new directions in chemical research. "Don't shoot the piano player, he's doing the best he can!" was how Söderbaum expressed his unease over having to evaluate work outside his expertise. In discharging its duties, the committee also had to tiptoe through a minefield of challenges from within the Academy. Disagreements over professorial appointments showed that Swedish chemists held diverging ideas about scientific accomplishment; a homogeneous view on who was "most worthy" was unattainable.

Although the elderly committee members remained gatekeepers for the prize, their direct influence on Swedish chemistry was waning. For them and for younger chemists, who gradually joined the committee, awarding or withholding prizes for particular specialties or orientations went hand in hand with striving to influence the nature of chemistry at home and abroad.

Committee members understood the prize's strategic importance. When choosing among nominees, they could be swayed by a desire to assist a particular milieu, to celebrate a national chemical tradition, or to improve scientific foreign relations. The varied interests that motivated nominators and committee members yielded results that occasionally ignored what many considered the most meritorious work.

## OVERCOMING ARRHENIUS: THE STRUGGLE FOR WALTHER NERNST

In the immediate postwar era, nowhere in the sciences were feelings of antagonism stronger than in the international chemical community. The Nobel Committee for Chemistry took sides with the Germans. The prize given to Fritz Haber in 1919 was a clear declaration of loyalty. But Swedish chemists faced a problem in their attachment to German science. The most prominent German chemist, the one person whom most German chemists and for that matter most chemists internationally believed to be most deserving of the prize, remained without one: Walther Nernst.

To remedy this omission, Swedish chemists had to overcome Arrhenius's long-standing opposition. What had begun twenty years earlier as a personal feud between these two talented physical chemists had grown into an obsessive antagonism. By 1918, Arrhenius could no longer distinguish between scientific evaluation and personal animosity. Would the debacle of war, including Nernst's personal tragedy of losing two sons, break the cycle of bitterness? This drama, spanning two decades, came to a conclusion in the immediate postwar years.

In the 1890s, Arrhenius and Nernst were both young rising stars of a new physical chemistry. They both borrowed from chemistry and physics in creating this new branch of science. Rivalry came quickly to the two researchers, who were as proud as they were creative. Although Nernst was always willing to assist Arrhenius professionally, he did so patronizingly. While visiting Arrhenius during the 1897 Stockholm Exhibition, Nernst tried to demonstrate a new apparatus. When, at the critical moment, he blew all the fuses in the laboratory, Arrhenius's guffaw was too self-satisfied. The break began. When one of Nernst's students challenged some of Arrhenius's findings, the break was final. In the last years of the 1890s, Arrhenius began a frenzied, defensive correspondence charging error in his opponent's work.

The news of Alfred Nobel's testament further estranged them. Learning of Arrhenius's desire to head a Nobel Institute, Nernst hurled abusive comments at a grab for riches that were meant for aiding international science. Name-calling accelerated; each used every opportunity to disparage the other's research. The always frugal and earthy Arrhenius resented, and no doubt envied, Nernst's growing personal wealth and professional success. The flamboyant Nernst was not afraid to indulge himself—especially his affection for automobiles—and was not afraid to let his sarcastic wit get the better of him. Arrhenius's success in getting a Nobel Prize and a Nobel Institute did not lessen his enmity.

Both men were dedicated to advancing physical chemistry, but they were not able to join forces. Although Arrhenius helped the cause by clearing the way for a prize to Ostwald, he doggedly worked against the growing number of nominations of Nernst. Privately, he insisted that Nernst had more than enough money from his patents; further riches would only have a further corrupting influence. Science and morals, Arrhenius thundered to a German friend who had nominated Nernst, do not always go hand in hand. Arrhenius became enraged upon learning that Nernst opened a café where students went drinking at night.

Arrhenius also picked at every weakness in Nernst's theories and experimental evidence. In the 1880s and 1890s, Nernst had provided important contributions to electrochemical theory, creating the foundation of the new physical chemistry. In recognition, Nernst was called to Berlin in 1905 as professor of physical chemistry. At this time, too, he proposed a heat theorem, which, with further modification and elaboration, in part through Planck's and Einstein's involvement, produced what has been called the third law of thermodynamics. Some observers found the experimental evidence for the theorem questionable; others were not able or willing to follow a path constructed with so much physical theory. Between 1900 and 1914, Nernst collected nominations. His earlier electrochemical contributions certainly deserved a prize, so said a broad spectrum of nominators from many nations. His heat theorem was yet more significant, but it was also more controversial.

Arrhenius was not a member of the chemistry committee, but he held it in his grip. As head of the preliminary Nobel Institute, for physical chemistry, Arrhenius had authority to judge works in this specialty. He intervened several times, even when the committee and Chemistry Section considered Nernst the strongest candidate. During the war, younger Swedish professors of chemistry, some of whom had studied with Nernst, were embarrassed by the omission and began to nominate Nernst themselves. Again, Arrhenius proved a formidable opponent. Nobody could match his depth of knowledge; he could amass lengthy exposés on every flaw, crack, and imperfection in Nernst's oeuvre. Nernst's heat theorem had problems; but nominators appreciated that even if not fully verified, its value was clear. As more and more data were published, Arrhenius charged Nernst's thriving laboratory in Berlin with becoming the epitome of what was wrong with German science. An autocratic professor was forcing assistants and students to produce results in support of his own theories; dissention and contrary evidence were suppressed.

Even if this charge held more than a kernel of truth, did such attacks demonstrate that Nernst was not worthy of a prize? Arrhenius's reports were at times selective, referring only to negative comments while ignoring general approval. Those closest to Arrhenius understood that he was incapable of separating objective assessment from personal feelings. To those who knew him best, it was clear that Arrhenius had convinced himself that Nernst was not worthy of a prize, and this premise was the starting point for all else. As his colleague Otto Pettersson noted, "A[rrhenius] couldn't stomach and couldn't forget a disagreement."

Following the war Arrhenius opened a new chapter in their relationship. Sharing the distress of German colleagues, and especially those who, like Nernst, had lost children, Arrhenius invited and warmly received his former rival in Stockholm. Arrhenius felt that Nernst had even lost some of his sharp tongue. He continued to strengthen ties to Nernst and to Berlin's chemistry community during the first years after the war. When Nernst returned as a leading candidate for a Nobel Prize in 1920, many thought that Arrhenius would now react favorably. Leading the case for Nernst, Haber—who was keyed in to the situation by von Euler and others—did what he could to counter Arrhenius's influence.

But Arrhenius dashed these hopes once again, issuing a new evaluation and recommending that Nernst not be given a prize. The committee withheld the previously reserved 1919 prize and reserved the 1920 prize as well. Perhaps the committee understood the implications of celebrating another German chemist who had been active in assisting the German war machine and accepted Arrhenius's conclusion. Nevertheless, Swedish professors of chemistry who were not on the committee, including the secretary of the committee, Wilhelm Palmær, were outraged. Although the committee voted to reserve the prize, Palmær led a rebellion in the Academy's Chemistry Section. He argued that Nernst had been nominated sixty-eight times by forty-two different nominators—fifty-eight times for a chemistry prize and ten times for a physics prize—including proposals from former prizewinners such as Fischer, Richards, Planck, and Haber. Palmær's report convinced committee member Söderbaum to break ranks. Section member von Euler declared that Haber's lengthy letter of nomination provided justification for giving Nernst a prize.

The debate brought swift results. By the end of the meeting, the only members of the section who still opposed giving Nernst a prize were the two committee members who most opposed the use of physical theory in chemistry, Hammarsten and Widman. They were delighted with Arrhenius's criticisms. But in the Academy the section's majority vote for Nernst was not enough. Arrhenius's opposition, coupled with committee chairman Hammarsten's reluctance to support "physicist" Nernst, prompted the Academy to back the committee's proposal for withholding prizes.

The following year, 1921, in a well-rehearsed plan, Nernst's Swedish supporters set about to break the opposition. Large numbers of nominations might help convince the Academy to ignore Arrhenius's criticisms. But the boycott of German science meant that no nominations could be expected for Nernst from the Allied nations; they organized a campaign themselves to swamp the committee with nominations. Virtually all Swedish professors of chemistry submitted proposals for Nernst; a few Norwegians joined in as well. Nernst amassed a clear, and unusual, plurality of nominations (twenty-two out of fifty-five).

Even more critical, the committee co-opted Hans von Euler as a special member to evaluate Nernst. Von Euler, a former student of Arrhenius and of Nernst in the 1890s, was an assertive Bavarian who chose to remain in Sweden after a sojourn in Arrhenius's laboratory. Now a professor of chemistry at the Stockholm Högskola and a member of the Academy's Chemistry Section, he was in close contact with Berlin and with German chemistry, to which he felt familial ties. He understood that Nernst's laboratory, the most important in German physical chemistry, was being strangled by economic difficulties.

At first, he was optimistic. In Berlin, he saw Arrhenius and Nernst hobnobbing together. It seemed that their friendship was truly restored. Von Euler submitted in late spring a preliminary, positive evaluation report. He assumed that Arrhenius might well let matters rest.

Suddenly, however, during the summer, Arrhenius submitted his own special report that countered von Euler's favorable assessment. Arrhenius knew of the tide of sympathy for Nernst; perhaps his next move was an effort to defuse this situation. He proposed an alternative way to help German chemistry: nominating for the prize a group, the German Chemical Society. Although an institution could in principle receive a prize, Arrhenius's nomination was disqualified, as it was based on potential future discoveries rather than documented accomplishments, as required by the bylaws. Widman and, to some extent, Hammarsten were deeply impressed by Arrhenius's "brilliantly written" evaluation of Nernst. It reinforced their skepticism. Although they could concede that Nernst's heat theorem had great significance for *physical chemistry* during the past decade, they thought that "for chemical [research it had]—none!" Widman stuck to his traditional organic chemist attitude and insisted privately that Nernst's work, and his recent physical chemistry, had its greatest significance in physics, not in chemistry. But he—and Hammarsten—wanted to support the majority in the committee, hoped to avoid antagonizing the younger chemists in the section, and aid German chemistry. Widman found a way. He added a note to the committee's deliberations, which claimed for the record that he had difficulty in accepting Nernst's work *as* chemistry but was willing to agree that it could have significance *for* chemistry. The committee voted to award the reserved 1920 prize to Nernst.

Although the committee unanimously endorsed Nernst, that was only the first step. The Academy awaited. Widman's support was shaky and Arrhenius was not one to give up a fight. Von Euler asked The Svedberg, a young popular physical chemist, to give a lecture at the Academy the day prior to the Nobel vote; "an energetic opposition" was expected. He cautioned Svedberg not to discuss the suitability of Nernst's heat theorem for a Nobel Prize but simply to portray its significance in general for chemistry.

In the Academy, Arrhenius threw a fit. When it seemed he could not win, he argued that laureate (and friend) T. W. Richards would be slighted if the formal justification for the prize specified the discovery of the heat theorem. Richards, not without his own envy, had been grumbling for years that actually he had all but discovered the same theorem. He informed Arrhenius that it would be more correct to award Nernst for the development of the theorem, but not its discovery. After many hours of debate, the Academy backed its committee. The formal justification was changed from the heat theorem to general contributions to thermochemistry. The debate and especially Arrhenius's intransigence were unseemly. Those present were aghast and expressed relief that "for the Academy's sake, luckily such a scene was rare."

No sooner had the Academy voted than Arrhenius changed his tune. He waited with open, welcoming arms for Nernst and made a special dinner in his honor. He now praised the imposing thermodynamical work of Nernst's laboratory. Arrhenius was filled with pride when, in the midst of continued international antagonism, the German shook hands at the Nobel ceremony with the French winner of the literature prize, Anatole France. Arrhenius failed in his campaign to prevent Nernst from getting a prize; he surely also lost prestige in the Academy. He seems to have been driven by the furies within him. Perhaps, in the end, Arrhenius actually also won;

his victory might well have been finding peace within himself after having been pressed to forgive and forget.

## IN SEARCH OF WORTHY CANDIDATES

Now that its last major "problem" candidate had been rewarded, the committee members could begin searching chemistry's landscape for new achievements to celebrate. What they saw was not encouraging. Chemists had been busy during the war, assisting on the battlefield and home fronts. To the extent that chemists—academic and industrial—had achieved new insights, developed significant techniques, and contributed to economic development, little registered in the nominations. Many of the same names reappeared year after year, generally representing a desire to celebrate past achievements of local heroes. But, other than Nernst and Haber, the committee had no candidate worthy of a prize between 1916 and 1921. Prospects remained dim. No candidate had sufficient allure. No candidate stood out from the rest. Committee members regularly dismissed them. These chemical heroes had made their reputations much earlier, generally through decades of first-rate research but without spectacular discoveries. More than ever, an advocate on the committee was essential to be considered for a prize.

In 1922, the situation seemed to be the same. Although nominations revealed no favorite candidate, committee member Söderbaum saw an opportunity to make a gesture to Allied chemistry. To do so, he would have to turn his colleagues' attention to innovations outside their own spheres of interest. When in 1908 the Academy gave a prize to physicist Ernest Rutherford for his work on the chemistry of radioactive elements, the action slighted his chemist coworker Frederick Soddy. The two had worked together on the transformation theory of radioactivity—the processes by which a parent element decays into daughter elements through the spontaneous expulsion of one or another kind of subatomic particle.

In an effort to make sense of the seemingly endless stream of new radioactive elements that could not be fit into the existing periodic table, Soddy articulated in 1911 the principle of chemical isotopes—substances that could not be further differentiated by chemical means (that is, having the same atomic number) but differed in atomic weight. Actually, his definition and insight were much narrower than the understanding of isotope that later emerged. During the decade after Soddy articulated the concept, isotopy had little significance for most chemists working with nonradioactive materials. This fact, coupled with the claim that the earlier pioneer work with Rutherford on radioactivity was too old to be considered, prompted the chemistry committee on several occasions to dismiss the sparse nominations for Soddy.

What changed in 1922 was a breakthrough by the Cambridge physicist Francis William Aston. Working under J. J. Thomson to try to explain anomalous spectral lines in samples of neon gas, Aston attempted to find by physical means a way of identifying a possible trace element in neon gas. Ultimately, he devised an instrument, called a mass spectrometer, that could distinguish atoms of a gas that differed minutely in mass. Neon, a nonradioactive element and chemically inert, which had been found to possess an atomic weight of 20.2, actually consisted of a mixture of two isotopes, one predominant form of mass 20 and another having mass 22. Aston demonstrated that Soddy's insight had application also to nonradioactive elements.

Although Aston and Soddy each received only one nomination, Söderbaum decided to back them. The rest of the committee did not seem particularly impressed. Widman and Hammarsten questioned whether a chemistry prize should be awarded to a physicist for a physical method. Did these innovations really have an impact on chemistry? But Söderbaum stood firm. Although he had been a professor of chemistry at the Royal Academy of Agriculture, Söderbaum's long and varied career gave him the broadest chemical experience of all the committee members. Put simply, he appreciated Aston's contribution to chemistry. Even more important for the outcome, Söderbaum was one of the original committee members who for over two decades had loyally served the committee. He rarely pounded the table and insisted; but he made his opinions known and spoke eloquently. He was also extremely well respected and liked in the Academy; in fact, he was soon to be elected perpetual secretary. It made no sense for the other committee members to oppose Söderbaum's claim that Soddy and Aston should be awarded, respectively, the reserved 1921 and 1922 prizes in chemistry.

At the very least, the Academy could thus extend a hand of reconciliation to British chemistry, which, other than the prize in 1908 to physicist Rutherford, had only Ramsay's 1904 prize to its credit. In these highly charged postwar years, those people voting understood the importance of making such a gesture after a string of German winners. Besides, no other candidate on the horizon was appealing. If the choice seemed an odd one for the chemists, it was even more surprising to Soddy, who could not believe the news.

## TIT FOR TAT

However, not all were made happy by the decisions—Oskar Widman for example. The recent prizes seemed to him to stress physics and physical methods. Widman was interested in assisting a candidate for whose work he had developed a passion. But it was a passion not shared by his colleagues. The Austrian physiological chemist Fritz Pregl had developed, together with a highly skilled German instrument maker W. H. Kuhlmann, an instrument for analyzing minute amounts of organic materials, so-called microanalysis. The problem for Widman was that few others seemed to be as excited as he was by this development, at least as reflected in the nominations. Colleagues on the committee considered the work too narrow in its overall significance for chemistry as well as problematic with respect to ascribing the achievement exclusively to Pregl. Equally telling against Pregl, the method had not produced any particular discoveries. Widman, however, felt it was now his turn to pick a winner. After trying to please others by allowing Nernst, Soddy, and Aston to receive awards, he now wanted his own preference. He understood the rules and the customs of the game. For better or worse, scientists were not always the most perfect of God's creatures; pettiness [*kitslighet*] was certainly part of their way to judge and decide matters. Why restrain himself?

The methods of analysis developed by the great nineteenth-century organic chemists, especially Justus Liebig, had served the science well. Problems arose, however, when substances, such as physiological secretions from organs, were available for analysis only in minute quantities. To purify such substances and then attempt a quantitative analysis could take months of painstaking labor. Pregl's slightly older colleague at the University of Graz, Friedrich Emich, was actually the founder of quantitative microanalysis, and his methods were those Pregl refined. Although highly skilled in laboratory practice, Pregl was forced to turn to instrument maker Kuhlmann to get the refinement he needed. In the immediate prewar years, Pregl had shown that the traditional methods of chemical analysis could be significantly reduced with respect to space, time, and quantity of sampling substances; even trace materials could be subjected to rigorous quantitative analysis. Although Pregl summarized his methods in a 1916 textbook, these were not easily learned. Still, when an Uppsala student returned from a sojourn in Graz and demonstrated the method, Widman became a devotee.

The first, and only non-Swedish, nominations for Pregl came in 1917 from two chemists working with enzymes, substances that called for microanalysis. Widman was ardent in his praise: "the greatest advance in organic chemistry since 1830." This said, he conceded that the method was of significance for a very limited area of chemistry. The rest of the committee agreed that this work, as valuable as it might be for particular subfields, did not have the broad significance the prize required. Pregl subsequently disappeared from the list of candidates.

In 1922, Widman decided to nominate Pregl. Again he evaluated Pregl's work, but still he could not convince his colleagues of the scientist's worthiness for a prize. However, he and other Swedish chemists raised funds to invite Pregl to Sweden. Direct experience with microanalysis and personal contact could soften opposition. Moreover, Widman and his colleagues were distressed over Austria's collapse: The Austro-Hungarian Empire had dissolved, the postwar Austrian economy was in ruins, and famine ravaged.

Widman got his chance in 1923. Pregl was well received in Stockholm and Uppsala. Politically and personally, he appealed to many in the Nobel establishment. He shared insights on his method, tales of economic catastrophe among Austrian academics, and complaints of "Gallic imperialism" among other Allied injustices toward the former Central Powers.

Whether the world of chemistry was too demoralized from the war, or whether few candidates excited the nominators, in 1923 a mere trickle of nominations again arrived in Stockholm. The committee had fourteen candidates from sixteen nominators. Three noncommittee Swedish chemists proposed Pregl (as one of several candidates), most notably von Euler. He in particular appreciated the precarious situation of Austrian chemistry. Von Euler was about to accept a professorial chair at the University of Vienna, but in 1922 he decided to stay put at the Stockholm Högskola when he understood how crippling the economic conditions were in Austria. Nobody else nominated Pregl. Neither the committee nor nominators showed much enthusiasm.

Responding to a rumor spreading to Berlin that some Swedish chemists were determined to give a prize to Pregl, Haber wrote a letter to Arrhenius, on behalf of himself and fellow laureate Richard Willstätter, urging Pregl's rejection. The two chemists had a principled opposition. They believed that the prize in chemistry should be awarded for work involving intellectual chemical achievement, theoretical or analytical finesse that revealed deep insight into chemical problems. Pregl had no doubt devised an important technique, but it was not in itself intellectually significant.

Indeed, committee members made just this point when taking issue once again with Widman, who had concluded his revised report with another endorsement of

Pregl. Söderbaum objected strongly to Widman's reasoning: Without a clear-cut discovery or practical improvement, invention of a new instrument could not in itself be eligible for a prize. Klason objected that if Pregl were to be considered, then he should share the prize with Kuhlmann, who actually achieved the technical specification that made Pregl's method valuable. But Widman was not about to allow money and prestige to go to an instrument maker. When the time came to voting, few committee members were convinced that Pregl was worthy. But it was Widman's turn to enjoy committee collegiality. Besides, most of the committee members had warm personal regard for Pregl. Söderbaum reserved his vote on principle and precedence. Pregl received a prize in 1923. The next year Pregl all but admitted his own debt by nominating his colleague Friedrich Emich, but of course the committee had little interest in making a similar award so soon.

## DIPLOMACY AND BIAS

In 1924 and 1925, the committee again staggered along, but this time unable to agree upon a candidate. True, the pickings were again slim. Again, few nominators responded; many candidates were the same senior chemists whom the committee had previously rejected. But a few recent candidates who enjoyed a following from at least national constituencies were also dismissed, not so much because of their inherent weaknesses but because of committee priorities.

For example, the committee eschewed giving prizes to potentially deserving scientists who were embroiled in priority disputes. It steered clear of Georg von Hevesy and Dirk Coster who painstakingly proved in 1922 that missing element number 72 fit Niels Bohr's prediction for its characteristics. In so doing, they also proved that the prominent French chemist Georges Urbain had erred in his analysis. That which Urbain had named in honor of Paris was unmasked; in its place came hafnium, after the Latin name for Copenhagen. A bitter feud resulted. The committee sidestepped nominations for both sides of the dispute, even when it was clear that the Danish-based results were incontestable.

Other features of the committee and Academy's propensities left their mark on the prize. Their bias toward Germany was never disguised. For the 1921 prize, Arrhenius nominated the German Chemical Society and for the 1923 prize the newly formed Emergency Fund for Aiding German Science [*Notgemeinschaft der deutschen Wissenschaft*], which was to serve as a research council to allocate grants. He worked constantly to ease the boycott against German chemists and to rehabilitate those who had been suspended or who had resigned from Allied chemical societies. Arrhenius's last nomination before his death, in 1927, called for a prize split between the two physiologically oriented German organic chemists, Adolf Windaus and Heinrich Wieland. Arrhenius had not previously shown any particular interest in this field or style of research, but he was in close contact with leading German chemists. Windaus and Wieland were regarded as having the best chances for bringing a prize to German chemistry. Their respective studies of the complex chemistry of bile acids and cholesterol had earlier been nominated by German colleagues.

Initially, their work had been evaluated as being too narrow in its significance for chemistry; a medical prize was seen as being more appropriate. Hammarsten noted in his 1924 evaluation that Wieland's painstaking organic analyses were masterful, but their importance lay in their application to biomedical questions. In

1926, the committee again worried about giving a prize to research that had such limited significance for chemistry as compared to its importance for physiology. Moreover, neither nominee had fully explained the chemical constitution of cholesterol or bile.

Once Arrhenius's last request was made, committee members understood they had a chance both to aid German chemistry and to pay tribute to Arrhenius's many years of assistance. Widman, who usually was more flexible in judging than Hammarsten, took over the evaluations. Söderbaum warned Widman that a very solid case was necessary to get the two physiologically oriented chemists accepted. Conferring closely with Söderbaum, Widman devised a strategy to make a strong statement in favor of German chemistry. In fact, he argued to reserve the 1927 prize first; then in 1928 each could receive a prize. Widman ignored Hammarsten's earlier objections about the limited significance this work had for chemistry. Instead, he incorporated positive statements about the quality of their analyses, and then he pointed out the growing interdependence of organic and physiological chemistry. He avoided the issue of the importance of the German chemists' work for but a limited area of chemistry and concluded by asserting that their high level of scientific achievement must be rewarded with a prize.

Actually, the full elaboration of the chemistry of bile and cholesterol was not forthcoming until the 1930s. Neither Windaus nor Wieland received support outside of a clique of German researchers. This double recognition of German research and double injection of Nobel money into German chemistry was a fine present just as the Allied boycott was ending. The following year Haber, on behalf of the Association of German Chemical Societies [*Verband Deutscher Chemiker Vereien*], sent a letter to the Academy's Chemistry Section expressing deep appreciation for all the help in bringing German science back into the international community.

## Keeping the Door Closed to Theory

The committee's willingness to stretch principles to assist German chemists did not extend universally, especially when questions of sophisticated theory were involved. In this respect, the committee shared Arrhenius's uneasiness with efforts to import more physics into chemistry. Although Arrhenius had been a pioneer in this respect, he was not willing, or able, to follow more recent developments in physical chemistry. So he dismissed the candidacy of an innovative and productive American physical chemist, Gilbert Newton Lewis, by claiming in a special report in 1924 that Lewis's work accomplished little beyond applying well-known principles of thermodynamics to chemistry. Arrhenius could not identify any new discovery or invention; instead, he saw simply a continuation along "thermodynamics' broad royal road," applying principles from researchers who were at least seventy years old. Therefore, Arrhenius concluded, although Lewis's overall contributions were great, they did not satisfy Nobel's conditions for a prize.

Committee members did not understand or care to understand Lewis's theories for chemical bonding and chemical reactions. They readily accepted Arrhenius's declaration that Lewis's physical chemistry was at the extreme physical end of the field; because of its heavy use of thermodynamics and mathematical physics, they surmised, it was more suitable for a physics prize, if any.

Of course, if all Lewis had done was to extend earlier principles, providing little new chemical insight or theoretical tools, why was he nominated? Arrhenius did not discuss Lewis's contributions to the theory of molecular and atomic configuration. No doubt, he felt threatened that physical chemistry had reached levels of sophistication beyond his grasp. For their part, committee members did not feel directly threatened; they merely refused to acknowledge such work as chemistry.

Arrhenius had an additional motive. Lewis was one of several physical chemists who many years earlier had attempted to modify Arrhenius's theory of electrolytic solutions. It was no secret that this theory, which was so influential in launching physical chemistry, was seen, by the 1900s, to be inadequate. Arrhenius's theory held only for dilute solutions; the results and processes governing strong solutions remained a puzzle. At national and international congresses, Arrhenius had stubbornly refused to acknowledge findings that offered answers that undercut his own position. Those who attempted to create a working theory of strong electrolytes received a cool reception in Stockholm. Arrhenius could, at one moment, seem to accept modifications of his theory but then, suddenly, stonewall opposition.

Such was the case with the Dane Niels Bjerrum. Committee members had little interest in physical chemistry, and as long as Arrhenius's opposition could be expected, nobody bothered to advocate candidates in newer physical chemistry. Arrhenius had not mellowed with age; those who contradicted his results were beyond the pale.

Given the dismal record in the early 1920s, how much longer could the committee continue operating as it did before losing credibility? It managed to remedy the embarrassing neglect of Nernst. Yet, it is hard to believe that the prizes to Aston, Soddy, and Pregl brought widespread acclaim. But these choices were certainly acceptable, if not inspired. And, no doubt, media acclaim in national communities kept interest simmering. The committee could be perceived as rising to the occasion by digging in its collective heels and refusing to award the several senior chemists who regularly populated the annual list of candidates. It seemed that little distinguished winners from losers other than luck and committee taste.

Difficulty in second-guessing committee priorities might have contributed to the generally low number of nominators who sent in proposals for most of the 1920s. But whatever the attitudes abroad, by mid-decade there was great disgruntlement in Stockholm. Von Euler reacted against what seemed a sluggish, out-of-touch committee by initiating a dogged campaign to remove senior committee members. He urged adoption of a statutory age limit for committee membership, and he tried to reduce the number of committee members as a means of eliminating senior men. Some did retire; others tried to hold onto their positions to prevent a takeover by individuals who threatened past traditions. But, after all, the committee did not have the last word on who was elected; the Academy had the final say. Four new members were elected to the committee between 1925 and 1929. Two of these men, in particular, influenced not just who received prizes but how the prize could be used to favor disciplinary growth. Svedberg and von Euler were emerging as major scientific leaders and as entrepreneurial researchers whose endeavors became connected with the prize. The committee had resisted their election, but once these two men gained membership, they tried to give the prize direction and purpose—for better or for worse.

# ELEVEN

## One Ought to Think the Matter Over Twice

Committee Renewal and Biochemical Bias

*W*hen The(odor) Svedberg studied at Uppsala University in the early 1900s, students and professors were astounded by his intelligence, energy, and enthusiasm. Botany and chemistry vied for his allegiance; Friedrich Nietzsche's philosophy, August Strindberg's drama, and modernism in art fascinated him. Philosophical questions drew him toward the newly developing field of colloids—the physics and chemistry of minute molecular substances suspended in solutions. Under a microscope, such particles exhibited so-called Brownian movement—what appeared to be perpetual, random motion that seemingly defied the laws of physics. Maybe the law of conservation of energy was not a universal principle; maybe these movements were the result of constant collisions by even more minute, invisible particles. Perhaps this phenomenon held direct proof of the existence of atoms.

Svedberg benefited from a sojourn with the German colloid specialist Richard Zsigmondy, who had devised an instrument, the ultramicroscope, that allowed closer study of these puzzles. In essence, the microscope featured a strong, focused light able to illuminate a solution. Using this instrument, young Svedberg began a series of studies that attracted attention. He claimed to prove the reality of atoms, but actually his analysis was riddled with error, as both Albert Einstein and, more vociferously, Jean Perrin noted. Still, Svedberg was soon included among the conquering heroes of atomism. He also cast himself into radioactivity, beginning what was to become a devotion to nuclear science. Together with a more senior Uppsala chemist, Daniel Strömholm, he discovered independently of Frederick Soddy the concept of the isotope.

### SVEDBERG'S RISE TO PROMINENCE

Uppsala scientists did not want to lose Svedberg; they managed in 1913 to create a personal professorship for him in physical chemistry. Supporting Svedberg had its costs, as Widman, who had backed this action, soon learned. Svedberg generated a considerable magnetic field, and the best students flocked to him. Within a few years, Widman's personal enthusiasm for Svedberg was eclipsed by his dismay over Svedberg's negative impact on organic and other branches of chemistry.

In a relatively small nation, few students entered science, and even fewer were outstanding. Widman suffered this sorry condition keenly. To complicate matters, rapid growth of Svedberg's colloid-based physical chemistry resulted in a dearth of candidates for organic chemistry. Svedberg and his growing enterprises soon monopolized Uppsala's chemistry building, much to the dismay of Widman and his successor. What made matters worse for Widman was that he did not consider Svedberg's research to be legitimate chemistry.

Widman feared losing a culture of chemistry that stood in clear differentiation from physics. Chemistry and physics, he wrote in 1903, were increasingly interchanging concepts and ideas, making it unclear which concepts belonged to which discipline. Widman's scientific lineage was based on a tradition of laboratory craftsmanship in purifying, identifying, and ultimately synthesizing complex organic substances. Physical theory and methods played almost no role in this tradition; rather, it was apprenticeship with a master chemist and a nose—figuratively and literally—for making sense of what was being analyzed or synthesized at the laboratory bench. What troubled Widman was not simply the lack of new recruits to organic chemistry but also a sense that the temple of chemistry was being desecrated. Individuals who worked under the banner of chemistry but in reality were doing physics subverted the science.

To withhold recognition to work that erased boundaries between physics and chemistry, Widman gladly supported Arrhenius's anti-Nernst position. To Widman, Nernst's theories were inadmissible as chemistry. When Widman heard rumors that Svedberg was moving yet further in the direction of physics, he breathed more easily. Now maybe, he thought, Svedberg might draw upon resources and students in physics rather than chemistry. Actually, Svedberg had just begun to make his mark on Swedish chemistry.

Svedberg represented not only a force pulling the methods and problems of chemistry in new directions but also, in the Swedish context, a new way of organizing science toward what was then sometimes disparagingly called the "Americanization" of science. Svedberg—full of energy, ideas, and ambition—discovered this new world of science when he accepted an invitation to spend 1923 at the University of Wisconsin. There he met a culture very different from Uppsala's. As he wrote years later in his unpublished autobiography, everything in Madison was different. In a world so alien to him, even the laws of nature might be different; at times he feared that laboratory demonstrations, which always worked during lectures at home, might fail. Openness and collegiality flowed with an exuberance unheard of in Europe. Wives of professors brought in lunch; all joined together for social pleasantries. But what struck him most was the American "can do" spirit that contrasted with the pessimism prevailing in the Old World. Whenever he asked whether it might be possible to acquire some costly instrument or for help with some other problematic concern, Svedberg was greeted with cheerful, optimistic assurances—sure, we can easily do that!

Equally important for his future career, Svedberg came into contact with American colloid chemists and biochemists. The University of Wisconsin hosted the First Colloid Symposium, where Svedberg presented his preliminary ideas for bringing greater exactitude to the study of these substances. Having worked with inorganic colloids, he learned that such rigorous quantitative methods could have great significance for exploring the nature of blood and other organic colloids. Little was then known about the physical chemistry of proteins. Whether they consisted of aggregates of smaller molecules, as most chemists believed, or were large molecules with definite structures remained a puzzle. Similarly, were molecules of a specific protein all the same size, or were they, as in the case of inorganic colloids, heterogeneous? Even such basic questions could not readily be answered.

Svedberg and Wisconsin graduate student J. B. Nichols began to devise a new instrument for studying these molecules. By rotating a vessel containing a colloid solution at extreme velocities, it should be possible to cause the minute substances suspended in solution to precipitate as sediment. By carefully controlling speed of rotation and the rate at which the particles sediment, chemists could determine the approximate size and weight of these molecules. To attain results, the instrument must tolerate extraordinary velocities for long periods of time without creating artificial currents in the solution—no easy or inexpensive task. On the voyage home to Sweden, Svedberg resolved critical design problems for this device, which he named the ultracentrifuge. Once home, he set out, together with colleagues, to construct preliminary models.

Early breakthroughs were achieved with their trial ultracentrifuges. In 1924, he and a junior professor of pathology, Robin Fåhraeus, tried to get results with blood. Because their earlier attempt at centrifuging egg albumin had failed, Svedberg had little faith that this procedure would work, as the molecular weight of albumen was assumed to be 34,000 and hemoglobin only 17,000. But as the machine whirled for hours into the night, suddenly particles of hemoglobin began separating. Fåhraeus pulled Svedberg out of bed with a phone call, "The, I see a dawn." Svedberg rushed to the laboratory and just before the chamber cracked, he witnessed the upper part of the cell clearing as the particles began to sediment. Additional trials pointed to the fact that the hemoglobin molecules in the sample were all the same size, and much larger than anticipated: a molecular weight of 67,000. Surprisingly, the Uppsala results revealed that all the molecules were of the same huge size, suggesting that they were not simply aggregates but held together by some binding force. "That sure ought to make the chemists 'sit up and take notice' as the American expression has it," was what Svedberg's colleague in Wisconsin wrote upon seeing these first results.

Svedberg and the German chemist Hermann Staudinger independently and with differences in method and theory both proposed at this time that chemists had to accept the existence of such macromolecules. Their respective studies were astonishing, but they hardly persuaded most organic chemists, who could not conceive how such huge structures could exist. But for those scientists hoping to bring more exact methods into biomedical research, the promise to measure the size and deduce the shape of organic molecules was a godsend. Svedberg dropped his inorganic colloid work and dedicated himself and his laboratory to using the ultracentrifuge to explore physiological substances.

Svedberg planned and dreamed as if he were in the United States. He conceived a major laboratory—a complete annex to the overcrowded chemistry building, the centerpiece of which would be an array of ultracentrifuges. In addition, Svedberg's fertile imagination conceived other physical methods for bringing quantitative and qualitative precision to the study of organic colloids, including that which became the method of electrophoresis developed by his student Arne Tiselius. Svedberg sought money, students, assistants, and technicians. Plans for his institute required previously unheard-of sums; they prompted disbelief in some, fright in others, and resolve to help from friends. Resistance, however, was also immediate: The university's top officials complained that it was not Svedberg's turn to build, nor was it feasible to request from the government such expensive equipment.

## A Prize to Make Dreams Come True

In 1924, the first of the elderly chemistry committee members, Åke Ekstrand, retired. Here was the first chance since 1913 to bring fresh blood into the evaluating process. Widman, Söderbaum, and Klason had been on the committee from its start; Hammarsten joined in 1905. Most members of the committee backed Wilhelm Palmær to be a member, as he had been serving as secretary for both committees and he combined expertise in physical chemistry, electrochemistry, and industrial chemistry. When the Chemistry Section voted, Palmær received five votes, including most of the committee's support (Ekstrand, Hammarsten, Widman, and Söderbaum). Svedberg received one vote, from medical biochemist Christian Barthel; Klason voted for fellow Germanophile and friend, Hans von Euler. In spite of the section's solid support for Palmær, Hammarsten feared that the full Academy would favor Svedberg. He knew that Arrhenius supported Svedberg and, no doubt, the physicists were prepared to back Svedberg, if for no other reason than to push him further in the direction of chemistry. Widman tried to use Svedberg's membership in the Physics Section as a reason to exclude him formally from consideration for the chemistry committee. Even Söderbaum did not want Svedberg on the committee just then; he prepared an impassioned speech on behalf of Palmær. Hammarsten's fears, in fact, were well founded. In the Academy Svedberg edged out Palmær by two votes.

Once on the committee, Svedberg tried to repay past debts. In 1926, he nominated Richard Zsigmondy for the ultramicroscope that had opened modern colloidal research. Previously, the committee had rejected Zsigmondy, whose candidature first arose in 1913 and regularly entered the annual list after 1921. Widman had led the opposition; he was not inclined to welcome this specialty into chemistry. Söderbaum had also questioned its identity as chemistry and its inexactness as compared to other branches of chemistry. Now Svedberg would have a chance to report on Zsigmondy. Turning to chairman Hammarsten, he asked for advice on how to overcome prior opposition and to prevent a humiliating defeat in his first try at promoting a candidate. Hammarsten invited Svedberg over to his laboratory to discuss strategy. With Hammersten's wisdom, and a dose of goodwill, Svedberg convinced his colleagues to give the previously reserved 1925 chemistry prize to Zsigmondy. The committee voted to reserve the 1926 prize.

The prize deliberations were running parallel to Svedberg's own campaign to raise over a million crowns for a new institute. He managed to convince Uppsala University's leadership to consider including his request in its budget proposal. Widman and the successor in the chair of organic chemistry, Ludwig Ramberg, supported Svedberg, if for no other reason than to liberate space and to create at least one truly up-to-date Swedish laboratory for physical chemistry. Widman recalled how he was embarrassed to request two hundred thousand crowns two decades earlier for a new chemistry building. How could anybody expect the government to fund such a huge request?

Carl Benedicks, one of Svedberg's old colleagues from student days, also considered this dilemma. Benedicks was aware of Svedberg's problem. He, among others, had been feeding the media with news related to Svedberg's inadequate conditions. Newspapers reported that the Rockefeller Foundation was willing to donate the equivalent of a million crowns to build an institute in the United States if Svedberg would come as its leader. Colloid chemistry is a branch of physical chemistry, *Dagens Nyheter* informed its readers, with many applications to medicine and industry, in which Americans have great interest. Tipped off that overcrowding in the chemistry building was extreme, the newspaper ran a front-page article: "Professor Must Work in Dark Coal-Cellar and His Students in a Toilet."

The media exposure was being orchestrated in advance of the Academy's Nobel meeting and the government's consideration of Svedberg's budget request. When members of the Academy had proposed Svedberg in 1918 and 1919, Arrhenius and Söderbaum supported him, but Widman and Hammarsten opposed. Widman held his younger colleague in high regard, but not as a chemist. For their part, Uppsala physicists also wanted little to do with Svedberg. When previously Svedberg and Perrin were proposed for a physics prize, they insisted that neither of the two had provided rigorous proof for atomism. They held that Max von Laue's and the Braggs' work with crystal lattices provided the first direct experimental evidence for the existence of atoms. When in 1926 Nernst proposed to divide a chemistry prize between Svedberg and Perrin, Svedberg, who was aware of past failure in winning approval, formally withdrew from consideration so that he could remain active on the committee and work for Zsigmondy.

Benedicks, as we know, used 1926 as his springboard for advancing alternative candidates, both in physics and in chemistry. First, he led a rebellion against the physics committee in order to give a prize to Perrin; then he rallied the Academy to reward Svedberg. The formal justification for the prize was vague. It referred generally to Svedberg's many investigations on colloidal ("disperse") systems and mentioned, erroneously, that he provided experimental confirmation of Einstein's theory of Brownian movement. It also mentioned that he recently devised the ultracentrifuge, "which enables highly interesting investigations to be made."

In fact, on prior procedure, it was not easy to justify a prize. Svedberg had not actually proven the existence of atoms, he had not alone invented the ultracentrifuge, and he had not been nominated for any particular significant discovery made by the new instrument. But it helps to play for a home-team crowd. The Academy did what it wanted, bylaws be damned. Svedberg was given the prize. Widman was personally happy for Svedberg, but he despaired at the ease with which the Academy had disregarded the principles by which its committees had worked for a quarter century. Still, traditions might have been trampled, but Sweden rejoiced.

Svedberg, no doubt like many other winners, experienced both depression and exhilaration. As he noted in his autobiography, his happiness was mixed with discontent. In reality, the only basis for a prize was his recently begun study of proteins. He chastised himself for not agreeing to wait a few years for the work to reach greater maturity. But getting a prize just then was critical to securing the resources for his grand research institute. Svedberg resolved to devote himself during the coming decade to justifying the prize. He accepted Söderbaum's advice not to hold his Nobel lecture until the spring. There was no need to inflame the bitterness and humiliation felt by his colleagues on the committee. The extra time would allow Svedberg a chance to report further findings, and, as it turned out, his presence alone at the Nobel festivities was in itself sufficient to achieve his immediate goals.

Rarely had any scientist a better opportunity for self-promotion. Because 1926 was the twenty-fifth anniversary of the prize, the media and the public were extra-hungry for celebrating a local hero. And Svedberg had, without doubt, charisma. He took the Nobel ceremonies by storm. Those in attendance applauded him, toasted him, and embraced him as few winners had experienced. It seemed as if by having one of their own among the winners, the Nobel establishment was able to break away from the recent pessimism and low-key celebrations with smiles, pomp, and nationalism. Through Svedberg, Swedes could celebrate their own cultural eminence. Highly patriotic songs and music ran throughout the ceremonies. When presenting the awards, Perpetual Secretary of the Academy Söderbaum underlined the hope that the prize would enable Svedberg to continue his important research at home, in Sweden.

The effort worked. The media kept up pressure on the government and parliament. How could a Swedish Nobel laureate be allowed to work in a damp and dark basement, denied adequate space and funding? The minister of church and education, whom Svedberg knew from student days, gave full support. If the university was not willing to apply for the funds, Svedberg should apply himself. The cabinet minister understood that the media would be raucous if the university or parliament hesitated. The government included in its 1927 budget over a million crowns for Svedberg. Although a parliamentary commission complained that the difficult economic situation made a grant of this size questionable, it nevertheless approved because of Svedberg's "extraordinary contributions" and the "splendid development physical chemistry has recently made."

Money flowed to Svedberg's project, as did yet more students and foreign guests. He and his laboratory became favorites of the Rockefeller Foundation. In 1925, Svedberg was scarcely noted in the diary of visiting officials; in fact, they could not even remember his name correctly. But, in 1927, Svedberg began a mutually enthusiastic, close relationship with the foundation that lasted over two decades. The first large grants following the prize helped to build and equip Svedberg's institute.

Svedberg's Institute for Physical Chemistry opened officially in 1932; a cluster of sophisticated ultracentrifuges attracted visitors from around the world. By this time, the Rockefeller Foundation decided to focus its aid to natural science so to reconfigure the biomedical sciences through injections of physics and chemistry: "more physics in physiology." Svedberg's laboratory, with its meticulous design, unique array of instruments, and over one hundred workers, became one of the foundation's showcases. Loyalty between Swedish and American members of the Rockefeller network was part of the enterprise. Svedberg, sitting in Uppsala, became a major player in international science.

## Svedberg's Agenda and the Prize

Svedberg's research and institution building consumed most of his time; work for the Academy's committee had to compete for attention. Maybe, he simply had sufficient political savvy to keep a low profile until he found candidates who could buttress his own interests.

Having nominated and evaluated Richard Zsigmondy, which constituted a celebration of his own colloidal heritage, he supported Adolf Windaus and Heinrich Wieland. Perhaps he was paying tribute to Arrhenius, who nominated them just before dying. But the two German chemists' contributions, which repeatedly had been thought more appropriate for a medical prize, appealed to Svedberg just then for other reasons. While Svedberg helped to promote candidates whose work exemplified that pure chemical research could be of great significance for physiology, he was negotiating a major grant from a private Swedish foundation for medical research, a foundation that had expressed reluctance to fund chemical research.

Discoveries that opened up relevant new lines of inquiry riveted his attention. When Svedberg saw that the 1932 discovery and isolation of heavy hydrogen (deuterium), an isotope that contains an extra neutron, held promise for studying biological processes, Svedberg took initiative. Harold Urey and his assistants at Columbia University had made the discovery; then, with Edward Washburn, they managed the difficult—and often thought impossible—task of separating the isotopes. Svedberg followed the start of Urey's research effort at Columbia using concentrated heavy hydrogen to study biochemical processes. In 1934, Svedberg proposed dividing a prize between Urey and Washburn. Nobody else nominated either of them. Svedberg evaluated their work and convinced the committee to give Urey the 1934 prize (Washburn had died and was, therefore, dropped from the proposal).

When Urey came to Sweden to receive his prize, he spent considerable time with Svedberg and informed the local press that his host's laboratory was the best-equipped chemistry laboratory in the world. Immediately after the Academy celebrated Urey, Svedberg applied to the chemistry committee's special fund and to the Rockefeller Foundation for grants to use deuterium in protein studies. Urey's prize arose from Svedberg's appreciation for how deuterium could help his own research interests. Svedberg's one nomination for Urey stood in contrast to a number of other candidates who received solid support from several nominators, generally over several years. Perhaps Urey would have received more support from nominators in subsequent years had Svedberg not jumped in as a solitary promoter and judge. Still, even if Urey had subsequently received greater support from nominators, that would not necessarily have led to a prize unless a strong committee member was willing to serve as advocate.

In a comparable manner, Svedberg intervened the following year to capture the 1935 prize for chemistry for the physicist couple Frédéric Joliot and Irène Joliot-Curie, for their discovery of artificially induced radioactivity. During the so-called Year of Miracles, 1932, a series of discoveries opened up the atomic nucleus for systematic study. The Joliots had produced neutrons but did not fully appreciate what they discovered; in Cambridge, James Chadwick, cued by his leader, Ernest Rutherford, zeroed in on this new, neutrally charged subatomic particle. Joliot and Joliot-Curie followed up with an announcement in 1934 that they had induced radioactivity in lighter elements. By radiating nonradioactive elements such as aluminum, they produced highly radioactive isotopes of the element—like making a piece of lumber suddenly sprout leaves. In 1934, the Joliots and Chadwick received substantial support for dividing the physics prize. When no prize was awarded in that year, the same three were proposed by a large number of nominators in 1935, again for a physics prize. Some nominators called for a division of the prize among the three; others kept to a single national candidate. The French couple enjoyed support for a physics prize from, among others, former prizewinners Niels Bohr, Louis de Broglie, Werner Heisenberg, and Jean Perrin. But Svedberg had his own design for them.

Svedberg nominated Joliot and Joliot-Curie for a chemistry prize. He defended his proposal by insisting that the couple had demonstrated "in a purely chemical" way reactions between atomic nuclei. He also submitted a nomination for Chadwick, alone, for a physics prize, which might favor his strategy to give the French couple a chemistry prize in spite of the overwhelming support for them in physics. Svedberg wrote to Perrin to inform him of his strategy and to ask him to do the same. He also asked Perrin to greet the couple for him and to ask for copies of their latest work, which he wanted to include in his evaluation. Svedberg knew he would have a "battle" to get the chemistry prize awarded to the two physicists. Once again, he threw himself into the task, evaluated their contribution, and, using his authority in the committee, convinced the rest of the members to accept his proposal, before the physicists came to a decision.

In effect, Svedberg declared that nuclear research was not exclusively a field of physics but was also an integral part of chemistry. Within Swedish science, he legitimated the right for a chemist, such as himself, to compete for funds to pursue nuclear research. Subsequently, Svedberg introduced a chemical nuclear research program in his laboratory and, when Sweden began a major drive after 1945 to become a nuclear power, he successfully competed with physicists. He established significant research facilities for nuclear chemistry and its biological applications, including what was for a period Europe's most powerful cyclotron.

Svedberg's strategic use of the prize included a show of loyalty. After receiving considerable aid from the Rockefeller Foundation, and hoping to receive more, Svedberg supported an appeal to award the chemistry prize to an American. In 1930, Karl Landsteiner of Rockefeller University came to Stockholm to receive the prize in medicine for his discovery of human blood groups. He discussed with Svedberg, and probably others such as von Euler, who was also a major recipient of Rockefeller largesse, that the chemistry committee's neglect of Americans was strongly resented. T. W. Richards had been the only American Nobel chemist, in 1915.

Both Landsteiner and Svedberg proposed Irving Langmuir. Previously, the physicists had refused to reward Langmuir for his splendid investigations of the physics of vacuum tubes. The chemistry committee had also refused to consider either Langmuir or his fellow American, G. N. Lewis, for their studies of the nature of chemical bonds. But Landsteiner and Svedberg turned to that aspect of Langmuir's enormous scientific production that spoke to their own biomedical research interests.

When studying the electric light bulb at General Electric's industrial research laboratory fifteen years earlier, Langmuir was drawn to consider the physics and chemistry of all aspects of this artificial little world. What happened on the surface of an intensely heated filament in a near-vacuum? What processes occur at the interface molecules? Langmuir explained a wide range of phenomena based on the chemistry and physics of surfaces no thicker than a molecule. Landsteiner and Svedberg appreciated that surface chemistry provided a valuable tool for studying cellular processes based on molecule-thick membranes.

They both proposed Langmuir for a prize in 1931. Von Euler came to Svedberg's aid by insisting that the time had come to acknowledge American chemistry, which could best be done by an award to Langmuir. But that year, a rebellion was underway. Disgruntled Academy engineers and industrial researchers demanded that a prize be given in technical chemistry. The committee acquiesced, although Svedberg made it clear to whom he wanted to give the prize. The next year Langmuir again received but token support; Von Euler and an American nominated him. By contrast, once again, G. N. Lewis, made a considerably stronger showing in the nominations, as did other candidates. Landsteiner also proposed Lewis, perhaps believing that the lack of a prize for Langmuir indicated difficulties in convincing the committee of his worthiness of a prize. Palmær took over the evaluation of Lewis and argued forcefully in his favor. Svedberg admitted privately that he agreed with Lewis's worthiness, but he was compelled to object—adamantly—as he was campaigning for another candidate. In the end, Svedberg prevailed, but not because of any suggestion of intrinsic superiority between Langmuir and Lewis.

Svedberg's authority was nearly impossible to overcome. Equally important, a committee majority had already taken a clear turn toward the application of chemistry to biology. Contributions to chemical principles were far less favored. Therefore, Langmuir received the 1932 prize for surface chemistry, but particularly because of his research's significance to the study of biological phenomena. His success was not an indication of its intrinsic value judged impartially; it reflected the interests that dominated the committee at the time.

Svedberg, like any other researcher, was not without his limitations. In fact, he emphasized instrumental innovation at the expense of theory. His neglect of advances in physical theory in chemistry influenced the work of the Nobel committee. Precisely because his opinion was held in such high regard, the one-sided and even flawed assessments could sway the committee and the Academy. But he was not the only strong-willed member, and he was not alone in drawing the boundaries of chemistry further into the domain of biomedicine.

## COMMITTEE BULLDOG: HANS VON EULER-CHELPIN

After years of rejection by the Academy, in 1929 Hans von Euler-Chelpin both joined the committee and got a prize. Von Euler was a man who overflowed with ambition, energy, and tenacity. He possessed an uncanny aptitude to secure professional and social advantage while remaining impervious to rebuke. Some feared that he would use the prize to advance his interests. His career gave reason to be wary.

Von Euler came to study and work with Arrhenius in the 1890s. He courted Astrid, the elder daughter of Uppsala chemistry professor P. T. Cleve, an active researcher in her own right. They married and produced several jointly authored papers. Von Euler moved increasingly away from physical chemistry and toward organic chemistry and biochemistry. In 1906, he was named professor of general and organic chemistry at the Högskola, where he studied the chemistry of enzymes and proteins. Once secure in his position, he suddenly accused his wife of infidelity and of having had a marginal role in their joint publications. Although there was no evidence for his scandalizing accusations, he divorced her and married into a Swedish aristocratic family.

Just before the outbreak of war, von Euler tried to get elected to the Academy's Chemistry Section. After the section had approved him, but before the full Academy vote, Hammarsten learned, through discreet channels, of von Euler's questionable conduct. Hammarsten regretted his support of von Euler, but he consoled himself with the knowledge that the section must vote purely on scientific merit while the Academy could consider personal merit as well. But nobody wanted to make public accusations and counteraccusations; von Euler was voted into the Academy's Chemistry Section.

The Academy's new member caused further commotion when he compromised Swedish neutrality during World War I by joining the German war effort. He spent one semester each year during the war teaching in Stockholm; the rest of the time he served as a fighter pilot in the German air force. After the war, he continued to ruffle feathers when he seemed ready to accept a chair in Vienna. Nobody knew what he intended. Frequent trips to Germany in the 1920s brought back news and information on local science and politics, but they also raised questions as to when he was going to make good on his repeated threats to jump ship. A loyal ally of the former Central Powers, he aided efforts to give prizes to Germans and Austrians.

Even though he was a well-informed and enterprising researcher in his own specialty, his single-mindedness of purpose made him less than ideal for the Nobel Committee for Chemistry. He was a natural candidate to replace Hammarsten in 1926, but he lost out to Ludwig Ramberg, who was less talented but more circumspect. When Svedberg was given a prize later that year, von Euler demonstrated his dissatisfaction by not attending the December festivities and instead lecturing in Germany. Arrhenius feared that von Euler would disparage the prize and in so doing hurt the Nobel establishment, which in turn would harm Sweden. But after being passed over three times, in 1929 von Euler was at last elected to the committee.

It seems that von Euler was not to be stopped. In the late 1920s, he single-mindedly pursued plans to expand his laboratory. Like Svedberg, he turned to the Rockefeller Foundation, moved increasingly into medically oriented biochemistry, recruited numerous doctoral students and assistants, and planned a major institute. His loyalty and friendship with German chemists helped. Richard Willstätter and others recommended that the Rockefeller Foundation's International Education Board support von Euler's proposal for a biochemical institute. Although they readily acknowledged that his research was not always of the highest quality—because he published far too much, too soon—they supported him to establish a second major European center for bioorganic chemistry in addition to that in Berlin. Knowing that von Euler was approaching the Rockefeller Foundation and Swedish sources, Willstätter and Planck, who knew von Euler's earlier efforts on behalf of German science, nominated him, even though his research had not as yet resulted in a major breakthrough. Similarly, local friends, who were certainly aware of his plans, regularly nominated him.

The pieces fell into place when the Rockefeller Foundation reversed its initial rejection and agreed to fund half of his biochemical institute. Von Euler scrambled to raise the remaining funds, turning to both local and foreign donors. He mounted a public relations program using supportive foreign colleagues to help sell "biochemistry"—the key to better health and better living—to Swedish chemists and potential donors. He took out patents and funneled profits from drugs, vitamins, and consulting into his research. He repeatedly threatened to move to Germany if he failed to achieve his plan, but his efforts succeeded and his new institute opened, with much fanfare, in 1929.

## Crowning Himself

Von Euler posed a problem to the chemistry committee. He had been nominated several times, and rejected several times, ostensibly on the grounds that because his work was not finished, its significance could not be assessed. Evaluations concluded that von Euler's work was not only incomplete but was also too narrow in scope. But they also noted that his research aimed at understanding the chemistry of "cozymase" in fermentation, a complex and difficult task.

After Buchner had identified zymase as the enzyme in cell-free fermentation, the British chemist Arthur Harden claimed in 1906 to have distilled zymase into two components, each of which was necessary for fermenting sugars. The notion that a minute, heat-stable "cofactor" was necessary for the activity of an enzyme was of great significance for metabolic biochemistry. Harden did little to push the investigation of this puzzling substance, and he moved on to other topics. In part, Harden won few followers to his claim. But von Euler and a German biochemist, Carl Neuberg, independently took up the study of co-zymase, proved its existence, and began to unravel its complex chemistry. From 1926 onward, von Euler was proposed annually by two or three nominators, including Willstätter, Planck, and one or another Swedish or Finnish friend. In 1929, Neuberg was nominated; but he also sent a proposal to divide the prize between von Euler and Harden (who also had not previously been nominated).

However, as long as von Euler was nominated and as long as he did not remove himself from consideration, he could not participate in the committee's work. This situation was awkward for the committee. Candidates could reappear annually for over a decade without ever winning, which raised the specter of a dismembered committee. And yet, with only four members available to evaluate a rapidly growing number of nominations, the committee could not afford to allow this situation to continue indefinitely.

Svedberg accepted the task of preparing special reports on four biochemical candidates: von Euler, Harden, Neuberg, and Otto Warburg. He had two problems. First, biochemistry could just as easily be considered for the prize in medicine; how could the Caroline Institute and the Academy avoid duplication? Although secrecy prohibited the two institutions from discussing evaluations, Svedberg contacted a colleague on the medical committee. Second, Svedberg admitted that he was no more than "an amateur in biochemistry." Svedberg accepted von Euler's evaluations of the others. The situation soon became messy when, to help Svedberg, von Euler sent reprints of scientific articles and commented on drafts of Svedberg's reports.

Svedberg was overwhelmed with a challenging task and a deadline. It would seem that he was determined to solve the von Euler problem; the question was, with whom should von Euler share a prize? The most important advances from von Euler's laboratory were achieved by teamwork. Moreover, Neuberg's contributions to co-zymase were significant. Harden had not contributed to the subject since the war. Complicating matters further, the research was not finished; many basic questions remained unanswered. Svedberg admitted that Neuberg's contributions merited a prize. But an equal three-way split was disallowed by the bylaws; to give one of them a half prize and to split the remaining half prize between the two others was permissible, but this would give priority to one over the others. It was not easy to assign priority. In the end, Svedberg endorsed the idea of a prize divided between von Euler and Harden. Maybe, von Euler thought that he could push for Neuberg, once biochemistry's "turn" came again, which was precisely what he tried.

Svedberg admitted privately that he was pushing for von Euler, in part to resolve the committee's dilemma. He certainly felt an affinity for a colleague who, like himself, was extending chemistry into biomedicine. Svedberg was actively engaged in testing the borders. He encouraged the Cambridge biochemist Sir Frederick Gowland Hopkins to send a nomination for the chemistry prize, much to the medical researcher's surprise. Von Euler would start doing the same once he was active on the committee. Although Svedberg's work differed from von Euler's significantly— the former more physical, the latter more chemical—Svedberg recognized the importance of an indebted ally.

But Svedberg's recommendation failed to convince the rest of the committee. Neither the candidates nor their work in chemistry won unanimous approval. Ramberg sympathized with Svedberg's concern to wrap up the von Euler problem sooner rather than later, but he noted that the proposal to divide a prize between von Euler and Harden was "perhaps not a recurrent nomination." Indeed, Ramberg was "very doubtful"; he urged that "one ought to think the matter over twice before allowing pure biochemical work to march into the *chemical* Nobel Prize's domain." Widman also hesitated, not least because he feared too many prizes were being given to Swedes without enough time in between. Although in principle opposed to Svedberg's recommendation, Palmær finally agreed to sign the committee report, "sighing and whining [*suckan och kvidan*]." He was swayed by the desire to be finished with von Euler's candidacy and thereby ensure the committee use of all five members. The result was two for and two against.

Even stiffer resistance was waiting in the Academy's Chemistry Section, where former committee members adamantly opposed the proposal. Some believed that Neuberg's contributions were more significant than Harden's and von Euler's. Others felt that the obvious winner in 1929 should be Hans Fischer, whose dramatic synthesis of porphyrin molecules, including hemin, offered important insights into the pigmentation of blood and other biological substances. Fischer, moreover, received considerably greater support from nominators.

In the debate, Widman questioned whether Harden and von Euler were of Nobel quality; he preferred Fischer. Hammarsten agreed, and argued that giving a prize to von Euler would lower the standard for a chemistry prize. He feared that nominators would follow this signal and propose others who, although notable, nevertheless had not attained a truly high level of accomplishment. Another member, S. G. Hedin, objected that Harden had been nominated by only one person. He also opposed the proposal because von Euler had not fully explained the chemistry of cozymase. It was not enough, he maintained, to claim progress. Moreover, he also believed that Neuberg had provided much of the explanation. Both he and Hammarsten preferred to see Fischer get the prize. But, if a prize were to be given for the chemistry of fermentation, it should be divided between von Euler and Neuberg.

Finally when the full Academy voted, Svedberg's proposal won. No doubt, it helped that it was Svedberg who made the case to the Academy. With a dose of goodwill and faith, many accepted his claim that it was only a question of time before von Euler would clinch the final solution to the chemistry of co-zymase. A prize in

1929 would not only ease the committee's workload but would also be a fitting capstone to von Euler's impressive entrepreneurial career. Again, playing for hometown supporters made a difference. Everyone understood that the prize money could fund research in the new institute; Swedish science would benefit. Von Euler sent a telegram to Svedberg: "Thanks for the prize and the roses, Your friend, Euler."

## Von Euler's Advocacy of Biochemical Collaborators

Although von Euler's prize meant that he would be available to evaluate future candidates, there were other consequences of his election. The committee would now have to deal with a highly ambitious man whose single-mindedness and sense of self-worth had just been upped a few notches. Von Euler was absorbed in his many research projects, which could cloud his judgment. No sooner had the Rockefeller Foundation approved plans for his biochemical institute than he suggested they also fund an institute for organic chemistry under his leadership. He was that kind of scientist.

One visiting Rockefeller fellow at his laboratory in the mid-1930s described von Euler as "a bundle of Prussian nerves." Apparently, few trusted him; he had a widespread reputation for allegedly stealing ideas and problems. Dictatorial and arrogant, he directed his enormous energy and drive to cover an enormous range of research topics and entrepreneurial activities. A master of organization, he kept close tabs on his small armies of assistants and students. They, in turn, frequently feared that their unfinished work, if not locked up at night, would be worked into a premature publication. Some colleagues doubted his originality; still, he and his laboratory produced results, massive results, some of high quality and some less so.

Von Euler's overreliance on his own opinions, and his inability to accept criticism, produced poor decisions; and his unquenchable desire for resources, coupled with an uncritical affection for Germany, led him into awkward situations. His loyalty to the Fatherland was unswerving. After complaining in 1932 that his new institute was too small, he positioned himself to become a leading candidate in 1934–1935 for appointment as Germany's "science dictator." Von Euler was determined to reverse the Nazi dismantling of German science through dismissal of Jewish researchers. He imagined that he could persuade physicist James Franck and other internationally prominent refugees to return. By his plan, they would pledge "that Jews would not monopolize the laboratories in which they were permitted to be director." He imagined that the Nazis would let some of them return. All was set for von Euler's move to Berlin when a German official visited his Stockholm laboratory and informed him that the offer was withdrawn. According to von Euler, the reason for the withdrawal was that he had given work to a number of Jewish refugees in Stockholm.

He continued to advance German interests in his own fashion, seeking accord with both Nazi authorities and those who were suffering from their policies. When the German dissident and pacifist Carl von Ossietzky received the peace prize in 1936, Hitler ordered a German boycott of all aspects of the Nobel institution. Von Euler tried to use his influence through diplomatic channels to ease the ban. For him, the prize was too valuable for Germany to accept passively Hitler's rage-filled decree; the propaganda value for science within the Reich and for the Reich in the world was too great. The anti-Nobel policy had the immediate effect of greatly reducing the number of Germans sending nominations. In 1938 and 1939, the Academy (and the Caroline Institute) confronted more directly the impact of the ban when considering how to distribute prizes. Many argued that it was best to respect Hitler's decree, as a prize could bring harm to a German recipient. For the most ardent supporters of Germany such as von Euler, the issue was couched differently. To accept the ban was to accept the introduction of politics into science and prize deliberations. The Academy must give the prize to those persons it finds most deserving, be they German or not.

Von Euler had been maneuvering on the committee to advance his colleague in vitamin research in Heidelberg, Richard Kuhn, and the Berlin biochemist Adolph Friedrich Butenandt for the 1938 prize. Unsure of how to proceed, the Academy paused in 1938, but the following year it accepted von Euler and his pro-German colleagues' arguments: Kuhn received the reserved 1938 chemistry prize, and Butenandt and Leopold Ruzicka, a Swiss researcher, shared the 1939 chemistry prize. The prize in medicine went to Gerhard Domagk, from Münster. Nazi officials and the Gestapo stepped in to make clear their government's displeasure; the three prizewinners wrote letters to Stockholm turning down their prizes. (After the war, they applied to reinstate their prizes; they received the medal and diploma but no money.) Von Euler's meddling and positioning finally resulted in his becoming "detested by most German scientists."

These same qualities plagued von Euler on the committee. Von Euler "thanked" Svedberg for his role in 1929 by supporting his candidates, but by the late 1930s the two were estranged. Von Euler openly disapproved of Svedberg's overreliance on physical methods at the expense of chemical procedures. Perhaps Svedberg's liberal and pro-American orientation and von Euler's uncritical support of Germany and its Nazi government did not mix. No matter; by then, the committee's recognition of biochemistry was already a fact.

Von Euler worked to give prizes to foreign researchers with whom he collaborated. He was not alone in promoting candidates whose work corresponded closely to his own, but he did so with unusual tenacity and lack of diplomacy. His work became ever more biologically and medically oriented, research on vitamins and cancer being among the most prominent. Even when he backed Svedberg by proposing Irving Langmuir for the prize in 1932, he also nominated his own collaborators, Paul Karrer, Carl Neuberg, and Gustav Embden. Karrer, with whom he worked on vitamin research, became his primary candidate. Von Euler vigorously advocated a prize for his Swiss colleague several years running, emphasizing Karrer's identification of carotene with vitamin A, of which von Euler was codiscoverer. A prize to Karrer would give von Euler a foothold as a contributor to Nobel-worthy vitamin research. Indeed, von Euler's own vitamin pills, of a size more fitting a horse than a human, were already being marketed with a label identifying him as a Nobel Prize winner; publicity for Karrer would only strengthen von Euler's position at home as a leading vitamin researcher. Similarly, he acted as an advocate for collaborator-colleagues Fritz Kögl, Carl Neuberg, and Richard Kuhn, whom he asserted were worthy of a prize. At times, his own enthusiastic conclusions in special reports on his favorites had to be moderated or eliminated in the general report. Rather than accept committee culture based on seeking consensus and on trying to reward work from different specialties, he did not hesitate to dissent from the committee majority to vote for his own candidates. This single-minded determination might well have hurt these otherwise noteworthy candidates.

Von Euler didn't see his vote as merely a matter of taste; he understood the prize as bestowing authority and legitimacy on new directions in chemistry. Just as much as he advocated chemical applications to biology and medicine, he resisted efforts to apply sophisticated physical theory to biochemistry and organic chemistry. His opposition was influential against Robert Robinson's physical organic chemistry.

Robinson was one of several highly talented British chemists who in the 1920s built conceptual models of organic reactions. For von Euler, as for most German organic chemists, these efforts were heresy. When one promising student refused to heed von Euler's advice to work with leading German chemists and instead chose to join Robinson in Oxford, von Euler angrily called the Englishman "a minor prophet." (Robinson used this epithet many years later in the title of his memoirs.) Von Euler noted in a 1933 Nobel evaluation that *he* did not agree with those who claimed Robinson "one of the greatest living organic chemists." In the case of Robinson, it did not help that Ludwig Ramberg, who otherwise worked to introduce greater physical theory into organic chemistry, had little enthusiasm for Robinson's particular theoretical methods. Not until von Euler (and Ramberg) left the committee could Robinson receive a balanced evaluation and finally get what many considered a long-overdue prize. Von Euler had little interest in recognizing theoretical achievements that enriched the broad underlying foundations of chemistry.

## OTHER VOICES, OTHER CHOICES: THE CHALLENGE TO DEFINE CHEMISTRY

When the biochemist Arthur Harden came to Stockholm to collect his half of the 1929 chemistry prize, he declared the award to be a sign that the contours of the discipline were changing:

> If I may for a moment yield to the Scandinavian atmosphere of saga and fairy tale, Biochemistry was for long the Cinderella of the Sciences, lorded over by her elder—though I will not say ugly—sisters, Chemistry and Physiology. But now the secret visit to the ball has been paid, the fur [sic] slipper has been found and brought home by the Prince, and Biochemistry, raised to a position of proud independence, knocks boldly at the door of the Palace of Life itself.

Not only had biochemistry entered the Nobel castle of chemistry, it was also trying to gain the throne. Historian of science Robert Kohler has shown that in the early 1900s biochemistry was nurtured in different institutional and disciplinary niches— in the United States and Britain in schools of medicine, in Germany in competition with organic and physiological chemistry. The term biochemistry meant different things. The recognition given the field through the prize reflected the particularity of the Swedish committee. Both Svedberg and von Euler used their chairs in physical and organic chemistry to cultivate biochemical research. Their positions on the committee supported candidates in this area. Although few critics would claim that biochemistry was not "worthy," the broader dilemma remained: How could the committee define the boundaries of chemistry, and how should significance be judged?

For the chemists in the committee and section who did not share an enthusiasm for biologically oriented chemistry, each year brought a difficult task. Although Wilhelm Palmær and Ludwig Ramberg, the other new members elected in the 1920s, brought expertise and advocacy in technical, physical, and theoretical chemistry, rarely could they sway their more powerful colleagues. Complicating matters, the built-in bias of the nominating system gave them little support for their cause.

Well into the 1930s, nominations continued to mirror the fragmented state of international chemistry. At the same time, many important developments in chemistry scarcely registered. The committee made minimal use of its statutory right to invite nominations from individuals and institutions to supplement the permanent nominators. Thus, the network of nominations was too limited to capture the diversity of chemical research. The British and Americans were fostering more schools of outstanding research than could be monitored by so few nominators. The results were stultifying. For 1928, for example, the committee invited two German universities (Kiel and Breslau) but only one university from each of the other major scientific nations: Oxford, Utrecht, Lyon, Milan, Toronto, Calcutta, Columbia, and the Technical University in Vienna. In contrast to the physics committee, the chemists generally did not invite individuals singly as allowed by the bylaws.

German chemistry enjoyed a disproportional representation among the nominators. German institutions not only were invited in greater numbers. They also enjoyed a permanent reinforcement of this pattern by the German prizewinners in physics and chemistry having the right to propose nominees. However, German chemists were largely marching to their own beat. Self-imposed ignorance or out-of-hand rejection of foreign developments in chemical theory and practice was common.

All these weaknesses in the nomination system resulted in the absence, or minimal showing, of many outstanding chemists. For example, Danish physical chemists J. N. Brønsted and Niels Bjerrum, famous for their pioneering work on the acid-base theory and electrolytic solutions, hardly registered. Yet, they were roundly acclaimed as among the few first-rate researchers by the Rockefeller Foundation's comprehensive evaluation of international chemistry. Significant British chemists such as Arthur Lapworth, T. M. Lowry, and Christopher Ingold were all but invisible. Moses Gomberg in the United States and S. P. L. Sörensen in Denmark, among others, were like fireflies in a dark night, suddenly appearing and disappearing from the annual list, despite their important contributions. Committee members either had to have privileged insight into the national chemical communities or else try to guess from the few nominations the extent of enthusiasm for individual candidates. And, of course, they especially needed knowledge of fields of chemistry in which they themselves did not work.

With little help from nominators, who rarely provided anything resembling a consensus, Palmær and Ramberg had to work against the biomedical interests of their colleagues. Although they did not always succeed, their election to the committee in 1926 and 1927 did allow dissenting voices to be heard and even, on occasion, heeded.

Only after Palmær entered the committee was the neglect of Carl Bosch remedied and industrial chemistry celebrated. Following the prize to Haber, Bosch received five nominations for the 1920 prize, but it seemed that little could be done to see that he won the prize. In 1926, the committee dismissed Bosch, claiming that it did not want to give a prize for work already rewarded. It justified its earlier decision by claiming that Haber alone had been nominated in 1919. Of course, most of those few nominations for Haber actually came from within the committee, and nobody on the committee was much interested in waiting to learn the facts about Bosch's contributions.

In 1929, Einstein, who only occasionally sent in nominations, and never for chemistry, recommended that a chemistry prize be given to Bosch. True, Bosch achieved great industrial success when in 1925 he orchestrated the merger of Germany's leading dye companies to create IG Farben. But this success did not detract from Bosch's critical role as scientist in turning Haber's theory into a successful industrial process. Since then, he had continued to develop high-pressure chemistry, creating processes for producing gasoline and methanol.

Palmær picked up Bosch's candidacy in 1929 and nominated him in 1930. When the committee and Academy backed Hans Fischer in 1930—making the fourth prize in a row to biologically significant chemistry—Palmær protested. He picked up support in the Academy. In 1931, the committee confronted nine Swedish nominations for Bosch, including one from Carl Benedicks, all the rest from the Academy's Technology Section. The committee had little choice but to accept Palmær's recommendation; Benedicks' record was known—and he had the Academy's Technology Section behind him, the same section that succeeded in giving Dalén a prize in 1912 against the physics committee's wishes. By adding Friedrich Bergius, whom one of the nominators proposed to divide the prize with Bosch, and by stressing general contributions to high-pressure chemistry, especially for producing synthetic fuels, the committee could defuse the movement to reward Bosch without having to admit error in the 1919 decision.

When the Academy gave its approval to this divided prize in 1931, it for the first time recognized industrial research. This time the chemistry committee was cornered; other efforts to reward industrially relevant chemistry failed. Although chemistry's prominence and growth since the late nineteenth century derived from its industrial applications, the academic-oriented committee members wanted to reserve the prize to work not associated with technology. A popular candidate such as metallurgist Henri Le Châtelier simply could not win support—this, despite Benedicks' campaigns in the late 1920s that helped bring many nominations including from former prizewinners such as Victor Grignard, Charles-Edouard Guillaume, Jean Perrin, and Frederick Soddy.

## Ramberg's Efforts on Behalf of Chemistry Proper

When Widman retired from the chair in organic chemistry, Ludwig Ramberg managed in 1919 to beat stiff competition. His research in pure organic chemistry had been limited, but in coming from Lund he brought fresh ideas and perspectives that had yet to infiltrate Uppsala. Ramberg once joked that he had one foot in organic, the other in inorganic, while his heart leaned toward physical chemistry. But he drew a sharp line between chemistry and the biological sciences.

In Uppsala, Ramberg was determined to put organic chemistry on solid physical-chemical foundations based on the mechanisms of reactions. Unlike other members of the Nobel committee, he was concerned with chemical bonding, acid-base theory, and even, in the 1930s, quantum chemistry. He championed chemistry and objected strongly against those who measured its significance against its assistance to other sciences. He preferred exact mathematical methods. That his understanding of chemistry and that of the chemists in Lund differed from Uppsala's came to clear expression when Ramberg wrote, "I'll be damned how dreary Uppsala chemists' writings are." Although a very social person, and for many years actively engaged in music and cabaret, his devotion to chemistry remained his greatest passion.

After arriving in 1919, when the Uppsala science faculty discussed reform, Ramberg advocated that both organic and inorganic (general) chemistry must include theory. He disapproved of how educational programs were defined by those who used chemistry rather than by the science's intrinsic problems. His efforts to include theory and physical chemistry in the undergraduate science degree failed because of objections from biologists. He then tried to squeeze theory into advanced students' training: "With time I hope . . . to be able to teach the lads to admit that physics has to be included in the major [*licentiat degree*] to consider becoming a chemist!" Ramberg understood that Swedish chemistry was lopsided in its expertise and priorities. Many new ideas had scarcely entered the curriculum and no less the research agenda.

In a lecture to the Swedish Chemistry Society in 1926, Ramberg advocated greater involvement with theoretical perspectives to stimulate experimental research. He warned that the new world of chemistry would only reach maturity if physics and physical chemistry were pulled into organic chemistry. This was also his remedy for inorganic chemistry, which "at present scarcely can be said to maintain [Sweden's] proud traditions."

When Ramberg was elected to the committee in 1927, he brought both his chemical sensibilities and a healthy measure of conscientiousness. On the committee, he judged fairly, declaring his biases. Ramberg tried to brake the biological drift, at times standing against von Euler. He saw no reason to give prizes to chemical works simply because they might have biochemical significance. He understood the extraordinary subtle analyses of complicated natural products and the great difficulties in synthesizing them, but their solution ultimately depended upon the state of general chemistry. His oft-repeated call to resist the move toward biomedical criteria was not heeded. His judgment was followed when it called for eliminating candidates whom other committee members had little interest in supporting. But when Ramberg urged rewarding chemists whom he believed were worthy, he rarely received backing.

Ramberg sought to reverse the committee's and Arrhenius's rejection of newer theories of electrolytic solutions. He accepted thermodynamics in physical chemistry and was keen to recognize chemists—such as J. N. Brønsted, Niels Bjerrum, Peter Debye, and G. N. Lewis—who used physical theory to solve basic problems related to reactions and bonding. But Ramberg found that Arrhenius's loyal followers kept vigil even after his death in 1927. In the case of Lewis, Ramberg added his support for the few members who gradually tried to recognize the American. After Arrhenius's negative evaluations of Lewis, Svedberg took responsibility for the American chemist in 1926 and expressed greater enthusiasm for him than did his predecessor, but he recommended monitoring developments. When Palmær subsequently evaluated Lewis in 1932, he explicitly disagreed with Arrhenius's view that Lewis's work was at the extreme physical end of physical chemistry's scientific spectrum. Palmær declared Lewis well worthy of a prize.

Ramberg was in a bind. Söderbaum supported Palmær, but Svedberg—with von Euler's support—wanted to give the prize to Irving Langmuir for his biologically relevant work on surface chemistry. Ramberg knew it would not be wise to oppose a fellow Uppsala colleague who was determined to have his way. No help could be had from the Chemistry Section, where old-timers such as Widman and Hammarsten were not about to join any rebellion in favor of a physics-oriented chemist. In an ironic action, the Academy in 1939 gave the first gold Arrhenius Medal for outstanding research in either chemistry or physics to Lewis for his "epoch-making contribution to dissociation theory, based on the application of exact thermodynamic methods."

In 1936, Ramberg seemed poised for victory. Among the candidates whom he advocated unsuccessfully, Niels Bjerrum and Peter Debye were his favorites. Debye's work, for over two decades related to numerous aspects of molecular science, was widely considered outstanding. Frequently nominated earlier for the physics prize, Debye had his chances halted by Oseen. Ramberg understood that few of the nominators in chemistry were among those able or willing to read sophisticated theories on molecular properties. For a while, it even seemed that Debye had fallen between the stools of physics and chemistry; nominators and committees for each prize assumed that the "other" prize would be more appropriate for the candidate.

In what appears to have been a move to repay Ramberg's earlier loyalty, Svedberg in 1936 accepted the evaluation of Debye. Perhaps Svedberg also understood the Academy's shame that Debye—one of the century's great researchers—had not been given a prize. Committee members surely must have also considered the international sympathy for this Dutch national who left his chair in experimental physics at Leipzig, to protest political interference, and moved to a Berlin chair in theoretical physics in 1934, where he tried to keep some level of civility amid Nazi harassment. Although Ramberg wanted a division between Debye and Bjerrum, in 1936 the Dane was not nominated; Ramberg accepted Svedberg's recommendation for an undivided prize to Debye.

The chemistry committee seemed poised to give the prize to Debye. But when the vote was taken, von Euler refused to back the majority. Instead, he insisted that the prize should be divided between his bioorganic chemist colleagues Paul Karrer and Fritz Kögl, whom he had several times declared prizeworthy. Knowing that these two candidates could appeal to the many medical and biological members of the Academy, Ramberg took up von Euler's provocation. In fact, Ramberg's patience with von Euler's single-mindedness had run out. He prepared an impassioned speech.

He told the Academy that the vote was one of principle. One researcher's work was of fundamental significance to the *general* solution of one of chemistry's basic problems: the structure of molecules. Another pair of researchers challenged this candidate. Von Euler declared discoveries of vitamins and hormones to have great significance not only for physiology and biology but also for applied medicine. But von Euler had not mentioned anything about these works' significance for *chemistry*.

> Shall the Nobel Prize in chemistry be given to those, who have afforded chemical science eminently effective theoretical and experimental tools for the *general* solution of its fundamental problems, or to those, who . . . partially . . . solve this problem, [but] reach results of great significance for physiology, biology, and applied medicine?

Chemistry, as he saw it, was being colonized by medically oriented biochemistry. Ramberg protested against the repeated awarding of the chemistry prize for the application of well-known concepts and techniques to biological and physiological problems, while fundamental innovations on basic chemical principles repeatedly went unrecognized. Acknowledging the enormous, and still growing, "orders" made to pure chemistry by industry, agriculture, hygiene, medicine, and physiology, among other fields, which have great impact socially and economically, Ramberg feared that only if chemistry was master of its own fundamental problems could it satisfy its clients.

Ramberg rallied the Academy to Debye, who received the 1936 chemistry prize. But the victory must have been bittersweet. Ramberg had informed the committee many times earlier that Debye should not be rewarded without sharing a prize with one of several candidates, including Bjerrum, Brønsted, or Lewis, but the Dutch chemist received it alone. As Ramberg expected, the committee majority returned to its biologically oriented agenda during the next few years: The 1937 prize went to Walter Haworth, for work with carbohydrates and vitamins, and Paul Karrer, for work with vitamins; the 1938 prize to Richard Kuhn, for work with vitamins; and the 1939 prize to Adolph Butenandt, for work with sex hormones, and Leopold Ruzicka, for work with sex hormones and terpenes.

## The Prize as a Reflection of the Committee

Some committee members tried to widen the boundaries of chemistry to encompass biomedical applications. Although opposed by a minority, the committee determined that this relatively new field, biochemistry, was firmly included in the scope of science eligible for a chemistry prize. However, to what extent hybrid specialties should be admissible was a question answered not in principle but by prejudice. The bias toward biologically oriented chemistry contrasted with the negative evaluations of some other hybrid chemical specialties.

The committee's evaluation of Victor Moritz Goldschmidt and his pioneering geochemical research was symptomatic. Goldschmidt's major accomplishments were made in the early and mid-1920s at the University of Oslo. His use of X-ray spectroscopy and other advanced physical methods, coupled with profound insight in chemistry and mineralogy guided him to a new theory for the distribution of elements in nature. As testimony to his standing, several German universities tried unsuccessfully to recruit him, although in 1924 the Munich faculty refused him, not wanting another Jewish professor. Finally, Goldschmidt accepted a call to Göttingen University.

Technically, Goldschmidt was a professor of mineralogy, but chemistry and physics were his tools of trade. From mineralogy he developed new theories and methods for crystal chemistry. Haber and Planck, among others, nominated him in the late 1920s and early 1930s. The point made to the committee was that Goldschmidt's establishment of a new geochemistry was a triumph for chemistry—the conquest of a new domain by chemistry and an intellectual enrichment of chemistry. But committee members refused to consider the work eligible. Using bogus reasoning, they claimed that his election as a foreign member of the Academy's Section for Mineralogy and Geology was proof that his work was not really chemistry. But Bosch and Langmuir had been elected to the Technology Section; did that make their work ineligible for chemistry prizes? Although von Euler and Svedberg willingly considered applications of chemistry to medicine and physiology, they saw no reason to enlarge the boundaries of their subject to include this specialty of geochemistry. Their own interests—and those of other members—determined which applications of chemistry were "significant," and these were limited, restrictive, and biased.

As the century progressed, chemistry had become a federation of states with little central governance. Commerce and trade in ideas could interconnect regions, but few chemists had the breadth to speak for chemistry as a unified science. Although the prize might well have contributed to a festive, symbolic unity, in reality the committee that awarded the prize was far from consensual, impartial, or omniscient. With but limited assistance from nominators, members negotiated solutions among themselves. In each case, the outcome reflected the "chemistry" of the committee and contingencies in the Academy. In consequence, to say that the chemistry prizes awarded between the wars represented a distinct class of excellence is mistaken.

# PART V

## SCANDALOUS TRAFFIC

Subverting Nobel's Legacy
in the Name of Science

# TWELVE

## Dazzling Dialects

### Withholding Prizes, Reserving Judgment

*J*ust prior to voting on the 1936 prizes, the Nobel committees and the Academy suddenly found themselves caught in the public light with their pants down. In the November 10 issue of the respected Gothenburg newspaper *Göteborgs Handels- och Sjöfarts- Tidning,* the physicist-oceanographer Hans Pettersson, son of former chemistry committee member Otto Pettersson, published an article demanding that the Academy and its committees explain themselves. Nobel's money was intended first and foremost for the prizes. Was it right for the Academy to declare in a period of spectacular scientific accomplishment that no candidate merited a prize, and then to divert the prize money to benefit local research? Pettersson conceded that the money was certainly going to good use—Swedish science was desperate for nourishment—but was it right to withhold prizes for this reason? He urged the Academy to stop the practice of withholding prizes when, undoubtedly, suitable candidates could be found.

Pettersson's charges came during a period of expanding ambitions and growing budgets in Swedish physics and chemistry. Attempts in the 1920s to invigorate local research had made limited headway, but the 1930s brought new possibilities. Even before World War II, a new style of organizing research—so-called Big Science—was coming into being. Increasingly, laboratories were becoming large installations, with teams of hierarchically structured researchers, engineers, and other technical staff and massive arrays of scientific machinery. New ways of thinking about research prompted the need for new ways of financing science.

For the members of the Nobel establishment, the 1930s saw changes in use of the prizes. As a culture of Big Science emerged in Sweden, some committee members confused narrow professional concerns with higher ideals, in the process blunting moral and ethical judgment.

### TO EACH HIS OWN NOBEL INSTITUTE

Physicists and chemists diverged in priorities, but they all understood the need for resources. Attempts to raise money precipitated charges of impropriety. In seeking to recruit the prize and its funds as a resource for research, committee members moved the institution farther away from the ideals of Alfred Nobel's testament. At least, that was according to the critics. The charges were serious, but were they accurate?

## An Institute Up for Grabs

Following Arrhenius's death in 1927, both the physics and chemistry committees discussed the fate of the building housing the Nobel Institute for Physical Chemistry. Both committees shared the expense, and they were trying to agree on an acceptable use for Arrhenius's modest Nobel Institute. Should the building be converted into an institute for chemistry or for physics, and for whom? Some saw a chance to establish a full-blown Nobel Institute for physics or one for chemistry. In an all too familiar tradition, subcommittees prepared report after report, which were in turn thrashed about in further rounds of memoranda. Amid prolonged bickering, the economic situation worsened.

Most chemists reasoned that to create a comprehensive institute would be too expensive and, besides, Swedish university laboratories were reasonably adequate for most scientists' needs. More pressing was how to use existing academic chemistry laboratories for research. For them, the committee's special fund held the greatest allure. A new, costly institute could siphon from this limited source of grants, and it might be impossible to agree on a leader for the institute. Most committee members preferred to cultivate the special fund as their own, private mini–research council.

By the early 1930s, von Euler and Svedberg obtained annual grants from the special fund, which they matched from other sources. Ramberg was more modest in his research needs, but he was embarking upon painstaking studies on rates of chemical reaction that needed round-the-clock technical assistance. For them and others, grants could help buy new instruments or keep young disciples in research. State-sponsored research councils and organized fellowship programs scarcely existed.

By contrast to the chemists, physicists Oseen and Siegbahn emphasized institutional expansion. University institutes of chemistry had two and even three professors, but those for physics had but one. In the late 1920s, Siegbahn was struggling to improve conditions in Uppsala by proposing a major new laboratory and an additional professorship. He was not making much headway and therefore began to consider other possibilities. But neither Arrhenius's existing institute nor the available funds were sufficient to establish a Nobel Institute for either physics or chemistry. Oseen and Siegbahn suggested that the modest building be used as a library, seminar, and conference locale until a better solution could be found.

## Oseen Makes a Pitch for Theoretical Physics

In 1930, all this changed. By the late 1920s, Oseen's doctoral students, the first generation of Swedish-trained theoretical physicists, had attained professional maturity. Some of these young scholars—David Enskog, Hilding Faxén, and Ivar Waller, among others—had already made their mark abroad. Oseen understood that his students—or the most promising of them—required jobs. The problem became more acute in 1930, when Oskar Klein, an outsider to Oseen's Uppsala school, edged out the competition for the professorship in mechanics and mathematical physics at the Stockholm Högskola. Oseen had been angling for his own students. Faxén at least had a job in Gothenburg, but Waller, who was working with such internationally leading figures as Wolfgang Pauli, Paul Dirac, and W. L. Bragg, had no prospects at home. Moreover, Oseen was tired of university life; he saw a means to escape that provided comfort and prestige as well as new opportunities for theoretical physics.

Oseen consulted with fellow committee members Siegbahn and Pleijel, who then wrote a proposal for transforming the existing building into a Nobel Institute for Theoretical Physics. The available funds would go mostly to Oseen, whose chair in Uppsala then could go to Waller. As head of this new Nobel Institute, Oseen could devote himself to research and evaluations for the Nobel committee. The chemists resisted, not wanting what had been Arrhenius's laboratory to fall into the hands of the physicists. They claimed that the physicists did not actually have sufficient funds for such an institute. But in the end, Oseen won. He remodeled the building and moved into his institute in 1933; Ivar Waller succeeded him at Uppsala.

Even this modest institute, based in an existing building, required money for salary, renovations, modernization of the living quarters, and other improvements. In the midst of the economic depression, these costs increased the strain on the committee's Nobel capital, especially when others also had plans for these funds.

### Siegbahn Gets By with a Little Help from His Friends—Again

In the meantime, Siegbahn struggled with his own aspirations. He lacked money, space, and time. Although his prize brought him additional funds, he watched enviously as Svedberg translated his prize into a stepping-stone for establishing in Uppsala a thriving internationally prominent research institute. Svedberg's prize opened the door for hefty grants from both the government and the Rockefeller Foundation; but Siegbahn's prize did not have the same effect.

The scarcity of means was doubly worrying, in that Siegbahn's research program was running out of steam. By the late 1920s, following the breakthroughs in quantum mechanics, Siegbahn's program in X-ray spectroscopy lost its glamour. His role as purveyor of precision measurements to atomic theorists in need of reality checks had been his key to international recognition. But by the late 1920s, not only was there less need for such measurements but also other laboratories competed with and even overtook Siegbahn's. He needed new programs of research and a well-equipped laboratory, if he and his students were to remain at the forefront of physics. These required instruments, space, and time.

Siegbahn was unhappy with his working conditions. He insisted that an additional professor in experimental physics must be appointed at Uppsala to relieve him of teaching and administration. The need for a second professor had been obvious for a half century, but petitions always failed. His petitions, even for an additional temporary junior lecturer, were all turned down. In 1929, Siegbahn's institute had sixteen advanced graduate students in place, five more had just registered. His laboratory was designed to accommodate between six and eight. For the academic year 1929–1930, he requested 32,500 crowns for his laboratory and received 13,640 crowns, which included extraordinary assistance on top of the ordinary appropriation. Physics was not yet the prestige field it would become. Swedish industry and government had not yet "discovered" physics.

Siegbahn's dream was to establish a large laboratory, well endowed and well equipped, that he could direct without heavy administration and teaching. On May 23, 1930, he discussed a plan for reviving the Academy's physics institution with banking financier Knut Wallenberg. As director of a richly endowed foundation,

Wallenberg was probably the only person in Sweden who could help. He agreed, but under the following conditions: Siegbahn must be leader of the institution; at least half of the 3 million crowns requested must come from other sources; and these matching funds must be pledged by the end of 1930.

When, in December, the Rockefeller Foundation rejected Siegbahn's proposal for matching funds, he tried to convince Wallenberg that a major laboratory would be in the national interest. Claiming that modern technology and industry relied upon physics, he argued that Sweden's university laboratories were solely for educating teachers and could not contribute to applied research. He sought an extension of the deadline and now also proposed to use Nobel funds for building a laboratory and paying salary. In addition, the Academy would be asked to contribute 20,000 crowns annually for fifteen years to cover operating costs. In short, by 1931, Siegbahn had to mobilize both the Academy and the Nobel Foundation to keep his plans alive.

Fortunately for Siegbahn, his most loyal colleagues had just then attained positions that could help his plan. Pleijel became permanent secretary of the Academy and chairman of the physics committee. Oseen accepted the presidency of the Academy. Gösta Forssell, who had backed Siegbahn's prize, was now vice-director of the Nobel Foundation. Together, they embarked on a coordinated offensive to secure massive funding, the likes of which had never been seen in Sweden.

When their efforts began, nobody could guess the outcome. Could the Wallenberg and Rockefeller foundations be persuaded to join? Could the new Social Democratic government make a contribution, especially as physics was now being pitched as a potential aid to industry? And how much of the various funds of the Nobel Committee for Physics could be diverted exclusively to Siegbahn's plan? After all, the Nobel Institute could have departments representing a spectrum of specialties.

Not knowing which pieces of the plan might tumble out of control, Siegbahn was not taking chances. He asked Uppsala's Academic Senate to petition the government for a new laboratory. Siegbahn would be freed from normal professorial duties except for supervising a few advanced students—a backup strategy resembling Svedberg's. Just like Svedberg, he could then add equipment and initiate ambitious research programs through astute use of the physics committee's special fund in conjunction with matching funds from the Rockefeller and Wallenberg foundations. In either case, he had his eye in 1934 and 1935 on the committee's economy. In the winter of 1934, things began to look up for Siegbahn. Forssell and Pleijel assured Siegbahn of Nobel funds for a new building and inventory. In addition, they and Oseen appealed directly to the government for a personal research professorship for Siegbahn. Success seemed almost in reach; but when Siegbahn contacted Wallenberg informing him that matching funds had been patched together, he learned that Wallenberg no longer had the money. At best, Wallenberg promised to examine the situation next year. Throughout 1934, the situation remained uncertain.

Early in 1935 Siegbahn received good news. The Nobel Foundation promised 800,000 crowns for a building, furnishings, and upkeep. He and his supporters still hoped to get the state to pay for a professorship (13,000 crowns a year) and some of the operational expenses (35,000 crowns a year). But, again, nothing was certain. Under the condition that everything fell into place, Wallenberg guaranteed his foundation's support (55,000 crowns a year). But typical of the problem facing Siegbahn, even when the large grants were falling into place, numerous smaller expenses remained. Nobody, for example, appropriated money to get the ball rolling; he had to go to one of the committee's funds to commission an architect (at a cost of 5,000 crowns).

As it turned out, the Nobel Foundation, the Nobel Committee for Physics, the government, and the Wallenberg Foundation all came through for Siegbahn. He established in 1936 what was at first called the Academy's Research Institute for Experimental Physics. In this respect, most of the physics committee's various funds were "drained [*bottenskrapat*]" for establishing an institute that not only was the domain of just one person but also turned out to be not even a formal Nobel Institute. What seemed to some people as "insider trading" by a few committee members to control all the relevant Nobel funds bred resentment. Given Siegbahn's and Oseen's well-placed supporters in high office, it was impossible to stop what Otto Pettersson called "the Firm Siegbahn, Oseen, & Co. from making an economic coup." But did they violate the trust they held as committee members to evaluate candidates for the prize?

## SECRECY, TRUST, AND DECEPTION

To those who were less than enthusiastic about Siegbahn, Oseen, von Euler, and Svedberg, the activities of the physics and chemistry committees were shameful. In particular, Otto Pettersson and Carl Benedicks believed that the committees were ignoring Nobel's intentions. When Pettersson looked into where the grants were going, he found that committee members who had a prize received most of the grant money. He discreetly took up the matter with elderly members of the Academy and the Nobel Foundation, who, like Pettersson, participated in earlier discussions on such issues. Said Pettersson: "In effect I believe I have brought to light a downright scandalous traffic, which if made more generally known will stop once and for all the economic gang that our Nobel Prize winners have established." Pettersson's son Hans then aired the festering discontent in public. He challenged the Academy in the November 10, 1936, issue of *Göteborgs Handels- och Sjöfarts- Tidning.*

In bringing the discussion to public notice, Hans Pettersson noted that nothing was wrong with using grants from the Nobel funds to aid Swedish research. This use was stipulated in the bylaws. But it seemed that most money was going to a few select committee members. He also questioned the reserving of prizes: The Academy must stop the practice of withholding prizes when suitable candidates surely could be found.

As expected, Oseen shot back a categorical denial of any impropriety. Pleijel followed suit. Pettersson, they insisted, had misunderstood and misrepresented the situation. Oseen regretted that the bylaws compelled him to secrecy. But, he reassured readers, the committee was not able to identify any worthy candidate in those instances when it proposed first reserving or then withholding a prize. Moreover, he added, the Academy was entitled to withhold prizes; it was required to award a prize only once every five years. No impropriety had been committed.

How, Pettersson wondered, was it conceivable that, in a period of such extraordinary scientific ferment, no candidate could be found for the 1931 and 1934 prizes in physics and the 1933 prize in chemistry? In each instance, the Academy had first reserved and the following year again declared that no merited candidates could be found. From behind his shield of secrecy, Oseen attacked Pettersson and asked the public to trust the Academy. The newspaper debate came to an end when Academy members Johan Forssman and Axel Enström, prominent members of the Technology Section, broke ranks and announced that Pettersson was right.

## Reservations About Reserving Prizes

A close examination of the charges, using more complete documentation than was available to Pettersson, supports his allegations. At this time, both committees made a significant change in the procedure by which their special funds could be augmented. From the beginning, when the bylaws were drafted, measures were included to discourage withholding prizes. A prize would have to be reserved one year and then withheld the following before its monetary value could be used for other purposes. It was not until World War I, when prizes began to be reserved, that an ambiguity appeared. Which prize should be awarded first, when a year after having reserved a prize, two were then available? If only one candidate was then declared worthy, which prize should be given?

Disagreement plagued the Academy whenever it tried to resolve this problem. Those who were interested in diverting prize money into research funds preferred to award the current year's prize first, and then to vote on the previous year's reserved prize. In this way, if only one candidate were proposed, that year's prize would be given out and the previously reserved prize would be permanently withheld. With goodwill from the Academy, the money could be directed to the relevant special fund. It would be awkward, the argument went, to choose the alternative, which could lead to a string of reserved prizes year after year.

What would be even more awkward, the Swedish attorney general proclaimed after World War I, would be the suspicion that prizes were being withheld while worthy candidates were waiting to be rewarded. He ruled that prize-awarding institutions were compelled first to vote on the reserved prize and then to take action on the current year's prize, which of course made it less likely that any money could be returned to the committees. (The physics committee repeatedly tried to reserve and withhold prizes for two consecutive years in the mid-1920s, but its own disagreement over Siegbahn's prize and the Academy rebellion in 1926 thwarted the plan.)

The ruling held for a decade; then the committees petitioned to reverse this practice. Again, those wanting enlarged funds argued for distributing the current year's prize first. Skeptics wondered. Oseen had just a year earlier urged the Academy to replace Carlheim-Gyllensköld on the committee with Bohr on the basis that one of the most significant revolutions in the history of science was underway: The committee needed assistance to evaluate the many worthy candidates in atomic physics. Now, a year later, Oseen and his allies claimed that they were facing the prospect of a long, embarrassing string of reserved prizes. Once a prize was reserved, they contended, the prospect of finding two worthy candidates the following year looked doubtful. Therefore, they wanted to change the practice to award the current year's prize first.

The real objective for wanting the change was transparent. Many in the Academy and the Nobel Foundation no doubt sympathized with the committee members who devoted time and energy to evaluate candidates. Here was an opportunity for committee members and other Swedish researchers to gain a few fringe benefits. Few were willing to protest. Yet, privately, as Hans Pettersson later noted,

> Quite a few members of the Academy with whom I spoke seemed surprised the first time the changed clause was applied. One of them, a man whose impartiality and integrity are above all discussion, reported to me that according to his understanding the clause in question indeed must have been changed to facilitate to a yet greater extent than hitherto future withholdings [of prizes]. And without question the change has now eliminated a serious impediment to unjustified use of withholding [prizes]. I have difficulty believing that even Professor Oseen's dazzling dialectics can offer another justification for the change than just this.

Here was an opportunity to help Swedish research. Formally, the practice of awarding the current year's prize first and not awarding reserved prizes did not violate the bylaws; at worst, it only skewed their intention. However, to declare candidates unfit, who otherwise might deserve a prize, was also a breach of trust.

### Refusing to See Merit

Did the committees really find it impossible to identify worthy candidates when they insisted that no worthy candidates could be found? Naturally, official evaluations served to justify the committees' conclusions; the very act of writing them was the act of erasing any and all extrascientific criteria. There were sometimes compelling reasons for not wanting to find worthy candidates, as during World War I. But sometimes, a simple declaration, dismissing a candidate, or a more elaborate rhetoric flourish was a tidy means of masking motives and intentions that could not be included in an evaluation claiming to be impartial and objective.

The Academy did not always have expertise to challenge committee evaluations. Regardless, to question the validity of an evaluation was to cast doubt on authority, to affront the expert and his claims; it was to provoke and invite quarreling. When in the 1930s the committees began to reserve and then withhold prizes, some protested against what they saw as a sleight of hand to eliminate worthy candidates for the sake of preserving funds for Swedish purposes.

The first clear-cut case of using the changed procedures arose in 1934. After heated debate, the Academy backed the chemistry committee's recommendation to first award the current 1934 prize (to Harold Urey) and then, after agreeing with the declaration that no other candidate merited a prize, to withhold the previously reserved 1933 prize. Otto Pettersson saw this as a prearranged event. The decision, he feared, foreshadowed a long-term strategy to reserve a prize every other year and then withhold the following year; first in chemistry and then in physics, for what purposes "we will certainly soon find out." Sarcasm mixed with scorn: "The big quarrel in the Academy . . . like a *business deal* decided upon in advance . . . : The '33 Nobel Prize shall be divided up, not distributed." Pettersson was not against withholding prizes in principle, but he believed that "the proposals should be evaluated in an impartial and proper manner, which is what the IVth [Chemistry] Section did not do this time."

Svedberg might well have pushed through the candidacy of Urey, but in both 1933 and 1934 several other outstanding candidates were solidly nominated. Some such as Paul Karrer waited only a few more years; others such as Otto Diels and Robert Robinson waited much longer. Some had been declared worthy of a prize earlier, such as Carl Neuberg and G. N. Lewis. And surely, few chemists would have objected had either of them, among others, received the 1933 prize.

When the physics committee reserved its 1934 prize and then withheld it the following year, Pettersson was aghast; his fears seemingly were being realized. First chemistry and now physics: Was this the start of the "cyclical permutation" he foresaw, in which during alternate years the prize money would be directed into committee funds?

Hans Pettersson's public charges were aimed at the physics committee. Did candidates during these years receive fair, unbiased hearings? A negative answer was implied by Oseen's evaluations of Werner Heisenberg and Erwin Schrödinger. As we have seen, Oseen countered the arguments put forth in 1931 and 1932 in favor of awarding. At first, Oseen had misgivings over the direction in which physics was heading. But, by 1932, when two prizes were available and a broad spectrum of the world's leading physicists amassed forceful evidence why the two quantum theorists deserved recognition, Oseen again refused to budge. His own evaluation actually showed that the two were worthy and deserved prizes. Did Oseen's plans for his own Nobel Institute, and Siegbahn's desperate search for funding, influence the outcome? As Hulthén and Svedberg noted, no compelling scientific reason had been given for *not* giving Schrödinger and Heisenberg prizes.

By 1934 and 1935, the list of first-rate candidates was crammed. Some were holdovers from the 1920s—including Otto Stern and Wolfgang Pauli—and some were the result of recent dramatic discoveries in particle and nuclear physics. Several new candidates stood out with respect to number and source of nominations for 1934 and 1935; these included James Chadwick (for the discovery of the neutron), C. D. Anderson (for the discovery of the positron), C. J. Davisson and L. H. Germer (for the discovery of electron diffraction), and Frédéric Joliot and Irène Joliot-Curie (for the discovery of the neutron and of artificially induced radioactivity). Other highly praised researchers almost fell through the grid but still made it onto the list of nominees: P. M. S. Blackett (for discovery of the positron), G. P. Thomson (for his work with electron diffraction), and Max Born (for his work with quantum mechanics), to name the most prominent. Committee members' insistence on the need to wait to assess significance conflicted with nominators' compelling arguments. Previously, calls for a liberal interpretation of the bylaws characterized efforts to squeeze committee favorites across the line—for example, Guillaume and Siegbahn. Such liberality was never applied uniformly. Similarly, committee efforts to deny the worthiness of compelling candidates were prompted by personal interests that had little to do with merit. To claim that, among all the candidates, only Chadwick—and first in 1935—warranted a prize in effect constituted a sudden, indiscriminate raising of standards.

Withdrawing two physics prizes during a period rich in accomplishment naturally had important consequences. Odd combinations—dividing prizes—were necessary to accommodate the growing backlog. Some candidates missed out on a chance for a prize, while others ended up waiting a decade or more to receive what, according to nominators, was already theirs in the 1930s. For example, Blackett, Born, Pauli, and Stern did not receive prizes until after the war. Was this simply an inconvenience for some, bad luck for others? The case of Stern is instructive.

## Betrayal?

In 1934, Otto Stern received massive support to receive a prize. Indeed, since 1927 he had been repeatedly nominated for experiments between 1920 and 1923 that gave conclusive evidence for early quantum theory. Sommerfeld had predicted that certain atoms would possess a magnetic moment of a particular magnitude, and that in the presence of a magnetic field, such atoms could assume only two distinct orientations. This so-called spatial quantization could not be accounted for by classical physics; so the idea was not at first well received. Stern saw a means to test the hypothesis experimentally. In 1919, he began to refine an earlier method by which molecules are projected in a vacuum chamber as a distinct beam moving along a straight trajectory. If classical theory held, a narrow beam of silver atoms passing through a nonhomogeneous magnetic field would broaden; according to the quantum theory, it would be split into two beams. Stern designed the experiment, but he needed more nimble hands to perform the delicate procedures. So, with colleague Walther Gerlach, he demonstrated spatial quantization and measured the magnetic moment of the silver atom. Subsequently, Stern was called to a chair in Hamburg and began setting up a sophisticated magnetic-beam laboratory, where more notable experiments were performed.

Some nominators called for a divided prize for Stern and Gerlach; most proposed Stern alone. Bohr, Franck, Heisenberg, and others pointed out that it was Stern who refined the method, saw the possibility of an experimental test for spatial quantization, and designed the procedure. Stern's subsequent record of research underscored his stature as one of the era's great experimental physicists. But when evaluating Stern, Oseen always found some reason for denying him a prize. In 1934, for example, he pointed to the disparity between Stern's recent measurements of the magnetic moment of the proton and the size predicted by Dirac. As usual, Oseen called for waiting, a wait that was hardly justified. True, the question of the magnetic moment of protons puzzled physicists, and indeed, Stern's measurements were later verified as being correct, but none of this had much to do with his earlier experimental achievements, whether judged alone or with Gerlach.

If the uncertainty and mistrust surrounding this recent work so tarnished Stern's record, why did so many nominators propose him? Stern outdistanced all other candidates in 1934. His worthiness for a prize was repeatedly spelled out year after year by several Nobel laureates, including Bohr, Franck, Heisenberg, von Laue, Planck, and Raman. A surprising acknowledgment of Stern's excellence came from Johannes Stark in 1929. Stern's Jewish background was well known, as was Stark's rabid anti-Semitism, support for the Nazi party, and loathing of quantum theory. Yet Stark, like many others, appreciated the beauty and intelligence of Stern's experiments. But Oseen, "who can write more disparagingly than anyone," was not persuaded, indeed he was not willing to be persuaded. The rest of the committee shared the desire to not want to find any candidate worthy. When Stern did receive a prize, a decade later, it was already too late.

Stern's laboratory in Hamburg was one of the great centers of European experimental physics. Physicists who crisscrossed the continent to attend conferences and workshops frequently stopped there; students and young researchers came from afar. All this came to an end with the Nazi regime. Stern left Germany in 1933, and he was fortunate to obtain a position at the Carnegie-Mellon University in Pittsburgh, where he tried to rebuild his laboratory and regain lost momentum. He could scarcely compete for the limited funds available to American physicists or attract the social support an émigré needs. First displaced by Nazi policy and then denied an award that most leading physicists considered well merited, Stern was unable to regain balance and self-respect in his new homeland. Winning a Nobel Prize would certainly have made a difference. Stern urged fellow refugee James Franck never to forget to mention his own Nobel Prize while in the United States. When Stern finally received the prize after the war, it mattered little, his research career was over. Not able to raise money for research, his spirit, battered by Nazi hatred, broke. After receiving the prize in 1945, he retired prematurely and cut himself off from the world of physics.

The committee had subverted precisely the ideal that Alfred Nobel sought for his fortune: to help those with ideas but not the means to realize them.

The chemistry committee's decision to reserve the 1933 prize, and then to withhold it the following year, had other damaging consequences. Candidate Carl Neuberg, earlier declared worthy of a prize, was passed over, despite continued support from laureates Heinrich Wieland and Richard Willstätter, among others. Neuberg was forced to leave his position as head of a Kaiser Wilhelm Institute, and he moved to Brooklyn, New York, where he languished at Brooklyn Polytechnic Institute before finding a job as an industrial chemist in New Jersey. His fate in the New World was enviable, as compared with that of many refugees whose careers came to an abrupt halt—not to mention the far more horrible fates awaiting those who could not flee. But, as his correspondence with major German-language chemists reveals, he was one of the era's leading bioorganic chemists. He was also frequently nominated for a prize in medicine for his biochemical studies. Had he received a prize, especially in chemistry—for which Svedberg and von Euler had evaluated him as deserving—he might have been able to rescue his research career in the United States. In any case, it would have been far closer to Nobel's explicit intention to have used these prizes, or the funds generated from them, for a purpose more altruistic and international than simply supplying more bricks for the edifice of Swedish science.

Injustice is a necessary concomitant part of any prizegiving. No committee is all-knowing; no committee can divorce itself completely from contextual concerns. A series of small, almost imperceptible adjustments in attitude and procedure can be as significant for committee actions as blatant, willful deviations. Injustice also occurs when institutions use prizes indiscriminately to assess the relative value of researchers and their specialties. Knowing full well the weight given—rightly or wrongly—to the prize, the Nobel committees and the Academy chose a poor time to act locally and ignore consequences globally. Precisely those persons entrusted to fulfill Nobel's legacy might have displayed greater sensitivity to international events. Did the fixation of some committee members with their own expanding budgetary needs numb their sense of moral responsibility? Perhaps the point is that they had no moral responsibility; they were supposed to be "objective." But, clearly, they were both amoral and subjective.

When the massive dismissal of "undesirable" scientists began in Germany in 1933, both prominent and not-so-prominent researchers suddenly found themselves without employment, without a safe haven, and without possessions. In Western Europe and North America, prominent scholars joined in creating emergency committees to find homes and jobs for displaced academics. Many leaders of international science understood that extraordinary times required extraordinary measures. Not all refugees were Jewish, and many who helped were not so much pro-Jewish or anti-German as simply concerned to help refugees, to show solidarity with displaced scientists. In the era of appeasement, many institutions treaded cautiously. The effects of the Great Depression added to the difficulties. Jobs and research grants were scarce; raising funds and creating positions required energy and imagination.

Discreet inquiries from relief organizations found their way to the Nobel committees and Nobel Foundation. When asked to lend their names, collectively or individually, members of the Nobel establishment claimed that this would compromise their impartiality and neutrality. When asked to assist financially, Nobel officials claimed that the bylaws prohibited them from doing so. If direct aid were precluded, would it be possible, Max von Laue asked, to propose giving a prize to one or more relief organizations? The answer was no.

At a time that called for compassion, the Nobel establishment found excuses rather than moral leadership. In the dark years of Nazism, gestures in support of the ideals of internationalism and openness in science—values embodied in Nobel's testament—would have been appropriate. The Nobel bylaws did not in fact preclude action. In 1919, pro-German sentiment resulted not only in at least one dubious politically inspired prize but also in a grant from the physics committee's special fund to Arnold Sommerfeld, whose Munich institute needed help. In 1939, when the war caused a cancellation of the Nobel ceremonies, the money that the Academy saved for its share of the festivities was donated to the Swedish Red Cross. Did their obsession with finding research funds cloud committee members' sensitivities? The monotonous annual toasts and speeches, praising the Academy members as the guardians of transnational ideals, proved to be effervescent fizz; once the festivities ended, the flat taste of self-interest dominated daily affairs.

It is possible that the Academy's reluctance to give a prize to Stern was in part motivated by a fear of angering the new regime in Germany. A prize to either Stern or Neuberg might have been criticized as being political. But not to give them prizes, which they otherwise merited, was also political. In fact, it violated Nobel's explicit wish that prizes should go to the most deserving, regardless of nationality.

Certainly, had the will been present, action could have been taken, effective or symbolic. An action as simple as canceling a banquet, or reducing its splendor, to raise funds for international relief organizations would have shown the Academy's and the Nobel Foundation's distress over the threat to the republic of learning. Use of the committees' special fund for supporting research did not have to be limited to members or Swedes, and indeed, as we shall see, two refugees were helped at the end of the decade and during the war. By the mid-1930s, nobody could hide behind the belief that the Nazi actions were temporary excesses. Racial laws and the

Nazification of universities with politically correct party members influenced academic practice on all levels. Of course, the reasons for the inability of the Academy and the Nobel Foundation to reveal moral leadership were obvious. There remained traditional strong links between Sweden and Germany. Was it merely a coincidence that Pettersson's attack was published in the one major Swedish newspaper that, already in 1933, had emerged as an uncompromising outspoken opponent to Hitler's reign of terrorism, much to the chagrin of the government and many elite groups in Stockholm?

"The world of science is a world without flags or frontiers" was how the Rockefeller Foundation tried to justify a grant in 1936 to a Kaiser Wilhelm Institute. In response, Felix Frankfurter, then a law professor at Harvard University, reminded the foundation that, by ignoring the question of whether Nazi Germany actually was such a world of science—without flags and frontiers—it helped "to adulterate the spiritual coinage of the world." He continued with words that could also have been aimed at the Swedish institutions, hiding behind a similar banner:

> In the central contest of the contemporary world between reason and antirationalism, between democracy and dictatorship, I have little doubt about the general direction of your thoughts and feelings. But those of us who believe in reason and democracy ought to be as firm and uncompromising and valorous in acting on our faith as are those who challenge everything that we hold dear.

Courageous is not the first word that comes to mind when considering the events that prompted Pettersson's criticism. The committees and the Academy were self-satisfied, secure in the international prestige of the prizes, and proud of the emergence in Sweden of several prominent schools of physics and chemistry. Arrogance grew with achievement; inflated authority in the Academy and in the national scientific community fostered indifference. The most powerful and influential committee members, Oseen, Siegbahn, von Euler, and Svedberg, differed greatly in outlook and scientific style. Some of them and their colleagues did care sincerely about the plight of refugees as well as the fate of those trapped in Germany and also in Stalinist Russia. Yet they also shared a common concern with increasing local institutional resources for research. They ignored the fact that they, the committees, and the Academy were caretakers of Nobel's legacy.

As Swedish scientists looked to ensure thriving national research in costly fields, their judgment became clouded. From today's perspective, this foraging for mere nuts and berries within the Nobel coffers may appear odd. The amounts generated by withholding a prize were modest—these were not even the sources for major institution building—but from their perspective still, nevertheless, vital. Thus, there grew a gap between declared ideals and actual practices.

The sharp and short-lived debate in the mid-1930s dampened the committees' subsequent enthusiasm for reserving prizes. But by this time the importance of the prize as an economic resource was already dwindling. Siegbahn's venture revealed that major experimental commitments required far more massive funding. Much more important uses of the prizes were soon to emerge. Scandalous traffic did not come to a halt.

# THIRTEEN

## Completely Lacking an Unambiguous, Objective Standard

### Big Science and Diminutive Morals: The Authority of Prizes

*I*n 1937, in a manner grander than he had ever imagined, Siegbahn became director of a major new Swedish research institution. From years of tense maneuvering, Siegbahn and his supporters had learned to think big. When, in 1930, he first contacted Knut Wallenberg, Siegbahn did not have a specific plan for his institution other than a chance to escape the drudgery of academic obligations and devote himself to research. Several years later, Siegbahn saw nuclear physics as a possible focus. Once again, he recognized an important niche where his particular talents could be applied.

In physics, bigger, costlier, and technologically sophisticated installations seemed to be the ticket to scientific success. Siegbahn looked to the United States. In the hunt for money prior to World War II, scientists with costly, ambitious plans had to win support from private foundations. During and after the war, when massive government funding of research began, they competed for slices of public largesse and for a voice in directing science policies. For members of the Nobel establishment in Sweden, and for nominators throughout the world, more than ever the Nobel Prize became a means of bestowing authority and prestige upon potential leaders—and for hindering them, as well. Two candidates—E. O. Lawrence, for his cyclotron, and Lise Meitner, for her contribution to nuclear fission—illuminate the role of the prize in the emerging international era of Big Science.

### CYCLOTRONISTS OF THE WORLD UNITE

#### Siegbahn's Nuclear Transformation

By the mid-1930s, the field of nuclear physics was no longer the province of just a few groups of scattered researchers. Following the "miraculous year" of 1932 an increasing number of physicists looked toward nuclear physics as a field ripe for theoretical and experimental cultivation. The discovery of the neutron gave scientists a particle without charge that could be directed at a nucleus without having to overcome repulsive electrical charges. Previously, the projectile of choice for exploring the inner atom had been Ernest Rutherford's alpha particle—two positively charged protons—which, because of its charge, was easily deflected by the numerous protons in heavier nuclei. In 1934, Frédéric Joliot and Irène Joliot-Curie's discovery of artificially induced radioactivity produced a smörgårdsbord of new isotopes, some characterized by powerful radiation. In Rome, Enrico Fermi followed up on this sensation by complementing his own pathbreaking theoretical studies on nuclear processes with experimental research, systematically investigating the periodic table for radioactive isotopes. Having begun bombarding one element after another with neutrons, he discovered that by putting a piece of paraffin between the radiating source and target, he could slow the neutrons, which then became even more effective as bullets for penetrating atomic nuclei. By the mid-1930s, several centers for physics were churning out a stream of publications that showed an ever more complicated picture of nuclear processes.

All these studies struggled with the problem that had plagued researchers from the earliest studies of radioactivity—the need for energetic sources of radiation. Whether based on processed radium or artificially created radioactive isotopes, these sources were costly and frequently provided disappointingly weak radiation. Laboratories often had to rely upon discarded radioactive materials that no longer were sufficiently potent for medical use. To investigate nuclear processes and the world of subatomic particles, considerably greater energies were needed. There were two avenues. First, by the 1930s it was clear that highly energized subatomic particles from outer space—cosmic rays—constantly bombard the earth. With great patience, researchers using a Wilson cloud chamber and photographic records could study particle collisions. Meanwhile, others sought to create their own high-energy bullets.

Under Ernest Rutherford's leadership, the Cavendish Laboratory in Cambridge made many of the experimental breakthroughs in nuclear science. It was here in 1932 that the first artificially accelerated particle smashed into a nucleus, thereby transforming an atom of one element into another. To achieve this result, John Cockcroft and E. T. S. Walton convinced their chief to allow them to build a modest particle accelerator. Following this and other accomplishments, Cambridge certainly had an edge over all other centers of physics in the development of nuclear physics. But Rutherford did not believe that bigger meant better. He did not see the scientific need to scale upward in size; moreover, he preferred to work with relatively simple equipment under his own control. Chasing patrons for massive grants and establishing ties with local industry for collaborative projects to erect huge machines was not his style. Although Rutherford's disciples continued working with accelerators, they could not compete with a man who believed in the credo of big, bigger, biggest—E. O. Lawrence of the University of California at Berkeley.

Lawrence devised the machine that made him and Berkeley physics famous—the cyclotron. In 1930, an article by the Norwegian engineer Rolf Wideröe had inspired Lawrence to consider the problem of accelerating a particle to great energy levels using relatively modest electric and magnetic fields. Wideröe suggested a scheme by which a reversing electric field could push and pull a charged particle in a tubular electrode. Lawrence understood that to achieve significant energy levels this design would be prohibitively expensive. Instead, he transformed the basic idea into a more feasible scheme: Accelerate charged particles in an electric field around a circular orbit with the aid of a powerful magnet. Each time a particle passed one of the poles, it could be sped up further through attractive or repulsive force. Beginning with a tiny prototype, Lawrence and his chief collaborator, graduate student

M. Stanley Livingston, announced their first success in January 1931. Lawrence built bigger machines during the 1930s while also assembling larger teams of assistants. Money was crucial to scaling up; Lawrence's optimism and entrepreneurial gusto kept the enterprise moving through the Depression years.

Single-mindedly obsessed with building ever-larger cyclotrons, Lawrence had no clear program for research. Consequently, his laboratory missed discoveries they were in a position to make—the first artificially disintegrated nucleus and artificially induced radioactivity, being among the most prominent. Soon after reading the Joliot and Joliot-Curie results, the Berkeley team produced a jungle of new isotopes. But, although a new world of nuclear chemistry and physics was coming into view, Lawrence focused his attention elsewhere: medicine. Producing isotopes for use in therapies and biomedical research could help justify the big grants needed to build larger cyclotrons. By 1935, Lawrence's cyclotrons were mass-producing radioactive isotopes for cancer clinics and medical research laboratories. Lawrence traveled across the United States, extolling his new miraculous wonder isotopes as the cure for all sorts of diseases. Emphasizing the value that the cyclotron and nuclear science had for medicine and biology, he pressed private foundations for grants.

In the second half of the 1930s, Lawrence helped several laboratories around the United States and Europe build their own cyclotrons. These highly complex machines could not be erected and switched on simply by reading a blueprint and instruction manual. The tacit know-how had to be learned through direct contact. Lawrence and his "boys" made house calls to help colleagues; and they opened their laboratory to foreign guests who needed to learn the fine points of building and running a cyclotron.

Lawrence's enterprise inspired Siegbahn; for a scientist of his temperament, it was irresistible. Siegbahn decided that his new institute must have a cyclotron. Although his budget could scarcely accommodate this costly addition, he saw the instrument as his ticket to the emerging research frontier. Siegbahn did not have any particular research plans or problems in mind. But his experimental acumen told him that a cyclotron was a sort of "ante" if a laboratory wanted to join in nuclear and particle research. Siegbahn saw that the project's technological demands would promote interaction between physics and industry, which he considered lacking in Sweden. More to the point, for a physicist such as himself, who was more at home with instruments and machines than with complex theories, a cyclotron and an array of auxiliary instrumentation could direct his talents productively. Just as X-ray spectroscopy was once in vogue, the cyclotron now offered similar opportunities.

Siegbahn invested his funds in a cyclotron in the basement of his new physics institute. Although he had a tiny number of assistants from Uppsala and a small research budget—the money going to infrastructure—Siegbahn was determined to have his laboratory take a lead in European physics, by being one of the first to have a cyclotron. He established ties with Lawrence, and he sent an assistant to Berkeley and other American cyclotron locations to learn the necessary know-how. Californians were sent to Stockholm. Bonds of collegiality and friendship quickly grew.

## A Friend in Need

By 1939, the Stockholm cyclotron had been planned and was under construction. Around the United States and Britain, a number of these machines were already hurling subatomic particles into atomic nuclei. In Berkeley, Lawrence intended to outdo himself with plans for a giant cyclotron, one that would dwarf his previous machines in size and cost. All he needed to do was convince a reluctant Rockefeller Foundation to donate a million dollars. He knew full well that a Nobel Prize could tip the Rockefeller scales in the right direction. He knew how the prize enabled Joliot and Joliot-Curie to attract substantial funding for their nuclear projects. Closer to home, down the coast, Robert Millikan and the California Institute of Technology had comparable success with the prize. All Lawrence needed was for those who had benefited from his generosity to help him get one of his own.

Prior to 1939, Lawrence had received a mere sprinkling of nominations; he was a man to watch, but others had achieved more results. Some felt that John Cockcroft and E. T. S. Walton deserved a prize before Lawrence because they were the first to show that a nucleus could be disintegrated with an artificially accelerated particle. They had seen what other experimenters had not. They were inspired to take a chance by a young Russian theoretician, George Gamow, who used quantum mechanics to predict that the energy needed for a particle to collide with a heavy nucleus was much less than previously calculated. Rather than requiring power all but beyond reach in a laboratory, a tiny number of particles could, probabilistically, bore into a nucleus at a much lower level of energy by a tunneling effect. The Cambridge group understood that it was worth trying; they succeeded. Although Lawrence could have achieved this result, he was thinking about technology and machines; the Cambridge group was doing physics.

When, in 1937 and 1938, Cockcroft and Walton were nominated, Siegbahn was already California dreaming. A cyclotron was to be in his future. He dismissed the candidacy of the two physicists by claiming theirs was not the definitive solution to the smashing of the atom. But this missed the point. They did not attempt to create an instrumental solution; rather, they set out to demonstrate that atoms beyond the lightest elements could be smashed without the need for tremendous energy to accelerate the particle "bullet." Siegbahn was, like Lawrence, too focused on the technological innovation. He also was too attached to the California effort to appreciate the Cambridge achievement. It was not until 1951 that the committee reversed itself and gave Cockcroft and Walton a prize—motivated mainly because of Cockcroft's leadership in British and European nuclear activities.

Nevertheless, in 1939, Siegbahn was ready to give a prize to Lawrence. For the American physicist, and for those who understood his plans for the gigantic cyclotron, the prize could confer authority and prestige, which were more important than the actual money involved. A prize to Lawrence could underscore the significance of the cyclotron; every laboratory director who was chasing after funds to build or improve a local cyclotron could benefit. But one vital question remained: Did Lawrence meet the formal requirements?

When Niels Bohr and O. W. Richardson discussed whom they might propose for the 1939 prize in physics, Richardson balked at nominating Lawrence. First, he noted, most of the American's impressive work was accomplished with coworkers. A problem that was to become common after the war was already apparent: How could the prize properly recognize teamwork? Was it right to celebrate Lawrence alone? Also, Richardson raised an even more serious question: Had Lawrence or his assistants made a significant discovery with the cyclotron? The invention of a new instrument did not appear to be enough to merit a prize; something of great significance had to be discovered with it, either by the inventor or by someone else. Richardson believed it only fair to let Lawrence wait until the all-but-inevitable discoveries came; the list of physicists worthy of a prize was already long.

Bohr, who was sojourning at Princeton, and well plugged into the gossip of American physics, certainly knew what a prize *sooner* rather than later could mean. Lawrence's application, pending at the Rockefeller Foundation, was the talk of the discipline. Bohr could sympathize with him. Having spent much time and energy raising funds to bring his Copenhagen institute into the nuclear age, Bohr appreciated Lawrence's need for recognition from the international community. Bohr himself benefited from Lawrence's willingness to share Berkeley know-how; his own Institute for Theoretical Physics was adding a cyclotron. So he acknowledged Richardson's points, but nevertheless he proposed Lawrence, based on the great significance cyclotrons were having for nuclear physics.

In the tense atmosphere in 1939, before the outbreak of World War II, there were few nominations. Hitler's ban kept Germans from participating. Elsewhere, those who chose to propose saw that most of the obvious candidates, of older and newer vintage, had been repeatedly passed over by the committee: Why continue to nominate them? Too few nominators responded to give the committee a sense of which newer or less obvious achievements could win broad approval. Richardson proposed E. V. Appleton for his discovery of the ionized layer of the upper atmosphere—the ionosphere—which is responsible for reflecting radio signals, and so permitting long-distance radio communication. W. L. Bragg, like Richardson, compared Appleton with Lawrence, and he put the former first in the list of most prizeworthy, as he "originated a new branch of science." Professors from the Massachusetts Institute of Technology and Harvard sent in collective letters of nominations for Harvard's P. W. Bridgeman for his continuing contributions to extreme high-pressure physics, which the committee had repeatedly found not sufficiently meritorious. Some Swedish nominators again proposed Vilhelm Bjerknes, alone or with collaborators, for a physics of the atmosphere. With so few proposals for relatively so many candidates, Lawrence suddenly stood out, with a surprising eleven nominations.

## Siegbahn's Strategy for Lawrence

Siegbahn evaluated Lawrence. Questions about Lawrence's originality, whether the cyclotron led to a significant discovery, and of his assistants' role, all required attention. Siegbahn was aware of the similarities between Lawrence's candidacy and his own. Just as he had received help from allies who pushed for a liberal interpretation of the statutes, Siegbahn tried for the same strategy. Sharing Lawrence's predicament—of being constantly in need of larger grants—Siegbahn appreciated the social credit a prize could bring. And what could better cement the relationship between the two laboratories than having Lawrence come to Stockholm? Siegbahn's own plans would, of course, also benefit from media attention given to Lawrence and the cyclotron.

He failed, however, to make a firm case for Lawrence. After reviewing the evolution of the cyclotron, his strongest claim pointed to the cyclotron's use in creating isotopes. The Lawrence team had produced a myriad of new isotopes of virtually all the elements. This discovery promoted the study of nuclear phenomena and all but assured future discoveries, as Siegbahn noted in the closing of his seventeen-page evaluation. "The great possibilities" that the cyclotron offers experimental research prompted some twenty-five laboratories around the world to build similar machines in spite of the great costs and construction difficulties. But no particular breakthrough could be named. In fact, the most sensational new discovery in the field, nuclear fission, was just made without a cyclotron.

Siegbahn did not directly confront these awkward questions. Instead, he diverted attention by playing up the benefit to humankind. In part, this was a strategy to defuse the campaign for Bjerknes that played upon the many lives saved and economic-social gains derived from improved weather forecasting. The Bjerknes camp appealed to Alfred Nobel's vision of benefiting humankind; but Siegbahn trumped. He pitched his case for Lawrence with a heavy dose of the same: the cyclotron's significance as a producer of radioisotopes for medical therapies. This also drew attention away from the problem of whether the cyclotron had led to a significant discovery *in and for physics.*

Nobody doubted, as one nominator stated, the cyclotron's great importance for medicine, and that it "has proven to be by far the most effective tool for the experimental study of the structure of atomic nuclei and for the production of artificial radioactive elements." But it was difficult to claim that these constituted a singular discovery worthy of a prize. The committee often dismissed proposals in which the reputation of a researcher was based on a series of important findings, none of which could boast to being a single great discovery. Rejections were also normally the case when physics, or the instruments of physics, yielded significant results outside of "physics proper," such as in astrophysics and geophysics. Medicine surely fell in the same category. Richardson's insight was correct.

However, goodwill in the committee could overcome such barriers. Evidence of the need to bend the traditional measuring stick to give Lawrence the 1939 prize entailed the comparable evaluation of Cockcroft and Walton. Again, the committee rejected them, saying that, because of the intense ongoing developments [*rastlösa arbetet*] in this field, it did not want to take a stand on the nomination of the two British researchers. But why then award Lawrence if developments were coming fast and furious? In 1938, the committee similarly concluded that it would be best to wait for "research results" from the many cyclotrons under construction before assessing Lawrence's merit. This was still the case in 1939. What had changed from 1938 was the acute desire to bestow authority on Lawrence to assist him with his enterprise.

Siegbahn counted on the committee's support. Oseen's critical voice was all but assured. Two potential rivals had little chance. Oseen adamantly rejected Bjerknes, and thundered, once again, that meteorology was not physics. Oseen's previous emotional outbursts during Academy debate had served only to generate sympathy for the Norwegian. Similarly, Appleton's contribution was, well, up in the air. This work came too close to Oseen's long-standing antagonism to "terrestrial physics," which, because it lacked controlled laboratory experiments, must remain inexact and, therefore, to his thinking, inferior science. Oseen needed a strong alternative candidate. Lawrence might not have satisfied the committee's traditional criteria, but he made an attractive choice, not least by being an American.

Oseen, and many others in the Academy, who had reservations about American culture, could not ignore the fact that the world's leading scientific nation was now across the Atlantic. The influx of many refugee scientists—including some of the most gifted—from Germany, Italy, and Austria, on top of impressive homegrown advances, ensured America's dominance. Some Swedes, such as Siegbahn and Svedberg, had developed a taste for American science in the 1920s, but by the late 1930s even the most reluctant and envious Europeans had to look west. Good relations with American colleagues and institutions were imperative. Here, then, was an added incentive to create yet stronger links between the two scientific communities. A prize to Lawrence would no doubt be very popular among American physicists.

Although Oseen and the rest of the committee backed Siegbahn's recommendation, a minority of the Physics Section refused Lawrence. Five members voted for Lawrence, three for Bjerknes. Given the formal problems with Lawrence's candidacy, Oseen feared the worst—a prize for meteorology. So he suggested postponement; war, too, had broken out a few weeks earlier. But the supporters of both Lawrence and Bjerknes were determined to win a prize for their respective candidate. In the Academy, a bitter debate broke out. Those in favor of Bjerknes insisted that Lawrence should wait. But the American enjoyed the backing of members who also saw the advantage of promoting the cyclotron: the influential medical radiologist Forssell, as well as the ever-popular Svedberg, who was considering research programs in nuclear chemistry. Svedberg envisaged either a cyclotron of his own or use of the one planned for Stockholm. He was in contact with Lawrence, aware of Lawrence's pending grant, and cognizant of the Berkeley enterprise's importance for his own plans.

In spite of this support, the Academy still had qualms about backing Lawrence. The final vote showed a reluctance to accept the committee's unanimous recommendation. Those who called for waiting until Lawrence could fulfill the formal requirements were sufficiently persuasive to put a few dents in Siegbahn's proposal; the final vote was Lawrence, fifty, Bjerknes, nineteen, and several for not awarding the prize.

Happily for Lawrence, the prize did come sooner rather than later. He was hailed in the press, and the Rockefeller Foundation swallowed its reluctance and gave him a million dollars. When the American effort to build an atomic bomb began, Lawrence's laboratory became one of the critical centers of research. It did not hurt his ability to win important contracts and missions for his laboratory that, in addition to his other honors, he could boast a Nobel Prize. Lawrence could not travel to Stockholm in 1939, because of the war in Europe, but the delay made his eventual trip in 1951 all the more valuable.

Siegbahn used the excellent relations with Berkeley to further his own endeavors. His first cyclotron, which was completed in 1942, was a dud. It was not as powerful as planned, or very effective. Still, Siegbahn was able to follow Lawrence's track; he prioritized the use of the machine for producing radioisotopes for medicine and biological research. This profile, and the example of Lawrence, enabled Siegbahn to approach the Wallenberg and Rockefeller foundations with plans for improvements and for a new, larger cyclotron. The new machine would provide radioisotopes to all the Nordic countries. Officials at the Rockefeller Foundation turned to Lawrence for an expert assessment of Siegbahn's plans. Lawrence urged support:

> I think it would be splendid in every way for the Foundation to make a grant to Professor Siegbahn along the lines indicated. . . . Apart from scientific value, it

would seem to me obviously in the best national interests to do what we can along this line to preserve cordial relations with our Swedish colleagues.

When the war at last came to an end, Siegbahn wanted yet more. He conceived an even larger machine, but by this time all things nuclear had taken on new significance. Among the many repercussions of the detonation of atomic bombs in August 1945, nuclear research became a major concern of the neutral Swedish government. But who—other than Siegbahn—was going to provide authoritative advice and who was going to claim major portions of the expected bonanza of government funding of nuclear research?

## DENYING LISE MEITNER A PRIZE: THE SWEDISH NUCLEAR STAKES

### Postwar Promise

In the United States, in particular, the war boosted Big Science. Money flowed to research as never before, in both the belligerent and some of the neutral nations. Large teams of scientists tackled myriad problems touching all aspects of total war (at home and on the battlefront). Whether to predict the height of waves for landing amphibious troops, to create substitutes for rubber, or to produce vaccines against contagious diseases, large collaborative projects under the direction of prominent scientists became commonplace. Enormous amounts of money for critically important projects—not the least for the gigantic efforts to build the atomic bomb and develop radar systems—accelerated changes in the conception of the organization, and even in the nature, of research. Scientific leadership increasingly entailed management of large teams, large budgets, and large responsibilities.

Science helped shape the outcome of this war. Radar helped win it, the atomic bomb helped end it, and innumerable other accomplishments underscored the importance of research. Nobody quite knew what was about to happen to the funding of science in the postwar era, but nobody expected a return to the past.

During the war, the neutral Swedish government greatly increased its research budget, for defense and for national self-sufficiency. Academic leaders such as Siegbahn and Svedberg, took the lead in forging links among science, industry, and the military. They helped create and lead new military research advisory committees and defense research organizations. What Siegbahn had earlier advocated—physics as a resource for national strength—was now widely accepted.

Soon after the shock of the American atomic bombs, the Swedish government and scientific leadership began discussing policy about how to mobilize to exploit nuclear energy. Swedish uranium deposits, it turned out, were among the largest in Europe; although the ore was of low-grade quality, the amount of it was estimated to be enormous. Some people saw a Swedish atomic bomb as the best means to guarantee future neutrality. Others discussed the potential of inexpensive nuclear energy to power a materially secure and socially just welfare state. Although the specifics of Sweden's nuclear future were unclear, scientists understood that a new golden age for research was dawning. The immediate postwar climate—heated by the realization that the energy in the atom could be released—was one of optimism, especially for those who looked to establish large, expensive installations for the study of the atom. Siegbahn's day had arrived.

Only one fact clouded Siegbahn's prospects of securing a major share of the anticipated resources and playing a pivotal role in policy. He was not a nuclear physicist. He knew how to plan, build, and operate a cyclotron; he knew how to raise money for such machines, but—as critics asserted—he had little insight into what could be done with them. In Sweden, there were other researchers with greater competence in nuclear science, and they also wanted facilities.

In particular, Sweden had become the home of an exiled nuclear physicist, one of the leading researchers in this field, an Austrian of Jewish lineage, Lise Meitner. In 1938–1939, Meitner played a key role, together with her former Berlin colleagues Otto Hahn and Fritz Strassmann, in the discovery and initial explication of nuclear fission. After the news of the American atomic bombs, journalists converged upon her. A few months later, she visited the United States where physicists gave her a V.I.P. treatment. She met with President Harry Truman and leaders of American nuclear agencies; she was interviewed on radio by Eleanor Roosevelt, and she received a number of awards. Many believed that she should share the honors for the discovery of fission, including a Nobel Prize. But, to give a prize to Meitner would inevitably elevate her from a powerless, despairing, and dependent refugee in Siegbahn's laboratory into a recognized, leading authority in nuclear physics, even in Sweden. That scenario made Siegbahn and others uneasy.

## The Meitner–Hahn Team and Fission

In July of 1938, Lise Meitner was forced to flee Berlin where she had lived and worked for thirty years. Only the second woman to receive a doctorate in physics from Vienna, Meitner came to Berlin in 1907, where she had to overcome Prussian restrictions of women's academic activities. Those who came into contact with this shy but intelligent and determined woman generally tried to assist her. The politically conservative Max Planck helped her into Berlin physics. She became his assistant for a short time and his friend for life. She was accepted socially and intellectually into the circle of prominent Berlin-oriented physicists that included, among others, Albert Einstein, James Franck, Max von Laue, Max Planck, and Otto Stern.

Although it was unthinkable to give a woman a university position, the new privately funded Kaiser Wilhelm Institute offered an alternative. She and chemist Otto Hahn teamed up in the new chemistry institute just prior to World War I. Hahn was also young and fresh to Berlin; he had worked abroad in the new field of radioactivity. To continue his work, he needed help from a physicist, while she wanted greater experimental collaboration than Planck could provide. Meitner and Hahn's collaborative efforts brought results. In 1918, they announced the discovery of a missing radioactive element, number 91 in the periodic table: protactinium. They were given their own respective departments and professorial posts at the Kaiser Wilhelm Institute for Chemistry—Meitner in physics and Hahn in chemistry. Their discovery of protactinium, and subsequent accomplishments, brought them scattered nominations for a Nobel Prize throughout the 1920s and 1930s. In the meantime, Meitner moved away from radioactivity and started working in nuclear physics.

Meitner's initiatives in this nascent specialty made Berlin an international center in nuclear physics. When in the 1930s large numbers of physicists turned to nuclear questions, Berlin and Meitner's laboratory were already prominent points of reference. It was Meitner who drew Hahn into collaboration on nuclear processes; and it was this line of research, begun in 1934, that resulted four years later in the discovery of fission.

Fission produced much embarrassment, even before it became explosive. In the mid- and late 1930s nobody could quite make sense of the complex and confusing results of bombarding heavy elements, including uranium, the heaviest, with neutrons. Although the nuclei of light elements might be shattered by collisions with particles, nobody quite knew what to expect from the heavier elements. From his team's systematic studies moving up the periodic table, Enrico Fermi announced in the spring of 1934 that the heavy elements tended to absorb neutrons and decayed by emitting beta rays, or highly energized electrons. This would favor creating an isotope of an element one place higher on the periodic table. It seemed that as irradiated uranium (number 92 on the periodic table) decayed through a series of transformations, new elements were created—number 93 and number 94—beyond what had been thought the end point of the periodic table.

Fermi initially announced his finding tentatively, but soon scientists and the public accepted it as fact. Although a German chemist, Ida Noddack, responded by advising that the discovery of alleged new elements heavier than uranium should not be declared until every last alternative explanation was explored, nobody seemed to heed her admonishment. The Fermi team was propelled into the limelight. Mussolini's propaganda machine transformed them into heralds of fascist Italy's return to scientific greatness. As the fervor spread, a number of teams joined in the endeavor to make sense of what happens to irradiated uranium.

These results riveted Meitner's attention and encouraged her to begin a new, intense research program, in which she enrolled Hahn and a gifted, young analytic chemist, Fritz Strassmann. They, and others working in this field, such as Irène Joliot-Curie and her collaborators, were aware that Fermi's results were awkward. The chemical properties of the new isotopes differed from what was expected, the radioactive half-lives of the transforming isotopes, once painstakingly identified, only added to the pattern of confusion. But they all worked feverishly onward, spurred by the hope that new physical theory and chemical principles were waiting to be discovered once the tangle was sorted out.

In Berlin, Meitner led the team. Hahn and Stassmann performed the painstaking chemical analyses to separate and analyze the minute radioactive specimens using a number of techniques. Working with well-accepted theory, Meitner attempted to make sense of the puzzling results, proposing schemes for how uranium and the new transuranium elements transformed into innumerable isotopes. Their jointly authored papers were widely read; they entered into a competition with the Joliot-Curie Paris team, each hoping to find and explain a definitive sequence of events ensuing from uranium's irradiation. Following a claim from Paris that again called the whole explanatory scheme into question, Meitner once more pushed her chemist colleagues to redo all their measurements even more carefully. But just as the collaboration entered this new intense phase, Meitner had to flee Germany.

For all three scientists, it had been important to stay on top of this controversial and prestigious line of research. All of them were suspect in Nazi Germany. Hahn refused to join the party; Strassmann was staunchly and openly anti-Nazi, even cutting off professional advancement by refusing to join the Nazi-dominated German

Chemical Society. Although she had been baptized as a child, Meitner was a Jew under German law. But only when Austria entered the greater Reich was she subject to the racial laws that ensured that she would be fired—and threatened with worse. Her friends tried to organize a move for her to another country, but when it became clear she would be prohibited from leaving, Meitner had to escape on short notice in July 1938 with but summer clothes, ten marks, and a ring from Hahn that could be sold for emergency money. At age fifty-nine, she was smuggled out of the country, leaving behind all her possessions, including research notebooks and designs for her customized instruments. Niels Bohr and Dutch scientists Peter Debye, who worked in Berlin, and Dirk Coster, in Holland, helped clear a path from Germany to Holland and then finally into Sweden. Just as Siegbahn was setting up his new laboratory in Stockholm and planning a cyclotron, he was asked to accept one of the world's leading authorities on nuclear research.

Meitner's circumstances did not stop her from sharing in the discovery of nuclear fission; but they did limit her ability to gain recognition for her achievement and to follow it up. Careful historical research, first by Fritz Krafft and then in immaculate detail by Ruth Sime, using unpublished correspondence, diaries, and drafts of publications, provides insight into the collaboration that is not readily grasped from published scientific articles related to fission. They have also been able to reconstruct the origins and propagation of the myth that the discovery was achieved by pure chemical means, without the assistance of physicist Meitner, who in fact has even been erroneously identified in popular history as Hahn's junior assistant. A summary of the events in the winter of 1938–1939, and then a review of Meitner's relations with Siegbahn, provide valuable background to the chemistry and physics committees' evaluations of the discovery of fission. More than a tale of injustice, this episode reveals an important role given the prizes in the emerging world of Big Science.

From July 1938, Meitner, the intellectual leader of the team, was separated from Hahn and Strassmann. Hahn continued to correspond with her, even though contact was a clear political liability. Before Meitner left Germany, however, Hahn and Strassmann had been confidently separating and neatly analyzing the various products of irradiated uranium. From Hahn's perspective as a chemist, the existence of the transuranium elements was "no longer in doubt," and "no further discussion" was necessary on their distinction from other elements. It was Meitner who was disturbed by the findings. She could not reconcile the confusing pattern of nuclear processes these findings implied with prevailing theories of nuclear structure. Her repeated assaults on this problem prompted her calls for redoing the chemical analyses.

Hahn reported to Meitner by letter on the team's work during the autumn of 1938, asking her advice on matters of interpretation and direction for further studies. They met secretly in Copenhagen on November 13 and 14—just after *Kristallnacht* marked an escalation in Nazi measures against Jews and dissidents. In Denmark, according to Hahn's diary, they discussed the uranium problem for many hours. Meitner knew that the results could not be correct and urged yet further refinement of the most critical experiments. Strassmann noted many years later that it was fortunate her opinion and judgment carried so much weight that he and Hahn immediately set about with the new round of measurements.

It was this latest effort at distilling and identifying specific products arising from irradiated uranium that led to Hahn and Strassmann identifying a substance resembling the element barium, roughly half the atomic number of uranium. Greatly agitated, Hahn wrote to Meitner on December 19 asking her if she "can come up with some fantastic explanation" for why the isotopes acted like barium. He conceded that they know that uranium cannot actually "burst into ba[rium]." He repeated his plea for guidance a few days later, afraid to publish this seemingly absurd physical result. Without her guidance on the physical interpretation, he was increasingly losing confidence in his and Strassmann's gift for chemical analysis.

In the article Hahn began writing, he concluded that the chemical results prompted them to say that the products from irradiated uranium were not forms of radium, but barium; however: "As 'nuclear chemists' fairly close to physics we cannot yet bring ourselves to take such a drastic step that contradicts all previous experience in nuclear physics. There could still perhaps be a series of unusual coincidences that has given us deceptive results."

Meitner, who was by now driven to frustration with all earlier attempts to make sense of the transformations of uranium, shot back a response to Hahn's letter. She gave him renewed courage to interpret the results. She agreed that radium acting like barium was puzzling, and the thought of a "large-scale breakup" of the uranium nucleus was very difficult, "but in nuclear physics we have experienced so many surprises that one cannot unconditionally say, 'It is impossible.'" Meitner gave Hahn a green light to think freely; she would not at all rule out some sort of previously unknown process by which the nucleus of a heavy atom could break apart. But how could a neutron result in a smashed uranium nucleus? How could it have an effect like the impact of a marble on an office building, causing the structure to collapse into rubble? Meitner was ready to accept the possibility; she urged Hahn to reason his way to an explanation of the chemical findings. Four days after receiving her reply, on December 23, Hahn added a short paragraph to the page proofs of his article suggesting, in a rather muddled manner, that what previously had been thought of as transuranium elements might actually be forms of elements much lighter than uranium. He offered no explanation or possible process that could account for this result.

In the meantime, Meitner continued to ponder the curious results. She visited her friend Eva von Bahr-Bergius over Christmas in Kungsälv, outside of Gothenburg. Here, she was joined by her nephew, Otto Robert Frisch, a young physicist at Bohr's institute. The two went out into the snow-covered landscape to discuss the uranium puzzle. They talked as he cross-country skied and she walked alongside. Using an earlier theory by Bohr that conceived the atomic nucleus to behave like a drop of liquid, they arrived at a model for how the heavy uranium atom could split after absorbing a neutron: not a shattering of the nucleus, rather a splitting analogous to an unstable drop of liquid in which the surface tension can scarcely hold it together. The added neutron induces the final instability that provokes the drop splitting into two parts and releasing an enormous amount of energy that shoots the fragments outward. Even without reference works, they could make some quick calculations and theoretical assumptions.

Back in Copenhagen and Stockholm, respectively, Frisch and Meitner continued their exploration of this new process. Frisch was able to conduct a series of experiments to confirm their conclusions. Inspired by a biologist colleague in Copenhagen, Frisch suggested the term nuclear fission, by analogy with the process by which a living cell splits in two. Meitner concluded that the other product of fission must be an isotope of the element krypton, which would decay into a series of other elements. She communicated all these results to Hahn, who had not identified the krypton, had not suggested any mechanism for splitting, and had not considered the release of energy.

Frisch discussed their conclusions on January 3 with Bohr. Although Bohr had been working on a liquid-drop model of the nucleus for years, he immediately and excitedly accepted their results and allegedly exclaimed: "Oh, what fools we have been! We ought to have seen that before." Bohr urged them to publish this sensational insight as quickly as possible. This they did, sending a manuscript on January 16 to the British journal *Nature*. But it was not quick enough. Bohr left immediately thereafter for the United States to spend two months at the Institute for Advanced Studies and to attend a number of physics congresses. Although he had promised not to say a word until the paper was published, he could not keep quiet. Before the paper was in print, word spread like a chain reaction. Bohr's leak jump-started widespread efforts to repeat and extend the experiments and measurements. Many physicists began devising theoretical and experimental research programs on fission. The original insight, so difficult to arrive at, was taken to be almost intuitively obvious—once understood. Meitner and Frisch's contribution to fission was soon buried in an avalanche of articles, reports, notes, and letters to the editor of scientific journals. By the time Meitner and Frisch's article appeared, it was no longer sensational news. Soon came the self-imposed secrecy among Allied nuclear researchers in an effort to keep the Nazi war machine from creating a bomb.

Even more disturbing for Meitner, Hahn's subsequent publications and personal communications showed the start of a process by which he cut her out from any role in the discovery. After she related to him in early January the result of her and Frisch's efforts to make sense of the chemical findings, he incorporated these insights, including the krypton decay series, into his and Strassmann's next article. Again, though, he did not provide any hint of his communications with Meitner. Hahn was, of course, under great pressure and stress. Physicists at the institute, members of the Nazi Party, were annoyed that they were kept in the dark. But just as the agitation against Hahn was reaching a peak, it also became clear that the discovery of fission was the "miracle," as Hahn described it to Meitner, that could protect him and his laboratory. To the physicists, he insisted that the discovery was purely chemical—there was no physics at all involved—and by claiming to have resolved a problem that pointed to important military and economic applications, he could remain securely in charge of his department. But when he started to retell this version of the story to Meitner, she despaired.

Just a few weeks earlier, while Hahn was still unsure of himself and unable to make sense of his and Strassmann's analyses, he turned, as he had done for years, to Meitner for assistance. Hahn even asked Meitner at Christmas if she had something she could publish at the same time so the three of them who had always been a team could still be seen as being behind this work. Naturally, they could not publish together; Meitner, a Jew who illegally fled Germany, was an enemy of the Reich. But once the phenomenon of fission was clear, and once Hahn understood that fission was critical for his professional and personal security under the Nazi regime, he began, first haltingly and then more decisively, to take credit for himself. He claimed that the discovery had nothing to do with physics and that Meitner, the physicist, would probably not have allowed him to make the discovery.

Meitner was left holding the bag—with the transuranium elements. She understood, even in January while she was exhilarated by Hahn and Strassmann's chemical results and her and Frisch's explanation, that the course of events made her look bad. It could easily appear that, as long as she had been in Berlin, they were mired in confusion; upon her leaving and without her interference, the chemists found the truth. Hahn made almost no reference to Meitner in his articles, and he made no mention of their communications leading up to the crucial publications. This made good sense while he was in danger, but as historians Sime and Krafft have shown, he continued to propagate this myth even after the war, when he had ample opportunity to set the record straight. Meitner feared that her new Swedish colleagues would interpret events in this manner, especially Siegbahn. As it turned out, this was the case, but what she could not have guessed was that some of her new colleagues simply did not want to find out what she had contributed. At first, she was denied the chance to continue her research, then she was to be denied a share in the Nobel Prize and in the leadership of Swedish nuclear research.

## Adding Insult to Injury

In 1938, Enrico Fermi had also interrupted his work. He used the occasion of his Nobel Prize in 1938—in part for his discovery of the alleged transuranic elements—as a means to escape, along with his Jewish wife, from fascist Italy via the prize ceremony in Stockholm to the United States. There, he was treated as a precious gift to American science, which was certainly the case. Meitner had also just become a refugee, but she received quite a different welcome in Sweden.

In the 1930s, Sweden was by no means a friendly haven for refugees from Germany. In the university towns of Uppsala and Lund, massive student demonstrations, supported by some professors, demanded the doors be kept shut to Jewish academics and medical doctors. Oseen used the term "our Jewish problem" with respect to two refugees who would need aid in the form of grants from the committees' special fund: Arrhenius's German Jewish nephew, Ernst Riesenfeld, and Lise Meitner. In spite of this remark, Oseen was actually one of the few in Sweden who extended hospitality and friendship to Meitner. Siegbahn was another matter.

Both Meitner and Siegbahn were used to being in charge. Both were strong and silent, reserved but determined, proud and stubborn. Siegbahn acquiesced to make room for Meitner after pressure from Bohr, Coster, and Oseen. Unfortunately, bad personal chemistry was compounded by their strong differences in scientific style. He focused on instruments and perfecting their operation; she engaged experiment for developing and confirming theory. Siegbahn expected all members of his scientific staff to be responsible for their own experiments. His was a hands-on laboratory culture; a researcher designed instruments and conducted experiments without an assistant. Meitner was used to being the brains who conceived and designed experiments, but she almost always needed someone to carry them out. In the hastily written correspondence, those who tried to aid Meitner neglected to mention to Siegbahn that she would need an assistant.

Siegbahn considered her a burden from the start. He did not know what to do with her: "She can do nothing with her hands." But that was only a minor part of the problem. Her requests for a budget, apparatus, and an assistant quickly resulted in a strained relationship. After working for years to raise funds for his new institute, he considered her demands on his laboratory an imposition, a breach of his moral order of things by which she, who did not work to build up the laboratory, had little right to its resources.

After the trauma of a dramatic escape from Germany, and after breaking away from highly significant research, she might well have been insensitive to Siegbahn's proprietary feelings and more generally to Swedish culture. But, having been a professor and head of a prominent research unit in Berlin, Meitner did not expect to be treated as a dependent student. Siegbahn had no intention of giving her either authority or resources. She had to ask him for even the smallest of appropriations and for permission to use any instrument. When Meitner expressed her disappointment to her old friend Eva von Bahr-Bergius, a highly talented physicist but without a professional position, she in turn contacted Oseen to see if anything could be done. Oseen arranged that Meitner be given a small annual grant from the Nobel committee's special fund to support an assistant. But this plan also faltered. The sudden demand for physicists in industry and schools resulted in difficulties in finding a recent graduate who could serve as her assistant. When a potential assistant did come to the institute, Siegbahn recruited him for other purposes.

Already dejected by her abrupt move to a new country, Meitner became thoroughly depressed. She was given a room in a new institute where very little research was being done or being planned. She had no access to funds, instruments, or assistants; she was completely dependent upon Siegbahn. He was absorbed in plans for building a cyclotron and not especially concerned with securing resources for a talented refugee. She, of course, had her own definite ideas on how to organize and run a laboratory. It was clear from the start that a rocky path was in store for her. As Oseen observed, "It is as yet unclear how this can be straightened out. Perhaps Professor Meitner must seek another home and another institute."

Within a short time, she felt more like a windup doll, as she put it, acting mechanically, rather than an internationally prominent researcher. She understood that Siegbahn was not a nuclear physicist, but he was the chief. And now she was stymied, and worst of all she was absent from Berlin just when years of joint research were yielding significant results. She knew Hahn could not mention her contribution in a German publication, but she feared that the emerging version of the discovery would make it even more difficult to restore her reputation while she was stranded in Sweden. Although those in the know and those who closely followed developments in this field understood that Meitner was a major contributor to the discovery of fission, those responsible for evaluating this work for a Nobel Prize refused to accept that she could have played any role of significance.

### Prizes to Anoint Leaders

Confusion over just what had been discovered and by whom in nuclear research began in 1939. The clandestine communication between Hahn and Meitner was just the start; secrecy became the norm as significant results were withheld from publication during the war years. Even after 1945, the record of accomplishments was not immediately obvious. The Royal Swedish Academy of Sciences and the Nobel committees had ample reason to pause until well after the war and to wait until greater certainty was at hand. A mere trickle of nominations during and immediately following the war, difficulties in following foreign research journals, and finally an understanding that an enormous amount of research had been kept secret should have made it clear that even the most conscientious committee might have trouble making fair judgments. But more than ever, what was at stake did not entail identifying the most worthy contributors to physics and chemistry.

During the war, the physics and chemistry committees evaluated the few nominations that arrived in Stockholm. Although potential candidates were identified, as long as the international situation was so unstable, the prizes were postponed and reserved. In general, Sweden's close relations with Germany—and fears of being invaded as Norway and Denmark had been—prompted vigilance and an avoidance of unnecessary friction. The wartime coalition government allowed the German military concessions that were in violation of Sweden's declared neutrality. For many members of Sweden's cultural, scientific, and commercial elites, the traditional affinity with Germany continued, as it had after 1933, regardless of their feelings about the Nazi regime. Historians have noted that after November 1942, Sweden began redirecting its political course westward, toward the Allied nations: German defeats at Stalingrad and El Alamein shattered the common belief in Sweden that Germany's victory was certain.

Little is known about the political beliefs of the Academy's members, but some clearly continued to promote relations with German science. Visits from and to Germany were not uncommon. But in the Academy, 1943 also marked a change in attitude. Even though the government desired, for neutrality's sake, that the Nobel prizes could continue to be withheld in 1943, the committees made it clear that they wanted to award them. Old-timers, such as Hjalmar Hammarsköld and Ragnar Sohlman, who had been through the debates during the First World War, were skeptical whether the committees could obtain impartial proposals and a sufficiently large sampling of nominations. Undoubtedly, some committee members became restless after reserving prizes since 1940; the backlog of potential candidates was growing. The case of erroneous evaluation based on inadequate information—such as of the Haber-Bosch process—however, was clearly forgotten, or ignored.

Errors in the evaluation of Meitner and Hahn for the chemistry prize beginning in 1939 provided a basis for subsequent treatment of their candidacy; there was opportunity later to redress misunderstandings, but the will to do so was lacking. For its part, the physics committee tried to avoid evaluating fission; it conveniently used the chemistry committee's negative assessment of Meitner's role to ignore the question of fission's importance in physics. That was the pattern prior to 1945. After the war, more overt measures for hindering Meitner's recognition became necessary, once the stakes and the available information grew.

Svedberg's deep interest in nuclear science resulted in the chemistry committee taking up the issue of fission soon after the announcement of its discovery. Previously, Svedberg had been evaluating Fermi's work on the transuranium elements for the chemistry committee, but he allowed the physicists to claim him. In January 1939, immediately after the initial announcements related to fission, Svedberg nominated Hahn and Meitner for the chemistry prize. Svedberg then evaluated their contributions. In his report, he concluded that the work on the transuranium elements was now understood to have been erroneous. He noted that Meitner and Hahn were not the only ones who had been mistaken; highly qualified researchers such as Nobel laureates Enrico Fermi and Irène Joliot-Curie had made the same error. But he cut Meitner out of any role having to do with fission. Pointing out that Hahn's latest studies were made after the collaboration allegedly ended, these "are without doubt of fundamental importance for the science of atomic nuclei." He implied that Hahn could make the discovery of fission once Meitner left. A divided prize based on the discovery of fission would be out of the question. Furthermore, he claimed that the role of elucidating the theory of fission belonged largely to Bohr. Svedberg concluded that because of the rapid developments in the field, the committee should wait before making any decision on awarding a prize for fission.

In 1940, nominators provided a corrective. This time the physics committee had to take a stand. Former prizewinner Arthur Compton proposed Hahn and Meitner for a divided physics prize: "As I understand the matter, Professor Hahn and Fräulein Meitner should be included in the award for their work respectively in identifying the fission process and in showing the tremendous energy liberated when the fission occurs." He admitted that it was "difficult for me to judge" whether their respective collaborators, Strassmann and Frisch, should be included. Compton considered the "discovery" of fission to consist of both Hahn's chemical and Meitner's physical results. Had he been unsure whether his proposal was justified, he would have added a note to this effect. That was his style. Compton did not nominate without careful reflection; he tended to confess reservations. Typically, he added in this same letter that the only other recent discovery worthy of consideration would be the new subatomic particle, the mesotron, which was not fully understood and therefore "too early to make the award intelligently."

In response, the physics committee simply referred to Svedberg's earlier evaluation that concluded it would be best not to make a decision as yet. The physicists added that nothing new has happened to change that view; the committee would not evaluate them.

In 1941, Compton's opinion received further confirmation. In a detailed letter of nomination, laureate James Franck presented the case for dividing a physics prize between Hahn and Meitner. Franck was one of the more morally upright physicists of his time. Although he had not been immediately threatened, he resigned his professorship in Göttingen as soon as the Nazis started dismissing less prominent Jewish and dissident physicists. After some years at Johns Hopkins, where he found anti-Semitism prevalent—as was the case in most elite American universities of the time—he again resigned, moving to the University of Chicago. Franck claimed in his letter of nomination: "I do not need to emphasize the importance of this discovery which is certainly the greatest in physics in the last ten years, but I would like to explain why I think that Hahn and Meitner should be honored together." He reviewed their thirty years of teamwork, and especially the tight collaboration up to the very last step prior to the discovery. But because Meitner had to leave Germany,

> she was not co-working in the paper which Hahn published with Strassmann, which actually contained the solution, but Hahn himself did not draw the consequence. Lise Meitner did it in collaboration with Frisch, and she was the

first to see the whole importance of the result and drew the consequences that the fission products should fly from another [sic] with tremendous energy. She and Frisch were also the first to observe this fact experimentally.

The physics committee again asserted that the discovery was more suitable for chemistry and refused to take it up for evaluation.

Although the chemistry committee also received nominations in 1941 calling for Hahn and Meitner to share a prize, Svedberg again concluded that Hahn alone deserved to be rewarded for fission. He now began specifying Hahn's work as having great significance for nuclear chemistry, and he claimed that Hahn's group continued to produce excellent work. Obstinately, as if to emphasize that Meitner could not possibly have played a role in fission's discovery, he underscored that, in contrast to Hahn, she had not produced any work of great significance during the past two years. He did not mention why she had not shown continued creativity. Although she and Frisch had, in March 1939, performed in Copenhagen the definitive analyses showing that the transuranic products were indeed nothing more than the products of fission, Svedberg erroneously gave credit for this confirmation to Americans. He also mistakenly claimed that the work she or she and Frisch published immediately after Hahn's publication has "not exerted to any significant degree an influence on the development." Svedberg concluded that only Hahn and, in part, a number of Americans had been leaders in fission research. Theoretically, Bohr "has shown the way." Svedberg had spoken; his colleagues now knew what he thought.

One chemistry committee member, who had at first himself nominated Hahn alone, began to have doubts. In 1942, Wilhelm Palmær openly expressed his uncertainty over the committee's one-sidedness; he wanted a second look, and he nominated Hahn and Meitner: "It seems to me clear that it would be in accordance with the demands of fairness to let, if possible, both researchers divide an eventual prize." He was to submit his own evaluation, but he died before it could be written. Instead, Svedberg's long-term friend and ally Arne Westgren assumed the responsibility, even though he had already made his opinion clear by having nominated Hahn alone.

Westgren's report revealed an unwillingness to consider Meitner's case. He did not accept Palmær's and others' nominations hinted at a case where Hammarsten's principle applied: Wait until certainty can be attained, rather than risk leaving out a deserving researcher from an award. Instead, he wrote his evaluation based upon the supposition that Hahn alone should be given a prize.

Admitting that had Meitner remained in Berlin she certainly would have been part of the discovery, Westgren wrote that, unfortunately for her, she did not. He ignored the issue raised in some of the nominations that it was Meitner's work that actually identified the physical process of fission along with the enormous amounts of energy released. By and large, he repeated Svedberg's evaluations, except for one point, which indeed Svedberg had also changed after his original 1939 evaluation. At first, Svedberg rightly maintained that several leading nuclear scientists propagated the error of the transuranic elements. But now Meitner remained alone, a scapegoat. No mention was made of the others. Westgren claimed that Hahn and Meitner's transuranic "mistakes" should be kept separate from Hahn's experiments on splitting uranium. Even without knowing the private communications between the two of them, an unprejudiced review of the literature would suggest otherwise. In asserting that the earlier work on the transuranic elements in no way led to, or made possible, the discovery of fission, Westgren ignored or missed the intimate linkage of Hahn and Meitner in terms of the research problem and experimental methods. The report contributed to the illusion that Meitner was the sole reason fission had not earlier been discovered. The committee agreed once again that Hahn should receive an undivided prize—but because of international tensions, it recommended reserving the prize.

In 1943, the situation remained the same with respect to the chemistry prize. Westgren alone nominated Hahn and simply repeated his conclusion. For the physics prize, Siegbahn nominated Hahn. Franck repeated his proposal of Hahn and Meitner again and reminded the committee that Hahn and Strassmann's article only hinted at fission; Meitner and Frisch were the first to identify the process and confirm it. In an odd decision, the physics committee again refused to evaluate. It wanted the chemists to judge, even though Meitner was nominated only for the physics prize and for the significance of fission for *physics*. Of course, the physicists knew full well that the chemists had not previously evaluated the physics of fission; in fact, the chemists had completely ignored all aspects of fission related to its being part of physics. Siegbahn and the rest of the committee also knew that the chemists had no intention of recognizing Meitner's contribution. She had already been eliminated. The physicists made no effort to bring into focus what Franck had called the most important discovery during the past decade in physics.

In 1944, Westgren again nominated Hahn, and Meitner was not mentioned in the report. Hahn had visited Sweden in 1943; being a German patriot but anti-Nazi, Hahn made a good impression. Westgren praised Hahn in the report for the continued excellence of his work, and in the spring meeting the committee proposed him for the 1944 prize. The committee added that if, when the Academy voted, the political situation still prohibited Hahn from accepting the prize—that is, if the Nazi regime and its ban were still in place—then it recommended reserving the 1944 prize. Once again, the only proposal for Hahn was Westgren's. In contrast, several strong nominations continued to arrive for the leader of British organic chemistry, Robert Robinson. Why the rush? Were the Swedish chemists already considering Hahn as a possible leader for German science once a de-Nazified Germany began rebuilding? Was Svedberg getting ready to launch a major effort to claim nuclear research as part of chemistry? But the Nazis' ban held; the 1944 prize was reserved.

At a meeting in June 1945, the committee agreed to propose giving the reserved 1944 prize to Hahn and the 1945 prize to the Finnish chemist A. I. Virtanen. The European war had ended in May 1945; then the atomic bomb helped bring the Pacific conflict to an end in early September. Soon it became clear how significant fission had been, as well how much secret research on fission had been done. After the war ended, Westgren and Svedberg received a double "aha" experience. First, they obtained missing American scientific journals, which, surprisingly, contained few contributions related to fission; then they learned these were but a tiny visible tip of a gigantic secret research endeavor mobilized to create the atomic bomb. The chemistry committee made an about-face—it would be best not to rush a decision.

The committee agreed on September 10 to propose reserving the 1945 prize and giving Virtanen the reserved 1944 prize. The reason given was that some of the

Allied work could possibly compete with the discovery of fission for worthiness of a prize. Guidance could come with the next year's nominations, especially from former prizewinners who had direct contact with wartime developments, such as Bohr (who fled from Denmark to the United States), James Chadwick, G. P. Thomson, and Harold Urey. In the committee protocol, the next sentence originally stated that this postponement would shed light on whether Meitner should divide the prize with Hahn. But this was certainly not to Svedberg's liking; Meitner's name was subsequently crossed over and replaced with "other researchers." The Chemistry Section agreed with the committee to wait for more information.

But when the Academy deliberated, a member of its Medical Section, Göran Liljestrand, objected. He argued that the Academy must now give the prize to Hahn. The reasoning was convoluted, but effective: By not awarding Hahn it would appear as if the Academy were being influenced by the Americans, who ostensibly would not look kindly on giving a prize to a German, especially one who was being held in isolation along with other captured nuclear researchers.

It is hard to know what really happened at the meeting and what guided a majority of the Academy's members to give Hahn the prize in 1945 against the recommendations for postponement. Liljestrand could not alone have mobilized the Academy to disregard an unambiguous and unanimous proposal from its authoritative members that fairness required waiting. Other powerful persons in the Academy certainly had reason for wanting to see Hahn alone rewarded for the discovery of fission. Siegbahn and his followers had ample reason for wanting a quick decision on fission that gave Hahn all the credit.

### Fallout: 1945–1946

Lise Meitner and Manne Siegbahn disliked each other. She made no secret of her feelings. She had little patience with his obsession with bigger—and more difficult to operate—instruments at the expense of research. Money that could go to personnel went instead to technicians and technology. Whatever Siegbahn actually thought of her as a researcher and person, he did see Meitner as a potential threat to his plans in that she knew more nuclear physics and enjoyed greater prestige abroad. He treated her shabbily and without respect. She was kept in the dark about events in the laboratory, such as the procurement of instruments important to her work. Once the cyclotron finally started working in 1942, she had to apply for time to use it, but virtually none was made available to her. Siegbahn preferred to prioritize the production of isotopes for medical use. She had insight and experience with nuclear physics, but he had power, authority, and unbridled ambition. The Nobel Institute was from its first visionary birth an institution where internationalism was supposed to flourish. Again and again, committee members had spoken of Nobel laboratories where foreign researchers could come as guests or staff. But as with so much of the idealism surrounding the Nobel legacy, the harsh realities of narrow professional interests at times undermined these noble aspirations.

Against this background, the prospect of massive funding that commenced following the dropping of the atomic bombs made Meitner a problem for Siegbahn. A new generation of physicists was coming into positions of authority; some had contacts with Meitner and held no allegiance to Siegbahn. Some were much closer to the emerging postwar Social Democratic political power base. Central among these were Lund University physicist Torsten Gustafson and Tage Erlander, who had studied physics in Lund, served as the wartime cabinet minister for education and research, and became in 1946 Sweden's prime minister. Gustafson was Erlander's informal adviser on nuclear matters; during the war, he put the politician in touch with Bohr, Oskar Klein, and Meitner. After the war, Gustafson continued to have access to Erlander and tried to guide policy. That Klein and Bohr, with whom Meitner had close relations, had access via Gustafson to the highest levels of decision making could not have pleased Siegbahn. The stakes soon became clear.

Following the bombing of Hiroshima and Nagasaki, the supreme commander of the Swedish Armed Forces gave top priority to research on the atomic bomb. An immediate half million crowns were put at the government's disposition during the fall of 1945; the Defense Research Institute followed up with a petition for 1.2 million crowns in the 1946–1947 budget for research related to atomic weapons. The government also established in November an advisory group, based largely of scientists, the Atomic Committee [*Atomkommittéen*]. Officially, little was mentioned with respect to either atomic weapons or energy; recognizing the sensitivity of the matter, not the least diplomatically, the expression "different practical applications" was commonly the given goal for creating a program for research and development. From the start, the heavy representation of academic scientists on the committee resulted in exaggerated claims for the need to shore up basic research. Political scientist Stefan Lindström, who pioneered a historic study of early Swedish atomic energy policy, noted scientists' efforts at the origins to downplay the technological problems and to exploit the opportunity to support academic-based "pure" nuclear research.

Researchers began positioning to claim shares of the expected bonanza of state support to the nuclear effort. Things were about to happen, but just what, was unclear. Siegbahn had been chairman of the wartime committee to coordinate defense-related research; he now became a member of the new committee (as was Gustafson and soon also Svedberg). In the fall of 1945, he was once again anxious over prospects for raising large-scale funding for a new, huge cyclotron. Certainly, Siegbahn's institute was already the best-equipped laboratory in Sweden for nuclear research. Critics feared that it could easily dominate to the extent that would not be healthy for the field. It was just this prospect that prompted Gustafson to express concern that critically important nuclear research will be "monopolized" by Siegbahn and his institute.

Competitors challenged Siegbahn's position. Several researchers had designs for major initiatives in nuclear research and for claiming a say in national policy. Svedberg had plans for Uppsala, although he was in principle willing to begin by trying to gain access to Siegbahn's laboratory. Gustafson wanted to create opportunities in Lund. And leaders of military research institutions were also making noises about needing adequate facilities. Oskar Klein hoped to create a nuclear physics research unit connected with his institute at the Stockholm Högskola. He wanted to recruit Meitner as a professor and senior adviser; given her age, she could not be expected to take on major administrative tasks of leading the initiative. In discussing plans with Bohr and some Swedish colleagues, Klein suggested that maybe Frisch, who was a refugee in Britain, could be recruited as leader. Here, then, was a challenge to Siegbahn's dominance. Not only would this research unit possess real expertise in nuclear physics, it would have very close ties to Bohr and his institute as well as potential support from several members of the Atomic Committee, including Gustafson.

Siegbahn had good reason to be worried. Meitner's enormous international reputation finally became clear to Swedes after the end of the war. Media focus upon her at home and abroad, as well as the now well-known stories circulating within the scientific community of the poor treatment she received in Sweden, undoubtedly rattled Siegbahn. For him, she was a ticking bomb. Her closest relationships in Sweden were with persons—such as Klein and Hans Pettersson—outside Siegbahn's network, and even hostile to him. Now she and those who appreciated her expertise were in contact with Erlander. Klein supported Meitner for a prize and for spearheading a rival center for nuclear physics. Moreover, Klein was finally elected to the Academy's Physics Section early in 1945. And if that was not troubling enough, the Swedish parliament indicated its reluctance to provide the extra funding Siegbahn requested for his massive cyclotron.

In 1945, the physics committee had to face Meitner and Frisch as candidates. Nominators Klein and Bohr hoped Hahn and Strassmann could divide the chemistry prize and Meitner and Frisch the physics prize. This would be the fairest distribution of recognition for this epoch-making discovery. Again, the physics committee refused to accept the responsibility of evaluating work on the physics of fission. It again claimed that the chemistry committee had this responsibility.

Moreover, Ivar Waller, Oseen's former student, who replaced him on the committee, was now in a position to undo his former mentor's unfair treatment of a candidate. Waller overturned Oseen's intransigent, antagonistic refusal to award a prize to the brilliant but arrogant theoretician Wolfgang Pauli (see Chapter 14). The committee agreed on Pauli; it was a choice that Klein would certainly embrace. But a prize to Pauli did not eliminate the threat of empowering Meitner through a subsequent prize for fission.

Indeed, Klein was heartened at first when the chemists recommended waiting another year before reconsidering Hahn and fission. Maybe next year, he mused to Bohr, both prizes could still go to those responsible for fission. But when the Academy suddenly disregarded the chemists' recommendation and gave the chemistry prize to Hahn, Klein understood that the task of achieving fairness would not be easy. The physicists could continue to ignore fission.

Meitner did not have just Siegbahn's opposition to worry about. Committee member Erik Hulthén, Klein's colleague at the Högskola, was a problem. Hulthén was well aware of Klein's plans for Meitner and for his own institute, and he opposed them. Hulthén made an agreement with Siegbahn in November 1945 to prevent the introduction of nuclear physics at the Högskola. But even if Hulthén and Siegbahn blocked this plan, other schemes were under discussion for rescuing Meitner from neglect in Siegbahn's institute and using her expertise to establish nuclear physics elsewhere. A number of physicists discussed possible scenarios—also with Erlander—to create a professorship for Meitner at the Royal College of Technology or at Lund, or maybe to create a position for her at the Defense Research Institute. In the meantime, Siegbahn did not know whether he would get the money for his huge cyclotron. Perhaps that was why he went to the press and announced, in November, that if he were given enough money, his institute could build a Swedish atomic bomb within weeks.

When the Nobel Committee for Physics met in February to begin assessing the nominations for the 1946 prize, Meitner and Frisch were again among the candidates. Klein and Bohr proposed them for their role in clarifying the physical process of fission and especially for being the first to discover the great release of energy during the fission of heavy atoms. Among others, Max von Laue, James Franck, and the prominent Norwegian theoretical physicist Egil Hylleraas also proposed Meitner or, alternatively, a prize shared between Meitner and Frisch.

Although they knew the prejudices against Meitner, Bohr and Klein tried to set the record straight. They knew from direct contact with the participants the actual events leading to the articulation of fission during the winter of 1938–1939; they tried to explain why the published record was inadequate for assessing credit. Although the committees had claimed that Bohr was mainly responsible for the theory of fission, they revealed to the committee the informal communications that enabled Bohr and American colleagues to begin work on fission in January 1939 before Meitner and Frisch's pioneering article appeared. They also countered in detail the charges that others also had produced the same results almost simultaneously. Franck's proposal further provided a corrective to the illusion that Meitner had little to do with the discovery of fission. The committee could no longer claim ignorance and simply hide behind the misleading wartime record of publication or rely exclusively on Hahn's version.

In even greater detail, Hylleraas argued against cutting Meitner out from recognition for the discovery of fission. Aware that the committee would be tempted to claim that Hahn's chemistry prize already took care of fission, Hylleraas wrote that fission was of such great significance for physics that "it would be striking [*påfallende*]" if it was not rewarded with a Nobel Prize in physics. In reviewing the literature on fission, he pointed out that even the published record showed Meitner's integral role in the research program that culminated with fission. Reaching back to 1934, he claimed that clearly it was Meitner who had mobilized the Kaiser Wilhelm Institute to enter uranium research so quickly and effectively. Then he followed developments to 1939, elaborating the circumstance by which Bohr's visit to the United States in 1939 set in motion America's rapid entry into the field of fission studies. Said Hylleraas: "It has become all in all very clear what significant historic role Lise Meitner in this case has played." Both Hahn and Meitner merited prizes. Then, alluding to news and rumors from the Academy, he let his annoyance spill into the letter: While conceding that awarding Nobel prizes in the natural sciences certainly differs from those in peace and literature, he was under no illusion about bias here as well: "I am nevertheless aware that natural science is completely lacking an unambiguous, objective standard [*målestokk*] for awarding prizes. It can therefore be that my own subjective view on the significance of Lise Meitner's contribution will not gain the requisite support [*fornödne tilslutning*]."

The committee allowed Hulthén to write a special report on Meitner and Frisch. Hulthén understood full well what was at stake; in the midst of positioning and bargaining over the future of Swedish nuclear research (new laboratories, new research programs, and new authorities), a prize to Meitner along with the inevitable media attention would propel her into a position that would be hard to assail. Could she then be kept off commissions, denied significant funding and ability to lead research projects, and even be excluded from the Nobel committee? Previously, Siegbahn's, Svedberg's, and von Euler's prizes had shown how local laureates garnered immediate prestige and authority. Klein and others were hoping to capitalize on Meitner, and even Frisch, to establish a significant research unit in nuclear physics, if not at the Stockholm Högskola, then elsewhere. Moreover, by winter of 1946, it was no secret that Meitner held strong ethical convictions. She opposed nuclear research for military purposes; she sided with Bohr and a growing international movement to ban nuclear weapons. Hulthén and Siegbahn both had connections with the military's research efforts. And, of course, Siegbahn was not only nervously waiting the outcome of his petition for an extra appropriation but also looking ahead for further funds for staffing his institute and embarking on actual research.

In addition, Hulthén had something in common with Meitner. He was also left holding the bag of transuranic elements. He had written the evaluation in 1938 for Enrico Fermi that recommended a prize for him, in part for the transuranic elements. Fermi could have been given a prize for any number of extraordinary contributions in both theory and experiment. It was Hulthén who had reversed the committee's earlier cautionary conclusions to wait for further confirmation before using this discovery to justify a prize. Hulthén's evaluation resulted in an embarrassment for the committee and Academy. He might have used the opportunity in 1946 to argue that virtually all major researchers in nuclear studies had tentatively accepted the reality of Fermi's transuraniums. He could also have set the record straight based on the leads provided in the nominations that the discovery of fission grew out of efforts to understand the physics of nuclear processes. Instead, Hulthén dug in. He conveniently emphasized that Meitner had been a hindrance to fission. He added that Meitner and Frisch's work had no special significance, but was merely one of many simultaneous contributions, in fact of less importance than many others. He scarcely addressed the points raised in the letters of nomination. He insisted that he was compelled to evaluate based on the published record; he disallowed Bohr's testimony, he refused to look any further. No grounds existed, he concluded, for giving a prize to Meitner or to Meitner and Frisch. Whether out of conviction or of expediency, Hulthén understood full well that in defense of his claims, he could rely on the committee's strongman, Siegbahn.

The committee agreed. Although the Academy had just declared it did not want to appear to be getting too cozy with the United States, in 1946 it did just that. After having rejected Harvard professor P. W. Bridgeman for more than fifteen years, the committee decided that 1946 was a good time to heed the Cambridge, Massachusetts, lobby. But when the Physics Section took up the measure, Klein was waiting. He attacked Hulthén's report, pointing out errors, distortions, and omissions. Klein stood alone. In the Academy, he lost his temper.

Perhaps compounding the shame Klein felt over how shabbily his countrymen were treating Meitner, and the committee's refusal to give her a fair hearing, his own frustration with the Academy's physics establishment fueled his outburst. In spite of his strong international reputation, he was repeatedly denied a place on the committee and, until recently, in the Academy's Physics Section. Indeed, he would soon be passed over again, as Siegbahn and others preferred to recruit less quarrelsome and less intellectually gifted colleagues. Klein no doubt felt the absurd imbalance between reputations in the Academy as opposed to those internationally. For example, when the prestigious Solvay Conference soon met for the eighth time since its founding in 1911—this time to discuss new results and problems in the study of subatomic particles—Sweden was for the first time since 1921 invited to send delegates: Klein and Meitner. But probably more to the point, Klein was seeing firsthand how inconsequential factual knowledge and a sense of fairness were in the face of—to use Ibsen's phrase—a "compact majority" whose own interests were threatened.

Klein blasted the errors and misconceptions in Hulthén's report; he attacked the committee's prejudice. But even if Klein was right in his charges, he could not expect the Academy to turn against its committee in this case. First, he was all too emotional and antagonistic; second, to give a prize to Meitner—an outsider in its midst—the Academy would thereby insult its professor of physics and national leader of the discipline, Siegbahn.

Over the next few years, Klein and others continued to try getting Meitner and Frisch a prize. Bohr understood the impossibility of overcoming Siegbahn's and others' opposition; he mistakenly thought the chemists might be more open to consider Meitner. Actually, that was not the case: The chemistry committee members washed their hands of the matter; Meitner's or Meitner and Frisch's work belonged to physics. Moreover, the chemistry committee declared that through the Academy's decision to award Hahn alone, "a clear definitive position" had been taken on the question of awarding the discovery of the process of nuclear splitting.

The case was closed. Both committees refused to take up the question again. For those who might have waited, the following year brought nominations for Meitner again from Compton, Maurice and Louis de Broglie, and Planck, among others. Perhaps what mattered more was encapsulated in a *New York Times* article not long after the Academy's vote: "Sweden Aids Atom Study," which reported that the Swedish government allotted $1,748,000 mainly to enlarge the atomic facilities of Professor Manne Siegbahn.

Hahn came to Stockholm in 1946 to claim his prize. He basked in the attention that journalists and colleagues showered on him. Meitner remained in the background, a shadow. Hahn largely stuck to his version of the story. He did, however, nominate Meitner for a prize the following year. The chemists were so intent on awarding Hahn—not Hahn and Strassmann, not Hahn and Meitner—that Westgren renominated Hahn alone for the 1946 prize, in case he could not collect his 1945 prize in time. If the achievement was one of pure chemistry—which it wasn't—then Strassmann certainly deserved to share the prize. He was the chief analytic chemist and coauthor of each article. For the committees, the stakes were too high to accept that there was honor enough for all.

The mid-1940s Nobel prizes were not awarded on the basis of recognizing merit; instead, they had become to a great extent instruments in the politics of science. Hahn, whom Academy members all knew personally from his earlier visit in 1943 to Sweden, was to be one of the "good" Germans to lead the rebuilding of the once-proud national scientific community. Without the taint of party membership, and with the prestige of a Nobel Prize, he could help represent and guide German science under the Allied occupation. Hahn's great patriotism was well known; he long fretted over the future of German science once the Nazi debacle ended. His nationalism was of the sort that devoted friends of German science happily supported: no wallowing in apology, only focusing on the injustices done to Germany and claiming the moral high ground for not having managed to make an atomic bomb. "Good" was a relative term in the 1940s.

Meitner was promised a professorship and other inducements to remain in Sweden. Some younger physicists, along with Klein and Gustafson, pleaded her case. Somehow, petitions to the government got misplaced, or delayed, or hindered by formal rules. Finally, after further gaffes and postponements, she was given a small research unit in nuclear matters at the Engineering Science Academy's laboratory. Although she subsequently received many prestigious prizes abroad, her fate as a researcher was first crushed by Nazi hatred and then sealed by Swedish scientific leadership's insensitivity and self-interest. She eventually moved to Cambridge, England, where Frisch was working; she died there in 1968. The shibboleth that the search for truth transcends political realities and personal prejudices might be comforting for some, but rarely are such "truths" pure and clean, especially once money and authority become critical resources in the quest to know.

# FOURTEEN

## The Knights Templar

### Into the Age of Nobel "Geniuses" and the Banality of "Excellence"

When The Svedberg visited the United States in the autumn of 1946, he was astounded by the differences he saw since his last visit a decade earlier. Svedberg was no stranger to America's fascination with the Nobel Prize; nevertheless, he was not prepared for what he witnessed. America's Nobel laureates, he reported to an Uppsala newspaper, were treated as the "Knights Templar." Svedberg heard stories of how they had become a class unto themselves. Everywhere, they and their opinions were sought—on the buildup of scientific capabilities, but also on a multitude of military, social, and political issues. Publishers and documentary filmmakers were beginning to churn out portraits of prizewinners, drawing lessons for learners. And that was only the beginning.

During World War II and into the postwar era, scientists, and especially physicists, entered public awareness as never before. With the emergence of national security states, in which the sciences remained in a permanent state of mobilization, the amount of money directed to research and to training on both sides of the Iron Curtain, as in neutral Sweden, grew at astronomical rates. Historians are now trying to make sense of the influence of the Korean War and the cold war on the scientific enterprise. These entailed an explosive growth in numbers and research budgets, as well as the rise of massive hierarchical teams of researchers in elaborate laboratories. Whether in universities, private corporations, or government installations, laboratory directors became research managers. Major scientists consulted with military and political leaders, working together to define priorities for "basic" and "applied" research missions involving hundreds of millions of dollars. All the while, new, exotic, and increasingly fragmented research fields flourished as never before. Scientific journals grew in size and in number; and academic science departments either swelled to the point of splitting or became unwieldy institutional umbrellas for semi-autonomous specialized units. Science seemed to have the answers to all problems. But who spoke for science? Who represented physics or chemistry as a mass society within science began to grow?

The postwar culture of science mobilized the Nobel Prize for many purposes. With a massive scientific work force and enormous public awareness came "Nobel fever." An early symptom of this syndrome occurred with the American-arranged "Nobel festivities" at the Waldorf-Astoria hotel in New York in 1944 to celebrate the large number of Americans who received prizes in the first awards since 1939. Americans set the tone for the postwar infatuation with the prize. The language of "genius" and "excellence" spilled into media accounts of laureates and their institutions. Prominent in the past, the prize became an icon.

## BUSINESS AS USUAL IN AN UNUSUAL ERA

After the war, things became more complicated with respect to the prize, but not better. Parallel with the eruption of largely uncritical enthusiasm, the Academy and its committees resumed the work that had abated but not stopped during the war. During the first five years after the war, nominators, including national leaders and former prizewinners, provided no clear mandate for any candidate. An unambiguous consensus "voice" of international science—or even of the respective national leaders—was rarely heard. A wealth of worthy candidates was out there—but who should be given prizes? In the end, it was still the respective Swedish committees and the Academy that decided the outcome. The prize remained their prerogative. The war did not mark a dividing line in the history of awarding the prize.

Although post-1950 awards are still beyond the grasp of detailed study, owing to the fifty-year secrecy clause, indications suggest that, in spite of sincere and meaningful efforts at improvement, the process of selection remains flawed. The following paragraphs, a brief excursion into the history of the prizes since 1945, show continuities with the past and reveal new problems that have emerged, accompanying developments in science more generally. The cases of E. O. Lawrence and Lise Meitner underscore the continued growth in awareness of the prize's importance for the internal politics of science. The first years after the war show a number of "political" themes that entered into committee members' deliberations; these were not new, but now they seemed more evident. Subsequent efforts to better equip the committees with greater insight helped eliminate mistakes based on factual errors, but the task of selecting winners can never attain that objectivity and consensus the public assumes. In short, there remained a highly subjective and personal "politics of excellence."

### Changes in the Committee, Changes in Fortune: Pauli, Robinson, and Appleton

As always, the committees' composition proved critical. Oseen's death in 1944 allowed important changes to enter the physics committee. The election of Ivar Waller brought a theoretical physicist onto the committee who differed from his mentor. Waller was much more in contact with foreign researchers; he combined talent with far greater openness. Just before dying, Oseen again attacked Wolfgang Pauli's contributions to quantum mechanics, declaring that the so-called uncertainty principle and other interpretations of atomic phenomena were not physics but merely logical postulates, philosophical constructions. When Waller joined the committee, he took up Pauli's candidacy and argued in 1945 for rewarding his colleague. In the Academy, at least one member of the Physics Section questioned Pauli's suitability with respect to the bylaws, given that his contributions were almost two decades old and had long been accepted by the international physics community. There were few nominations for him. One was in a short telegram from Einstein, who was with Pauli at the Institute for Advanced Studies in Princeton. But no new discoveries were mentioned, no new revelations about the significance of the work, and no new outpouring of support entered into the decision. It was the change of committee membership that proved critical.

In chemistry, the story was no different. Robert Robinson, whom some nominators called the greatest living organic chemist, had long been a strong candidate. But, while he collected broad support, which continued into the war years, the committee had favored in its evaluations candidates whom its own members themselves were nominating, such as Georg von Hevesy, Otto Hahn, and A. I. Virtanen. During the war, von Euler again dismissed Robinson as "not worthy." In fact, he called Robinson's work unoriginal. Von Euler had long found it hard to accept an Englishman as a leader in organic chemistry—traditionally, the pride of Germany—and found Robinson's use of theory particularly distasteful. When the war ended, it became known that Robinson's distinguished career now also included work on penicillin. In 1946, von Euler grudgingly declared that he found Sir Robinson "worthy," based largely on his recent contributions to pharmacological chemistry.

But when von Euler stepped down from the committee—his past coziness with Nazi Germany was certainly a liability—a younger committee member, Arne Fredga, broke the prejudice against Robinson. In 1947, Fredga, a talented and attuned chemist, reached back to Robinson's earlier accomplishments and rebutted the committee's earlier, negative evaluations. Referring to Robinson's work on alkaloids and steroids as experimental "virtuosity," Fredga even found grounds to praise Robinson's previously maligned theory of organic reactions, although other committee members made sure that this accomplishment did not enter the formal justification for the prize. Robinson possibly understood why he had to wait so long; the title for his autobiography borrowed the term von Euler once applied to him: *Memoirs of a Minor Prophet*. Robinson laughed last.

The committee and the Academy continued also to see the prize as a private affair. In 1947, Henning Pleijel announced his retirement from the physics committee. He had long been chairman; four years earlier he stepped down from his position as the Academy's perpetual secretary. Professor of theoretical electrotechnology at the Royal Technical College, Pleijel wanted to celebrate his own specialty, by rewarding E. V. Appleton for his work in using radio waves to explore the highest reaches of the earth's atmosphere—the ionosphere—work that had led to long-distance telecommunications.

Appleton and his collaborators had discovered and investigated layers of electrically conducting ions that were inaccessible to direct observation. Understanding the physics of these layers, including their relations with solar activity, was necessary to hemispheric and polar communications during the war.

Although a few British colleagues—such as W. L. Bragg and O. W. Richardson—had nominated him, Appleton was repeatedly passed over. In 1935, the committee noted that too many persons had contributed to the field to single out any one or two researchers. In 1939, it admitted that even though Appleton could be declared the leading figure in ionospheric research, the overall significance of the work did not merit a prize. Nothing changed in the committee's view until 1947, when Pleijel decided to back Appleton. The one nomination that year for Appleton—from K. G. Emeleus of Belfast, who had once worked under him—added little new information.

Although Pleijel had left the committee, he asked to prepare a special report on Appleton. Bringing into new focus past details and evaluations, he concluded that Appleton deserved a prize. Even though a prize to Appleton would be seen as opening the door to geophysics, committee and Academy culture insisted upon respecting their dedicated retiring member's last wish. Siegbahn especially owed much to Pleijel. Even if he harbored doubts about Appleton, he knew who had helped him become the most powerful physicist in Sweden. Appleton's prize might have come as a surprise to many, but upon reflection most could accept the decision—at least no formal protests were forthcoming. At the Nobel ceremony, Appleton was extolled as the inheritor of the mantle of British genius from Isaac Newton, Michael Faraday, and James Clerk Maxwell. But the reason he won had more to do with circumstances in Stockholm than with international opinion; indeed, Appleton had been decidedly rejected prior to Pleijel's retirement gift.

## The Prize as Diplomacy: Virtanen and Blackett

Spokespeople for the Nobel establishment have always denied that nationality has ever played any role in the selection of winners. More often than not, this has been true. Nationality rarely plays a dominant role, but nationalism is more subtle. The chemistry committee rushed to give a prize to Fritz Haber in 1918 precisely because he was a German; nationality overshadowed evaluation and judgment. Svedberg and von Euler urged their colleagues in the early 1930s to give a chemistry prize to an American; the choice of which American reflected various interests. Prizes to Swedes were not color-blind to the blue and yellow flag. Sensitivity to the international gaze appeared when the Academy overturned its own committee and gave Otto Hahn a prize in 1945 in order to appear free from American influence.

After the Second World War, Americans became the hope and the problem. Earlier in the century, Swedish scientific relations with Germany dominated Nobel deliberations. Now, committees feared becoming American satellites, but they also wanted access to the gigantic American scientific enterprise.

In 1944, when Allied victory was assured, and the prizes again were given out, Americans dominated. Svedberg gave a lecture extolling American science and pointing to the importance of democracy and freedom in the advance of knowledge. True, it was hard to escape American influence. During the war, Americans began to dominate the nominating, and they continued thereafter. A whole colony of prize-winners from Europe, all having permanent nominating rights, had moved to the United States. Without question, the size and wealth of American science made for a hothouse for research. Svedberg took great pains to reject any suggestion that the choices for the 1943 and 1944 prizes had anything to do with diplomacy. He implicitly denied that Swedish researchers might have wanted to diminish the memory of Sweden's pre-1943 accommodation with Nazi Germany.

As much as Svedberg plied American relations, he was always wary of the prize losing its claim to international impartiality. Such statements had to be made, but they also had to be taken with a grain of salt. It is difficult to believe that Svedberg and his committee were blind to considerations of nationality. In fact, Svedberg feared that the prize might become lopsided. He seems to have wanted to balance the scales when in 1946 he suddenly nominated a Russian, Peter Kapitzka, for the physics prize and another Russian, N. N. Semonov, for the chemistry prize (to be shared with the British scientist Cyril Hinshelwood). His choice of a Russian for both prizes in the same year would suggest that, not withstanding public proclamations to the contrary, he did not ignore nationality. In this case, the committees did not back Svedberg. Nationality might not have been as important as other considerations, in the committees' decision, but possibly some members' strong anti-Soviet feelings could have entered the picture.

The case of A. I. Virtanen shows the prize as both prerogative and diplomacy. Although there was in the mid-1940s a backlog of first-rate nominees for the chemistry prize, the committee, led by von Euler, chose to pay homage in 1945 to a close friend, Virtanen, a Finnish biochemist. Virtanen was not a strong candidate. Having begun his career working in the Finnish dairy association's laboratory after World War I, he continued studying dairy problems when he became a professor in Helsinki. His training included a sojourn in the early 1920s in Stockholm where he studied bacteriology and enzyme chemistry. He remained in close touch with several Swedish chemists.

To produce high-grade milk, cows need fodder with a high protein-to-carbohydrate ratio. By studying the various conditions under which nitrogen fixation in protein-rich plants occurs, Virtanen hoped to find the optimal conditions under which bacteria in the roots convert atmospheric nitrogen into useful chemical compounds. While trying to resolve this complex problem, he developed a process for preserving the nutritional value of grain silage. During long winters in northern climates, stored fodder loses protein, vitamins, and carbohydrates; it also develops a taste repulsive to cattle. Using field and laboratory tests, Virtanen created a method that produced nutritious silage; through the winter, livestock remained healthier and produced better-quality milk.

Virtanen had few nominations. In the 1930s, the same two Finnish colleagues had occasionally proposed him. But even though his study of nitrogen fixation was incomplete, and the silage methods were still being tested, von Euler and Christian Barthel, who had trained him, declared him "worthy." During the war, von Euler continued to do so, based on his own nominations. After the war, several nominations came from Finland. While admitting that Virtanen's research was incomplete, the committee agreed that the silage method had been so important in Finland and abroad that it was itself worthy of a prize. Virtanen received the 1945 chemistry prize, although nobody outside a tiny circle of close Swedish and Finnish colleagues ever nominated him.

In some respects, this case evokes memories of Haber's candidacy in 1918—an overwhelming desire to recognize a national research tradition. When the Soviet Union declared war on Finland during the winter of 1939–1940, many Finns expected Sweden to help. Swedish neutrality prevented official intervention, but thousands of Swedes volunteered to fight alongside Finnish troops. Popular campaigns raised humanitarian aid. Members of the Nobel Committee for Chemistry initiated a donation, and the physicists followed suit. Those committee members, such as von Euler, who had clear German sympathies considered Finland an anti-Bolshevik, pro-Teutonic ally. Finland suffered and lost territory to the Soviet Union. Regardless of political orientation, many Swedes felt shame or guilt over being unable to do more for their neighbor.

The committee seems to have wanted to extend a hand of reconciliation and friendship. It was willing to overlook von Euler's admission that even though scientifically prepared silage was used in many countries, not all of these processes were based on Virtanen's method. Although he provided no evidence, von Euler asked his colleagues to consider Virtanen the originator of the many varieties. An undivided prize to Virtanen, according to von Euler, would aid Finnish research.

Virtanen became Finland's Nobel laureate, and the authority and prestige that come with the prize created a powerful scientific leader. Other chemists wondered. When the Rockefeller Foundation resumed on-site visits in Europe to assess prospects for support, its field officers inquired about Virtanen. The Cambridge biochemist A. C. Chibnall discussed the Finn's work at length. Although the silage method was useful in Virtanen's country, he saw nothing phenomenal about it. The method did not even work in England. He considered Virtanen's prize the result of lobbying—it was "nothing less than fantastic. . . . But by God, we are supposed to treat him as a great savant." As related in the Rockefeller diary, "[The] sad result is that everyone, and chiefly V. himself, concludes that V. is a great biochemist." The criticisms may have been extreme, but Warren Weaver of the Rockefeller Foundation nevertheless recommended that support had to be terminated, "for we will . . . just be compounding the illusion that the Nobel Committee has created."

Diplomacy could take different forms. Some of the prizes in the 1940s appear to be exercises in making amends for neglecting researchers who had done significant work, who during and after the war emerged as national and international figures. In part, Appleton and Robinson might be understood in this light. Both had contributed toward winning the war, and they were leaders in Britain's postwar scientific community. Another in this category was Patrick M. S. Blackett.

Blackett was conspicuously passed over in 1936 for the discovery of the positron, when the honor went to Cal Tech's C. D. Anderson. True, Anderson's publication appeared just prior to Blackett and Guiseppe Occhialini's announcement. Anderson had serendipitously discovered a subatomic particle while trying to verify a speculative theory on the origins of cosmic rays, a theory advanced by his mentor, Robert Millikan. In contrast, Blackett had set out to find the antielectron, which had been predicted theoretically by Paul Dirac. Blackett and Occhialini designed ingeniously an instrumental scheme that actually detected the particle before Anderson, but they waited to verify that this was the particle Dirac had predicted. For the committee to claim that Anderson alone deserved the prize because he "beat" Blackett and Occhialini into publication by a month surely evoked surprise among cosmic-ray researchers.

Blackett's candidacy was kept alive during and after the war, but the committee repeatedly turned it aside. Anderson had received credit for the positron; therefore, the committee claimed, there was no need to make a new award. As to Blackett's work on cosmic rays more generally, the committee noted that others were also contributing to this field. Nominations had arrived for a shared prize between Blackett and others. But the committee showed no interest in considering Blackett prior to 1948. Then, however, it made a complete about-face.

Blackett was not just a talented researcher. In 1948, two nominations that underscored another side of Blackett's reputation arrived. In a lengthy letter the American laureate Arthur Compton described Blackett's distinguished wartime contributions, including the creation of "operational research," which allowed for a rational use of Britain's military resources, including the use of convoys during the Battle of the Atlantic. Blackett was a war hero, but he was also a left-wing critic of postwar militaristic nationalism and of the budding cold war nuclear diplomacy. He was also an advocate of organized science policy and of using the natural sciences to plan social development.

In a letter to Siegbahn, the Cambridge crystallographer and prominent leader of the left-wing British scientists J. D. Bernal proposed Blackett for a physics prize. Faced with conservative antagonism in British government and scientific elites, Blackett, and others who shared a vision of state and science collaborating to plan a more just society, needed help. In Sweden, the Social Democrats, with support from leaders of the scientific community such as Siegbahn and Svedberg, were themselves trying to shape a new science policy based on comparable ideals. Moreover, Swedish scientists were increasingly fearful of the escalation of American and British nuclear militarism; voices such as Blackett's, which appealed for international controls, found a sympathetic audience. It might well be that the committee saw Blackett as a highly gifted researcher, a wartime hero, and a spokesman for democratic science policy. Here was a reason for not dividing a prize between him and his collaborator, Occhialini, or with any other active researchers in cosmic-ray studies. An undivided prize could give the greatest possible authority and prestige to Blackett. Although Blackett received the 1948 prize, he and other like-minded scientists could not stop the escalation of international nuclear militarism.

### The Prize as Alliance: Stanley, Seaborg and McMillan, and Cockcroft

By degrees, the Academy and its committees became integrated into the world of transnational research. Allegiances to such communities competed with traditional loyalties and identities. To a greater extent than before, committee members did not sit in distant judgment. They could at times have direct stakes in the awarding of prizes to members of their own networks. While never without attachments to foreign researchers and their institutions, some members increasingly worked on shared-research programs. Prior to the war, Svedberg and von Euler tended to promote candidates whose work could bring prestige or support to their own. After the war, these tendencies surfaced more sharply.

In the late 1930s, Svedberg's American Rockefeller network colleagues, J. H. Northrop, J. B. Sumner, and W. M. Stanley, all began to be proposed for the Nobel chemistry prize, either singly or paired. Sumner's 1926 paper, claiming to have crystallized an enzyme in pure form and to have shown it to be a protein, came under attack from German chemists. Svedberg's institute provided further evidence, but not until the early 1930s, when Northrop extended this work, did the identity of enzymes as proteins win broad international support. Working with Northrop's methods at the Rockefeller Institute, Stanley announced in 1935 to have crystallized a virus and to have identified it as a protein. Stanley's sensational discovery claimed that the virus was not a tiny, submicroscopic organism, as biologists claimed, but a giant chemical molecule.

Svedberg and his protégé Arne Tiselius used their unique instruments in Uppsala to provide measurements for their Rockefeller network colleague. In 1936, they performed measurements that established the size, shape, and weight of

Stanley's crystallized plant virus, and Stanley continued investigating plant viruses using Uppsala-developed instruments. All these investigations received support from the Rockefeller Foundation. The dramatic findings seemed to vindicate the belief that chemical and physical methods could transform inexact biology into an exact science. Stanley's claims fit the Uppsala–Rockefeller program's focus on proteins as the key to the riddle of life. In jumping to the false conclusion that a virus is a protein, though, he overlooked warnings that there was more to the puzzle.

Although many people believed that the work on enzymes by Northrop and Sumner deserved a medical prize, Svedberg and von Euler repeatedly promoted the two for a chemistry prize. Svedberg also tried to advance Stanley. In 1939, he proposed a shared prize between Stanley and two British virus investigators, Frederick Bawden and Norman Pirie, who, having found nucleic acid in the same type of virus, questioned the completeness of the American's results. Perhaps Svedberg nominated all three so as to have a chance to review in detail all the relevant literature and respective claims. Subsequently, however, Svedberg acknowledged only Stanley.

In 1946, Svedberg and his Swedish colleagues were alone in nominating the three Americans. Perhaps because Svedberg was in the United States and preoccupied with supervising a large number of projects, the committee gave the task of evaluating the three to Tiselius, who also enjoyed close working relations with them. Tiselius concluded his special report by stating that it would be best to divide the prize between Northrop and Sumner for their crystallization of enzymes and that Stanley should wait. He hoped that either the medical committee would give Stanley one of its prizes or, better yet, by waiting, Stanley could subsequently divide a chemistry prize with other investigators once further insight into the nature of the virus might be gained. This outcome would be as wise as it would be fair. Tiselius noted the many British objections to Stanley's claims, but he insisted that none of these studies "threaten our belief that Stanley crystallized truly pure virus." At its early September meeting, the chemistry committee tentatively accepted Tiselius's recommendation.

Svedberg did not attend that meeting. But when he came to the next one, two weeks later, he objected to the choices. Stanley was his man; Stanley's finding was a crown jewel in the Rockefeller-supported efforts to transform biomedical research. Svedberg had proposed Stanley and intended to see him with a prize. He suppressed the part of Tiselius's text that claimed "waiting was best." Instead, Svedberg recommended giving half of the prize to Sumner and dividing the other half between Northrop and Stanley. It is possible that in promoting Stanley, Svedberg was not only bringing honor to his own laboratory but also applauding Tiselius, whose instrumental innovation—the electrophoresis apparatus—played a major role in establishing Stanley's widely discussed claims. Tiselius had, that same year, received his own institute of biochemistry. By bringing all three Americans to Sweden, Svedberg could celebrate the full blossoming of the Uppsala–Rockefeller network. In fact, Tiselius received the prize two years later.

Unfortunately, Svedberg's enthusiasm on behalf of his network clouded his judgment. He was so completely committed to the belief that they captured the inner lane in the race for unraveling the virus that he refused to take seriously the objections of British and other researchers. Stanley's pure crystals contained impurities; they were not even three-dimensional crystals. More critical, Bawden and Pirie were right; the virus was not pure protein. Stanley missed a critical component that made up about 6 percent of the virus mass: its nucleic acid (RNA). This was soon understood as the actual key for understanding virus self-replication and not, as Stanley maintained, enzyme action and crystal growth. Ultimately, the rise of molecular biology eclipsed the protein model of virus; nucleic acids, and not proteins, opened new avenues for exploring the secret of life. In the 1950s, Stanley conceded his mistake: "I have had to swallow a lot." By then, Svedberg had moved on to matters nuclear and was devoting himself increasingly to nuclear chemistry.

Svedberg was already well connected with the Berkeley nuclear community; he planned yet closer ties. Svedberg once again revealed an almost magical ability to conjure support. He obtained a huge donation from a wealthy Swedish industrialist, which led to the establishment in 1949 of the Gustaf Warner Institute for Nuclear Chemistry. By the end of the decade, his own cyclotron, being built in Uppsala, was going to be the largest in Europe. Svedberg, as often was the case, saw an opportunity to mark the opening of this new installation in 1951.

That same year was the fiftieth anniversary of the Nobel Prize. For Svedberg and Siegbahn, the two leaders of Swedish chemistry and physics, this was an opportunity to celebrate Swedish nuclear research. Both Svedberg's new cyclotron and Siegbahn's new cyclotron were being readied for opening. Both men and their disciples were deeply involved in government policy for an internationally neutral, Social Democratic, nuclear-based future. In 1950, Svedberg contacted Lawrence and suggested that he would like to see Lawrence's two protégés, Glenn Seaborg and Edwin McMillan, share a chemistry prize.

The two Berkeley scientists had, together with several colleagues, used the cyclotron to discover and produce several transuranic elements. In 1940, after Fermi's claims for new elements were shown to be the result of fission, McMillan and Philip Abelson created the first true transuranic element, number 93, called neptunium. As they began searching for the next possible element, McMillan left Berkeley for other wartime research. A group led by Seaborg continued the painstaking work, ultimately producing and isolating element number 94, plutonium. This discovery in late 1940 was kept secret; the world learned of it in the form of a bomb dropped on Nagasaki.

After the war ended, and the record of wartime nuclear research emerged, Seaborg began receiving nominations, both singly and together with various collaborators (namely, J. W. Kennedy and Emilio Segrè). By 1950, Seaborg had established himself as the leading nuclear chemist, as he continued to extend the periodic table by producing a string of transuranic elements. He was a natural for a prize, if nuclear chemistry was to be acknowledged. (During these years, the committee ignored important contributors to theoretical general principles, such as Linus Pauling. Eventually, Pauling did receive a prize, in 1954.) To give a chemistry prize to McMillan, a physicist, was largely Svedberg's idea. True, McMillan had contributed elegantly to the chemical and physical research on neptunium, but it was his nonchemical expertise that made him so attractive. McMillan helped design the uranium and plutonium bombs; even more relevant, he turned to the physics of particle accelerators to solve the problem of keeping spiraling particles stable at very high energy levels.

Lawrence was perhaps surprised by Svedberg's suggestion, but he was also pleased. He accepted Svedberg's suggestion to nominate Seaborg and McMillan for 1951 and allowed Svedberg to word his letter of nomination. Just in case, Svedberg sent in his own proposal. Svedberg successfully steered the two candidates through the committee and Academy.

Once the shared prize was voted on, Svedberg suggested that Lawrence join his two protégés at the ceremony. In advance of the Berkeley visit, he began discussing technical problems arising in trial runs of the cyclotron; help from the Berkeley team would be most appreciated.

Meanwhile, the nuclear component of the prize's fiftieth-anniversary celebration was being discussed in the physics committee. Having repeatedly rejected John Cockcroft and E. T. S. Walton for their pioneering accelerator work in 1932, the physics committee suddenly gave them the 1951 prize. Although the records remain closed, the actions point to the diplomatic alliance-building nature of the award. The two had been thought "dead" candidates; after declaring that their work was overshadowed by Lawrence's, the committee treated their reappearance in 1943 and 1946 as not requiring further evaluation.

But some nominators believed otherwise. In 1947, Cockcroft, alone, received several nominations from, among others, W. L. Bragg and Blackett. Cockcroft also had been a major figure in wartime research; he was one of the heroes of radar, and was then leader of the British–Canadian atomic energy laboratory. In 1946, he was named head of Britain's new Atomic Energy Research Establishment at Harwell. Bragg and Blackett might well have been interested in helping him in his politically delicate position.

This time, the physics committee did comment. It noted with interest that Blackett, having been present at Cambridge in 1932, had dropped Walton from the nomination, and it concluded that Cockcroft was the leader of the experiment. Nevertheless, there was no need to split hairs: The committee again concluded that Lawrence's cyclotron was more effective and the 1939 prize precluded giving another award. Five years later, this position dramatically changed. What had happened?

By 1951, Cockcroft had established himself as not only a British but also a European leader in nuclear activities. He was supervising several British particle accelerators and programs for fundamental and applied nuclear research. More than that, he was a spokesman for the peaceful use of nuclear power. Swedish nuclear policy at home and abroad favored reaching out for better relations with Cockcroft. Furthermore, Siegbahn desired to bring further attention and prestige at the Nobel festivities to nuclear physics. But why share the prize with Walton, who had turned away from research? Perhaps some eloquent spokesmen in the committee or Academy pointed to the possible damage if Walton were ignored; or perhaps the division made less obvious the underlying politics for resurrecting Cockcroft's candidacy in Stockholm.

In 1951, a nuclear Nobel celebration began. Setting the stage prior to the Swedish events, Norway's nuclear reactor—built in close collaboration with the Swedes—was scheduled to open in late November. Here, Cockcroft joined Lawrence and other European nuclear leaders to help the Norwegian king and dignitaries open the reactor. Then they moved the festivities to Sweden. Two days prior to the actual Nobel ceremony, Svedberg inaugurated his cyclotron. Svedberg transformed not just elements but also the traditional Swedish festivity of St. Lucia, which for the occasion was moved from December 13 to the 8th. Cockcroft, Lawrence, and other distinguished guests sat around candlelit tables deep in the pit of the cyclotron, while a procession of girls dressed in white, singing Lucia carols, circled around them. Then

Lawrence pushed the button to start the cyclotron. The Nobel ceremonies were almost anticlimatic, but the message came through clearly: Science was the shield of Western civilization; the peaceful atom was to lead the way. Having for almost two decades rejected Cockcroft and Walton's work as not being significant, the Swedish nuclear establishment now hailed it as opening "a totally new epoch in nuclear research."

As the cold war era began, Nobel business was, to a large degree, business as usual.

## A TURN FOR THE BETTER?

Beyond the immediate postwar years, some changes were subsequently made to the selection process. In 1949, each committee sent out approximately three hundred invitations to nominate. In that year, fifty-two nominators proposed thirty-seven candidates for the physics prize; for chemistry, the corresponding numbers were fifty-eight and thirty-six. The numbers continued to rise in the ensuing years. (Currently, more than two thousand invitations to nominate are sent out, and approximately two hundred fifty nominees are proposed.) The number of institutions invited to nominate on a rotating basis has been increased. Both the number per major nation and the number of nations have also increased. No doubt, the addition of this finer-grained mesh to the net of nominations has yielded a more comprehensive picture of significant developments in physics and chemistry. Many more scientific communities around the world, as well as a broader range of opinion within the leading nations, are heard. This change has certainly contributed to a lessening of the problem that arose during the interwar years, when many important chemical discoveries and innovations scarcely registered on the committee's agenda.

At the start of the postwar era, nominators continued to interpret their mandate in different ways. Voting "within the family" predominated, whether the nominator's own research specialization, home institution, or nation. Some nominators, especially former prizewinners, urged adding to the list of winners elderly researchers who had been passed over by the committees. Both the increasing numbers of candidates and the multitude of rationales posed an enormous challenge to the committees. But several changes in procedure have enabled them to perform their duties more efficiently and to avoid "mistakes."

For example, although the Academy still preserves the tradition of electing only Swedes to the five-person committees, recently these committees have been willing to seek expert advice from outside their ranks. The committees now turn on occasion to foreigners to write special reports on candidates or developments about which they lack expertise. The Academy's Physics and Chemistry Sections have also grown in number, and a guaranteed number of members under the retirement age has ensured greater participation by active researchers. The procedure by which a section acts on the committee recommendation seems to be more formalized, and it allows greater opportunity for discussion. Bringing the sections more actively into the process might reduce the influence of a strong-willed committee member, but it could also mean greater quarreling.

The committees are also more sensitive to the inadequacies of the published record for assessing priority. Publications cannot always be relied upon to show sources of inspiration and influence. Sometimes the committees have asked foreign evaluators for insight. Having those evaluators' inside knowledge in the form of statements, even if not fully impartial, could help committee members sift among numerous claims for priority.

What Oskar Widman said in 1919, that scientists are not always God's nicest children, still holds true, with a vengeance. The stakes are now higher, the frenzy for renown is greater, and priority disputes have been known to get nasty. To sort out conflicting claims and to avoid unjustly overlooking contributors to an important innovation, committee members are becoming increasingly more mobile. They attend international meetings and specialized workshops around the world. They may even escape the Swedish winter at important research centers in more pleasant climates. To a greater extent than ever before, they can hear informal discussions on who did what, and how; they can hear, firsthand, claims and challenges. Far from perfect and far from comprehensive, these personal contacts have greatly improved the flow of information.

As some committee members themselves admit, the task of selecting "the best" is nearly impossible. The sheer number of specialties and impressive accomplishments in chemistry and physics precludes any precise determination. In 1948, the chemistry committee declared that it was impossible to compare two so totally different contributions as Tiselius's instrumental innovation, the electrophoresis apparatus, used in biochemistry, and William Giauque's theory of chemical thermodynamics at extreme low temperatures. The committee's choice was based on its own priorities. Tiselius received the prize; and then Svedberg and Tiselius and two other Swedish colleagues advanced Giauque the very next year. Such impossible or near-impossible comparisons have become more the rule than the exception. Few achievements are so brilliant and so consequential for an entire discipline that they eclipse all other nominated work.

During the first half-decade after the war, committee prerogatives remained vital. Most members tended to nominate and evaluate their own favorite candidates, which does not necessarily mean that the practice has continued. The physicists nominated candidates less frequently, but the many sudden reversals on the worthiness of candidates show that committee members' judgments, tastes, and priorities all entered in decisions. Even in an idealized situation in which nominators and committee members are able to transcend interests, there will almost always be too many worthy candidates and too few objective criteria with which to make an impartial choice. Committee members' opinions necessarily remain crucial; their group dynamics, decisive.

But what can be said about their practices in more recent times? Because recent actions are not open to examination, a few glimpses into more recent committee work, garnered from correspondence, are all we have as evidence. However, these examples do support the contention that there is no simple line that divides an earlier imperfect Nobel past from a perfect present.

## A Chemical Judgment

In 1962, the Swedish organic chemist Holger Erdtmann turned to Robert Robinson to express his disappointment over the committee's recommendation. Although members agreed that one of the candidates, Dorothy (Crowfoot) Hodgkin, merited a prize for her X-ray techniques of studying biochemical substances, they did not want "to let her share the prize with the more macromolecular X-ray chaps" (Max Perutz and J. C. Kendrew). They insisted that she should get a prize alone, "at a later date." Erdtmann was not happy with "a later date." He saw that his colleague Fredga was already favoring the prize for an "Italian macro-man" next year. But Erdtmann saw no reason why the committee could not give the prize two years in a row in the same subfield. Whether thinking aloud or hinting at points for Robinson to include in a proposal, he noted that Hodgkin's recent work on the vitamin B12 molecule was exciting and could well strengthen her case. He reminded Robinson that as many people as possible must nominate "DH." Since nobody had done so, he and some colleagues had to propose her. Sending a hint, he saw "no reason why L. T-d [Lord Alexander R. Todd, prizewinner in 1957] should nominate only W-d; why not both W-d [R. B. Woodward] and DH, not to be shared but—as is often done—propose by way of several candidates." Academy members continued to advocate candidates and occasionally solicited aid from like-minded nominators.

In this case, the work involved could be evaluated fairly soon after discovery. Dramatic and significant discoveries have generally been more the exception than the rule in chemical science, and chemists often must wait longer than physicists to be sure whether a new process or method of analysis possesses long-term significance. This situation may be reflected in the long lag between the time at which innovative work is made and the time when a prize is awarded; of course, the vast number of achievements also creates a backlog. Physicists, however, have had to move quickly, to sort out a cluster of dramatic research claims coming from several communities, all being rapidly developed. To wait could otherwise lead to a new cluster of significant developments overshadowing the old.

## A Tale of Promotion, Triumph, and Sadness in High-Energy Physics

Committee members and nominators might well have tried to learn from the past, but scientists are like the rest of us. Even though values vary, the human drama of science seems always populated with a mix of attitudes and abilities. In the paragraphs that follow, we can see complexities and difficulties of more recent prize decisions, even if the historical record is still incomplete.

The Pakistan-born particle physicist Abdus Salam, who first worked in England and then headed an international center for theoretical physics in Trieste, was an active and gifted theoretician and administrative leader. His contributions to theories concerning elementary particles and their interactions were part of the ferment, confusion, and triumphs of physics in the 1960s and 1970s. Because confirmation of theoretical claims and predictions in this field often challenged the limits of existing high-energy accelerators, only one or two giant installations in the world, if any, could attempt a confirmation. Both theoreticians and experimenters felt great commitment to their ideas. Often, several physicists contributed to developing each new theoretical scheme, sometimes leading to parallel breakthroughs by different investigators. Salam, at different times, was running with the pack—sometimes taking the lead, alone or with others, and sometimes chasing others.

Not surprisingly, when the stakes are so high, supporters organize campaigns. Although this phenomenon is not new, the magnitude and drive of the campaign for Salam in the autumn of 1968 prompted strong reactions. Werner Heisenberg was miffed over the number of appeals he received urging him to add his vote for Salam. Although he was quite willing to understand why others might want to nominate Salam, whom he considered a competent physicist, Heisenberg objected to the pressure tactics. Perhaps this was simply part of the new way things were being done in physics, but he felt obliged to inform the committee of his dislike. Others also registered their annoyance over being strong-armed into joining a campaign, which in this case continued for some years. Some considered such tactics the introduction of "politics" into science.

A few of Salam's supporters claimed that he had been unfairly passed over by the committee a decade earlier. Some advanced reasons why it was important to reward a great Asian physicist. It is possible that these tactics might have hurt Salam's candidacy. They prompted Tsung Dao Lee at Columbia University to express privately not only his dislike for the organized campaign but also his belief that Salam had not as yet made significant contributions to physics other than his excellent efforts to promote international collaboration. Lee shared the 1957 physics prize for work on nonconservation of parity, precisely the discovery for which some supporters claimed Salam had deserved some credit. Lee asserted that the committee's alleged oversight was perfectly justifiable: Salam's earlier contributions on parity were themselves derived from other physicists. He then registered his opinion that Salam's more recent and original theory on another aspect of particle physics had not as yet led to a significant discovery.

But this was not simply a popularity contest. In the late 1960s, some theoretical physicists, such as laureate Paul Dirac, were convinced that Salam indeed had a right to be considered worthy of a prize. Dirac wrote to longtime colleague and veteran member of the committee, Ivar Waller, that he considered Cal Tech's Murray Gell-Mann and Salam to be the most deserving contributors in theoretical high-energy physics. Several candidates from this specialty were vying for a share of the prize. For Waller and the committee, the problem was that even the members of this theoretical community found it extremely difficult to establish priority for originality. Moreover, the recent rush of exciting claims made it difficult to assess which of these was the most deserving. For physicists outside the clan of theoreticians, some of the theories were too fanciful and could not be taken seriously until they received experimental verification, even if that meant waiting until more powerful atom smashers were built. In the end, Gell-Mann received in 1969 an undivided prize.

In the mid-1970s, when the committee was evaluating additional contributions from Salam, his supporters renewed their drive. Waller, who was sympathetic to Salam, was in contact with other leading theoreticians at conferences and with leaders of the giant European particle accelerator in Geneva (CERN) on progress in confirming Salam's predictions. Waller confided to Dirac in 1975 that although he was convinced Salam's recent theoretical work was worthy of a prize, he was hoping to see still further contributions from Salam. These might help convince those committee members who were still hesitating. Waller failed to rally the committee or Academy that year for his candidate, but he was not giving up on Salam.

A major problem entailed the makeup of the committee. To one dispirited observer on the scene, the committee's recent work had deteriorated in quality. Two of its members were experimentalists who understood very little theory, and they had no desire to reward theoretical achievement. Another member fancied himself an expert in particle theory, and as long as he continued on the committee, Salam's candidacy would have great trouble. Soon thereafter, dramatic new confirmations of Salam and other theoreticians' predictions made with the SLAC accelerator at Stanford removed further resistance. The Academy could no longer ignore him. In 1979, Salam shared the prize with Steven Weinberg and Sheldon Glashow.

Salam was not the only physicist who felt slighted by not sharing the 1969 prize with Gell-Mann. In 1970, the Japanese laureate Hideki Yukawa, who won the 1949 physics prize, discreetly wrote to Waller about his colleague Soichi Sakata. Yukawa, a pioneer in theoretical particle physics, was disturbed by the committee's decision. Sakata had been a leading figure in the theory of particle physics since the late 1930s. He had predicted the existence of significant processes and particles that were later confirmed; he initiated models of elementary particles (hadrons) that set in motion highly significant theoretical work, including that of Murray Gell-Mann.

In fact, the committee had indeed tried to sort out the priority claims of several theoretical physicists who had worked on these and related problems. The committee had commissioned a special report from a researcher active in this field. While keeping his purpose secret, he interviewed persons present at critical seminars and who were in contact with participants in the developments. Information and insight did not necessarily make the task any easier; sometimes no clear answers for priority could be found.

Waller, who had known Yukawa for over twenty years, and who had promoted him for an undivided prize in 1949, responded candidly. He noted that the committee was well aware of the great importance of Sakata's work for the theory of elementary particles, for which Gell-Mann was rewarded. Waller confessed his admiration for Sakata's many contributions. But when the committee had to decide which of several physicists should rightly share the prize with Gell-Mann, it simply could not make a "just selection."

In September, Yukawa politely wrote to Waller informing him that Sakata had been ill when the nomination was written; since then, his condition had worsened significantly. This year might be the last to propose his colleague. Three weeks later, Sakata died. Yukawa informed Waller that a prize to Sakata would have brought him much honor and encouragement. He, then, in the name of leading Japanese particle physicists, asked to know what the Nobel committee thought of Sakata's merits, for that would perhaps bring them consolation.

Waller expressed his regrets over Sakata's death and sympathized with Japanese scientists over the loss of this outstanding physicist, who also embodied so many fine personal characteristics. Although unable to reveal details of the committee deliberations, he admitted to the overall high opinion held for Sakata's scientific contributions. But the letter upset Waller.

Why should Yukawa be so concerned with what five Swedish physicists thought of his colleague? Sakata had been held in high esteem around the world by all who knew his work. Of course, for Yukawa and his colleagues, the prize—regardless how imperfect it might be—was a shorthand symbol of undisputed excellence; it would give Japanese particle physicists great prestige in the eyes of the mass media and the political leadership.

Waller saw things differently. After a quarter century of committee deliberations, he knew there was no magic, no suprahuman quality to the Nobel decisions. He might well have even become immune to the lavish festivities and inflated rhetoric. He was sorry that Sakata had not received a prize but chided Yukawa for giving this matter unjustified importance. Both he and others, he pleaded, must often make mistakes no matter how hard they try to select the most deserving candidates. As the amount of research continued to grow, the task was becoming more difficult. Waller then stated, simply and directly, that which should be obvious: There were usually equally deserving candidates among those who did not get the annual prize.

# Further Reflections

*W*hy do people venerate the Nobel prizes? There is no easy answer. This book has examined but one part of this complex question: How during the first half-century have those entrusted with implementing Alfred Nobel's testament, in the particular domains of physics and chemistry, discharged their duties? By examining the process by which choices were made, disputed, and resolved, it becomes possible to replace illusion and myth with understanding. Without such a history, critical debate and reflection have lacked a fulcrum.

## GROUNDS FOR BELIEF

Winning a prize has never been an automatic process, a reward that comes for having attained a magical level of achievement. As much as we might want to believe that merit, like a well-blown bubble, floats upward, effortlessly rising to its natural level of recognition, the actual world of the prize has been turbulent. This study asks that we take seriously the kaleidoscope of human agency involved. Changing patterns of calculation and probity, bias and insight, arrogance and prudence preclude capturing in a simple statement the committees' attitudes and actions. But the net result has been that the list of winners and the research specialties represented were not natural or inevitable choices. Choosing winners has entailed judgment and volition, compromise and insistence.

There are no grounds, based on history, for assuming the laureates constitute a unique population of the very best in science; even less so to impute to them, as a class, the status of genius. The winners have, by and large, been highly talented and in many cases gifted. But their elevation to the status of a peerless elite, standing qualitatively apart, itself is puzzling. The attention, authority, and privilege afforded them, and their particular research interests, have sometimes been well deserved, but having won a prize in itself was—and undoubtedly remains—a questionable standard. It is perhaps asking much too much of any selection process to define a single group of the alleged "best" in science. Moreover, it has become clear that many important branches of science are not addressed by Alfred Nobel's testament, and many branches of physics and chemistry have been either passed over or rewarded on the basis of little more than local Swedish and committee preference. There is nothing wrong with wanting "heroes" in science, but we must understand the criteria used to select those whom we are told to revere.

The reasons for "believing" in the transcendent values ascribed to the prize are to be found not in the list of winners but in the broader history of science and culture in the twentieth century. The oft-repeated claim that the prize's prestige has reflected the skill of the Royal Swedish Academy of Sciences in picking the right winners simply does not hold up to inspection. A hypnotic fascination with the prize and its rituals has been with us from the very start, from even before the first awards were announced. Similarly, to say that the money involved has not really been important is quite untrue and misses the point. The hold of the prize in the twentieth-century imagination arises from the cumulative history of its uses, why influential groups were willing to grant the prize significance.

Many people have wanted to believe that the annual prize giving reflects competition on a level international playing field, on which the truly best compete on an equal basis. Ritualized by an annual feast of news coverage, and sanctified by legend, myth, and celebration, the prize has had a life of its own largely irrespective of the choices made. True, a long series of howlers sometimes dented the prestige and raised questions, but the margins of choice were always wide. The idea of having a truly objective international prize, free of personal, religious, political, methodological, or epistemological prejudice, was simply too valuable a resource. With time, the number of interest groups and institutions with a stake in maintaining the "creed" of the prize swelled. Let us pull together a few bits and pieces of this history before turning to the meaning of the prize today.

## CULTURES OF COMPETITION

Nobel's testament played upon sensibilities and aspirations common in nineteenth-century Europe. National and regional academies had a long tradition of prize-giving. Whether to honor individuals or to give financial aid for research, such prizes, and the competition they involved, was, by the late 1800s, a common feature of the cultural landscape. Controversy and accusations of cronyism were never far behind the winners. Still, interest—and hope—rarely faltered. Although it was not unusual for a national, provincial, or colonial academy to honor a great savant of another country, prizes were understood to be for the encouragement and recognition of one's own. What made Nobel's vision so exciting was the idea of bestowing objective, international recognition in a culture given to international commercial, military, and political rivalry.

Civilized nations did not go to war to prove their strength; the idea was to get away from warlike competition. This "Age of Progress," beginning in the 1880s, presumed a competition in the marketplace and in colonial expansion; sport and intellectual competition in this context had a natural role. Exhibitions and fairs where crafts, manufactures, and other manifestations of material culture enabled nations to display their industrial prowess could be complemented by cultural tournaments as well. As such, prizes reflected the evolutionary fitness of competing ideas and nations. The victors in the "race" were, of course, European, not Asian, and North American, by exception.

At a time of growing insecurity, for colonial empires, naval supremacy, and economic prowess, Europeans were ready to believe in an impartial Olympics of culture. This competition allowed nations to prove their right to honor by showing their ability to make supreme contributions to the collective advance of civilization. German declarations of cultural superiority came quickly; some Britons wondered whether their poor showing early on pointed to a national hereditary decline.

By 1900, widespread literacy in Europe and North America and the development of inexpensive means of publishing created vast public forums for news and propaganda. The mass media were especially attuned to issues of national honor; to competitive exploits such as heroic expeditions. From the first announcement of Nobel's testament, media in major nations played a dominant role in generating interest in the prize and transforming it into a public event. Speculation crackled: Who might win? Newspapers provided extensive coverage of the winners and their culture. Equally, the prize prompted grandiose expectations and miracles.

With so much money involved, and with so much at stake, surely those involved with selecting winners could only act most nobly. No doubt Mittag-Leffler's snide comment that the bourgeois mentality translated greatest amount of money into the highest honor helps in part to explain the immediate elevation of the prize. But equally important, the extravagant ceremony, and royal participation, appealed to *fin-de-siècle* sensibilities. Pomp defined the circumstance: a celebration of progressive civilization's and idealistic culture's most worthy contributors.

The prize became an emblem of the "gold standard" well before a long list of winners was at hand. Illustrations show that scientists and their patrons took the prize seriously from the earliest years. Arrhenius's 1903 Nobel Prize in chemistry prompted a call to Berlin and the Swedish king's insistence that Sweden must keep him at home. After Lenard received the 1905 physics prize, the Prussian minister for culture not only promoted him but also started to seek his advice. In 1911, Berlin professors filed a complaint to the Academy: They should be invited more often to nominate and thereby influence the outcome. They knew the importance of bringing home a prize.

World War I dampened belief in the capacity of competition to drive civilization's progressive evolution, but in its wake remained nationalistic competition, only with less idealistic packaging. Once the war ended, commentators spoke of the expected economic competition for national survival. Similarly, after the loss of millions of soldiers and great sacrifices at home, politicians and establishment media needed to reassure their people that national survival was worth the sacrifice. Olympic medals, Nobel prizes, and daring expeditions were among the easily displayed internationally recognized honors that fueled national pride. By keeping alive the belief in the prizes as an indisputable sign of excellence, the press, as well as the scientific and cultural leadership, could elevate the announcement of a prizewinner to a national event, sanctioned by fair competition with the rest of the world.

It has been claimed that the resilience of the prize in science during the post-1918 period of boycott and antagonism indicates a triumph for the transnational spirit in science. The argument holds that, in spite of their quarreling, scientists from the Allied and Central Powers kept the prize out of their cold war. But not only did the boycott largely extend into nominating, except for the case of Einstein, closer inspection shows that European communities saw the prize as an important resource to be parlayed into funding and public recognition at home. Everyone had a stake in maintaining the cult of the prize in spite of hostilities. The experience of the literature prize underscores this point.

Although French and British newspapers openly criticized this prize for being biased, the same writers devoted inordinate attention to rumors and results. Why hadn't H. G. Wells won a Nobel Prize in literature? Has Stockholm again insulted French culture by not awarding the prize to a Frenchman? The great attention given the prize had little to do with universal approval. It soon became a ritual to muse and gossip. And as soon as a local hero did receive a literature prize—for example,

Anatole France in 1921—the French press erupted into a frenzy of applause for French culture, which had just been impartially crowned. Five years later, Perrin received the 1926 Nobel Prize in physics, and then used his new-won prestige to campaign for a renewal of French science. Having a prize from one's own national academy did not bestow the same charisma.

German scientists celebrated their 1919 sweep of the science prizes. They might have lost the military war, but they remained supreme in culture. They and the press played upon a deep chord of cultural heritage and legend. When, over a hundred years earlier, Napoleon's troops ravaged the German states, the Prussian emperor instituted major reforms in higher education, including the creation of Berlin University, that opened the door to original research as an academic duty. The emperor was determined to have Prussia compensate for what was lost on the battlefield through honor gained in the seminar room and laboratory. Now, again, although vanquished in battle, Germany had its victory in the prize.

The interwar German scientific community appreciated the value of maintaining the prestige of the prize. Faced with a hostile climate both at home and abroad, it understood that the prize and its myth held great promise. Scientists could appeal to the government and private patrons for support based on the claim that German research could spearhead national efforts to regain international respect. Having been snubbed by foreign colleagues, German scientists relished the recognition given their national science by the prize. Even if they had doubts about the competence of the committees, they knew that a prize could help enlarge a local laboratory or invigorate a research specialty.

In the United States, emerging after World War I as a world economic power, there still lingered a sense of cultural and scientific inferiority among those who cared about such things. Young scientists who intended careers in physics and chemistry still did their utmost to complete their training in Europe. But the tide was turning; an American doctorate would soon compete with a German one. Robert Millikan's Nobel Prize in physics in 1923 was not only hailed as proof of America's prowess but taken by Californians as an indication that the nation's cultural center of gravity had moved westward. Millikan and his cronies had declared southern California the last stronghold of a pure Anglo-Saxon culture. They appealed to wealthy local patrons to create a temple of science at Cal Tech that could help the region take its rightful place. The Norwegian Vilhelm Bjerknes was visiting the school at the time, and he was amazed by what he witnessed. Millikan was hailed in the press as the greatest "electrician" in the world since Benjamin Franklin. Bjerknes quipped that the prestige of the prize seemed to increase geometrically with the distance from Sweden: On the American West Coast, Millikan was treated as genius and sovereign.

During the interwar years, as Americans began to collect greater numbers of prizes, they served as grist for the mills of self-interest. California versus the eastern establishment, Berkeley in the north versus Cal Tech in the south, private research laboratories versus academic institutions, and so forth—the prize was treated as an indicator of excellence. A culture of academic competition gathered strength in the United States as universities promoted themselves to enroll students, recruit staff, and exploit their alumni. It did not hurt to have one of the world's "best" physicists, chemists, or medical researchers on campus. Simmer turned to boil. By the end of World War II, science seemed to be full of innovations, from war-winning atomic bombs and radar to synthetic fabrics, wonder drugs, and other miracles. America had won the war; America reigned supreme. Winning more Nobel prizes, and winning more than any other country, it seemed to some, was part of the world's recognition of American exceptionalism.

In the overheated and politicized economy of research after the war, prizes played many roles, such as helping to define leadership, to steer science policy, and to create a scientific culture based on reverence for individual genius. Too many institutions, research specialties, and mass media had a stake in keeping the faith in the prize to allow serious challenge. Moreover, too little was known about the prize to generate anything more than cranky comment. The cult of the prize was not a conspiracy, but the result of an evolving accommodation to shifts in cultural climate. Cold war propaganda did not miss the opportunity to underline American world scientific leadership. The annual rituals in Stockholm provided opportunities for science and the media to comment on the state of national research. Did a momentary eclipse of American science in any given Nobel season suggest a weakening of national resolve in basic science? Could American prestige let a European particle accelerator win the race to find a critical subatomic particle and scoop the prize? The list, depressingly, goes on.

In other nations, the cult of the prize took on other forms. After World War II, West German laureates used their status to assume leadership roles in rebuilding their national scientific community. In the early 1950s, annual public gatherings at Lindau brought national and foreign laureates together to bring science to the public and to win goodwill. Swedish friends of Germany, such as von Euler, urged the Nobel Foundation to give official support. But the foundation became wary of lending its name to high-profile events and products. Beginning in the late 1940s, film, television, and publishing companies wanted to produce portraits of the winners and reports from the ceremonies; many sought to find some tie-in with the magical Nobel name. The foundation did its best to curb such activities. It understood that by its giving official approval, prizewinners would feel a duty to comply with requests for their time.

Now that we know that the selection process has a human face, it should be easier to make comparative national studies of the prize and to explore the reasons for wanting to believe in the prize as an indisputable mark of excellence. Our culture's obsession with choosing the "best" each year, be it in films, popular songs, or universities, predisposes it to give undue attention and respect to the prize. Indicative that a will to believe precedes a critical basis for acceptance, the so-called Nobel prize in economics is one such example.

Not really a Nobel Prize at all, the award in economics in memory of Alfred Nobel came about as the result of efforts by a powerful member of the Royal Swedish Academy of Sciences. After his intensive lobbying, the Academy accepted that if he raised the money, it would administer such a prize. The new award was immediately promoted as a Nobel Prize. Some commentators quickly began ascribing meaning: Here was proof that economics was the most important of the social sciences. Without understanding the limitations and weaknesses of the process, the recipients were afforded instant prestige as part of the Nobel cult.

Similarly, when the MacArthur Foundation launched its high-profile and bountifully endowed fellowships to allow creative individuals in the sciences, humanities, and arts opportunity to free themselves from other obligations, the media dubbed them "genius awards." Newspapers and university administrators paid attention; lists of winners received enormous attention. This hullabaloo preceded any insight into the means of selection or into the outcomes expected from being treated as a "genius." Even if claims of genius were soon pushed into the background, and even if academics understand that the selection of winners is far from perfect, few forgo harboring the wish to be chosen (including this author).

Now, more than ever, the Nobel ceremony is the stuff of which dreams are made. After fumbling and stumbling in the early years, the Nobel Foundation achieved an institutional expertise, handed down and perfected, with skill equal to any royal or presidential staff, to create a dazzling annual event. The Nobel festivities are the Swedish social event of the year. Journalists devote space to the seating, the fashions, and the food. Live coverage of the ceremony, dinner, and after-dinner dancing is part of Swedish television tradition. Even hard-core Social Democrats have been known to fall under the spell of this bourgeois dream: The prize lets the people share in the elegance, pomp, and decorum of the upper classes. Whatever messages these ceremonies convey, they are relics of past patriotic display, an assertion of Swedish pride, an event full of symbolic cultural-political meanings within Swedish society. It matters little for what achievements prizes are given; it is enough that, as they see it, the cream of Swedish society is circulating with the cream of science and culture.

All this matters little to the prizewinners, overwhelmed with their hosts' hospitality, geniality, and generosity. The young Ernest Rutherford in 1908 roared approvingly that he and his wife had the time of their lives. Rutherford warned Harold Urey in 1934 to be prepared to have his democratic principles tested by the Stockholm ceremony. Urey wasn't daunted. Indeed, as science became increasingly democratic, recruiting from less privileged groups and becoming a sprawling mass society full of its own social hierarchies, the image of being singled out, whisked to a captivating but not too familiar European capital, and treated like royalty—in addition to collecting a substantial sum of money—well, some may ask, what more could any scientist want? In fact, not too long ago most freshmen at the California Institute of Technology had no problem answering the query about their goals in science: to win a Nobel Prize.

Why the Nobel Prize? By and large, there just is no competition—and once the obsession began, cultural momentum, like an advancing glacier, ground down criticism in its path. But should winning a prize be the goal of a career in science? Should the gratification of ego be the ultimate reward? Is competition the driving force that elicits results? Many see nothing wrong with this. Why not, as in the case of the Olympics, why not just enjoy the ritual? Regardless of whether or not they manage to clean up the business—with or without the scandal of steroids and single-minded devotion to winning—why not have some fun, cheer the winners, and take it all as game? Why the curmudgeonly tone?

## RECLAIMING NOBEL'S LEGACY

Much has been written about the loss of the Olympic ideals and the larger sea change in attitude concerning international sport. Ideals and values evolve. As early as 1900, the great Norwegian polar explorer and early promoter of skiing, Fridtjof Nansen, bemoaned that "athletics" was turning into "sport." Rather than developing character and allowing individuals to test and challenge their own abilities, sporting competition was turning into a soulless quest for medals and records. We largely take for granted today that sports is about winning; nations work themselves into frenzies over Olympics and World Cup Football competitions. For Americans, at least those old enough to remember, that credo was succinctly put in the 1960s by the legendary coach of the Green Bay Packers, Vince Lombardi. In popular memory he was to have said: Winning is not only the most important thing, it is the only thing. And yet, there exist historically, and at present, other ways of thinking about athletics. What about science?

Science, as a rational enterprise, is not an activity devoid of its own culture. Values touch it at almost every point. Here, too, it is difficult for most researchers and academic administrators to conceive of other ways to think about the life of science than that which they experience. As with so many other aspects of the scientific enterprise, the way things are often seem to be natural and inevitable.

The prize has played a role in shaping popular images of science. Once an award is announced, a gifted researcher is elevated to the level of exaltation through many channels. Directly and indirectly, the prizes communicate images and messages as to what is important in science; as such, they both reflect and propagate values that underpin modern science. The prize is, of course, just one of many means by which such processes occur.

The prize has bequeathed us a pantheon of heroes based upon romantic notions of genius-inspired individuals. Is this actually the reality, even for the population of prizewinners? Regardless, is this the only ideal for which we may want to select those deserving our highest respect in science? Many candidates were nominated over long periods of time for careers of important achievements and for having founded significant schools of research, but because they did not have a particular discovery that caught the committee's fancy, they were passed over. Is it wise to equate a scientist's worth by the number and type of prizes that scientist collects?

Whom should we honor, and for what actions? Does Richard Willstätter deserve to be remembered simply because of his brilliant studies of chlorophyll, for which he received a Nobel Prize in 1915, or also for his resigning his prestigious chair of organic chemistry at Munich in 1925 in protest against his university's blatantly anti-Semitic hiring practices? And how should Hans von Euler be remembered? In his single-minded striving for prestige and professional success, he, and others in this history, displayed moral reflexes that charitably can be described as reptilian. Just what is being celebrated when awarding a Nobel Prize?

The prize creates a false impression of discovery as an individual researcher's eureka experience; the prizes reward in "quanta" when in reality they should recognize a more continuous spectrum of accomplishment by many investigators over time. A greater recognition of both the team aspect of research and the importance of communal enterprise to reach successful results is essential. It might even foster healthier cultures of daily laboratory life and promote a republic of science that treats all its citizens with dignity.

In the 1960s, the molecular biologist James Watson, who shared the Nobel Prize in medicine for unraveling the structure of the DNA molecule and showing its significance for heredity, published an account of how the race to unlock the secret of life was won. His book, *The Double Helix,* caused a great stir, as it portrayed the daily practice of science as something quite different than a collective enterprise with members who pursue the truth with gentlemanly disinterest. He portrayed an enterprise populated instead by many different types of characters, some highly competitive and even willing to engage in unseemly behavior to gain an upper hand. Whether or not it was his intention, the tale was received by some as a guidebook for how to succeed in science. Winning the race to unravel the big questions is the main thing; how you do it is less important. The reward is a place in the Nobel Hall of Fame. It would be wrong to say that Watson's book spawned the present culture of science. The United States thrives on competition and makes winning a dominant goal. But generations of young science students did read the book and took its message to heart. Competition is the way to "best" science; the Nobel Prize is the Holy Grail that inspires the quest. Is science just a glorified goose? Is squeezing out more golden eggs all we should aim for?

The cult of the prize thus touches the soul of the modern scientific enterprise, but it speaks for only part of the enterprise. The heritage of science is far richer.

## Benefit on Mankind

Perhaps no clause in Nobel's testament has been more contentious than the one referring to persons who "shall have conferred the greatest benefit on mankind." Are the interests of humankind to be defined by a small group of academic specialists who contend that their esoteric knowledge ultimately benefits all? Some in the Academy insisted that greater attention be given to contributions that satisfy both a quest to know and a desire to serve society. Some even wondered whether the academic self-interest reflected in the choice of winners had anything to do with Nobel's intentions.

At times, committee members invoked the phrase as a rhetoric device. Thus, Gullstrand argued against the Academy's rebellious desire to award Perrin by asserting that the Frenchman's work had not "benefited mankind." In the early years, nominators did not know how to interpret this phrase. Others used it to focus upon technical and applied achievements. Occasionally the committees accepted that argument, such as when they awarded the prize to Marconi and Braun. Not satisfied, Swedish practitioners of technical chemistry and engineering regularly lauded achievements in technological and industrial processes. They understood well the value of the prize in conferring prestige. Rebellions resulted in prizes for Dalén and for Bosch and Bergius, frightened the physics committee into rewarding Richardson, and brought Le Châtelier regularly to the committees' attention. Champions of Bjerknes repeatedly implored the Academy to accept that his school's innovative research in developing a physics of weather change went to the essence of Nobel's intentions: to celebrate those who contribute knowledge that has a direct positive impact on society.

No matter how hard they tried, those who wanted to respect Nobel's wishes faced ambiguities and quarrels. As Oseen confessed: "The interpretation of words changes from person to person. What do we mean by physics? What do we mean by 'during the preceeding year' or 'recently'? . . . Do any words have a fixed significance?" As soon as one finds an interpretation "consistent with one bylaw, one finds another inconsistent." Those with the authority to do so defined what Nobel meant.

Moreover, the interpretations chosen were not innocent of deeper purpose. Academic scientists defined "benefit on mankind" in their own ways. They generally underscored the belief that science advances best when state and industry give both money and complete freedom. An uncritical "trickle-down" model repeatedly surfaced. In more recent times, as long as the cold war and economic prosperity stimulated massive investments in research, the consensus remained largely unchallenged. That is, academic research provides reliable knowledge that can be applied by others to economic, social, and military advantage. To its critics, this model became an apology for self-interested science, especially when presented as the *only* way to organize national research. In the past decade, as many nations have begun to scale down support for research, it becomes relevant again to ask: How is research legitimized as a social activity? Why is it good for a society to support research and celebrate individuals who devote their lives to science?

Although we can never know with certainty what Alfred Nobel intended, we can try to appreciate what he meant by "benefit on mankind" in historical context. In a society marked by an increase of commercial and materialistic values, some members of the Nobel committees and Academy defined "benefit on mankind" in terms of the pursuit of academic knowledge for its own sake: celebrating the selfless example of a researcher who, through hard work, contributed to knowledge without seeking personal gain in the form of profit or fame. Such a person benefits humankind by offering a counterexample to the vulgar pursuit of wealth and aggrandizement. At least, so it seemed to an academic elite, which felt its social, cultural, and moral leadership and legitimacy being undermined. For these members of the Academy, benefit to humankind was written into the very act by which the king presents the academic researcher his prize: a laurel wreath to the seeker of truth who stands apart from the world of markets and who—by remaining pure—advances civilization.

This theme goes to the foundations of modern academic research. Whether for the university reformers in early nineteenth-century Prussia, who first introduced original research as a criterion of professorial appointment, or for those visionaries in Baltimore who established the Johns Hopkins University in 1876 as the first American university embracing these German ideals, research for its own sake entailed a moral vision of the professor insulated from commercial values. The importance was placed not on the results of research but on the life devoted to research. Daniel Coit Gilman, first president of Johns Hopkins, noted: "I can never rid myself of the belief that the essential value of the university does not depend upon the discoveries it makes, or the knowledge it accumulates and imparts, but in the character which it develops. . . . In the hunt for truth, we are not first hunters and then men; we are first and always men, then hunters."

Gilman did not separate the intellectual growth of the disciplines from the development of moral character. Research, teaching, and civic concern were bound together in an effort to produce genteel leaders. The methods and practice of the sciences in the seminar and laboratory were themselves a form of experimental method of molding proper citizens. Research offered an alternative or supplement to liberal arts for cultivation. Learning the methods of research went hand in hand with learning the conduct of being a researcher. Those with Ph.D.s would then teach in colleges and other schools, spreading through their example a morally uplifting way of life. In Ira Remsen's chemistry laboratory at Hopkins, shirtsleeves remained unfolded, the gas was turned off at 5 P.M.

According to what became a prevalent nineteenth-century ideal, one did not engage in research for the purpose of making money or seeking fame. It was morally good to devote oneself to research—for its own sake—and such an individual benefited society through his example. Confronting the mysteries and enormity of nature would foster awe and humility; seeking truth through disciplined hard work ennobled the researcher. This was perhaps aristocratic, but the moral effect was powerful. Other views also circulated and competed for allegiance, including research as a means to modernize and improve society. Regardless, we cannot turn the clock back. We are not simply free to choose values in the way that we select soft drinks and to define cultures based on taste. These are complex processes. But it is important for us to see that there have been, and there remain, many attitudes and ideologies of research. History reveals science's buried traditions and paths not taken. The dominant voices in any period are never the only ones, and rarely are they uncontested.

The pursuit of research as a means of personal cultivation seems to have been lost in the postwar era of Big Science, big bucks, and competition for prestige and resources. What happens when those who claim to seek knowledge for its own sake can do so only by bringing marketplace values into the practice of science? Where is the moral basis for research in entrepreneurial science "for its own sake" that must engage in cutthroat competition for resources? In such a culture, the value and worth of the scientist and her products become commodities both in the academic marketplace and in society at large. That is, just what the nineteenth-century academic reformers sought to avoid. During the cold war, all too many scientists and university administrators were willing to cast aside moral identity in the chase after extramural "big bucks."

In contrast to Gilman, the cold war era might be typified by Stanford University's Frederick E. Terman—professor, dean of engineering, and, finally, provost—who summed up his success in helping transform a private regional California institution into a university of international fame and prominence through astute use of military and industrial grants: "This game of improving an educational operation is great fun to play because it is so easy to win. Most of the competition just doesn't realize that education is a competitive business, like football, only with no conference rules." And just as in football, so in science winning has become the only thing. In the highly competitive and bureaucratic world of contemporary science, prizes such as the Nobel Prize provide a simple means for declaring excellence and high achievement. Unlike Gilman, many of today's university presidents and funding agencies do not want gentlemen but hunters—damned good ones, too. Readily evaluated results that can be counted and displayed, that bring prestige and grants, seem to dominate the ethos of science.

There is an irony in this development. Many enthusiastic boosters of the Nobel cult saw the prize as a means to celebrate individual "genius," yet the prize and the culture of prizes today enable universities and government agencies to weigh and measure "quality" on these most superficial indices. What does the number of laureates at a university—regardless of how they are selected—say about the overall quality of research, of the teaching programs, and of the quality of life at the institution? Whatever Alfred Nobel might have meant by benefit on mankind, he certainly did not have in mind the promotion of narrow professional interests, institutional boosterism, and careerist advancement.

As the fury of competition increases, a buzz of discontent is heard. A new post–cold war era for science has begun. New voices are rethinking and reclaiming what might be considered Nobel's legacy in the new century. They express sentiments that can best be summed up by a classic Calvin and Hobbes cartoon. Having beaten his stuffed tiger, Hobbes, in a game of checkers, six-year-old Calvin characteristically goes berserk, shouting, "I won! I did it! I won! I won! . . . I'm the champion! I'm the best there is! I'm the top of the heap!" Suddenly he looks around, senses the emptiness of his jubilant hysteria and comments, "Is this all there is?" Is science or society served by a fixation on prizes and on nurturing cultures of extreme competition? Perhaps once the mystery of the Nobel Prize is reduced, and with a bit of reflection as to what is truly significant in life and in the life of mind, we might also ask of the prize, "Is this all there is?"

# Appendix A

Winners of the Nobel Prize in Physics and Chemistry, 1901–2000

| Year | Physics | Chemistry |
|------|---------|-----------|
| 2000 | Zhores I. Alferov, Herbert Kroemer, Jack S. Kilby | Alan J. Heeger, Alan G. MacDiarmid, Hideki Shirakawa |
| 1999 | Gerardus 't Hooft, Martinus J. G. Veltman | Ahmed H. Zewail |
| 1998 | Robert B. Laughlin, Horst L. Störmer, Daniel C. Tsui | Walter Kohn, John A. Pople |
| 1997 | Steven Chu, Claude Cohen-Tannoudji, William D. Phillips | Paul D. Boyer, John E. Walker, Jens C. Skou |
| 1996 | David M. Lee, Douglas D. Osheroff, Robert C. Richardson | Robert F. Curl, Jr., Harold W. Kroto, Richard E. Smalley |
| 1995 | Martin L. Perl, Frederick Reines | Paul J. Crutzen, Mario J. Molina, F. Sherwood Rowland |
| 1994 | Bertram N. Brockhouse, Clifford G. Shull | George A. Olah |
| 1993 | Russell A. Hulse, Joseph H. Taylor, Jr. | Kary B. Mullis, Michael Smith |
| 1992 | Georges Charpak | Rudolph A. Marcus |
| 1991 | Pierre-Gilles de Gennes | Richard R. Ernst |
| 1990 | Jerome I. Friedman, Henry W. Kendall, Richard E. Taylor | Elias James Corey |
| 1989 | Norman F. Ramsey, Hans G. Dehmelt, Wolfgang Paul | Sidney Altman, Thomas R. Cech |
| 1988 | Leon M. Lederman, Melvin Schwartz, Jack Steinberger | Johann Deisenhofer, Robert Huber, Hartmut Michel |
| 1987 | J. Georg Bednorz, K. Alexander Müller | Donald J. Cram, Jean-Marie Lehn, Charles J. Pedersen |
| 1986 | Ernst Ruska, Gerd Binnig, Heinrich Rohrer | Dudley R. Herschbach, Yuan T. Lee, John C. Polanyi |
| 1985 | Klaus von Klitzing | Herbert A. Hauptman, Jerome Karle |
| 1984 | Carlo Rubbia, Simon van der Meer | Robert Bruce Merrifield |
| 1983 | Subramanyan Chandrasekhar, William Alfred Fowler | Henry Taube |
| 1982 | Kenneth G. Wilson | Aaron Klug |

*(continued)*

| Year | Physics | Chemistry |
|---|---|---|
| 1981 | Nicolaas Bloembergen, Arthur L. Schawlow, Kai M. Siegbahn | Kenichi Fukui, Roald Hoffmann |
| 1980 | James W. Cronin, Val L. Fitch | Paul Berg, Walter Gilbert, Frederick Sanger |
| 1979 | Sheldon L. Glashow, Abdus Salam, Steven Weinberg | Herbert C. Brown, Georg Wittig |
| 1978 | Pyotr Leonidovich Kapitsa, Arno A. Penzias, Robert W. Wilson | Peter D. Mitchell |
| 1977 | Philip W. Anderson, Nevill F. Mott, John H. van Vleck | Ilya Prigogine |
| 1976 | Burton Richter, Samuel C. C. Ting | William N. Lipscomb |
| 1975 | Aage Bohr, Ben Mottelson, James Rainwater | John Warcup Cornforth, Vladimir Prelog |
| 1974 | Martin Ryle, Antony Hewish | Paul J. Flory |
| 1973 | Leo Esaki, Ivar Giaever, Brian D. Josephson | Ernst Otto Fischer, Geoffrey Wilkinson |
| 1972 | John Bardeen, Leon N. Cooper, J. Robert Schrieffer | Christian B. Anfinsen, Stanford Moore, William H. Stein |
| 1971 | Dennis Gabor | Gerhard Herzberg |
| 1970 | Hannes Alfvén, Louis Néel | Luis F. Leloir |
| 1969 | Murray Gell-Mann | Derek H. R. Barton, Odd Hassel |
| 1968 | Luis W. Alvarez | Lars Onsager |
| 1967 | Hans Albrecht Bethe | Manfred Eigen, Ronald George Wreyford Norrish, George Porter |
| 1966 | Alfred Kastler | Robert S. Mulliken |
| 1965 | Sin-Itiro Tomonaga, Julian Schwinger, Richard P. Feynman | Robert Burns Woodward |
| 1964 | Charles Hard Townes, Nicolay Gennadiyevich Basov, Aleksandr Mikhailovich Prokhorova | Dorothy Crowfoot Hodgkin |
| 1963 | Eugene P. Wigner, Maria Goeppert-Mayer, J. Hans D. Jensen | Karl Ziegler, Giulio Natta |
| 1962 | Lev Davidovich Landau | Max Ferdinand Perutz, John Cowdery Kendrew |
| 1961 | Robert Hofstadter, Rudolf Ludwig Mössbauer | Melvin Calvin |
| 1960 | Donald A. Glaser | Willard Frank Libby |
| 1959 | Emilio Gino Segrè, Owen Chamberlain | Jaroslav Heyrovsky |

| | | |
|---|---|---|
| 1958 | Pavel Alekseyevich Cherenkov, Il'ja Mikhailovich Frank, Igor Tamm | Frederick Sanger |
| 1957 | Chen Ning Yang, Tsung-Dao Lee | Alexander R. Todd |
| 1956 | William Shockley, John Bardeen, Walter Houser Brattain | Cyril Norman Hinshelwood, Nikolay Nikolaevich Semenov |
| 1955 | Willis Eugene Lamb, Polykarp Kusch | Vincent du Vigneaud |
| 1954 | Max Born, Walther Bothe | Linus Carl Pauling |
| 1953 | Frits (Frederik) Zernike | Hermann Staudinger |
| 1952 | Felix Bloch, Edward Mills Purcell | Archer John Porter Martin, Richard Laurence Millington Synge |
| 1951 | John Douglas Cockcroft, Ernest Thomas Sinton Walton | Edwin Mattison McMillan, Glenn Theodore Seaborg |
| 1950 | Cecil Frank Powell | Otto Paul Hermann Diels, Kurt Alder |
| 1949 | Hideki Yukawa | William Francis Giauque |
| 1948 | Patrick Maynard Stuart Blackett | Arne Wilhelm Kaurin Tiselius |
| 1947 | Edward Victor Appleton | Robert Robinson |
| 1946 | Percy Williams Bridgman | James Batcheller Sumner, John Howard Northrop, Wendell Meredith Stanley |
| 1945 | Wolfgang Pauli | Artturi Ilmari Virtanen |
| 1944 | Isidor Isaac Rabi | Reserved; in 1945 awarded to Otto Hahn |
| 1943 | Reserved; in 1944 awarded to Otto Stern | Reserved; in 1944 awarded to George de Hevesy |
| 1942 | The prize money was allocated one-third to the main fund and two-thirds to the special fund | The prize money was allocated one-third to the main fund and two-thirds to the special fund |
| 1941 | The prize money was allocated one-third to the main fund and two-thirds to the special fund | The prize money was allocated one-third to the main fund and two-thirds to the special fund |
| 1940 | The prize money was allocated one-third to the main fund and two-thirds to the special fund | The prize money was allocated one-third to the main fund and two-thirds to the special fund |
| 1939 | Ernest Orlando Lawrence | Adolf Friedrich Johann Butenandt, Leopold Ruzicka |
| 1938 | Enrico Fermi | Reserved; in 1939 awarded to Richard Kuhn |
| 1937 | Clinton Joseph Davisson, George Paget Thomson | Walter Norman Haworth, Paul Karrer |
| 1936 | Victor Franz Hess, Carl David Anderson | Petrus (Peter) Josephus Wilhelmus Debye |

*(continued)*

| Year | Physics | Chemistry |
|---|---|---|
| 1935 | James Chadwick | Frédéric Joliot, Irène Joliot-Curie |
| 1934 | Reserved; in 1935 the prize money was allocated one-third to the main fund and two-thirds to the special fund | Harold Clayton Urey |
| 1933 | Erwin Schrödinger, Paul A. M. Dirac | Reserved; in 1934 the prize money was allocated one-third to the main fund and two-thirds to the special fund |
| 1932 | Reserved; in 1933 awarded to Werner Heisenberg | Irving Langmuir |
| 1931 | Reserved; in 1932 the prize money was allocated to the special fund | Carl Bosch, Friedrich Bergius |
| 1930 | Chandrasekhara Venkata Raman | Hans Fischer |
| 1929 | Louis-Victor de Broglie | Arthur Harden, Hans K. A. S. von Euler-Chelpin |
| 1928 | Reserved; in 1929 awarded to Owen Willans Richardson | Adolf Otto Reinhold Windaus |
| 1927 | Arthur Holly Compton, Charles Thomson Rees Wilson | Reserved; in 1928 awarded to Heinrich Otto Wieland |
| 1926 | Jean Baptiste Perrin | The(odor) Svedberg |
| 1925 | Reserved; in 1926 awarded to James Franck, Gustav Hertz | Reserved; in 1926 awarded to Richard Adolf Zsigmondy |
| 1924 | Reserved; in 1925 awarded to Karl Manne Georg Siegbahn | Reserved; in 1925 the prize money was allocated to the special fund |
| 1923 | Robert Andrews Millikan | Fritz Pregl |
| 1922 | Niels Bohr | Francis William Aston |
| 1921 | Reserved; in 1922 awarded to Albert Einstein | Reserved; in 1922 awarded to Frederick Soddy |
| 1920 | Charles Edouard Guillaume | Reserved; in 1921 awarded to Walther Hermann Nernst |
| 1919 | Johannes Stark | Reserved; in 1920 the prize money was allocated to the special fund |
| 1918 | Reserved; in 1919 awarded to Max Planck | Reserved; in 1919 awarded to Fritz Haber |
| 1917 | Reserved; in 1918 awarded to Charles Glover Barkla | Reserved; in 1918 the prize money was allocated to the special fund |
| 1916 | Reserved; in 1917 the prize money was allocated to the special fund | Reserved; in 1917 the prize money was allocated to the special fund |

| | | |
|---|---|---|
| 1915 | William Henry Bragg, William Lawrence Bragg | Richard Martin Willstätter |
| 1914 | Postponed; in 1915 awarded to Max von Laue | Postponed; in 1915 awarded to Theodore William Richards |
| 1913 | Heike Kamerlingh-Onnes | Alfred Werner |
| 1912 | Nils Gustaf Dalén | Victor Grignard, Paul Sabatier |
| 1911 | Wilhelm Wien | Marie Curie, née Sklodowska |
| 1910 | Johannes Diderik van der Waals | Otto Wallach |
| 1909 | Guglielmo Marconi, Carl Ferdinand Braun | Wilhelm Ostwald |
| 1908 | Gabriel Lippmann | Ernest Rutherford |
| 1907 | Albert Abraham Michelson | Eduard Buchner |
| 1906 | Joseph John Thomson | Henri Moissan |
| 1905 | Philipp Lenard | Johann Friedrich Wilhelm Adolf von Baeyer |
| 1904 | Lord Rayleigh (John William Strutt) | William Ramsay |
| 1903 | Antoine Henri Becquerel, Pierre Curie, Marie Curie, née Sklodowska | Svante August Arrhenius |
| 1902 | Hendrik Antoon Lorentz, Pieter Zeeman | Hermann Emil Fischer |
| 1901 | Wilhelm Conrad Röntgen | Jacobus Henricus van't Hoff |

# Appendix B

Committee Members, 1900–1951

| Member | Years of Service | Affiliation | Specialization |
|---|---|---|---|
| | | Physics Committee | |
| Knut Ångström | 1900–1910 | Uppsala University | Experimental physics |
| Svante Arrhenius | 1900–1927 | Stockholm Högskola/Nobel Institute | Physical chemistry, cosmical physics |
| Bernhard Hasselberg | 1900–1922 | Academy of Sciences | Astrophysics, metrology |
| Hugo Hildebrand Hildebrandsson | 1900–1910 | Uppsala University | Meteorology |
| Robert Thalén | 1900–1903 | Uppsala University | Experimental physics |
| Gustaf Granqvist | 1904–1922 | Uppsala University | Experimental physics |
| Vilhelm Carlheim-Gyllensköld | 1910–1934 | Stockholm Högskola | Cosmical physics, mathematical physics |
| Allvar Gullstrand | 1911–1929 | Uppsala University | Ophthalmology |
| Carl Wilhelm Oseen | 1923–1944 | Uppsala University/Nobel Institute | Theoretical physics |
| Manne Siegbahn | 1923–1962 | Uppsala University/Nobel Institute | Experimental physics |
| Henning Pleijel | 1928–1947 | Royal Institute of Technology | Theoretical electronics |
| Erik Hulthén | 1929–1962 | Stockholm Högskola | Experimental physics |
| Axel Lindh | 1935–1960 | Uppsala University | Experimental physics |
| Ivar Waller | 1944–1971 | Uppsala University | Theoretical physics |
| Gustaf Ising | 1947–1953 | Stockholm Högskola/Private research | Experimental physics, geophysics |

## Chemistry Committee

| | | | |
|---|---|---|---|
| Per Theodor Cleve | 1900–1905 | Uppsala University | Inorganic chemistry |
| Peter Klason | 1900–1925 | Royal Institute of Technology | Industrial chemistry |
| Otto Pettersson | 1900–1912 | Stockholm Högskola | Inorganic chemistry, physical oceanography |
| Henrik Söderbaum | 1900–1933 | Academy of Agriculture | Agricultural chemistry |
| Oskar Widman | 1900–1928 | Uppsala University | Organic chemistry |
| Olof Hammarsten | 1905–1926 | Uppsala University | Physiological chemistry |
| Åke Ekstrand | 1913–1924 | Government service | Industrial chemistry |
| Theodor Svedberg | 1925–1964 | Uppsala University | Colloid chemistry, biophysical chemistry |
| Wilhelm Palmær | 1926–1942 | Royal Institute of Technology | Physical chemistry, electrochemistry |
| Ludwig Ramberg | 1927–1940 | Uppsala University | Physical-organic chemistry |
| Hans von Euler-Chelpin | 1929–1946 | Stockholm Högskola | Organic chemistry, biochemistry |
| Bror Holmberg | 1934–1953 | Royal Institute of Technology | Organic chemistry |
| Arne Westgren | 1942–1965 | Stockholm Högskola/Academy of Sciences | Metallurgy, crystallography |
| Arne Fredga | 1944–1975 | Uppsala University | Organic chemistry, stereochemistry |
| Arne Tiselius | 1947–1971 | Uppsala University | Biochemistry, biophysical chemistry |

# Appendix C

## Money Matters

| Prize Amount (in Swedish crowns) | | Value In | |
|---|---|---|---|
| **Year** | **Amount** | **U.S. $** | **U.K. £** |
| 1901 | 150,782 | 41,800 | 8,300 |
| 1910 | 140,703 | | |
| 1915 | 149,223 | | |
| 1920 | 134,100 | 36,250 | 8,252 |
| 1923 | 114,935[a] | | |
| 1930 | 172,947 | 43,236 | 9,348 |
| 1935 | 159,917 | | |
| 1940 | 138,570 | | |
| 1945 | 121,333 | 28,900 | 7,137 |
| 1947 | 146,115[b] | | |
| 1950 | 164,304 | 31,700 | 11,300 |
| 1953 | 175,293[c] | | |
| 1960 | 225,987 | | |
| 1969 | 375,000[d] | | |
| 1980 | 880,000 | 190,500 | 88,900 |
| 1990 | 4,000,000 | 678,000 | 377,400 |
| 2000 | 9,000,000 | 1,000,000 | 647,500 |

Variations in the Funds, 1901–1940 (in Swedish crowns, × 1,000)

| | 1901 | 1910 | 1920 | 1930 | 1940 |
|---|---|---|---|---|---|
| Main fund[e] | 27,800 | 28,900 | 30,191 | 31,548 | 33,300 |
| Organization funds[f] | | | | | |
| Physics | 322 | 204 | 293 | 499 | 10 |
| Chemistry | 322 | 204 | 293 | 499 | 1,040 |
| Special funds[g] | | | | | |
| Physics | — | — | 140 | 177 | 482 |
| Chemistry | — | — | 419 | 681 | 1,024 |

## Notes

*a* The lowest amount for a prize.

*b* The Swedish government removed most remaining taxes on the Nobel Foundation's funds in 1946. Much of the variation after World War I was related to changing rates of taxation. Some of the extreme post-1918 taxation was relieved in 1925–1926.

*c* The foundation's investment rules were liberalized. The original stipulation that Nobel's estate had to be invested in "safe securities" received broader interpretation during the post–World War I economic crises and then again during World War II and its aftermath. Further liberalization and exceedingly insightful investment policies by the Nobel Foundation's Executive Director Stig Ramel in the 1970s and 1980s yielded bountiful gains, reflected in the rapid growth of the prize.

*d* The prize in economic science is added; main fund is augmented.

*e* Alfred Nobel's estate generated more than 31 million crowns ($8.6 million or £1.6 million) from which the initial funds were created. Interest generated from the main fund provides the annual money for all the Nobel prizes and operating expenses.

*f* The five committees (physics, chemistry, medicine/physiology, literature, and peace) each received an equal share of the 1.5 million crowns set aside for organizing the Nobel Institutes. Both the physics and chemistry committees used a portion of their funds in 1904 to establish Svante Arrhenius's Nobel Institute for Physical Chemistry. In the 1930s, the physics committee used most of its funds to create institutes for C. W. Oseen and Manne Siegbahn.

*g* The special funds originated from the capital and interest on prize money that was not distributed. Prizes were first withheld during World War I. During the interwar years, the committees used some of the yields generated annually from these funds for research grants and for improvements to the Nobel Institute. The differences arose from the number of prizes withheld and the willingness of the committee members to use or save the annual yields. Their importance changed after World War II, when the prizes were always distributed and government-sponsored research councils provided new means for financing research.

*Sources:* Nobel Foundation Calendar, Nils K. Ståle, "Administration and finances of the Nobel Foundation," in *Nobel: The Man and His Prizes,* 1st ed. (Stockholm and Norman, OK: Sohlmans Förlag and University of Oklahoma Press, 1950), pp. 578–581; Sveriges Riksbank, *Historisk Statistik.*

# Notes

### A Note on the Archival Documents

Subsequent to the Nobel Foundation's relaxing in 1974 of the provision in the bylaws that imposed secrecy on the deliberations about prizes, each of the institutions that award prizes opened its archives to qualified scholars. To grant access to documents related to the prizes in physics and chemistry at least fifty years old, the Royal Swedish Academy of Sciences created a Nobel Archive Committee [NOAK] and deposited the relevant materials in the archives at its Center for History of Science.

The relevant documents fall into two groupings: the respective archives of the Nobel Committees for Physics or Chemistry and the archives of the Royal Swedish Academy of Sciences related to Nobel matters. The former archives entail the protocol of committee meetings and the annual reports in which candidates are evaluated. These documents frequently include early drafts of the reports that were discussed at meetings prior to the committee's vote; changes in wording from draft to draft at times can provide insight into the process of arriving at a final decision. Affixed to the annual reports as appendixes are a number of individual special reports providing detailed descriptions and assessment of a select number of candidates. The physics committee tended to prepare detailed reports of candidates who were considered serious contenders. The chemistry committee gradually accepted a practice in which most new candidates received a special report along with those candidates who were thought to be in contention that year. Individual committee members accepted assignments to write special reports for one or more candidates, so that this single evaluator's own insight and judgment played a critical role in assessing whether the candidate was found worthy of a prize. One or two committee members wrote the more general report that provides a short description of each candidate's work and a short assessment of its merit. Also included are excerpts from the special reports. The general reports close with a recommendation for how the prize(s) for that year should be allocated, including a short clause containing the formal justification for awarding the prize to the candidate. Each committee member signs the report; if one or more members dissent from the majority, their reasons and alternative choices are attached as an appendix.

The final draft of the report and its appendixes of detailed assessments were sent to the Academy after the committee voted. The Academy's Nobel archives consist of protocol over the Academy's meetings related to Nobel issues as well as the meetings of the respective Physics and Chemistry Sections [*klasser*]. The votes of the respective ten-member sections are recorded; dissenting statements are included as appendixes to the protocol of the meeting. The deciding vote in the full Academy is not recorded, only the result is provided. No summary of the discussion is included; on occasion a comment such as, "after long discussion in which different opinions were expressed a vote was taken," might be noted. Insight into the actual discussion in the full Academy can only be gleaned through private correspondence, diary notations, and, once in a while, vote tallies made by the Academy's permanent secretary on his copies of the protocol. The Academy's protocol for each year includes a copy of the final version of the physics and chemistry committees' evaluation reports as well as all the letters of nomination that were eligible to be considered. (Letters from persons who did not have a right to submit were not included.)

The language of all the reports and protocol is Swedish; letters of nomination were accepted in Swedish, Danish, Norwegian, English, French, or German.

## References to Nobel Documents

In the Notes, the following abbreviations are used to refer to the Nobel archival documents.

Reference to meetings of the Nobel Committees for Physics or Chemistry:

*Protokoll vid Kungl. Vetenskapsakademiens Nobelkommittéen för fysik/kemi sammanträde* [Minutes of meetings of Nobel Committee for Physics or Chemistry of the Royal Swedish Academy of Sciences; hereafter, *Protokoll*, NKF/NKK]; followed by the date. Documents attached as appendixes are identified as *Bilagor* A, B, C . . . to the protocol of the meeting. An early draft of an evaluation report is cited as a *Bilaga* to the given meeting in which it was discussed; where relevant, the date of writing the draft is provided.

Reference to the final draft of the annual evaluation report is referred to either as a *Bilaga* to the meeting in which the committee voted or simply, for example:

*Kommittéutlåtande, Nobelkommittéen för kemi* [Committee report, Nobel Committee for Chemistry], followed by the year. Abbreviated as KU, NKK, 1903.

Reference to the Academy's meetings related to Nobel matters:

*Protokoll vid Kungl. Vetenskapsakademiens Sammankomster för Behandling af Ärenden Rörande Nobelstiftelsen* [Minutes of meetings of the Royal Swedish Academy of Sciences, for discussion of matters concerning the Nobel Foundation]. Abbreviated as *Protokoll*, KVA/N, followed by the date.

Reference to meetings of the Physics Section (III klassen) or Chemistry Section (IV klassen):

*Protokoll,* KVA, IV klassen, 15 okt 1910 [Protocol, Royal Swedish Academy, IVth Section, October 15, 1910].

Letters of nomination follow the Academy's annual protocol. These are referred to by the abbreviation KVA/N [Royal Swedish Academy of Sciences protocol on Nobel matters] and the year. Occasionally the date of the letter is included.

## Abbreviations

**Unpublished Sources**

| | |
|---|---|
| AG | Allvar Gullstrand Papers, Uppsala University |
| AR | Arthur Rindell Papers, Åbo Academy, Turku, Finland |
| BF | Bjerknes Family Papers, University of Oslo |
| BH-KVA | Bernhard Hasselberg Papers, Royal Swedish Academy of Sciences, Stockholm |
| BH-LUB | Bror Holmberg Papers, Lund University |
| BSC | Niels Bohr's Scientific Correspondence, Microfilm, American Institute of Physics, Niels Bohr Center for History of Physics, College Park, MD (originals and supplementary correspondence in NBA) |

## Notes 291

| | |
|---|---|
| CA | Christopher Aurivillius Papers, Collection No. 2, Royal Swedish Academy of Sciences |
| CB | Carl Benedicks Papers, Royal Library, Stockholm |
| CC | C. V. L. Charlier Papers, Lund University |
| CWO-KVA | Carl Wilhelm Oseen Papers, Royal Swedish Academy of Sciences |
| CWO-LAL | Carl Wilhelm Oseen Papers, Regional Archives in Lund [*Landsarkivet i Lund*] |
| EH | Edward Hjelt Papers, National Archive (Finland), Helsinki |
| ER | Ernst Riesenfeld Papers, Royal Swedish Academy of Sciences |
| GE | Gustaf Ekman Papers, Regional Archive in Gothenburg [*Landsarkivet i Göteborg*] |
| GH | George Ellery Hale Papers, California Institute of Technology, Pasadena |
| GL | Göran Liljestrand Papers, Royal Swedish Academy of Sciences |
| GR | Gustaf Retzius Papers, Royal Swedish Academy of Sciences |
| HH | Hjalmar Hammarskjöld Papers, Royal Library, Stockholm |
| HHH-KVA | Hugo Hildebrand Hildebrandsson Papers, Royal Swedish Academy of Sciences |
| HHH-UUB | Hugo Hildebrand Hildebrandsson Papers, Uppsala University |
| IEB | International Education Board, Rockefeller Archive Center, North Tarrytown, NY |
| IW | Ivar Waller Papers, Royal Swedish Academy of Sciences |
| JF | James Franck Papers, University of Chicago |
| JL | Jacques Loeb Papers, Library of Congress, Washington, DC |
| KP | Kristian Prytz Papers, Royal Library, Copenhagen |
| KVA | Royal Swedish Academy of Sciences [*Kungliga vetenskapsakademien*] |
| LM | Lise Meitner Papers, Churchill College, Cambridge University |
| LR | Ludwig Ramberg Papers, Uppsala University |
| ML-KB | Gösta Mittag-Leffler Diaries, Royal Library, Stockholm |
| ML-MLI | Gösta Mittag-Leffler Correspondence, Institute Mittag-Leffler, Djursholm |
| MS | Manne Siegbahn Papers, Royal Swedish Academy of Sciences |
| NBA | Niels Bohr Institute Archives, Copenhagen |
| NBP | Niels Bohr Personal Correspondence, Niels Bohr Institute Archives, Copenhagen |
| OA | Ossian Aschan Papers, Åbo Academy, Turku, Finland |
| OK | Oskar Klein Papers, Niels Bohr Institute Archives, Copenhagen |
| OW | Oskar Widman Papers, Uppsala University |
| RF | Rockefeller Foundation, Rockefeller Archive Center, North Tarrytown, NY |
| SA | Svante Arrhenius Papers, Royal Swedish Academy of Sciences |
| TS | Theodor Svedberg Papers, Uppsala University |

| TWR | T. W. Richards Papers, Harvard University |
| UBO | University of Oslo Library |
| UUB | Uppsala University Library |
| VB | Vilhelm Bjerknes Papers, University of Oslo |
| WC | W. W. Campbell Papers, Mary Lea Shane Archives of the Lick Observatory, University of California, Santa Cruz |
| WO | Winthrop Osterhout Papers, American Philosophical Society, Philadelphia |
| WP | Wilhelm Palmær Papers, Royal Swedish Academy of Sciences |

**Published Sources**

| DN | *Dagens Nyheter* |
| DSB | *Dictionary of Scientific Biography* |
| NDA | *Nya Dagligt Allehanda* |
| SvD | *Svenska Dagbladet* |

## Author's Note and Acknowledgments

**xi** in the British science journal *Nature*: "Nobel physics prize in perspective," *Nature*, August 27, 1981, pp. 793–798.

**xii** published in specialized historical journals: "The prizes in physics and chemistry in the context of Swedish science" (with E. Crawford), in Carl-Gustaf Bernhard, Elisabeth Crawford, and Per Sörbom, eds., *Science, Technology and Society in the Time of Alfred Nobel*, Nobel Symposium, Björkborn, Karlskoga, Sweden, August 17–22, 1981 (Oxford: Pergamon Press for the Nobel Foundation, 1982), pp. 311–331; "The Nobel prize and the history of scientific disciplines: Preliminary thoughts and principles" (in Swedish), in Gunnar Broberg, Gunnar Eriksson, and Karin Johannisson, eds., *Kunskapens trädgårdar: Om institutioner och institutionaliseringar i vetenskapen och livet* (Stockholm: Atlantis, 1988), pp. 136–152; "Americans as candidates for the Nobel prize: The Swedish perspective," in Stanley Goldberg and Roger Stuewer, eds., *The Michelson Era in American Science, 1870–1930* (New York: American Institute of Physics, 1988), pp. 272–287; "Text, context, and quicksand: Method and understanding in studying the Nobel science prizes," *Historical Studies in the Physical and Biological Sciences*, **20**, 1 (1989), pp. 63–77; "The Nobel prize and disciplinary conflicts of interest: The cases of Vilhelm Bjerknes and Viktor Moritz Goldschmidt" (in Norwegian), *Forskningspolitikk*, **12**, no. 2 (1989), pp. 18–21; "The Nobel prizes and the invigoration of Swedish science: Some preliminary considerations," in Tore Frängsmyr, ed., *Solomon's House Revisited: The Organization and Institutionalization of Science*, Nobel Symposium 75 (Canton, MA: Science History Publications and Nobel Foundation, 1990), pp. 193–207.

**xiv** the most important monograph on the subject: Elisabeth Crawford, *The Beginnings of the Nobel Institution: The Science Prizes, 1901–1915* (Cambridge: Cambridge University Press, 1984).

## Introduction

**2** If somebody were to catch cold: Letter, Hugo Theorell, to Sune Bergström, October 11, 1950, copy, Hugo Theorell Correspondence, KVA. I thank Professor Anders Lundgren for bringing this letter to my attention. Additional references recur regularly in correspondence from members of the medicine committee referring to the fact that the responsible members of the Caroline Institute are decisive for selecting prizewinners. For example, Göran Liljestrand to Statsråd [Cabinet Minister] Ragnar Edenman, December 12, 1964, copy, GL, complaining that Swedish colleague Ragnar Granit had not as yet received a Nobel Prize in medicine because of personal antagonisms within the institute. Granit finally did receive a prize in 1967.

**4** for better or for worse scientists are not always the most perfect of God's creatures: Letter, Oskar Widman to Bror Holmberg, February 17, 1919, BH-LUB.

**5** The period witnessed an almost explosive growth of national ceremonial traditions: Arno J. Mayer, *The Persistence of the Old Regime: Europe to the Great War* (London: Croom Helm, 1981), esp. pp. 189–273; Eric Hobsbawm and Terence Ranger, eds., *The Invention of Tradition* (Cambridge: Cambridge University Press, 1983), esp. pp. 263–307.

**6** shared a common biological heritage: On German-Nordic relations in politics and culture, *Tyskland og Skandinavia, 1800-1914: Impulser og Brytninger* (Oslo: Norsk Folkemuseum, 1997); on racial theories, see Gunnar Broberg, "Rasbiologi og antisemitism," in ibid., pp. 172–178; and Gunnar Broberg and Mattias Tydén, *Oönskade i folkhemmet. Rashygien och sterilisering i Sverige* (Stockholm: Gidlund, 1991).

**7** the Academy's greatest accomplishments: Tore Frängsmyr, "Introduction: 250 years of science," and "Swedish polar exploration," in Tore Frängsmyr, ed., *Science in Sweden: The Royal Swedish Academy of Sciences, 1739-1989* (Canton, MA: Science History Publications, 1989), pp. 1–20, 178–198; Gunnar Eriksson, *Kartläggarna: Naturvetenskapens tillväxt och tillämpningar i det industriella genombrottets Sverige, 1870-1914,* Umeå Studies in the Humanities 15 (Umeå: Umeå Universitetsbibliotek, 1978), pp. 33–39; E. W. Dahlgren, *Kungl. Svenska Vetenskapsakademien: Personförteckningar, 1739-1915* (Stockholm: Almqvist & Wiksells, 1915), pp. 94–101; Gunnar Broberg, "The Swedish Museum of Natural History," in Frängsmyr, ed., *Science in Sweden*, pp. 148–176; Urban Wråkberg, *Vetenskapens vikingatåg: Perspektiv på svensk polarforskning, 1860-1930* (Stockholm: Royal Swedish Academy of Sciences, 1999).

The Academy's nine sections in 1900: 1. Pure Mathematics (six members); 2. Applied Mathematics (six members); 3. Practical Mechanics (eight members); 4. Physics (six members); 5. Chemistry and Mineralogy (twelve members); 6. Zoology and Botany (sixteen members); 7. Medicine and Surgery (fifteen members); 8. Economic Sciences (fifteen members); 9. General Scholarship (sixteen members).

**7** the Academy's overreliance on established authority: Letter, Svante Arrhenius to Arthur Rindell, January 2, 1902, AR.

**7** ruled autocratically by the senior professor: Historian of universities Thorsten Nybom called the Swedish "an extreme version of the German *Lehrstuhl* system," quoted in Svante Lindqvist, ed., *Center on the Periphery: Historical Aspects of Twentieth-Century Swedish Physics* (Canton, MA: Science History Publications, 1993), pp. xxxii–xxxiii. Sten Lindroth, *Uppsala Universitet, 1477-1977* (Uppsala: Almqvist & Wiksells, 1976).

**8** ". . . the one professor dragging along the next one with him": Sonja Wassiljewna Kowalewsky, "Autobiografisk skizze," *Samtiden: Tidsskrift for literatur og samfundsspørsmaal,* **12** (1901), p. 327. A highly talented Russian mathematician, Sonja Kowalewsky was recruited to the Stockholm Högskola, where she remained from 1884 to 1891.

**8** fear of appearing fallible: Otto Pettersson used the occasion of Arrhenius's Nobel Prize in 1903 to attack the Uppsala academic culture: "Svante Arrhenius," *Svensk kemisk tidskrift,* **15** (1903), pp. 206–207.

**8** claiming that its bylaws specified only "men" as members: Frängsmyr, ed., *Science in Sweden*, pp. 17–18.

**9** the Högskola's life in its first several decades: Sven Tunberg, *Stockholms Högskolas Historia före 1950* (Stockholm: Norstedts, 1957). The Norwegians in question were geologist W. C. Brøgger, botanist Nordahl Wille, and physicist Vilhelm Bjerknes.

## ONE The Stupidest Use of a Bequest That I Can Imagine

**13** "a retiring, considerate person...": Biographical details are based on Henrik Schück, "Alfred Nobel: A Biographical Sketch," and Ragnar Sohlman, "Alfred Nobel and the Nobel Foundation," both in Nobel Foundation, ed., *Alfred Nobel: The Man and His Prizes*, 2nd ed. (New York: Elsevier, 1962). This official history was first published in 1950. Also, Sohlman's more detailed account, *Ett Testamente: Nobelstiftelsens tillkomsthistoria och dess grundare* (Stockholm: Norstedts, 1950). Quotation is from Schück, p. 3.

**15** familiarity with Stockholm science: Elisabeth Crawford, *The Beginnings of the Nobel Institution: The Science Prizes, 1901–1915* (Cambridge: Cambridge University Press, 1984), pp. 62–63.

**15** king-in-council: The term used for the government: the prime minister and cabinet (heads of the administrative ministries) as arms of the king. The government proposes laws and budgets that are sent to the Riksdagen (Parliament) for approval. On the machinations, see Sohlman, "Alfred Nobel and the Nobel Foundation," p. 46.

**15** the Swedish institutions expressed less than enthusiasm for Nobel's plans: Sohlman, *Ett Testamente*, pp. 195–200, 231–240.

**15** decline the task: Letter, Oskar Widman to Edward Hjelt, January 9, 1897, EH-RF.

**16** "permanent battles will surely be waged for every prize distribution": Letter, Svante Arrhenius to Hjelt, January 4, 1897, EH.

**16** "... the stupidest use of a bequest that I can imagine!": Letter, Otto Pettersson to Gustaf Retzius, January 4 [1897], GR-KVA. [*"Den dummaste användning af donationsmedel jag kan tänka mig! Att begärd premimium för sitt arbete är ej tilltalande för vetenskapsmän."*]

**16** "in a great mess": Arrhenius to Hjelt, January 4, 1897. Also expressed in Widman to Hjelt, January 9, 1897.

**16** Sohlman contemplated direct negotiations: Sohlman, *Ett Testamente*, pp. 109–112.

**16** willing to cooperate: Ibid., pp. 231–240.

**17** Af Wirsén managed to rally members: Sohlman, "Alfred Nobel and the Nobel Foundation," p. 109. On af Wirsén and the Swedish Academy, see Staffan Björck, *Heidenstam och sekelskiftets Sverige* (Stockholm: Nature och Kultur, 1946), pp. 7–69; Kjell Espmark, *Det litterära Nobelpriset. Principer och värderingar bakom besluten* (Stockholm: Norstedts, 1986), pp. 15–36.

**17** reimbursement equal to a professor's annual salary: Letter, Carl Lindhagen to Ragnar Sohlman, January 16, 1897, cited in Crawford, *Beginnings*, p. 65.

**17** "very large sums [of money] for scientific purposes": Letter, Arrhenius to Wilhelm Ostwald, December 25, 1896; reprinted in Hans-Günther Körber, ed., *Aus dem wissenschaftlichen Briefwechsel Wilhelm Ostwards*, II Teil (Berlin: Akademie-Verlag, 1969), p. 146.

**17** intrigues and spread malicious gossip in the media: Sven Tunberg, *Stockholms Högskolas Historia före 1950* (Stockholm: Norstedts, 1957), esp. pp. 86–87. The disputes even became the source of satirizing poems and cartoons in the media; see, for example, *Söndags-Nisse,* January 27, 1895. Letter, Arrhenius to Hjelt, July 26, 1895, EH, describes how protagonists were willing to soil the name of the institution by bringing quarrels into the public. Otto Pettersson tried to convince himself and his colleagues that the local conflicts did not influence Nobel's decision, but a heavy shadow of doubt remained. Letter, Pettersson to Retzius, January 4 [1897]. The Nobel puzzle about why no prize in mathematics was stipulated might well be related to Nobel's awareness that Mittag-Leffler would necessarily gain control over such a prize.

**18** invited prominent foreign scientists: Göran Liljestrand, "The prizes in physiology," in Nobel Foundation, ed., *Alfred Nobel: The Man and His Prizes*, 2nd ed. (New York: Elsevier, 1962), pp. 334–337.

**18** "help dreamers, who find it difficult to get on in life": Nobel reportedly made this statement a few months prior to his death. Fritz Henriksson, *The Nobel Prizes and the Founder Alfred Nobel*, New Sweden Tercentenary Publications (Stockholm: Albert Bonniers, 1938), p. 12. Also, Liljestrand, "The prizes in physiology," p. 143.

**18** Arrhenius began rehearsing a life of research leisure: Letters, Arrhenius to Ostwald, December 25, 1898, January 21 and 24, 1899, in Körber, *Aus dem wissenschaftlichen Briefwechsel*, pp. 151–154.

**19** to join his meetings with scientists: Sohlman, "Alfred Nobel and the Nobel Foundation," pp. 62–63.

**19** a spokesman for the soul: Ibid., p. 40.

**19** separate statutes for each prize-awarding institution: Liljestrand, "The prizes in physiology," pp. 336–337.

**19** several rounds of negotiations: Crawford's account in *Beginnings* builds upon the official histories, supplementing these versions with additional insight gained from private correspondence among the negotiators. My abbreviated narrative is indebted to her text as well as to Sohlman's *Ett Testamente*. The main source for all accounts is the official published record of the negotiations: *Protokoll hållna vid sammanträden för öfverläggning om Alfred Nobels testamente* (Stockholm: Nobelstiftelsen, 1899).

**20** should both propose and evaluate candidates: Crawford, *Beginnings*, p. 75.

**20** testimony against Arrhenius and for an alternative candidate: Letters, Otto Pettersson to Hjelt, June 24, 1895, EH; Arrhenius to Hjelt, July 26, 1895, EH; Pettersson to Retzius, n.d. [1895], GR.

**20** considerable expansion of foreign participation: Letter, Arrhenius to Lindhagen, February 15, 1899, quoted in Crawford, *Beginnings*, p. 82; *Protokoll . . . om Alfred Nobels testamente*, pp. 44–46.

**22** Stockholm on the verge of becoming an international center for research: Letters, Vilhelm Bjerknes to Fridtjof Nansen, January 18, 1902, Fridtjof Nansen Correspondence, UBO; Arrhenius to Hjelt, December 26, 1898, and January 18, 1900, EH; Arrhenius to Ostwald, January 24, 1899, in Körber, *Aus dem wissenschaftlichen. Briefwechsel*, p. 154.

**22** greater prestige and authority: Kristian Birkeland reveals his eagerness to move to Stockholm in letters to Mittag-Leffler, 1900–1905, ML-MLI. Arrhenius hoped Bjerknes would remain in Sweden, thinking that the Norwegian would soon become a committee member and enjoy the added salary and prestige. [Letter, Arrhenius to Nordahl Wille,

January 23, 1902, Nordahl Wille Collection, UBO.] Use of Nobel committee salary and professional benefits as a means to interest prospective candidates can be seen especially around 1905 after Arrhenius relinquished his professorship at the Högskola, as exemplified in correspondence from Mittag-Leffler to Kristian Birkeland, as well as from Bjerknes to Philipp Lenard. On being received with greater prestige, Arrhenius to Hjelt, January 8, 1903, EH, and Richard Abegg to Arrhenius, December 23, 1901, and January 23, 1902, SA.

**22** ambiguous and nonbinding: Liljestrand, "The prizes in physiology," pp. 135–139; Crawford, *Beginnings*, pp. 79–80; letter, Widman to Lars Fredrik Nilson, February 19, 1899, L. F. Nilson papers, UUB.

**22** 30 million Swedish crowns: At prevailing rates of exchange, equal to about $8.6 million or £1.6 million.

**24** Deliberations are kept secret; no protest or appeal is permitted: Nils K. Ståhle, "The statutes of the Nobel Foundation and the prize-awarding institutions," in Nobel Foundation, ed., *Alfred Nobel: The Man and His Prizes*, pp. 645–665.

## TWO Coming Apart at the Seams

**27** Nineteenth-century chemistry itself became fragmented: John W. Servos, *Physical Chemistry from Ostwald to Pauling: The Making of a Science in America* (Princeton, NJ: Princeton University Press, 1990), pp. 3–19ff.; Owen Hannaway, "The German model of chemical education in America: Ira Remsen at Johns Hopkins," *Ambix*, **23** (1976), pp. 145–164; William H. Brock, *The Fontana History of Chemistry* (London: Fontana Press, 1992), pp. 173–354.

**28** Widman also asked this Finnish colleague: Letter, Widman to Hjelt, September 12, 1900, EH; Widman to Hjelt, November 7, 1883, EH, on their discovering one another as being two rare Nordic synthetic organic chemists "on the outskirts of civilization."

**28** Widman urged Hjelt to submit: Letter, Widman to Hjelt [n.d.] December 1900, EH.

**28** van't Hoff appeared to be the strongest candidate: Twenty responses came, of which four were from committee members. About three hundred invitations were sent; clearly, most did not know what to make of the request to submit a nomination.

**29** Consensus was the most important goal: Letter, Widman to Hjelt, January 12, 1902, EH.

**29** important for his making the case for Fischer: Letter, Widman to Hjelt, January 11, 1901, EH.

**29** own circle of acquaintances: Fellow Uppsala chemist and committee member P. T. Cleve also nominated Fischer, as did member of the Academy's Chemistry Section Karl Mörner and, of course, Hjelt and Widman. The only nomination outside Widman's circle of acquaintances was from the German physical chemist Georg Bredig.

**29** left deep scars in Arrhenius: Arrhenius's unpublished autobiography ["*Levnadsrön*," SA] reveals the persistence of his antipathy toward Uppsala and especially the Uppsala scientists who failed to appreciate his early achievments. He also used the occasion of his Nobel Prize to chastise the Uppsala scientists and their philosophy of science (*Les Prix Nobel en 1903*, p. 49). On Arrhenius's early career and personality, see Elisabeth Crawford, *Arrhenius: From Ionic Theory to the Greenhouse Effect* (Canton, MA: Science History Publications, 1996), pp. 1–108. For some of the machinations and positioning by which Arrhenius and his supporters endeavored to get him, preferably, a physics prize and even half a physics and half a chemistry prize, see correspondence with Hjelt from Widman and Arrhenius, 1902–1904. Further details available in E. Crawford and R. M. Friedman, "The prizes in physics and chemistry in the context of Swedish science," in Carl-Gustaf Bernhard, Elisabeth Crawford, and Per Sörbom, eds., pp. 312–317; and Elisabeth Crawford, *The Beginnings of the Nobel Institution: The Science Prizes, 1901–1915* (Cambridge: Cambridge University Press, 1984), pp. 116–123. Arrhenius wanted a physics prize—to make life more miserable for the Uppsala physicists—and he tried to mastermind a bizarre plan by which he would receive half a physics prize and half a chemistry prize. The plan failed when nobody felt it right for Lord Raleigh (J. W. Strutt) and William Ramsay, as proposed, each to share the other half prize. In the end, Arrhenius had to be satisfied with a chemistry prize. Much significant research has been written on the early history of physical chemistry; for an overview, see Erwin N. Hiebert, "Developments in physical chemistry at the turn of the century," in Bernhard, Crawford, and Sörbom, eds., *Science, Technology, and Society at the Time of Nobel*, pp. 97–114; William H. Brock, "On the dissociation of substances dissolved in water," Chap. 10, *The Fontana History of Chemistry* (London: Fontana Press, 1992), pp. 355–395. Insight into the cabal on Arrhenius's behalf in letter, Widman to Hjelt, January 15 and April 20, 1903, EH. Arrhenius to Arthur Rindell, January 2, 1902, AR, notes the need for a "mass demonstration" of support so that even if the physics committee claims that his work is not eligible, the Academy will not be able to ignore his candidacy. Moreover, he thinks general opinion favors giving a prize to a Swede or a Scandinavian. After settling for a chemistry prize, Arrhenius visited Berlin to thank his friends there "who helped me get the Nobel Prize" [letter, Arrhenius to Gustaf Ekman, December 31, 1903, GE].

**31** "Strength through unity": Letter, Henrik Söderbaum to Widman, August 15, 1905, OW, expressed his sentiment in French, *L'unité fait la force*; he resorted to Latin and French for making important points.

**31** had not first received prizes: Letter, Emil Fischer to Hjelt, December 23, 1902, EH.

**31** he then compared notes with others who had access to the committee: Letters, Friedrich Kohlrausch to Arrhenius, November 8, 1901, SA-KVA; Widman to Hjelt, September 12, 1900, and January 11, 1901, EH; Fischer to Hjelt, December 23, 1902, EH, on confusion as to how to interpret the bylaws.

**31** a tacit rule of thumb rather than formal regulation: Letters, Fischer to Hjelt, December 23, 1902; Charles Guillaume to Bernhard Hasselberg, December 30, 1910, BH-KVA.

**32** under the guise of preparing for an eventual obituary: Letter, Adolf von Baeyer to Widman, June 8, 1905, OW, refers to the contents of Widman's request of June 4.

**32** the international chemical community favored an award: *Kommittéutlåtande, Nobelkommittéen för kemi* [Committee report, Nobel Committee for Chemistry. Hereafter abbreviated as KU, NKK], 1903, 1904, 1905.

**32** "Rebellion is in the air": Letter, Widman to Hjelt, October 21, 1905, EH.

**33** confirmed its full significance: KU, NKK, 1906.

**34** "... same sort of humbug that more fittingly characterizes himself": Letter, Otto Pettersson to Carl Benedicks, November 8–9, 1937, CB on Arrhenius's opposition (Pettersson was reminiscing about the past); and Diary, Gösta Mittag-Leffler, November 12, 1906, ML-KB, on characterization of Arrhenius during the debate.

**34** But soon these men died: Similarly, the elderly William Crookes, whom Ramsay enthusiastically supported, also died, after appearing several years with a smattering of nominations. Letters, William Ramsay to Arrhenius, October 1, 1907, and March 15, 1908, SA; Pettersson to Gustaf Retzius, n.d. [November 10, 1908], GR.

**34** if they at all could reach Stockholm in the winter: Letter, Arrhenius to J. H. van't Hoff, December 18, 1905, van't Hoff Correspondence, Special Collections, Milton S. Eisenhower Library, Johns Hopkins University, Baltimore.

**34** Most candidates received two or three: Two candidates (Walther Nernst and Wilhelm Ostwald) received four nominations, two received three nominations (Stanislao Cannizzaro, Theodor Curtius), and seven received two nominations each (Marcelin Berthelot, Eduard Buchner, Giacomo Ciamician, William Crookes, S. M. Jørgensen, Dmitry Mendeleyev, and Paul Sabatier).

**35** his support was critical: See, for example, Ramsay to Arrhenius, October 1, 1907, who followed Pettersson's advice on consulting with Arrhenius to try to win support for Crookes.

**35** Rutherford for the 1907 chemistry prize: Widman advocated splitting the prize between Rutherford and his chemist collaborator, Frederick Soddy, which would certainly have made the whole matter more palatable for chemists and put the focus on the chemistry of radioactivity. For more about the 1908 prize and its relation to Arrhenius's efforts to celebrate atomism through Rutherford's chemistry prize and a physics prize for Planck, see Elisabeth Crawford, "Arrhenius, the atomic hypothesis, and the 1908 Nobel prizes in physics and chemistry," *Isis*, **75** (1984), pp. 503–522.

**35** cofounder of modern physical chemistry, awarded the prize: Crawford and Friedman, "The prizes in physics and chemistry in the context of Swedish science," pp. 317–319.

**35** Ostwald's moment had come: Arrhenius relates Ostwald's conversion in a letter to Georg Bredig, September 6, 1908, copy, SA.

**36** but he had inside help: Among other candidates who were capable of raising an eyebrow or two in the committee, the Norwegian engineer and physicist team of Sam Eyde and Kristian Birkeland appealed especially to Klason for their preliminary efforts in creating an industrial process for artificial fertilizer, the beginnings of the industrial giant Norsk Hydro. But all their support came from Norwegians; moreover, no clear winner seemed to have emerged as yet in the fierce competition among several chemical concerns to establish a commercially viable means for relieving the world's growing shortage of fertilizer. Competitors O. W. Schönherr and Adolf Frank had also each received a nomination. Moreover, among the nominees the committee members also had to note pioneering French organic chemists, such as Gabriel Bertrand, Victor Grignard, Paul Sabatier, and J. B. Senderens. True, none of them received very many nominations, but it was important that the leading German organic chemist, von Baeyer, nominated Grignard, and the president of the French Academy, Gaston Darboux, suggested Bertrand.

**36** Widman asked him to prepare the report: On Ostwald's formal problems in satisfying the bylaws, letter, Widman to Hjelt, November 25, 1909, EH and KU, NKK, 1906–1908. On the strategy for 1909, letter, Widman to Arrhenius, April 22, 1909, SA.

**37** the only changes highlighted yet more clearly the catalytic studies: On writing the evaluation and discussing fine points of strategy, letters, Widman to Arrhenius, June 9 and July 12, 1909, SA.

**37** key questions concerning catalytic reactions: Quotations respectively from historians of chemistry J. R. Partington, Erwin Hiebert, and H.-G. Körber, in Crawford and Friedman, "The prizes in the context of Swedish science," p. 319.

**37** ". . . having solicited your [Arrhenius's] assistance": Letter, Olof Hammarsten to Arrhenius, April 28, 1914, SA.

**37** That was the situation in 1910: Letters, Arrhenius to Winthrop Osterhout, October 20, 1910, WO; Widman to Hjelt, November 25, 1909, EH.

**38** ". . . The committee couldn't agree on anybody": KU, NKK, 1910; letter, Widman to Hjelt, January 3, 1911, EH [*I år höll det på att gå alldeles sönder*].

**38** ". . . but then met solid resistance": *Protokoll,* KVA, IV klassen, October 15, 1910; Widman to Hjelt, January 3, 1911.

**38** nobody had voted for Wallach: On comparing Wallach and von Baeyer, KU, NKK, 1907 and 1910.

**39** but many in the world of chemistry were left nonplussed: On the change of attitude leading to voting for Wallach, letter, Åke Ekstrand to Ossian Aschan, December 17, 1910, OA; on reaction, letters, Bredig to Arrhenius, January 22, 1911, SA, and Hjelt to Widman, December 30, 1910, OW. Among those mentioned as more fitting to receive the 1910 prize were Richard Willstätter, Alfred Werner, even the elderly Theodor Curtius, but above all, Nernst remained waiting for what seemed to some a well-deserved prize.

**39** ". . . sooner or late Willstätter will receive a prize": Letters, Hjelt to Widman, December 5, 1912, OW; Widman to Hjelt, November 20, 1912, EH.

**39** Grignard and Sabatier were brought together to split a prize: Sabatier, an inorganic chemist, worked on the hydrogenation of hydrocarbons by catalysis with finely divided metals. This endeavor spanned two decades and was done with the younger organic chemist J.-B. Senderens. Grignard's pioneering research on organomagnesium synthesis began as his dissertation, defended in 1901, in Lyon under P. A. Barbier. Mary Jo Nye provides a detailed study of these chemists' careers and their controversies in *Science in the Provinces: Scientific Communities and Provincial Leadership in France, 1860–1930* (Berkeley: University of California Press, 1986), pp. 137–187.

**39** criticized the Academy's choice more openly: On the rationale, letter, Widman to Hjelt, November 20, 1912; on reactions, Nye, *Science in the Provinces,* pp. 150–151; letter, Hjelt to Widman, December 5, 1912.

### THREE Sympathy for an Area Closely Connected with My Own Specialty

**40** inaugurating a new era: On nineteenth-century physics, see Stephen F. Mason, *A History of the Science* (New York: Collier Books, 1968), pp. 468–502; Lawrence Badash, "The completeness of nineteenth-century physics," *Isis,* **63** (1972), pp. 48–58; J. L. Heilbron, *"Fin-de-siècle* physics," in Carl-Gustaf Bernhard, Elisabeth Crawford, and Per Sörbom, eds., *Science, Technology and Society in the Time of Alfred Nobel,* Nobel Symposium, Björkborn, Karlskoga, Sweden, August 17–22, 1981 (Oxford: Pergamon Press for the Nobel Foundation, 1982), pp. 51–73; opening chapters in Daniel J. Kevles, *The Physicists: The History of a Scientific Community in Modern America* (New York: Knopf, 1978); Helge Kragh, *Quantum Generations: A History of Physics in the Twentieth Century* (Princeton, NJ: Princeton University Press, 1999); and Christa Jungnickel and Russell McCormmach, *Intellectual Mastery of Nature: Theoretical Physics from Ohm to Einstein,* Vols. 1 and 2 (Chicago: University of Chicago Press, 1986).

**41** Swedish physics had a much narrower perspective: Robert Marc Friedman, "Nobel physics prize in perspective," *Nature,* August 27, 1981, pp. 793–795; Sven Widmalm, "Vetenskapens korridorer: Experimentalfysikens institutionalisering i

Uppsala 1858–1910," *Lychnos: Årsbok för Idé-och Lärdomshistoria* (1993), pp. 35–70; Thomas Kaiserfeld, *Vetenskap och karriär: Svenska fysiker som lektorer, akademiker och industriforskare under 1900-talets första hälft*, Stockholm Papers in the History and Philosophy of Technology (Arkiv Förlag, 1997), pp. 33–39 ff. On experimenticist philosophy of science, see Gerald Holton, *Thematic Origins of Scientific Thought: Kepler to Einstein* (Cambridge, MA: Harvard University Press, 1973), pp. 275–277. A masterful, in-depth analysis of the Uppsala school of physics was published too late for use as a reference for this book, but will be essential reading for further study of the early Nobel prizes in physics, Sven Widmalm, *Det öppna laboratoriet. Uppsalafysiken och dess nätverk 1853–1910* (Stockholm: Atlantis, 2001).

**41** if not American science more generally: These sentiments can be found both in popular works such as Bernard Jaffe, *Michelson and the Speed of Light* (Garden City, NY: Doubleday, 1960), and in scholarly books such as Kevles, *The Physicists*, pp. 27, 79.

**41** linked with Albert Einstein's elaboration in 1905 of the special theory of relativity: As Gerald Holton originally showed, the connection between the experiment and Einstein's theory is, at best, indirect in contrast to the popular version of the former directly prodding the latter; see Gerald Holton, "Einstein, Michelson, and the crucial experiment," *Isis*, **60** (1969), pp. 133–197.

**42** a Harvard University astronomer: Harvard was invited to send nominations that year; although neither a physicist nor a chemist, E. C. Pickering was the only Harvard faculty member to respond. This discussion on Michelson's prize is based in part on my "Americans as candidates for the Nobel prize: The Swedish perspective," in Stanley Goldberg and Roger Stuewer, eds., *The Michelson Era in American Science, 1870–1930* (New York: American Institute of Physics, 1988), pp. 272–280.

**42** at the forefront of contemporary physical research: *Kommittéutlåtande, Nobelkommittéen för fysik* [Committee report, Nobel Committee for Physics; hereafter abbreviated as KU, NKF], 1904.

**42** considered for the 1904 prize: KU, NKF, 1904, Philipp Lenard (for cathode rays), Ernst Abbe (for optical instruments), James Dewer and Karol Olszewski (for low-temperature physics), and J. W. Strutt [Lord Rayleigh] (for the discovery of argon).

**42** "to do all in my power to procure the prize for him [Michelson]": Letter, Bernhard Hasselberg to G. E. Hale, July 5, 1907, GH-CIT (also copy, Microfilm, Niels Bohr Center for History of Physics, American Institute of Physics, College Park, MD).

**43** but no one received a clear mandate: German physicist Hermann Ebert indicated that his first choice would be William Thomson (Lord Kelvin) or for a divided prize between Julius Elster and Hans Geitel (atmospheric electricity); as a second choice, he indicated sharing a prize between Michelson and Augusto Righi and Ernest Rutherford. Hasselberg correspondence with Prytz in Kristian Prytz papers, Royal Library, Copenhagen. Those who received larger numbers of nominations were Ernest Rutherford and Gabriel Lippmann, each getting seven nominations; J. D. van der Waals, six nominations. The nominations for the latter two, however, were largely "favorite son" nominations from their French and Dutch countrymen. Other candidates who received notable support included Max Planck, Wilhelm Wien, and Henri Poincaré.

**43** might constitute a "discovery": Letter, Hasselberg to Hale, July 5, 1907.

**43** "... *can* contain the seed of new discoveries": KU, NKF, 1907, Supplemental special report on Michelson.

**43** Arrhenius did not frustrate his campaign: Recalled in Hasselberg to Gösta Mittag-Leffler, January 24, 1910, ML-MLI.

**43** "... metrological investigations carried out with their aid": KU, NKF, 1907, Supplemental report on Michelson.

**44** But little of the ferment in European theoretical physics: Topics of papers delivered and discussed at the respective Stockholm and Uppsala "Physics Society [*Fysiska sällskapet*]" reveal considerable appreciation for recent experimental advances but limited appreciation of Continental or British theoretical work. Summaries of the meetings were provided in local newspapers; collections of clippings on the Stockholm meetings in Nils Ekholm Papers, KVA and SA. On Uppsala's Physics Society, Arne Eld Sandström and Arne Haglöf, *Fysiska Sällskapet 100 År* (Uppsala: Uppsala University, 1987). Stockholm professor Bjerknes noted in a letter to H. A. Lorentz [September 4, 1904, Lorentz Collection, National Archives, The Hague, Netherlands] that he could find maybe one or two persons in all of Sweden with whom he could discuss advances in electron theory and the electromagnetic world picture, which prior to the acceptance of relativity and quantum theories were considered the cutting edge in European theoretical physics.

**44** "I cannot but prefer works of *high precision*": Letter, Hasselberg to Hale, December 29, 1907, GH.

**44** "... our only way to new discoveries": Hasselberg's presentation speech in *Les Prix Nobel en 1907* (Stockholm: Norstedts, 1909), p. 14. Hasselberg actually did not have a chance to deliver his comments, as King Oscar II died two days before the ceremony, requiring a simpler, less public affair.

**45** "theoretical speculations" concerning the spectra of elements: Discussion of Rydberg based on Arvid Leide, "Janne Rydberg och hans kamp för professuren," *Kosmos. Fysiska uppsatser*, **32** (1954), pp. 15–32; and Paul C. Hamilton, "Reaching out: Janne Rydberg's struggle for recognition," in Svante Lindqvist, ed., *Center on the Periphery: Historical Aspects of Twentieth-Century Swedish Physics* (Canton, MA: Science History Publications, 1993), pp. 269–292.

**45** that others had overlooked: Hamilton, "Reaching out," pp. 279–280.

**45** tried to keep him from being elected to the Academy's Physics Section: Svante Arrhenius, Ms. "Levnadsrön [Scientific autobiography]," SA.

**45** that brought prestige in Sweden: Letter, Arrhenius to Wilhelm Ostwald, December 26, 1890, reprinted in Hans-Günther Körber, ed., *Aus dem wissenschaftlichen Briefwechsel Wilhelm Ostwards*, II Teil (Berlin: Akademie-Verlag, 1969); see also letter of October 16, 1888. Throughout his life, Arrhenius never stopped taunting Uppsala scientists in his many autobiographical and historical writings.

**46** difficult to rely upon the committee for sound evaluations: Letter, Arrhenius to Gustaf Retzius, January 8, 1903, GR.

**46** "Hasselberg's results are = 0": Letter, Otto Pettersson to Retzius, March 8, 1901, GR.

**46** "A *truly* scientific cooperation with him is indeed almost impossible": Letter, Hasselberg to Hale, October 21, 1904, GH.

**46** depth of Hasselberg's feelings: Letter, Hale to Hasselberg, November 14, 1904, copy, GH. Hale resolved the problem by asking Arrhenius to speak on meteorological issues.

**47** Hasselberg resolutely opposed allowing another one into the committee: Letter, Hasselberg to Hugo Hildebrand Hildebrandsson, November 17, 1902, HHH-UUB. For Arrhenius's efforts to prevent Hasselberg from being reelected to the Nobel committee and other skirmishes between the two, see letters, Arrhenius to J. H. van't Hoff

1903–1905, in van't Hoff Papers, Eisenhower Library, Johns Hopkins University. Gustaf Retzius tried to caution Arrhenius over expecting the inertial Academy members to reject one of their own members in favor of an outsider in a letter of December 25, 1904, SA. Arrhenius replied to Retzius, December 27, 1904, GR, pointing out that Hasselberg was simply too lazy to follow other branches of physics beyond his own narrow interest in spectral analyses; he was of little help to the committee and therefore must be removed.

**47** "places it on the highest level of contemporary scientific research": KU, NKF, 1907.

**47** "merit attention": *Protokoll vid Kungl. Vetenskapsakademiens Nobelkommitéen för fysik sammanträde* [Minutes of meetings of Nobel Committee for Physics of the Royal Swedish Academy of Sciences; hereafter, *Protokoll*, NKF], September 2, 1911. At the time "mathematical physics" was often used to designate both theoretical and mathematical physics, the former being more concerned with the unifying general principles of physics and the latter more concerned with the elegance of mathematical analysis generated by equations describing physical phenomena.

**48** without dividing the honor with an experimentalist: *Protokoll*, NKF, September 24, 1908.

**48** "... not for the sake of the instrument—as in certain other places": Letter, Vilhelm Carlheim-Gyllensköld to Arrhenius, June 7, 1907, SA.

**48** "... completely beyond the majority's horizon": Mittag-Leffler Diary, September 17, 1915, ML-KB.

**48** Nobel deliberations gave fuel to the flames: Celebratory anniversary works rarely note this aspect of institutional cultures. Private correspondence, diaries, and unpublished autobiographical manuscripts provide greater insight into daily life than the airbrushed portraits provided in ceremonious literature.

**48** "... sick from vexation and grief [*förargelse och ledsnad*]": Letter, Hugo Hildebrand Hildebrandsson to Emmy [daughter], March 8, 1904, HHH-KVA. See also, for example, HHH to Emmy, November 26, 1902, and to son Sven, February 28, 1910, HHH-KVA. On Ångström's many tribulations, see Widmalm, "Vetenskapens korridorer."

**48** "Nobel corruption": Letter, Vilhelm to Honoria Bjerknes, February 26, 1904, BF.

**48** turning around to vote the opposite: Mittag-Leffler Diary, January 27, 1909; September 17, 1915.

**49** "howl with the pack": Letter, V. Carlheim-Gyllensköld to C. V. L. Charlier, February 9, 1914, CC; see also Mittag-Leffler Diary, September 17, 1915.

**49** "... one foot in the coffin and the other in bed": Letter, Carlheim-Gyllensköld to Charlier, January 15, 1914, CC.

**49** formidable figures in the early years of the Nobel Prize: Crawford explored this rivalry in *The Beginnings of the Nobel Institution*, pp. 109–149. Further insight into Arrhenius's social and cultural attitudes is provided in her biography of Arrhenius. Former student and then colleague Hans von Euler-Chelpin describes Arrhenius's petit bourgeois tastes and habits in his unpublished autobiography, *"Minnen"* [n.d.; copy kindly provided by Professor Anders Lundgren, Uppsala University], p. 34.

**49** Academy turned down the committee's proposal: Mittag-Leffler Diary, November 10, 1908. The vote in the Academy was thirteen for Planck and forty-six for Lippmann, letter, Mittag-Leffler to Paul Painlevé, December 9, 1908, copy, ML-MLI.

Further insight on the debate recalled in letter, Nils Ekholm to Arrhenius, March 10, 1910, SA.

**49** campaign to overcome the opposition in Stockholm: Crawford, *Beginnings*, pp. 140–147.

**50** if Committee Chairman Ångström could be convinced to back Poincaré: Letter, Hasselberg to Mittag-Leffler, January 4, 1910, ML-MLI.

**50** committee tradition for interpreting "recent": KU, NKF 1910.

**50** When it became clear that there was little enthusiasm for doing this, Arrhenius was prepared: The one nomination for Ångström was sent by Wilhelm Röntgen, the first winner of the Nobel Prize in physics. Letter, Arrhenius to Gustaf Granqvist, September 8, 1910, CWO-KVA, shows Arrhenius trying at the last minute to find a justification for not awarding Ångström, once he had died, and once Arrhenius had understood that he could probably succeed in blocking Poincaré by advancing van der Waals as an alternate candidate. Arrhenius even included a few blows against Poincaré in his special report on the elderly Dutchman. [It is not clear why this letter is among Granqvist's correspondence to Oseen.]

**51** In the end, he successfully mobilized the Academy to back the elderly Dutchman: The Academy voted at eleven at night after four and a half hours of debate on the physics prize [letter, Arrhenius to Gustaf Ekman, November 6, 1910, GE].

**51** Other nominees also received respectable support: Other strong nominees that year included, again Planck (ten nominations), the frequently proposed Augusto Righi (five), and the collaborators Julius Elster and Hans Geitel (four).

**51** to keep Arrhenius from becoming professor of physics in Stockholm: See correspondence between Poincaré and Paul Appell with Mittag-Leffler, 1895, ML-MLI.

**51** his accomplishments were far too old: On assuming only experiment being eligible, letter, Philippe-Auguste Guye to Arrhenius, February 10, 1910, SA; on assumptions on age, letter, Charles Guillaume to Hasselberg, December 30, 1910, BH-KVA.

**51** his role was that of a general who had influenced . . . modern physics: Letter, Henri Delandres to Hasselberg, August 21, 1912, BH. Delandres reported Gullstrand's comments.

**51** "It is, as always, the human comedy": Delandres to Hasselberg, August 21, 1912.

**52** was singing Bjerknes's praises: Mittag-Leffler Diary, December 5, 1903. For evidence that also Ångström tried to keep Arrhenius's potential allies from getting elected to the committee, see Mittag-Leffler Diary, November 12, 1907. On Bjerknes as physicist, R. M. Friedman, *Appropriating the Weather: Vilhelm Bjerknes and the Construction of a Modern Meteorology* (Ithaca, NY: Cornell University Press, 1989), pp. 11–32.

**53** ". . . the gentlemen in the Academy of Sciences": Mittag-Leffler Diary, November 4, 1909.

## FOUR  Each Nobel Prize Can Be Likened to a Swedish Flag

**54** "Certainly it can be hoped that personal interests will not come to play": Letter, Hjelt to Widman, [n.d.] December 1900, OW-UUB. ["*Man kan väl hoppas att personliga interessen icke skola alltför mycket spela in i dessa frågor.*"] Hjelt responded on occasion to Widman's queries about the relative worth of various candidates, but he also tended not to comment on the intricate details of committee intrigues, which undoubtably he found distasteful.

**55** "... even though he has no chemical talent": Von Baeyer quotation in Joseph S. Fruton, "The interplay of chemistry and biology at the turn of the century," in Carl-Gustaf Bernhard, Elisabeth Crawford, and Per Sörbom, eds., *Science, Technology and Society in the Time of Alfred Nobel*, Nobel Symposium, Björkborn, Karlskoga, Sweden, August 17–22, 1981 (Oxford: Pergamon Press for the Nobel Foundation, 1982), p. 82. Christian G. Reinhard, "Eduard Buchner," in L. K. James, ed., *Nobel Laureates in Chemistry, 1901–1992* (Washington, DC: American Chemical Society and Chemical Heritage Foundation, 1993), pp. 42–48.

**56** he "resaddled" and wrote a private letter: Mittag-Leffler Diary, November 12, 1907, based on Olof Hammarsten's account told at *"Nachspiel"* after the Academy meeting. Hammarsten's personal papers were burned after his death, so the content of Arrhenius's letter can only be gleaned through the reply received: Hammarsten to Arrhenius, October 8, 1907, SA.

**56** When the matter reached the Academy, Arrhenius criticized Buchner's candidacy: The following account is based on the entry in Mittag-Leffler Diary, November 12, 1907, and on the text of Arrhenius's criticism delivered at the Academy [Mss., Unpublished Manuscripts, SA].

**56** Buchner received a prize, worthy or not: Decades later, Hans von Euler admitted that he considered this prize one of the Academy's mistakes. Buchner owed too much to the two bacteriologists who taught him the technique to have received an undivided prize. *Minnen* (unpublished autobiography), p. 333.

**57** "... even *Unser einer* [one of us] ... can get the prize": Mittag-Leffler Diary, November 10, 1908. German used in the original quotation.

**57** The Academy's sense of values ... did not always correspond with ... committees: Little has been written about the Academy's culture. Although the Academy's institutional activities in relation to society have been studied, how the Academy made collective decisions and defined its own values is relatively unanalyzed. Sten Lindroth's magisterial history of the Academy stops at 1818. Some of the articles in Tore Frängsmyr, ed., *Science in Sweden: The Royal Swedish Academy of Sciences, 1739–1989* (Canton, MA: Science History Publications, 1989), offer insights. Although it would be helpful to know more, correspondence and other archival documentation do provide many important insights.

**57** a comment allegedly made by Thomas Edison: Edison actually had not been nominated at the time. In 1915, he was nominated to split a physics prize as well as to split a chemistry prize. The engineers' disgruntlement and their rebellion in the Academy are discussed by Gunnar Eriksson, *Kartläggarna: Naturvetenskapens tillväxt och tillämpningar i det industriella genombrottets Sverige, 1870–1914*, Umeå Studies in the Humanities 15 (Umeå: Umeå Universitetsbibliotek, 1978), pp. 95–97.

**57** They mobilized a majority: The Academy's Nobel protocol does not record the votes; the vote was noted in a letter from Erik Ljungberg to Arvid Lindman, referred to in Eriksson, *Kartläggarna*, p. 96. KU, NKF, 1912, shows the committee's negative evaluation of Dalén.

**59** the effect being to reinforce royalist sympathy as a part of ... patriotism: Staffan Björck, *Heidenstam och sekelskiftets Sverige* (Stockholm: Nature och Kultur, 1946), p. 12.

**59** "so much injustice, oppression, and humbug": Quoted in ibid., p. 36 ["... tjänstgör som duperande skylt för så mycken orättvisa, förtryck och humbug"].

**60** to belong to the Germanic race, allegedly the crown of the human species: Eriksson, *Kartläggarna*, pp. 197–204; Gunnar Broberg, "Statens institut för rasbiologi-

tillkomståren," in *Kunskapens trädgårdar: Om institutioner och institutionaliseringar i vetenskapen och livet* (Stockholm: Atlantis, 1988), pp. 179–186.

**60** where he lent his presence to validate official culture: Josef Link and H. A. Ring, *Oscar II Sveriges konung, 1872–1907: En minnesskrift* (Stockholm: Svenska bokförlaget, 1908), pp. 450–451.

**60** tested a nation's abilities in technology, science, . . . and organization: Ibid., pp. 240–241ff.

**60** Swedish man of science could wear the laurels of the hero: Tore Frängsmyr, *Vetenskapsmannen som hjälte: Aspekter på vetenskapshistorien* (Stockholm: Nordstedts, 1984), especially on Nordenskiöld; also, Urban Wråkberg, *Vetenskapens vikingatåg: Perspektiv på svensk polarforskning, 1860–1930* (Stockholm: Royal Swedish Academy of Sciences, 1999), pp. 86–104.

**61** faith, love, hope and humility: Link and Ring, *Oscar II Sveriges konung*.

**62** The exhibition brought to a focus national self-confidence: My account is indebted to Björck, *Heidenstam*; Link and Ring, *Oscar II Sveriges konung*; and especially Anders Ekström, *Den Utställda Världen: Stockholmsutställningen 1897 och 1800-talets världsutställningar*, Nordiska Museets Handlingar 119 (Stockholm: Nordiska Museet, 1994).

**62** ". . . This is Swedish": Björck, *Heidenstam*, p. 53, quoting Academy Director Biskop Rundgren.

**62** Swedes were the aristocrats of the Germanic peoples: Ibid., pp. 53–57, 65; quotation from conservative professor Vilhelm Lundström.

**62** The conservative newspaper *Svenska Dagbladet* proclaimed: *SvD*, April 30, 1897, quoted in Ekström, *Stockholmsutställningen*, p. 240.

**62** the Swedish people's industriousness and diligence: Ekström, *Stockholmsutställningen*, pp. 282–287.

**62** "Thanks to the important duties entrusted to them": Göran Liljestrand, "The prizes in physiology," in Nobel Foundation, ed., *Alfred Nobel: The Man and His Prizes*, 2nd ed. (New York: Elsevier, 1962), p. 339; quotation from *Göteborgs Handels- och Sjöfartstidning*.

**63** "The unprecedented record": Ibid. Quotation from *Uppsala Nya Tidning*.

**63** each seemingly embodying its own national traits: Hildebrand speech at Nobel festivities, 1909, *Les Prix Nobel en 1909*. The comparison with the Olympic Games was already boldly declared by *Svenska Dagbladet* as soon as Nobel's testament was announced [January 4, 1897].

**63** nothing compared with this astronomical sum: Elisabeth Crawford, *The Beginnings of the Nobel Institution: The Science Prizes, 1901–1915* (Cambridge: Cambridge University Press, 1984), p. 64.

**63** the national spirit of impartiality: *DN*, June 28, 1912.

**64** ". . . all that which makes us cherish our fatherland": Hildebrand quoted in Ekström, *Stockholmsutställningen*, p. 285. [*"Det är uppifrån och från fåtalet, som det högre ljuset kommer, som skall genomtränga massorna. Det är detta fåtal med sann bildning, som lifligast känner, som djupast fattar både det verkligt nationella och värdet af allt det, som gör fäderneslandet för oss dyrbart."*]

**64** ". . . shall year for year receive a prize from Swedish hands": *Les Prix Nobel en 1901*, p. 40.

**64** through the entire universe [*est porté dans tout l'univers*]: *Les Prix Nobel en 1905*, p. 60. Lord Rayleigh's toast, *Les Prix Nobel en 1904*, p. 57.

**65** Silence descended, as the exhibition grounds were stripped bare: Ekström, *Stockholmsutställningen*, pp. 312–313; Allan Pred, "Spectacular articulations of modernity: The Stockholm Exhibition of 1897," *Geografiska Annaler*, **73**, 1 (1991), pp. 75–80.

**65** would the King be willing to take part in "such a comedy"?: Letter, Arrhenius to Retzius, November 20, 1905, GR.

**65** "great patriotic ceremonies for glorifying intellectual culture": *Les Prix Nobel en 1909*, pp. 8–9. [*"stora fosterländska högtidligheter för andliga kulturens förhärligande."*]

**66** he received hardly a notice: Referred to in handwritten minutes by H. H. Hildebrandsson from April 15, 1908, meeting, cited in Julia Lindqvist, "Det krönta huset. Kungl. Svenska Vetenskapsakademiens stamhus i Frescati uppförd, 1911–1915" (work in progress); see also Mittag-Leffler Diary, January 27, 1909.

**66** Swedish purpose for setting in motion the festivities: Quotations and account from Mittag-Leffler Diary, December 10, 1904, and December 10, 1908; Rutherford's reactions in letter to Arrhenius.

**66** "charming French ladies": Letters, Hildebrandsson to daughter, December 16, 1903, HHH-KVA; Middagen Å Grand Hôtel den December 10, 1908, in HH.

**66** whenever something went amiss: Quotations and descriptions from Mittag-Leffler Diary, December 10, 1906, December 10, 1908, December 11, 1909. On the yellow journalists, Hildebrandsson to Emily 1903, HHH-KVA.

**67** The power of the word over thought: Mittag-Leffler Diary, December 10, 1903. [*Ordens makt öfver tanken. Högsta = materiellt största.*] Arrhenius makes the declaration of greatest award in his acceptance toast, *Les Prix Nobel en 1903*, p. 49.

**67** only the most noble sentiments would enter into selecting winners: Letters, Richard Abegg to Arrhenius, December 23, 1901, and January 23, 1902, SA. Abegg had been a postdoctoral researcher in Arrhenius's laboratory, and although he knew that Arrhenius and his supporters were active in furthering physical chemistry at home and abroad, he still considered van't Hoff's prize to be the expression of pure impartiality.

**67** "The Land of Smörgårsbord": Björck, *Heidenstam*, p. 11.

**67** "The Royal Swedish Royalty": Poet Gustaf Fröding used this expression [*kungl. svenska kunglighsten*] in a newspaper debate, noted in ibid., p. 12.

**67** burden for those not in full party form: Gustaf Retzius to Arrhenius, November 29, 1906, SA, relates how his wife needs to rest up prior to the Nobel disturbances [*oroligheterna*].

**68** "High-minded and noble struggle": *Les Prix Nobel en 1908*, p. 9. [*att ägandet af denna lyckligare ställning skulle göra människorna benägna att i högsint och ädel kamp sträfva efter det stora målet att göra den högra odlingens frukter til tillgängliga för och möjliga att nås af hela mänskligheten.*]

**68** "... the destiny of people": *Les Prix Nobel en 1905*, p. 10. [*folkens öden.*]

**68** "What will he say this time?": Letter, Fredrik Wachtmeister to Hjalmar Hammarskjöld, November 3, 1911, HH.

### FIVE   Should the Nobel Prize Be Awarded in Wartime?

**71** At every year's Nobel ceremonies, Count Fredrik Wachtmeister: *Les Prix Nobel en 1905*, p. 10; *Les Prix Nobel en 1908*, p. 9; *Les Prix Nobel en 1909*, pp. 8–10ff.

**72** national unity behind its royal navy and army: On German academic propagandists for naval expansion, see Fritz Ringer, *The Decline of the German Mandarins: The German Academic Community, 1890–1933* (Cambridge, MA: Harvard University Press, 1969), pp. 139–140. Wachtmeister's comments are in *Les Prix Nobel en 1912*, pp. 8–9. On the Swedish political conflict regarding defense and especially the Dreadnought question, are "Försvars- och utrikspolitik i stormakts- alliansernas skugga 1905–1914," *Den svenska historien*, **14** (Stockholm: Albert Bonniers, 1968–1969), pp. 46–65. Prestigious members of the Academy, such as explorer Sven Hedin and Gösta Mittag-Leffler, agitated openly for defense.

**72** a Nordic paradise which everybody will long to visit again: *Stockholm Dagbladet*, June 21, 1914; *Aftenbladet*, June 22 and 28, 1914.

**72** ". . . the new German state! The ideas of 1914!": See Ringer, *Decline of the German Mandarins*, pp. 180–182; the quotation is from economist Johnann Plenge, p. 181. It was a Swede, Rudolf Kjellen, who popularized the theme of "the ideas of 1914" in Germany and Sweden. On the natural scientists' enthusiasm, see J. L. Heilbron, *The Dilemmas of an Upright Man: Max Planck as Spokesman for German Science* (Berkeley: University of California Press, 1986), pp. 69–72. Numerous letters from German to Swedish colleagues such as Fritz Haber to Arrhenius, September 23, 1914, SA, sharing the bubbly intoxication of enthusiasm for united patriotic action.

**72** Should the Nobel prizes be awarded during wartime?: On Sweden and the First World War, W. M. Carlgen, "Svensk neutralitet 1914–1918 och 1939–1945," *Historisk tidskrift* (1979), pp. 377–383, 388–395; and "Sverige under första värdskriget," *Den svenska historien*, **14** (1969), pp. 74–91. *Aftonbladet*, September 7, 1914.

**73** Hammarsten asked Aurivillius how to proceed: Letter, Olof Hammarsten to Christopher Aurivillius, September 7, 1914, CA.

**73** wiser and worldlier men must act: Letter, Hammarsten to Retzius, September 8, 1914, GR.

**74** Widman grudgingly backed the proposal: In 1913, Ramsay sent a seventeen-page report and more than one hundred reprints of Richards's work (Ramsay to NKK, January 9, 1913, KVA/N-1913; Arrhenius to NKK, January 20, 1913, KVA/N-1913. Arrhenis began his campaign for Richards with a report to the committee in 1912. Widman had initially reported on his reluctance to award Richards in 1910. Although accepting the proposal now, he reveals no enthusiasm while noting that he would prefer to back Willstätter. *Protokoll*, NKK, August 29, 1914, Bil. D.

**74** looked to replace Hammarsten: Letter, Widman to Hjelt, November 20, 1912, EH. Subsequently, Widman took over the evaluation of Willstätter.

**74** Richards had been waiting longer, the committee proposed him for the 1914 prize: Bohuslav Brauner (Prague) had proposed Richards whenever his university was invited to nominate (1909, 1912). He was then invited to nominate (based on the bylaws' category §6 allowing any number of individuals to be invited on an ad hoc basis) in 1913 and 1914, undoubtedly to keep Richards in contention. Brauner emphasized that Richards's work was fundamental for all chemistry, and most importantly, if the prize was to be truly international, the time had come to bestow an award in chemistry to the United States, especially to a candidate whom he characterized as "first-rate, original and ingenious, as well as prolific [*so ausgezeichnete, originelle und genialer, sowie fruchtbarer Forscher*]," KVA/N-1913 and 1914. Brauner had worked on the periodic table and was therefore sensitive to the importance of attaining new levels of precision for atomic weights. On the 1914 deliberations, *Protokoll*, NKK, September 21, 1914, Bil. A & D.

During August and September, Henrik Söderbaum, most likely with Arrhenius coaxing him, sent a stream of letters to his committee colleagues bringing new arguments to add to his evaluation of Richards, which helped clinch the case.

**74** quantum physics in some form had most definitely come to stay: Wilhelm Wien to NKF (January 22, 1914) for Planck insisted that the latter's quantum theory was incontestably worthy of a prize; the same insistence was indicated in letters from, among others, Wien's fellow Nobel laureates H. A. Lorentz, Pieter Zeeman, J. D. van der Waals, Kamerlingh Onnes, and the respected Italian physicist Vito Volterra ("the hypothesis of the quanta is today one of incontestable usefulness"). KVA/N-1914. No Swedish scientist was invited to the Solvay Conference.

**75** achieved by a young researcher: Nobody, however, seemed to consider the critical roles of the even younger, highly talented researchers Walter Friedrich and Paul Knipping, who took von Laue's suggestion and actually designed, constructed, and conducted the difficult experimental procedures for revealing the diffraction pattern. Paul Forman, "The discovery of the diffraction of X-rays by crystals: A critique of the myths," *Archive for the History of the Exact Sciences*, **6** (1969), pp. 38–71, analyzes how the discovery actually came about in contrast to the layers of heroic myth-making and time-eroded memories that had already begun in von Laue's 1920 Nobel lecture.

**75** without reservation, work of the greatest benefit to mankind: NKF, September 15, 1914, KU, Bil. 1.

**75** noted the growing instability in Europe: See, for example, letter, T. W. Richards to Arrhenius, June 21, 1913, SA: "... the so-called civilized countries of the world continue to waste energy in constructing engines of destruction. Perhaps governments are right in believing that poor human nature has not outgrown the need of war, but it certainly seems a sad pity that the preparation for it should be so wasteful."

**76** "... no less holy than hearth, home, and soil": Slightly different versions have been quoted or reproduced over the years. Danish and Swedish texts were sent to Scandinavian academics; these are found among the preserved papers of members of the Royal Swedish Academy of Sciences. Quotation from Danish, *"Til hele Kulturverden!"* VB; similar version in Swedish in OW. Signatories also included leading professors of the humanities, social sciences, and theology, as well as cultural personalities such as Engelbert Humperdinck and Max Reinhardt. The explicitly racial paragraph is not always reproduced in reprints of the manifesto. On background, see Otto Nathan and Heinz Norden, eds., *Einstein on Peace* (New York: Schocken, 1968), pp. 1–14ff.

**77** "... we may safely trust that race to persist in vitality and intellectual activity: William Ramsay, "Germany's aims and ambitions," *Nature*, October 8, 1914, quoted in Lawrence Badash, "British and American views of the German menace in World War I," *Notes and Records of the Royal Society of London*, **34**, 1 (1979), pp. 98–99.

**77** "... all peoples which respect the Law of Nations": *The Times* (London), October 21, 1914, quoted in Badash, "British and American views," p. 104.

**77** syphilis must be at the bottom of this German delirium: Badash, "British and American views," p. 102.

**78** a higher level of organization and culture, under the guiding hand of Germany: "Det kulturella Tysklands sändebud. Ett samtale med Wilhelm Ostwald...," *DN*, October 28, 1914.

**78** "... may be directed against our whole country": Anders Österling, "The literary prize," in Nobel Foundation, ed., *Alfred Nobel: The Man and His Prizes*, 2nd ed. (New York: Elsevier, 1962), pp. 102–103.

**78** even though no political consideration motivated these choices: Letter, K. A. H. Mörner to "His excellency, the Prime Minister [Till Excellensen, Statsministern]," October 17, 1914, HH-KB. Mörner was perhaps stretching the facts. Laue, a German, was working in Zurich, but he had already accepted a call to the new university in Frankfurt.

**78** rumor had it that the Academy of Sciences was against postponement: *Protokoll*, KVA/N, October 28, 1914. "Nobelprisen," *DN*, October 26, 1914.

**79** international treaties were not always so readily translated into practice: Carlgren, "Svensk neutralitet," pp. 378–382.

**79** impartiality called into question: *DN*, October 31, 1914.

**79** "politically motivated actions": Carlgen, "Sverige under första värdskriget," pp. 78–82; Mats Kihlberg and Donald Söderlind, *Två studier i svensk konservatism 1916–1922* (Stockholm: Almqvist & Wiksells, 1961), pp. 11–28ff, on the oppositional conservative movements. *NDA* and *Svenska Morgenbladet,* October 31, 1914.

**79** By having acted in a cowardly fashion, Sweden had failed: *SvD,* November 1 and 4, 1914.

**80** the Swedish academies had capitulated in the face of nationalism: *Social-Demokraten*, October 31, 1914. Branting was clearly pro-Entente, but did not campaign openly for the Allies. The paper's editor, and other younger radicals, held pro-German sympathies congruent with their solidarity with their counterpart Social Democratic Party, the largest in Germany.

**80** equated the cancellation of the Nobel prizes with cowardice: "Heidenstam om dagens stora fråga," *SvD*, December 7, 1914. He refers to the spirit of *karolinerna:* the soldiers who had served through thick and thin two centuries earlier under warrior King Charles XII and who had become a symbol for a nationalist identity embodying courage, determination, and loyalty.

**81** the advance of knowledge, uninfluenced by passion: Brigitte Schroeder-Gudehus, "Nationalism and internationalsim," and D. E. M. Edgerton, "Science and war," in R. C. Olby, G. N. Cantor, et al., eds., *Companion to the History of Modern Science* (London: Routledge, 1990), pp. 909–919, 934–945; Paul Forman, " Scientific internationalism and the Weimar physicists: The ideology and its manipulation in Germany after World War I," *Isis*, **64** (1973), pp. 151–180; and Brigitte Schroeder-Gudehus, "Division of labour and the common good: The International Association of Academies, 1899–1914," in Carl-Gustaf Bernhard, Elisabeth Crawford, and Per Sörbom, eds., *Science, Technology and Society in the Time of Alfred Nobel*, Nobel Symposium, Björkborn, Karlskoga, Sweden, August 17–22, 1981 (Oxford: Pergamon Press for the Nobel Foundation, 1982), pp. 3–20.

**82** ". . . second-rate Teutonic learning": Sir James Chrichton-Browne, quoted in Badash, "British and American views," p. 100.

**82** ". . . German instruments of control and monopoly of science": Ibid., pp. 100, 104–105.

**83** German response to the hostile English attacks: Lenard's comments, ibid., p. 105; "*Aufforderung*," copy in VB. The signatories included Arnold Sommerfeld and Johannes Stark.

**83** ". . . ardent love and energetic work for one's own country": Planck's letter to the Dutch physicist H. A. Lorentz, February 27, 1915, which was then sent to physicists in the Allied countries, is quoted and analyzed in Heilbron, *Dilemmas of an Upright Man*, pp. 75–80ff. See also Russell McCormmach's document-based "novel" on German theoretical physics, *Night Thoughts of a Classical Physicist* (Cambridge, MA: Harvard University Press, 1982), for a depiction of contemporary sentiments and actions. Letters, Planck to Arrhenius, November 15 and December 6, 1914, SA.

**84** a book that received considerable attention: *Värdenskulturen och kriget: Huru återknyta de internationella förbindelserna?* (Stockholm: Albert Bonniers, 1915).

**84** It was part of German cultivation [*Bildung*] to appreciate other cultures: Ibid., pp. 5–6, 36ff.

**84** Sweden, could help out later by mediating the resumption of international cultural cooperation: Ibid., pp. 20–22, 26–28.

**84** "We are fighting for mankind, for the sake of divine peace": Ibid., pp. 23–25.

**85** that cultural work belongs to mankind . . . respected across national boundaries: Ibid., pp. 83–91.

**85** human comprehension was sorely taxed: Ibid., pp. 6–7, 169–171, 191–193.

**85** their efforts would be doomed: Ibid., pp. 167–169, 173–177.

**86** debased the idea of the prizes as expressions of impartial judgment: Ibid., pp. 6–7, 167–188.

**86** Scientific congresses that included Germans and Austrians would be boycotted: Ibid., pp. 5–6, 117–118.

**86** unable to achieve even the most paltry intellectual effort: Ibid., pp. 194–195, 191–193.

**87** "From night till night and now": Isaac Rosenberg, "Dead Man's Dump," in *Collected Works of Isaac Rosenberg* (London: Chatto & Windus, 1922).

**87** maybe even embrace peaceful coexistence: Letter, Arrhenius to Winthrop Osterhout, August 23, 1915, WO. Arrhenius's many letters to Ousterhout, Richards, and Jacques Loeb (LC), among others, reveal his descent into despair. Arrhenius considered both generals and merchants evil, as both were profiting from the war.

**88** ". . . intercourse with Germany will be practically stopped for this generation": Letter, Ernest Rutherford to Arrhenius, June 1, 1915, SA.

**88** advance the progress of culture better than the one huge prize: Letters, Arrhenius to Bjerknes, May 11, 1915, VB; Arrhenius to Hjelt, March 7, 1916, EH, wonders whether it would be wiser to make several smaller prizes out of the Nobel in order to assist the advance of science.

**88** if they were to be awarded at all: Letters of nomination, KVA/N-1915. H. A. Bumstead, an American with close Cambridge ties who recently visited England, proposed splitting between father and son Bragg.

**88** News came of his death: Also named in sharing the prize with Moseley were C. G. Darwin and W. H. Bragg. Chemists' evaluations in KU, NKK, 1915; on Moseley, see J. L. Heilbron, *H. G. J. Moseley: The Life and Letters of an English Physicist, 1887–1915* (Berkeley: University of California Press, 1974).

**89** deprive deserving scientists and writers their rightful reward: "Skola Nobelprisen utdelas i år?" *SvD*, October 20, 1915. Actually the committee reports offer little support for this claim.

**89** went on record in opposition to the vote: *Protokoll*, KVA/N, October 27, 1915.

**89** "the cultural solidarity within the entire civilized world": "Uppskof med fysik- och kemiprisens utdelande," *SvD*, October 30, 1915; "Nobelprisen och kriget," *Stockholms Tidningen*, October 18, 1915.

**90** bringing the hostile factions together again: "Nobelprisen och värdskriget," *Stockholms Dagbladet*, October 13, 1915. The line of reasoning seems influenced by Arrhenius.

**90** it would be awkward to allow another to postpone: "Nobelprisens utdelning," *Aftonbladet*, November 4, 1915.

**90** were too politicized to be useful: "De väntande Nobelprisen," *SvD*, November 5, 1915.

**90** then it must go to von Laue: *Protokoll*, KVA, III klassen, October 30, 1915.

**90** Arrhenius did not need further proof: On hopes for better relations, Letter, Arrhenius to Hjelt, October 25, 1915, EH. On reaction to Rolland, *Stockholms Dagbladet*, November 9, 1915, and *DN*, November 10, 1915.

**91** ended at eleven o'clock in the evening: Mittag-Leffler Diary, November 11, 1915, ML-KB.

**91** literary worth was strongly criticized in France: Letter, Arrhenius to Hjelt, December 11, 1915; Östling, "The Literary Prize," p. 103; Kjell Espmark, *Det litterära Nobelpriset. Principer och värderingar bakom besluten* (Stockholm: Norstedts, 1986), pp. 41–44; the literature committee actually split, with a majority voting for the Spanish writer Benito Pérez Galdós and a minority for Rolland. 1914's candidate, Carl Spitteler, was now eliminated because of his subsequent anti-German and Austrian comments.

**91** "... old friends can have acted as they did except through ignorance": Letter, Ramsay to Arrhenius, December 23, 1915, SA.

**91** sunshine in the midst of European horrors?: Letters, Richards to Arrhenius, December 13, 1915, and February 16, 1916, SA; Arrhenius to Richards, February 21, 1916, TWR; Max von Laue to Arrhenius, November 30, 1915, and March 10, 1916, SA.

### SIX   While the Sores Are Still Dripping Blood!

**94** the bylaws regulating the prizes needed to be changed: Mittag-Leffler Diary, November 12, 1915.

**94** declaring annually a lack of worthy candidates: On plans, *Aftonbladet* and *Stockholms Tidningen*, October 31, 1915; DN, November 4, 1915.

**94** no candidates could be identified as being eligible for a prize: Göran Liljestrand, "The prizes in physiology or medicine," in Nobel Foundation, ed., *Nobel: The Man and His Prizes*, 2nd ed. (New York: Elsevier, 1962), p. 171. The Caroline Institute managed in this manner to avoid awarding Nobel prizes until 1920. But by then, inflation, followed by further economic problems in the early 1920s, hampered plans for a new Nobel Institute.

**95** therefore, the prize would be reserved until 1917: *Protokoll*, NKF, September 13, 1916, Bil. D, and September 21, 1916, Bil. A.

**96** two, Danes, Karl Gjellerup and Jakob Knudsen: A minority of two had proposed an undivided prize to Gjellerup.

**96** the baptism in blood of a new, invigorated Europe: *Världskulturen och kriget: Huru återknyta de internationella förbindelserna?* (Stockholm: Albert Bonniers, 1915), pp. 189–190.

**96** "incomprehensible position" against Germany: On the literature prize, Kjell Espmark, *Det litterära Nobelpriset. Principer och värderingar bakom besluten* (Stockholm:

Norstedts, 1986), pp. 43–46. Quotation from Wilhelm Wien's complaint to *Svenska Dagbladet* that the neutral Dutch and Swiss-Germans were not as friendly as could be hoped for. *Världskulturen och kriget*, p. 89. Carl Spitteler was awarded the reserved 1919 prize in 1920.

**97** blunt the edge of German submarine warfare: A. J. P. Taylor, *The First World War: An Illustrated History* (New York: Putnam's, 1972), pp. 165–181, 187–195.

**97** Finland after the end of Russian imperial rule: "Sverige under första värdskriget," *Den svenska historien*, **14** (Stockholm: Albert Bonniers, 1968–1969), pp. 85–91. Arrhenius's letters to Loeb (LC), among others, contain much detail on domestic social problems.

**98** Amid the growth of democracy: In correspondence among Arrhenius and T. W. Richards, Jacques Loeb, and Winthrop Osterhout, they increasingly bemoan the tendency to organize and direct research.

**98** "... forecasting weather for the Zeppelins": Letter, Vilhelm Carlheim-Gyllensköld to Carl Störmer, April 21, 1916, Carl Störmer papers, UBO.

**98** Not much enthusiasm ... for the annual Nobel rituals in 1917: Letter, H. H. Hildebrandsson to Sven [Hildebrandsson], October 18, 1917, HHH. The prize in literature was divided between two Danish authors, Karl Gjellerup and Henrik Pontoppidan. Some members wanted an undivided prize for Gjellerup. Very few supported the great Danish writer Georg Brandes, whose literary realism did not fit the prevailing interpretation of "idealistic" in the Academy. Espmark, *Det litterära Nobelpriset*, pp. 45–47.

**99** Arrhenius assumed that the German defeat: Letter, Arrhenius to Loeb, August 4, 1918, JL.

**100** In his lecture "Science and War": Arrhenius, Mss, "Vetenskapen och krig," SA. Arrhenius, and many colleagues appreciated the growing danger in Sweden of supporting industrial research while letting students hemorrhage away from universities and science to technical colleges and engineering. If Sweden was to serve as a mediator in postwar international scientific relations, local research must be sufficiently potent to command respect.

**101** for their collective contributions to relativity: Former prizewinners von Laue and Wien proposed the division.

**102** Barkla dogmatically rejected well-received theory: Paul Forman, "Charles Glover Barkla," *DSB*, pp. 456–459.

**102** Carlheim-Gyllensköld's short report: *Protokoll*, NKF, September 4, 1918, Bil. B.

**103** His autobiography, written on the occasion of his sixtieth birthday: Peter Klason, "Några drag ur en kemikers lefnad," in *Festskrift tillägnad Peter Klason på hans sextioårsdag af Lärjungar* (Stockholm: Norstedts, 1908), pp. vii–cxxvi.

**105** By 1918, almost half of all German artillery shells fired on the western front were filled with gas: L. F. Haber, *The Poisonous Cloud: Chemical Warfare in the First World War* (Oxford: Oxford University Press, 1985); Everett Mendelsohn, "Science, scientists, and the military," in John Krige and Dominique Pestre, eds., *Science in the Twentieth Century* (Amsterdam: Harwood Press, 1997), pp. 175–185.

**105** "He plunges at me, guttering, choking, drowning...": Wilfred Owen, "Dulce et Decorum Est" [1920], in *Collected Poems* (London: Chatto & Windus, 1946).

**105** Klason had championed Haber's candidacy for the work on nitrogen fixation: For another account on the deliberations on Haber and its aftermath, see Sven Widmalm,

"Science and neutrality: The Nobel prizes of 1919 and scientific internationalism in Sweden," *Minerva*, **33** (1995), pp. 339–360.

**105** Indeed, what did this list represent?: The comparable statistics for physics shows that in spite of the war, a considerable number of nominators took the trouble to respond. Also in spite of the war, and in spite of the mobilization of physicists into military-oriented research, a stream of innovative studies related to atomic physics incited some nominators into action. French and Italian chemists continued to propose the Swiss researcher P. A. Guye for his precise determinations of atomic weights, but the committee did not believe his work possessed the significance of Richards's. Frederick Soddy returned to the list for his discovery of the chemical isotope. The only other significant candidate from the past decade who repeatedly reappeared on the list was Walther Nernst, but Arrhenius was set against him. Nernst was not on the 1918 list.

**106** critical information on the process was still lacking: *Protokoll*, NKK, June 3, 1918. See Henrik Söderbaum to Oskar Widman, September 7 and 15, 1918, OW.

**106** the Americans could not build a Haber-Bosch factory until the 1920s: I thank Professor Roy MacLeod for this information, which should be appearing in a forthcoming work he is publishing with Professor Jeffery Johnson.

**107** reject Haber because of the current political conditions: *Protokoll*, NKK, September 18, 1918, Bil. C; letter, Hammarsten to NKK, September 18, 1918.

**107** "Come gurgling from the froth-corrupted lungs": Owen, "Dulce et Decorum Est."

**107** The final vote was seven for Haber, two against, and one member not present: *Protokoll*, KVA, IV klassen, October 26, 1918, Bil. A & B, respectively, on Klason's allegations. Otto Pettersson also proposed, following Söderbaum's suggestion, that the reserved 1917 prize should go to fellow Swede The(odor) Svedberg for his studies of colloids. This proposal actually created more controversy than that for Haber. Some chemists were reluctant to admit Svedberg's work as belonging to their science; they hoped the physicists could consider it instead. Others saw the desirability of having a neutral Swede balance the prize to Germany. (On Svedberg, see Chapter 11.)

**107** but not without a certain amount of quarreling: KVA/N-1918, November 12, 1918. H. H. Hildebrandsson refers to "the chemical quarrel" in a letter to Aurivillius, November 20, 1918, CA.

**108** with the blue and yellow colors of the Swedish flag?: *Les Prix Nobel en 1911*, pp. 51–52.

**109** had no plans to publish in any other language: Letter, Allvar Gullstrand to Sebastian Finsterwalden, November 12, 1919, copy, AG-UUB. Gullstrand had declined to join formally Scandinavian commissions to help reestablish international cooperation so that he might be able to do more behind the scenes [Letter, Gullstrand to Ivar Bendixson, March 11, 1918, copy, AG].

**109** an honorary doctorate for Arrhenius: Letters, Johannes Stark to Arrhenius, November 8, 1919, SA (personal letter and formal institutional letter).

**109** Arrhenius could not fathom the: Correspondence with Richards, Bredig, and Leo Jolowicz, including copies of Arrhenius's letters to them in 1919–1920, SA. Letters, Arrhenius to Ossian Aschan, December 3, 1918, OA.

**109** in recognition of Alfred Nobel's desire to advance the cause of peace: Letter, P. A. Guye to Arrhenius, December 22, 1918, SA.

**110** "as a service for my less informed countrymen ...": On need for open eyes, letter, C. W. Oseen to Carl Benedicks, September 19, 1913, CB-KB; on need to educate colleagues, letter, Oseen to Niels Bohr, May 17, 1918, BSC [*till tjänst för mina okunnigare landsmän, icke minst för dem som ha sin plats i vetenskapsakademiens fysiska klass*].

**110** Planck was of temperament and outlook: J. L. Heilbron, *The Dilemmas of an Upright Man: Max Planck as Spokesman for German Science* (Berkeley: University of California Press, 1986), pp. 3–5, 64–67, and 74–81.

**111** he contrived a formulation that won assent: R. M. Friedman, "Text, context, and quicksand: Method and understanding in studying the Nobel science prizes," *Historical Studies in the Physical and Biological Sciences*, **20**, 1 (1989), pp. 63–77, analyzes Arrhenius's three drafts.

**111** He could help the committee overcome the formal problems: Acutely aware of Klason's difficulties in providing a sufficiently compelling evaluation, Hammarsten asked committee secretary Wilhelm Palmær to provide supplemental information.

**111** The committee had received few other nominations: French and French-speaking Swiss nominators tended to support the grand old man of French organic chemistry, Albin Haller; Americans backed the English W. H. Perkin, Jr.; Rutherford again proposed Frederick Soddy. A handful of Swedes, including committee member Söderbaum, again put forth the young, local whiz kid, Svedberg. As usual, the committee accepted that its own members nominate candidates and then evaluate others, such as Ekstrand, who vigorously proposed Haber and then evaluated the English organic chemist Perkin, Jr.

**112** "For all civilized peoples [*kulturfolk*] nitrogen ... is necessary ...": *Protokoll*, NKK, September 17, 1919, Bil. 5; written July 8, 1919; during the summer, Klason sent a stream of additional materials arguing for Haber.

**112** a factory in Sweden: Arrhenius's correspondence in 1919 (with Gustaf Ekman [*Statsarkiv i Göteborg*] and Ernst Risenfeld [KVA]) on plans for a Haber factory in Sweden.

**112** But he was willing not to oppose: *Protokoll*, NKK, May 7, 1919, Bil. 7 (Söderbaum's reservations).

**112** invited back into the circle of civilized peoples: Letter, A. J. Pressman to Arrhenius, October 14, 1919, SA. Statements to the fact that not until their German colleagues showed signs of true repentance, especially for signing the Manifesto, would many scientists from the Allied powers consider shaking hands with Germans, let alone invite them into international collaboration, were made in numerous letters to Arrhenius, such as from T. W. Richards. In Paris at this time, Mittag-Leffler recorded the hostile reactions from some French and Belgian scientists to Swedish initiatives on behalf of former Central Powers researchers [Diary, 1919, KB].

**112** Late at night, the Academy voted: The quarreling primarily entailed the other available chemistry prize—whether to reward Svedberg, Svedberg and the Frenchman Jean Perrin, or to reserve it. Söderbaum and Arrhenius argued that awarding Svedberg or Svedberg and Perrin would defuse the suggestion that the Academy was an instrument of German cultural politics. This argument did not generate much support. Widman and Hammarsten insisted that this work was not really part of chemistry. Arrhenius tried to postpone awarding prizes to Germans after learning that most of the money would be taken by the German state in taxes.

**113** "... Dulce et decorum est/Pro patria mori": "Sweet and fitting it is to die for one's country." The line is taken from Horace, part of the bread and butter Latin exercises for English schoolboys. Owen, "Dulce et Decorum Est."

**113** "... But because Haber is a Prussian professor we find it acceptable!'": "Akademi-politik," *Socialdemokraten*, November 17, 1919.

**113** a defense of the Academy ... which they feared were being undermined: *DN* November 18, 1919; *Göteborgs Posten* (as reported in *SvD*, November 18, 1919). See also *NDA, Aftonbladet, Aftontidning,* November 17–19, 1919; *DN* and *SvD*. November 19–21, 1919.

**114** to improve relations with Entente researchers: *Aftontidningen*, November 18, 1919.

**114** no protest had been registered against Haber: Letter, Helge von Koch to Mittag-Leffler, December 12, 1919, ML–MLI, claims no votes against Haber. A further blow against the Academy came from Svedberg, who charged in the press that Haber should not have received a prize based on scientific grounds. Only pure research was to be rewarded, and the pure scientific merit, as compared to the industrial merit, of Haber's discovery was limited (*DN*, November 21, 1919—from *Uppsala nya tidning*). Svedberg compared the prize to those to Dalén and Marconi, who were seen as scientifically unworthy of the physics prize.

**114** "... heard nothing, and understood nothing?": *DN*, January 25, 1920, quoting the French newspaper *Le Journal*.

## SEVEN   Einstein Must Never Get a Nobel Prize

**119** Physics changed dramatically between the two world wars: Much of the excellent detailed studies of the intellectual and institutional changes in physics is ably presented (along with comprehensive bibliographies) in Helge Kragh, *Quantum Generations: A History of Physics in the Twentieth Century* (Princeton, NJ: Princeton University Press, 1999); Mary Jo Nye, *Before Big Science: The Pursuit of Modern Chemistry and Physics, 1800–1940* (New York: Twayne, 1996); and, through the lens of American physics, Daniel J. Kevles, *The Physicists: The History of a Scientific Community in Modern America* (New York: Vintage Books, 1979, and later editions).

**120** Industrial research and science-based engineering created new markets: Spencer R. Weart, "The physics business in America, 1919–1940: A statistical reconnaissance," in Nathan Reingold, ed., *The Sciences in the American Context: New Perspectives* (Washington, DC: Smithsonian Institution, 1979), pp. 295–328.

**120** Bohr attained in Copenhagen: Peter Robertson, *The Early Years: The Niels Bohr Institute, 1921–1930* (Copenhagen: Akademisk, 1979); Finn Aaserud, *Redirecting Science: Niels Bohr, Philanthropy, and the Rise of Nuclear Physics* (Cambridge: Cambridge University Press, 1990).

**122** "small popes in Uppsala": Letter, Arrhenius to Vilhelm Bjerknes, February 22, 1925, VB.

**122** "common sense foundations of physics": The term was regularly used in the committee's evaluation reports prior to World War I when quantum theory and relativity theory were being discussed. Regardless of whatever triumphs these theories achieved, committee members had little desire to relinquish the classical foundations of their science or to celebrate publicly the fall of the cultural certainties grounded in classical physics.

**123** Einstein on the World Stage: An excellent overview based on recent international historical research on Einstein is David Cassidy, *Einstein and Our World*, Control of Nature series (Atlantic Highlands, NJ: Humanities Press, 1995); still valuable are

Philipp Frank, *Einstein: His Life and Times* (New York: Da Capo, 1953, 1989); and general introduction to Einstein and his thought: Jeremy Bernstein, *Albert Einstein and the Frontiers of Physics* (New York: Oxford University Press, 1996). Also, the most recent comprehensive biography, Albrecht Fölsing, *Albert Einstein: A Biography*, trans. Ewald Osers (New York: Viking Press, 1997, original German edition 1993); and relevant chapters in Fritz Stern, *Einstein's German World* (Princeton, NJ: Princeton University Press, 1999).

**123** Einstein was transformed into a celebrity: Jeffery Crelinsten analyzes the British and American reactions immediately following the announcement in "Einstein, relativity, and the press," and "Physicists receive relativity," *Physics Teacher*, **18** (1980), pp. 115–122, 187–193; questions related to how quickly the German press reacted are discussed in Lewis Elton, "Einstein, general relativity, and the German press," *Isis*, **77** (1986), pp. 95–103. More generally, on national differences of the scientific and cultural responses, see Thomas F. Glick, ed., *The Comparative Reception of Relativity*, Boston Studies in the Philosophy of Science 103 (Dordrecht: Reidel, 1987).

**124** but instead of gold, all he touched turned into newsprint: Einstein's lament in letter to Hedvig and Max Born, September 9, 1920, *The Born–Einstein Letters* (New York: Macmillan, 1971), p. 35.

**126** The two Nobel laureates placed all their scientific authority: They wrote to Swedes hoping to keep the prize from Einstein. Most correspondence was directed to Arrhenius, especially from Stark, Lenard, and Weyland, SA; Lenard and Stark also wrote to other influentials such as Gullstrand and Oseen.

**126** Einstein was no stranger to the Academy's physics committee: Nominations for Einstein came from: 1910, Wilhelm Ostwald; 1912, Clemens Schaefer and Wilhelm Wien (division with Lorentz); 1913, Bernard Naunyn, Ostwald, Wien (again division with Lorentz); 1914, Orest Khvol'son and Naunyn; 1916, Felix Ehrenhaft; 1917, Arthur Haas, Emil Warburg, Pierre Weiss; 1918, Ehrenhaft, Max von Laue (division with Lorentz), Edgar Meyer (together with Friedrich Paschen and Planck), Stefan Meyer (Einstein as alternative after Planck), Warburg, Wien (division with Lorentz as alternative choice after Planck); 1919, Arrhenius (for Einstein's work on Brownian movement, to include in evaluation of Svedberg and Perrin), von Laue (division with Lorentz as choice after Planck and Sommerfeld), E. Meyer (same as 1918), Planck, and Warburg; 1920, Niels Bohr, W. H. Julius, Heike Kamerlingh Onnes, H. A. Lorentz, L. S. Ornstein, Wilhelm von Waldeyer-Hartz, Warburg, Zeeman. Soon after learning the results of the eclipse expedition, Lorentz initiated efforts among Dutch physicists who were eligible to nominate to propose Einstein [Lorentz to Paul Ehrenfest, September 22, 1919, Microfilm, Ehrenfest Scientific Correspondence, reel 7, AIP]. It should be recalled that prior to 1919, many physicists who might have considered Einstein were still trying to impress the committee to reward Planck.

**127** Arrhenius's report was decidedly tilted: KU, NKF, 1920, Bil. 1 (written in August). Arrhenius was well aware of the opposition to Einstein and his theories as well as of the plans to attack these at the German Natural Science Association meeting in August, which he had intended to attend [Letters, Lenard and Weyland to Arrhenius, July 23, 1920, SA; Arrhenius to Ernst Riesenfeld, October 1, 1920, ER]. Arrhenius's views are hard to intepret; he seems to have been trying to gauge German opinion and was still willing to accept the importance of Lenard's and Stark's opposition.

**128** Guillaume at best was marked *by* physics: Letter, Bjerknes to C. W. Oseen, November 14, 1920, CWO-KVA. Bjerknes was not ready to accept relativity theory, but he was willing to accept that Einstein put his stamp on the advance of physics, in contrast to Guillaume's more passive and modest achievements. Oseen agreed on both counts [letter, Oseen to Bjerknes, December 14, 1920, VB]. On Guillaume, see *DSB*.

**128** Guillaume and Benoît's work warranted merely a simple notice: Letter, Hasselberg to Kristian Prytz, January 22, 1908, and draft of letter, Prytz to Hasselberg, December 29, 1907, KP. Prytz had also collaborated with Guillaume and the International Bureau. KU, NKF, 1908.

**128** "surely seems to fulfill Alfred Nobel's requirements": KU, NKF, 1912. Pietro Blaserna and Wilhelm Foerster alternated nominations in 1910, 1912, 1913, and 1914. Foerster was one of the original members of the International Committee (1875) and its second chairman (1891–1920); Blaserna, its secretary from 1901 to 1918. C. H. Page and Paul Vigoureux, eds., *International Bureau of Weights and Measurements, 1875–1975*, translation of the BIPM Centennial Volume, National Bureau of Standards Special Publication 420 (Washington, DC: U.S. Government Printing Office, 1975), pp. 233–236.

**128** Hale and Deslandre, who had considerably better chances of winning: In 1916, Hasselberg broke his own reluctance to nominate candidates when he proposed Hale and Deslandres. KVA/N-1916.

**129** Hasselberg wrote to the committee that he would be very happy to see Guillaume receive the prize: *Protokoll*, NKF, September 8, 1920, Bil. G. Carlheim-Gyllenskiöld's report, KU, NKF, 1920, Bil. 2.

**129** Sweden's wisdom and good taste: Letter, Arnold Sommerfeld to Oseen, September 26, 1921, CWO-LAL, related the confused reaction among German physicists. P. A. Guye to Arrhenius, November 20, 1920, SA, on the joy among the Swiss over their first native-born laureate and on the importance for supporting the International Bureau.

**129** In 1921, Einstein's place in physics received unambiguous confirmation: Those who nominated Einstein were C. L. Charlier, H. W. Dällenbach, A. S. Eddington, Arthur Haas, Jacques Hadamard, George Jaffé, Theodore Lyman, Erich Marx, Gunnar Nordström, C. W. Oseen, Max Planck, C. D. Walcott, Emil Warburg, and Otto Wiener. Most nominations were for relativity theory; some noted other contributions. Oseen, as will be discussed, mentioned only the law of the photoelectric effect.

**129** boycott of German science that lasted until 1926: In 1924, the Russian physicist Khvol'son proposed a split prize between Sommerfeld and Paschan (Russia was not formally part of the boycott); in 1925, Robert Millikan proposed Sommerfeld, and his fellow American W. W. Campbell followed suit in 1926. German chemistry received virtually no nominations from chemists from the Allied nations during the interwar years. Scientists from neutral nations generally disregarded Allied wishes and nominated Germans, with whom they traditionally had close relations.

**129** scattered votes for favorite sons: Elderly Danish physicists, for example, used their permanent rights to propose their compatriots Martin Knudsen, P. O. Pedersen, and Valdemar Poulsen (a total of six nominations) for precision measurement and technological achievements. Knowing the views of their Swedish colleagues, and following the 1920 decision for Guillaume, they gave themselves hope. Some French scientists proposed Jean Perrin, either as a first or second choice after Einstein. French chemists Albin Haller and Henri Le Châtelier, foreign members of the Swedish Academy of Science, cast their nominations respectively for Pierre Langevin and the Lumière brothers, whose inventive cinematic technologies could possibly appeal to the Academy, as had Lippmann's color photography, in 1908.

**129** if it proved impossible to give a prize to a theoretician: KVA/N-1921. Edgar Meyer (University of Zurich) to NKF.

**130** In Sweden, as elsewhere, academic philosophers and cultural commentators feared: Some aspects of the reception of relativity in Sweden are covered in Carl-Olov Stawström, "Relative acceptance: The introduction and reception of Einstein's theories in Sweden,1905–1965," in Svante Lindqvist, ed., *Center on the Periphery: Historical Aspects of Twentieth-Century Swedish Physics* (Canton, MA: Science History Publications, 1993), pp. 293–305; and Kjell Jonsson, "Einstein at the amusement park: The public story of relativity in Swedish culture," in J. L. Casti and Anders Karlqvist, eds., *Mission to Abisko: Stories and Myths in the Creation of Scientific "Truths"* (Reading, MA: Perseus Books, 1999), pp. 101–120; much more research is needed on the topic.

**130** he was unable to sit in judgment of physicists working with theoretical research vastly different from his own: On Gullstrand, C. W. Oseen, "Allvar Gullstrand. Minnesteckning föredragen på Kungl. Vetenskapsakademiens högtidsdag den 26 mars 1935," *K. Svenska Vetenskapsakademiens Levnadsteckningar*, Bd. 6, pp. 480–513. On the election to the committee, Mittag-Leffler Diary, September 17, 1915. My portrayal of Gullstrand is based in part on impressions and details from his extensive unpublished papers at Uppsala University, including correspondence, diaries, academic reports, and work for the Nobel committee.

**131** Gullstrand had to evaluate things he did not understand: Exchanges between Gullstrand and Oseen in 1921 in CWO-LAL, CWO-KVA, AG. Oskar Klein to Niels Bohr, March 9, 1921, BSC, relates Oseen's tribulations in correcting Gullstrand's criticisms of Einstein, and Oseen's confession that Gullstrand hindered giving the prize to Einstein; Oseen expressed his dismay to Arnold Sommerfeld, September 1, 1921 [Sommerfeld papers, Deutsches Museum, Munich], that Gullstrand was responsible for the evaluation of relativity.

**131** Gullstrand's exceptionally long fifty-page special report: NKF, September 7, 1921, Bil. A. Gullstrand's report is dated May 12, 1921.

**131** "... speculations such as these to be the object of his prizes": Letter, Bernhard Hasselberg to H. H. Hildebrandsson, September 9, 1921, HHH-UUB.

**132** rooted in Western civilization's classic Greek heritage: *NDA*, April 2, 1923, reprinted elderly Academy physics member O. E. Westin's protest letter to the Academy [*Protokoll*, NKF, June 6, 1923, Bil. C; written March 26, 1923]. He bemoaned the deaths of Granqvist and Hasselbert, who had managed, while alive, to keep the prize from going to Einstein. Westin called Einstein's fantasies "sickly and contagious phenomena." Sten Lothigius makes references to classical learning in his critique, *De Einsteinska Relativitetsteoriernas Ovederhäftighet* (Stockholm: M. T. Dahlström, 1922). Hasselberg to W. W. Campbell, July 22, 1920, WC, on fears for civilization in postwar Europe.

**132** that allegedly conflicted with physical reality: *Protokoll*, KVA, III klassen, October 29, 1921, Bil. B.

**132** for Einstein and against Gullstrand's evaluation: Mittag-Leffler alludes to the problem that, for the Academy, a vote for Einstein would mean going against Gullstrand's resolute attack [Diary, November 9, 1921]. Mittag-Leffler and others tried to convince the committee to seek nominations from individuals who might provide insightful reasons for giving the prize to Einstein, and even to seek a foreign expert to write the evaluation on relativity theory; the latter was soundly rejected by Gullstrand (see, for example, Mittag-Leffler to Gullstrand, January 4, 1922, and the response, January 5, 1922, AG).

**133** "Einstein must never receive a Nobel Prize...": Mittag-Leffler, Diary, July 27, 1922. Gullstrand felt that even if the whole world were throwing itself over these theories and thereby inflating interest in them, soon other discoveries would send Einstein's work back into darkness. Gullstrand compares the prize with the Swedish flag in a toast, *Les Prix Nobel en 1911*, p. 52.

**133** at the same time it might rescue Bohr from committee bias: What follows is an expansion of my original study of this episode, "Nobel physics prize in perspective," *Nature*, August 27, 1981, pp. 793–798, and "Text, context, and quicksand: Method and understanding in studying the Nobel science prizes," *Historical Studies in the Physical and Biological Sciences*, **20**, 1 (1989), pp. 66–67. Nominations for Einstein in 1922: Marcel Brillouin, T. E. de Donder, Felix Ehrenhaft, Robert Emden, Jacques Hadamard, Paul Langevin, Max von Laue, Edgar Meyer, Stefan Meyer, Bernard Naunyn, Gunnar Nordström, C. W. Oseen, Max Planck, E. B. Poulton, Arnold Sommerfeld, Ernst Wagner, and Emil Warburg.

**135** or the combined significance of the two: NKF, September 6, 1922, Bil. 1.

**135** Arrhenius concluded that it would be seen as "strange": KU, NKF, 1921, Bil. B. Arrhenius considered Einstein's work to be derivative of Philipp Lenard, J. J. Thomsen, and Max Planck.

**135** All he could pry out of the Physics Section: *Protokoll*, KVA, III klassen, October 29, 1921.

**136** conceptual alternatives to the light-quantum concept: Roger Stuewer, *The Compton Effect* (New York: Science History, 1975); Bruce Wheaton, *The Tiger and the Shark: Empirical Roots of Wave-Particle Dualism* (Cambridge: Cambridge University Press, 1983); Robert H. Kargon, *The Rise of Robert Millikan: Portrait of a Life in American Science* (Ithaca, NY: Cornell University Press, 1982), pp. 70–81.

**136** but not for relativity: Letters, Oseen to Bjerknes, December 14, 1920, VB; Oseen to Carl Benedicks, September 20, 1920, CB. Although Oseen had favorably discussed relativity theory when lecturing to young physicists in 1919, he confessed privately that in their present forms the theories were unsatisfactory.

**136** "the most beautiful of all the beautiful": Oseen's enthusiasm for Bohr's work is central in his 1919 lectures on theoretical physics, *Atomistiska Föreställningar i Nutidens Fysik: Tid, Rum och Materia. Femton Föreläsningar* (Stockholm: Albert Bonniers, 1919).

**136** Oseen identified a variety of scientific orientations and outlooks: KU, NKF, 1922, Bil. 2.

**137** Bohr deserved recognition: KU, NKF, 1922, Bil. 3.

**137** Oseen respected and liked the young Danish physicist: Their correspondence of over a decade reveals the warm personal and scientific relations. Although Oseen did not then or later uncritically accept Bohr's various theories, he respected the younger physicist's opinions. See BSC and CWO-KVA.

**138** Oseen's publications ... showed the extent of decadence in modern science: *NDA*, April 2, 22, and 29, 1923. After the Academy's vote, Oseen and Arrhenius were warned to restrict the citation on the Nobel diploma to what the Academy had voted on: no relativity to be mentioned [letter, Oseen to Arrhenius, November 20, 1922, SA].

**138** "Nobel Prize in the Firing Line": *NDA* and *SvD*, February 11, 1923.

**138** "clear Germanic intellect": *SvD*, February 11, 1923. The article is signed Rh, and from its content and tone it seems to reflect Arrhenius, but could also be Ragnar Holm, a young physicist who published favorable popularizations of Einstein's theories.

**138** while he was en route to Japan: Related in letter, Albert Einstein to Arrhenius, January 10, 1923, SA.

**139** Einstein was relieved to receive the prize: In his letter to Arrhenius [January 10, 1923], Einstein noted that he could now cease being harassed by those asking him why he had not received the prize. Fölsing, *Albert Einstein*, thoroughly discusses Einstein's economic and private predicaments after the war. When Arrhenius learned in 1919 that the postwar German government was going to tax the prize money, he initially proposed that the Academy not give out the prizes to the Germans, but then he found a way to keep the money in a Swedish bank for the foreign winners until the tax danger subsided.

**139** "... harvest gains [*skörda vinst*] from their work just like businessmen": *DN*, July 11, 1923 ["En upplevelse bara att se prof. Einstein"].

**140** as his prize was not awarded for that: *NDA*, August 11, 1923 ["Alarmsignaler blåser redan innan Einsteins besök"].

**140** Sitting in the front row of the audience: On Einstein in Gothenburg, see also Jonsson, "Einstein at the amusement park," and Aant Elzinga, "Einstein i Sverige," *Tvärsnitt*, no. 3 (1990), pp. 2–13.

## EIGHT  To Sit on a Nobel Committee Is Like Sitting on Quicksand

**141** Oseen was a man of great intellect: On Oseen see R. M. Friedman, "The Nobel prizes and the invigoration of Swedish science: Some preliminary considerations," in Tore Frängsmyr, ed., *Solomon's House Revisited: The Organization and Institutionalization of Science*, Nobel Symposium 75 (Canton, MA: Science History Publications and Nobel Foundation, 1990), pp. 193–207; Karl Grandin, *Ett slags modernism i vetenskapen: Teoretisk fysik i Sverige under 1920-talet*, Institutionen för Idé- och Lärdomishtoria, Uppsala Universitet, Skrifter, Nr. 22 (Uppsala, 1999), pp. 19–100; Einar Hogner, "Inför C. W. Oseen och hans verk," *Kosmos* (1946), pp. 5–16; Carl-Olov Stawström, *Idéhistoriska synspunkter på innehållet i vissa skriftliga arbeten av Carl Wilhelm Oseen*, Stockholm Papers in the History and Philosophy of Technology (Stockholm: KTH Library, 1986); Ivar Waller, "Carl Wilhelm Oseen. Minnesteckning," *Kungl. Svenska Vetenskapsakademiens årsbok* (1948), pp. 357–379. Much of the following is based on Oseen's scientific and popular articles, as well as several hundred scientific letters to and from Oseen; Grandin has subsequently covered the same materials and offers a useful survey of Oseen's thoughts and activities.

**141** "... to create an independent Swedish science, ...": Letter, C. W. Oseen to Carl Benedicks, March 25, 1914, CB-KB. [*"tyckte jag mig finna att det som för oss är viktigast är att försöka få fram en självständig, svensk vetenskap, som vågar ställa och bearbeta sina egna problem, oavsett om de för tillfället stå i kurs i utlandet eller ej."*]

**142** even funding for research could eventually be attained: On the Academy's favoring descriptive sciences, see letter, Carlheim-Gyllensköld to C. V. L. Charlier, December 6, 1917, CC; Manne Siegbahn's frustrations with Permanent Secretary Christopher Aurivillius expressed in letter, December 9, 1918, CA. On hopes for a national physics society, see letters, Oseen to Bjerknes, November 12, 1920, VB; Oseen to Arrhenius, September 29, 1920, SA; Siegbahn to Oseen, November 26, 1920, CWO-KVA. See also Thomas Kaiserfeld, *Vetenskap och karriär: Svenska fysiker som lektorer, akedemiker och industriforskare under 1900-talets första hälft*, Stockholm Papers in the History and Philosophy of Technology (Arkiv Förlag, 1997).

**142** Gustaf Granqvist squeezed space for doctoral students: Grandin discusses Oseen's inability to obtain space in the Uppsala physics laboratory and frustrations over not being able to conduct hydrodynamic experiments in *Ett slags modernism i vetenskapen*, pp. 43–48.

**142** Oseen conceived his ultimate goal to be a Nobel Institute for Theoretical Physics: Letters, Oseen to Bjerknes, February 20, 1921, and February 3, 1923, VB.

**143** if they worked in Uppsala: Letters, Siegbahn to Oseen, September 18, 1918, and February 8, 1919, CWO-LAL.

**143** characterized Siegbahn's subsequent career: On Siegbahn: Much insightful information in the historical articles in the sixty-fifth birthday Festscrift to Siegbahn, Lars Melander et al., eds., *Manne Siegbahn 1886 3/12 1951* (Uppsala, 1951); Olle Edqvist, "Manne Siegbahn," *Kosmos* (1987), pp. 163–176; R. M. Friedman, "Siegbahn, Karl Manne Georg," *DSB*, Suppl. 2, **18**, pp. 821–826; Axel E. Lindh, "En svensk nobelpristagare," *Kosmos* (1925–1926), pp. 5–63. Drafts of Hugo Atterling's biographical essay in MS-KVA are useful but must be used with care, as numerous errors also enter into the manuscript, subsequently published as "Karl Manne Georg Siegbahn," *Biographical Memoirs of Fellows of the Royal Society,* **37** (1991), pp. 429–444.

**143** The chaotic state of quantum theory was a boon for Siegbahn: On the importance of X-ray spectroscopy for atomic theorizing at this time, see John L. Heilbron, "Lectures in the history of atomic physics, 1900–1922," in Charles Weiner, ed., *History of Twentieth Century Physics* (New York: Academic Press, 1977), pp. 50–108; Paul Forman, "The doublet riddle and atomic physics *circa* 1924," *Isis,* **59** (1968), pp. 156–174; Thomas Kaiserfeld, "When theory addresses experiment: The Siegbahn-Sommerfeld correspondence, 1917–1940," in Svante Lindqvist, ed., *Center on the Periphery: Historical Aspects of Swedish Twentieth-Century Physics,* Uppsala Studies in the History of Science 17 (Canton, MA: Science History Publications, 1993), pp. 306–324.

**143** extra funding for equipment and technical assistance: On economic obstacles to increasing precision, see letter, Siegbahn to Oseen, October 26, 1920; see also November 27, 1920, CWO-LAL. [*"har det blivit klart för mig att den önskade ökningen av precisionen närmast är en ekonomisk fråga, d.v.s. kostnaden för en ny spektrograph à c:a 5000 kr. . . . Att finansiera en sådan apparat med årsanslaget har sina svårigheter (det ordinarie är just 5000 kr).*"] On seeking additional funding in Uppsala, see letter, Siegbahn to Oseen, September 24, 1922, CWO-LAL. See also letters, Siegbahn to Oseen, February 24, April 8 and 19, 1923, CWO-LAL.

**144** Oseen expressed an unusual degree of optimism: Letter, Oseen to Bjerknes, February 3, 1923, VB.

**144** "the prospects for the Academy's future are . . . quite hopeless": Letter, Arrhenius to Arthur Rindell, April 1, 1923, AR-ÅA; Arrhenius to Ossian Aschan, October 27, 1922, OA-ÅA; Oseen relates Arrhenius's economizing campaign to Bjerknes, January 6, 1923, VB; budgets and grants related to Arrhenius's institute, *Protokoll,* KVA/N, October 4 and December 5, 1923.

**145** What did come, in 1923, was a will to restrict the scope of physics: The question of changes in the scope of physics eligible for the prize was first discussed in R. M. Friedman, "Nobel physics prize in perspective," *Nature,* August 27, 1981, pp. 793–798; R. M. Friedman, "The Nobel prize and the history of scientific disciplines: Preliminary thoughts and principles" (in Swedish), in Gunnar Broberg, Gunnar Eriksson, and Karin Johannisson, eds., *Kunskapens trädgårdar: Om institutioner och institutionaliseringar i vetenskapen och livet* (Stockholm: Atlantis, 1988), pp. 136–152; and R. M. Friedman and E. Crawford, "The prizes in physics and chemistry in the context of Swedish science," in Carl-Gustaf Bernhard, Elisabeth Crawford, and Per Sörbom, eds., *Science, Technology and Society in the Time of Alfred Nobel,* Nobel Symposium, Björkborn, Karlskoga, Sweden,

August 17-22, 1981 (Oxford: Pergamon Press for the Nobel Foundation, 1982), pp. 311-331.

**145** The case of Hale and Deslandres is worth considering: First treated in R. M. Friedman, "Americans as candidates for the Nobel prize: The Swedish perspective," in Stanley Goldberg and Roger Stuewer, eds., *The Michelson Era in American Science, 1870-1930* (New York: American Institute of Physics, 1988), pp. 272-287.

**146** ". . . ought come to be regarded as deserving of the Nobel physics prize": KU, NKF, 1913. The stilted language is in the original Swedish. In 1909, the committee declared that it needed more time to see whether Hale's achievements would have a "thoroughgoing significance" in solar physics. Those who nominated Hale were William Hallock, A. A. Michelson, E. F. Nichols, H. C. Parker, M. I. Pupin, Frank Tufts, and Maximilian Wolf. Gustaf Granqvist had proposed Guglielmo Marconi and Ferdinand Braun that year and saw to it that they shared the 1909 prize. In 1913, O. H. Basquin (Chicago) proposed Hale; Alfred Pérot proposed Deslandres as a second choice after Emile Amagat.

**146** postponed 1914 prize for German von Laue: Support for Hale in 1914 came from C. R. Van Hise and T. W. Richards (KVA/N-1914). In 1915 (KVA/N-1915; KU, NKF, 1915), Carlheim-Gyllensköld broke ranks with the other supporters of cosmical physics and nominated the two Norwegian auroral researchers Carl Størmer and Kristian Birkeland, which further precluded forming a consensus around Hale and Deslandres.

**146** Hasselberg's earlier pairing of Hale and Deslandres, proposing them himself: Apart from the one nomination in 1913, the only proposals for Deslandres came from within the committee. French nominators did not propose Deslandres. Non-Swedish nominations for Hale came from the following persons: in 1916, the American C. R. Van Hise; in 1917, six Americans from Harvard; in 1922, Robert Emden, Antonio Lo Surdo, Vito Volterra, and C. D. Walcott; in 1923, J. B. Clark, and Gustav Jäger. In 1923, Carlheim-Gyllensköld proposed both Hale and Deslandres.

**146** only if the committee still wanted to consider astrophysics as part of physics: Letters, Allvar Gullstrand to Arrhenius, May 13, 1923, and Gullstrand to Vilhelm Carlheim-Gyllensköld, May 17, 1923, copies, AG. Gullstrand could discuss the issue directly with his Uppsala colleagues.

**146** ". . . deserves certainly a thorough consideration": Letter, Hasselberg to Simon Newcomb, October 26, 1900, Newcomb Papers, LC.

**147** ". . . the proposed new [Nobel] physical institution": Letter, Hasselberg to W. W. Campbell, December 24, 1916, WC; *Protokoll*, KVAs förvaltningsutskott, October 8, 1915, KVA; *Protokoll*, NKF, January 29, 1916, Bil. A. On plans for Nobel institutes more generally, see Nobel Foundation, ed., *Alfred Nobel: The Man and His Prizes*, 2nd ed. (New York: Elsevier, 1962), pp. 334-338, 521-523.

**147** as yet no department of cosmical physics at the Nobel Institute existed: KU, NKF, 1915, Bil. 3. On plans for a museum, Anders Carlson, "Bland kulturens finaste blommor: Vilhelm Carlheim-Gyllenskiöld och Museet för det exakte vetenskapernas historia," *Forskningsprogrammet Stella. Arbetsrapport 1. Avd. för vetenskapshistoria, Uppsala universitet* (Uppsala, 1994); and Wilhelm Odelberg, "Vetenskapsakademiens lärdomshistoriska samlingar," *Svenska museer,* **2** (1962), pp. 4-14.

**147** to be defined exclusively by astrophysical spectroscopy: On the need for a department of cosmical physics, see KVAs förvaltningsutskott, January 1916, *Protokoll*, NKF, May 18 and September 21, 1916; on skepticism, see letter, Arrhenius to Bjerknes, May 12, 1916, VB; Granqvist's concern in *Protokoll*, NKF, March 24, 1917, Bil. B.

**148**  committee members could request funds to help evaluate relevant proposals: Grants applications discussed, *Protokoll,* NKF, January 10, 1923; Abbott was proposing C. D. Walcott for a prize to be split with Hale.

**148**  Arrhenius's authority on this question: Oseen's opinions are included in letters from Gullstrand to Arrhenius, May 13, 1923, and Gullstrand to Carlheim-Gyllensköld, May 17, 1923, copies, AG.

**148**  Gullstrand baited Carlheim-Gyllensköld: Letter, Gullstrand to Carlheim-Gyllensköld, May 17, 1923. For two years, Svedberg had nominated Rutherford for having created the nuclear model of the atom, having identified the nature of the radioactive alpha particles, and having used these to split light atoms. Rutherford had received a chemistry prize for earlier work on radioactivity, but the leading physicist of the English-speaking world did not as yet have a physics prize.

**149**  Therefore, Hale and Deslandres had to be eliminated: *Protokoll,* NKF, September 5, 1923, Bil. F. Over and beyond the campaign against astrophysics, Arrhenius probably was not particularly happy with Hale, who had been a strong supporter of the anti-German International Research Council.

**149**  Moreover, Hale and Deslandres . . . were nominated after all by physicists: *Protokoll,* NKF, January 31, 1924, Bil. A. It seems that Carlheim-Gyllensköld's counter-arguments were not included with the papers sent from the committee to the Academy's Physics Section in September 1923, and turn up first as an appendix in 1924.

**149**  In Gullstrand's draft . . . he left blank the concluding comments: *Protokoll,* NKF, September 25, 1923, Bil. 1.

**149**  play a role of importance in the world of physics: Robert H. Kargon, *The Rise of Robert Millikan: Portrait of a Life in American Science* (Ithaca, NY: Cornell University Press, 1982), pp. 36–81.

**150**  forging closer ties with American scientific institutions: According to Atterling, "Siegbahn," mss., prior to being named professor in 1919 Siegbahn had considered moving to the United States. In 1923, Siegbahn was already planning a tour of American universities, which he made in 1924 and which led to many close connections. Two nominations for Millikan were from Nobel laureate T. W. Richards, who had a permanent right to nominate for both physics and chemistry prizes, and Henry Fairfield Osborn, who was a foreign member of the Academy and therefore also a permanent nominator for both prizes, even though he was neither physicist nor chemist. He generally sought out opinions from local colleagues as to who should represent the United States as a candidate.

**150**  In . . . Cal Tech's Athenaeum: I thank Robert Kargon and Judith Goodstein for helping me remember where the portraits hang. Possibly, but not necessarily, with time Millikan might have picked up greater support from nominators, as many new candidates began to attract attention.

**150**  "to act to make sure that we do not elect yet another weak meteorologist": Letter, Gullstrand to Arrhenius, December 25, 1923, SA; see Friedman, "Nobel physics prize in perspective," pp. 796–798, for further details on the antimeteorological turn. Additional support for the arguments made in that article have come to light in Carl Benedicks' papers, especially letters from V. W. Ekman and Otto Pettersson.

**151**  Work by Saha and Bethe: Although Saha was first nominated by his fellow Indian countrymen in 1930, in 1937 Nobelist Arthur Compton tried to impress the committee with Saha's importance for physics and astrophysics. Oseen claimed the work significant for astrophysics but not for physics [KU, NKF, 1937 and 1939]. Bethe received strong endorsements that were equally strongly rejected by Oseen as not having sufficient significance for physics [see, for example, KU, NKF, 1943].

**152** Arrhenius devoted himself to the task: Letters, Arrhenius to Winthrop Osterhut, December 20, 1924, WO; Arrhenius to Vilhelm Bjerknes, February 22, 1925, VB; Arrhenius to Georg Bredig, April 18, 1925, copy, SA.

**152** For Oseen, the physical world picture held consequence: His concern appears in numerous correspondence but comes most clearly to view in "Frågan om viljans frihet, betraktad från naturvetenskaplig synpunkt," *Heimdals småskrifter*, Nr. 6 (Uppsala, 1909); "Sambandet mellan ljus och elekricitet," *Vetenskapen och livet*, **4** (1918); "Vetenskapen och framtiden" and "Vad vet vi om relativitetsteoriens sanning?" *Vetenskapen och livet*, **10** (1925); "Determinism och indeterminism," *Religion och kultur*, **1** (1930); and "Empirismen och sannolikhetsbegreppet," *Spektrum*, **2** (1932). A series of historical and biographical writings by Oseen beginning in the late 1920s were in part motivated by efforts to make sense of the contemporary revolution in physics by comparison with earlier periods of intellectual upheaval.

**153** no candidate could be found who deserved the 1924 prize: KU, NKF, 1924. The Finnish theoretical physicist Gunnar Nordström noted in a letter to Oseen, February 28, 1923, that although many good candidates remained after Planck, Einstein, and Bohr, no single researcher stood over the others [CWO-LAL]. In contrast to those who called for an undivided prize to Franck, Einstein proposed a prize divided between Franck and Gustav Hertz.

**153** Actually, Sommerfeld was chanceless: Oseen began to register his dislike for Sommerfeld's efforts in atomic physics in his 1919 lecture series; his comments became more virulent in special reports on Sommerfeld for the Nobel committee.

**154** In 1925, Siegbahn returned to the list of nominees: KVA/N, 1925.

**154** Siegbahn perfected that which others had created: Lindh, "En svensk nobelpristagare," reviews the development of Siegbahn's instruments and experiments; see also Siegbahn's textbook, *Spektroskopie der Röntgenstrahlen* (Berlin: Springer Verlag, 1924), for development of techniques.

**155** "the small popes in Uppsala, especially Siegbahn": Letter, Arrhenius to Bjerknes, February 22, 1925. On the resistance to aiding Pettersson, Olof Hammarsten to Arrhenius, November 21, 1924, SA; correspondence, Benedicks and Pettersson in CB-KB and *Protokoll*, NKF, for 1924 and 1925. Several Stockholm scientists wanted to help Pettersson, whose controversial experiments on atom splitting had raised serious challenges to Rutherford's Cambridge school, challenges that were not put to rest until 1928. Before the denouement, Siegbahn would not consider for an instant Pettersson's claims; Marie Curie and Albert Einstein, among others, were willing to suspend judgment. The conflict between Vienna and Cambridge on the nature of splitting atoms is masterfully analyzed in Roger Stuewer, "Artificial disintegration and the Cambridge-Vienna controversy," in Peter Achinstein and Owen Hannaway, eds., *Observation, Experiment, and Hypothesis in Modern Physical Science* (Cambridge, MA: MIT Press, 1985), pp. 239–307.

**155** "Make the peaks higher" provided the guiding philosophy during the 1920s: Robert E. Kohler, *Partners in Science: Foundations and Natural Scientists, 1900–1945* (Chicago: University of Chicago Press, 1991). Some of Siegbahn's observations were recorded in a letter to Ivar Kalberg, September 19, 1925, found in the Arthur Compton Papers and reproduced in Nathan Reingold and Ida H. Reingold, eds., *Science in America: A Documentary History, 1900–1939* (Chicago: University of Chicago Press, 1981), pp. 387–389. (Ivar Kalberg remains unidentified.)

**155** Siegbahn requested $30,000 for instruments: Letter, Siegbahn to Wickliffe Rose, February 23, 1925, International Education Board (IEB), RF Folder 587.

**155** ". . . he will need something comparable in the way of equipment": Letter, Augustus Trowbridge to Rose, April 20, 1925, IEB Folder 587. Trowbridge based his observations on Siegbahn's visit to the IEB's Paris office.

**156** ". . . he is not regarded as at present doing work of very great originality": Letters, Trowbridge to Rose, May 15, 1925, Rose to Trowbridge, June 2, 1925; IEB, Sweden-3, Folder 587. The favorable personal traits are expressed in a letter, Rose to Simon Flexner May 6, 1924, IEB, Box 14, Folder 204.

**156** "the picture of atomic structure . . . is based on these measurements": KU, NKF, 1925.

**157** to eliminate candidates not to his liking: KU, NKF, 1925. See, for example, KU, NKF, 1928, where Oseen claims, again, that the stringent bylaws prohibit the committee from considering Paschen's experimental work as well as Sommerfeld's recent electron theory of metals because the latter was not specifically mentioned by nominators.

**157** The Physics Section voted six to three for Siegbahn: The committee used the argument that it needed to postpone judgment on Arthur Compton's merits, who had, in 1922, confirmed the particulate nature of light that Einstein had theorized. *Protokoll*, KVA, III klassen, October 29, 1925.

**158** ". . . evidence of the man's standing in his profession": Letter, Trowbridge to Rose, March 17, 1927, IEB, Folder 587. Trowbridge was enthusiastic about the narrow but still important research programs Siegbahn proposed, increasing one hundred-fold the precision of the spectral regions already studied and extending measurements to wavelengths in the gap between X-rays and ordinary light. Trowbridge understood that such measurements would be of great value for theoreticians such as Niels Bohr, Erwin Schrödinger, Max Born, and others.

**158** ". . . Siegbahn and C. W. Oseen form a sort of scientific-political clique": Warren Weaver and L. W. Jones, Diary Log, May 3–June 13, 1933: Uppsala, May 8, 1933, Natural Sciences, RF. One of Siegbahn's more prominent students, Hannes Alfvén, whose work on interplanetary plasma physics was rewarded with the 1970 Nobel Prize, allegedly claimed that Siegbahn did not really understand physics [outside his own narrow specialty]. Private communication, Professor Nicolai Herlofsen, September 21, 1986.

**158** none of the candidates merited a prize: KU, NKF, 1926.

**159** who surely deserved more careful consideration: Letter, Benedicks to C. E. Guillaume, December 22, 1924, copy, CB.

**159** Benedicks no doubt wondered how he might advance Perrin's candidacy: Letters, Benedicks to Guillaume, December 22, 1924; Benedicks to Hans Pettersson, November 12 and December 26, 1926, copies, CB.

**159** a lackluster celebration: Letters, Arrhenius to Osterhout, December 20, 1924, WO; Arrhenius to Ernst and Hanna Riesenfeld, November 14 and December 14, 1925, ER. Stockholm newspapers began chattering in October about the expectations for grand festivities.

**160** At the meeting of the Physics Section, Benedicks charged: *Protokoll*, KVA, III klassen, October 23, 1926.

**160** how to present Perrin's achievements . . . to rebut the committee's . . . judgment: Letters, Benedicks to Charles Maurain, October 27 and November 5, 1926, copies, CB; Maurain to Benedicks, November 1, 1926, CB.

**160** "I must thwart most resolutely this [Benedicks'] proposal: Letter, Arrhenius to Oseen, November 2, 1926, CWO-KVA.

**161** propose a division between Franck and Hertz: In 1925, Oseen was ready to follow the recommendation of the majority of nominators and consider Franck alone. At the last minute, Siegbahn, who was attending a meeting abroad, wrote to Oseen. He was told by Dutch physicist A. D. Fokker, who had worked for a while with Hertz, that the latter actually was the leading light of the now-separated duo. According to him, it was Hertz, and not Franck, who had suggested the experiment. Since this opinion was contrary to what was widely known at the time (and at present), Siegbahn asked whether Fokker was sure of his facts; Fokker sharply responded, yes. This came as a surprise to Siegbahn and apparently to Oseen; they were assuming that at some point Franck alone would eventually receive a prize. Now Siegbahn urged caution until they appreciated the situation better. [Letter, Siegbahn to Oseen, July 12, 1925, copy, MS, Perm 29:19.] Although only two of thirteen nominations for Franck called for a divided prize and nobody had nominated Hertz alone, Oseen backed a divided prize after the letter. Historians agree that Franck was the initiator and the one who carried the line of research further.

**161** the threat of rebellion averted: Letters, Oseen to Arrhenius, November 3, 1926, SA; Gullstrand to Arrhenius, November 3, 1926, SA.

**161** "Yesterday evening offered much of interest": Letter, Benedicks to Hans Pettersson, November 12, 1926, copy, CB ["*varför det naturligtvis skall festas*"].

**161** Again, Benedicks rallied the Academy: Letters, The Svedberg to Benedicks, November 14, 1926, and Benedicks to Svedberg, November 16, 1926, copy, CB.

**161** "To sit on a Nobel committee is like sitting on quicksand": Letter, Oskar Widman to Wilhelm Palmær, November 19, 1826 [sic], WP.

**162** Anything less than a magnificent reception: Newspaper accounts noted the extra opulence of the 1926 festivities, the missing prime minister, the optimistic chairman of the Nobel Foundation, all the toasts and speeches. Svedberg was the main attraction, receiving an ovation that lasted more than a minute.

**162** Perrin was a prominent leader among liberal and left-wing professors: On Perrin and the politics of French physics between the wars, see Spencer R. Weart, *Scientists in Power* (Cambridge, MA: Harvard University Press, 1979); and Dominique Pestre, *Physique et physiciens en France, 1918–1940* (Paris: Editions des Archives Contemporaines, 1984).

**162** he knew it was actually being given by Benedicks: Letters, Svedberg to Benedicks, November 14, 1926; Benedicks to Hans Pettersson, December 16, 1926, copy, CB.

## NINE  Clamor in the Academy

**163** "... that I am now overwrought": Letters, Oseen to Bjerknes, November 17, 1925, VB ["... *Nobelprisen utdelning, en bråd och ansträngande tid. Nu är det för denna gång över men attacken har denna gång varit så svår, att jag nu är överansträngd*"]; Oseen to Bjerknes, June 6, 1925, and November 17, 1937, VB, on his nerves and overwork. A letter from Eva von Bahr-Bergius to Oseen, October 10, 1923, CWO-LAL, urged less stringency.

**163** The events of 1926 put a stop, for the moment: Letter, Arrhenius to Arthur Rindell, November 22, 1926, AR.

**164** C. T. R. Wilson, deserved the prize: Benedicks related his conversation with Perrin in a letter to Hans Pettersson, December 16, 1926, copy, CB. Another letter from

Jean Perrin to Benedicks, January 29, 1927, CB, also contains a draft of Perrin's letter of nomination for Wilson which indicates that they had conferred. CB. Benedicks also began an effort to enlist foreign nominators on behalf of the elderly French chemist and metallurgist Henri Le Châtelier for a chemistry prize.

**164** Compton ... continued to receive broad support, now also from leading German physicists: Max Born, James Franck, Max von Laue, Max Planck, and Wilhelm Wien, among others, proposed Compton. The strong German support is noteworthy. The Allied boycott of German science formally ended in 1926; and although much of the German scientific establishment still opposed joining the despised International Research Council, relations between the two sides were improving.

**165** Both Wilson and Compton were found worthy of the prize: KU, NKF, 1927.

**165** The others voted for a split: *Protokoll*, KVA, III klassen, October 29, 1927. The only member of the Physics Section to support Benedicks' motion was Svedberg, who was already fascinated by nuclear research and, like Benedicks, was not particularly excited about quantum physics. Unlike that for James Frank and Gustav Hertz, who had worked together on the series of experiments that were rewarded, or H. A. Lorentz and Pieter Zeeman, whose theoretical and experimental work were closely linked, this divided prize was, for the first time, based on the physics committee's contingent opportunism.

**165** the sudden reversal of fortune: Letter, C. T. R. Wilson to Benedicks, November 15, 1927, CB. Impressions of the Academy meeting were based on Otto Pettersson's communication to son Hans as related to Benedicks, November 20, 1926, CB.

**165** Benedicks claimed that these candidates deserved renewed attention: *Protokoll*, KVA, III klassen, October 29, 1927. A letter from Bohr to Oseen, January 29, 1926, CWO-KVA, expressed support for Richardson and his willingness to accept sharing the prize with Irving Langmuir.

**165** It was Arrhenius who paired Fleming and Richardson: Arrhenius submitted a nomination on the last eligible day, January 31, after having read other nominations. He proposed that the committee examine carefully the merits of these two candidates. Sir Robert Hadfield's nomination also played a role in convincing Arrhenius that Fleming's candidacy should be examined. [Letter, Benedicks to Sir Robert (Hadfield), January 19, 1929, copy, CB.]

**166** For the 1928 prize, Benedicks repeated Arrhenius's proposal: In 1927, Fleming had also been proposed by W. H. Bragg and by two highly prominent technological researchers, Guglielmo Marconi and Sir Robert Hadfield, but in 1928 he received support only from Benedicks. Many nominators tended to send in a proposal to test the waters; when they sensed that the committee was not biting, they dropped the candidate. Others might continue to nominate for given contributions even when they appreciated that chances were marginal that the committee would respond favorably. Richardson also enjoyed only modest support that year, receiving just two additional nominations.

**166** Based on nominations, a fairly even match could be seen: Evenly matched were Friedrich Paschen and Arnold Sommerfeld, Paul Langevin and Pierre Weiss (for diamagnetism and paramagnetism), Otto Stern and Walter Gerlach (for quantization of the magnetic moment of atoms) or Robert Wood.

**166** Oseen had decisively and definitively rejected Richardson, as well as Fleming: Rejections were justified first by the alleged lack of sufficient significance for physics of the respective candidates' contributions. But from 1925, when radio broadcasting began in Sweden, the rejection was justified largely on the impossibility of deciding which of the several contributors should be rewarded. Neither Fleming nor Richardson's role was deemed so significant that it deserved acknowledgment over those of other investigators in this field.

**166** purists in Sweden believed, could only be created in an academic setting: Letter, Siegbahn to Ivar Kalberg, September 19, 1925, reproduced in Nathan Reingold and Ida H. Reingold, eds., *Science in America: A Documentary History, 1900–1939* (Chicago: University of Chicago Press, 1981), pp. 387–389. Academic prejudice in the treatment of industrial physicist Ragnar Holm is discussed in Thomas Kaiserfeld, "Professorstillsättningar i fysik under 1900-talets första del," *Lychnos* (1993), pp. 71–107. Although Siegbahn was always much more open to industrial relevance than Oseen, he also was not very favorably inclined toward Holm and Holm's impressive record in industrial research.

**166** Benedicks was pleased; others objected: Letter, Benedicks to Sir Robert (Hadfield), January 19, 1929.

**166** Gullstrand hoped that Söderbaum could bring greater wisdom: Letter, Gullstrand to Oseen, October 8, 1928, CWO-KVA.

**166** Oseen entered into a long discursion: *Protokoll*, KVA, III klassen, October 23, 1928.

**167** Gullstrand agreed to remain in place a bit longer: Letters, Oseen to Gullstrand, November 1, 5, and 8, 1928, January 13 and 22, 1929, AG; Gullstrand to Oseen, November 4 and 7, 1928, January 20 and 24, 1929, February 6, 1929, CWO-KVA.

**167** he was safe from being pressed into committee service: Letter, Hilding Faxén to Ivar Waller, September 14, 1930, IW.

**168** "Clamor in the Academy of Sciences!": Letter, Benedicks to Otto Pettersson, October 12, 1929, copy, CB. *Protokoll*, KVA, III klassen, October 6, 1929. The bylaws permit a Nobel committee to have three, four, or five members, based on approval of the respective awarding institution.

**168** "I am appalled by the coterie domination . . .": Letter, Otto Pettersson to Benedicks, November 14, 1929, CB.

**168** Gullstrand voiced his repugnance for the latest developments in quantum physics: A blow-by-blow account of the meeting, including the quotation, is provided in a letter, Benedicks to Otto Pettersson, November 20, 1929, copy, CB. The formal statements are included in *Protokoll*, KVA gemensamma sammanträdet av III & IV klasserna, November 16, 1929.

**168** Collegiality outweighed scientific expertise: *Protokoll*, KVA, III klassen, November 30, 1929 (vote five to three in favor of Pleijel and five to three for Bohr); *Protokoll*, KVA/N, December 4, 1929. Only about half of the Academy's present members voted; perhaps many did not want to participate in this unfortunate brawl.

**169** A number of repeat candidates . . . commanded greater support than did Richardson: Otto Stern and Walther Gerlach, Arnold Sommerfeld and Friedrich Paschen, Paul Langevin, and Robert Wood all received stronger endorsements. Among the few nominations for Richardson were those from Benedicks and Bohr (the latter once again suggested as a possible alternative a split between Richardson and Langmuir). Wilson and Rutherford sent in a joint proposal for Richardson, the former probably having been encouraged by Benedicks. Bohr also suggested, as a another alternate candidate, Wood, who was proposed by, among others, new committee member Hulthén. It was not unusual for some nominators, such as Bohr and Franck, to propose two, three, and even four alternate candidates.

**169**  Carlheim-Gyllensköld protested: *Protokoll*, NKF, October 9, 1929; *Protokoll*, KVA/N, November 12, 1929.

**169**  As Bohr and Rutherford both expressed: Letters, Bohr to Richardson, November 13, 1929, copy, BSC; Rutherford to Bohr, November 19, 1929, BSC.

**170**  "cultural relativism and pessimism": Oseen, "Vad vet vi om relativitetsteoriens sanning?" *Vetenskapen och livet*, 10 (1925).

**170**  This was not to his taste: Letter, Oseen to Bohr, March 6, 1926, BSC; Oseen reviewed the current quantum literature in his graduate seminar as well as in his evaluations for the Nobel committee. Letters from Bohr and others imply that Oseen's unhappiness was known to colleagues.

**170**  He had, in 1923, indicated a desire to devote but ten more years to physics: Letter, Oseen to Bjerknes, December 31, 1923, VB.

**171**  Oseen tried to hide behind the bylaws: Oseen seems to have been inching toward considering de Broglie and Schrödinger for the discovery of the wave properties of electrons. But when it became clear in 1929 that a prize to Schrödinger would appear to be endorsing his version of quantum mechanics, without acknowledging Heisenberg's, Oseen endorsed de Broglie alone while treating the two others together for their more general contributions toward a quantum mechanical depiction of the atom. [KU, NKF, 1929.]

**171**  To counter Oseen's intransigence, The(odor) Svedberg nominated Heisenberg: KVA/N-1930, Svedberg to NKF, January 31, 1930, noting Arnold Euken and K. F. Bonhoeffer's discovery of allotropic forms of hydrogen.

**172**  Raman had been campaigning for several years: Letter, Vilhelm Bjerknes to Oseen, February 13, 1925, copy, VB, relates Raman's desire to bring the first Nobel Prize in science to Asia. He wanted Bjerknes, whom he knew from Pasadena, to arrange a meeting with Oseen, clearly for the purpose of priming a key committee member so that when he did make a great discovery the path would be that much easier to Stockholm.

**172**  insisted on awarding Raman alone: KU, NKF, 1930. Letter, C. V. Raman to Bohr, December 6, 1929, asking to nominate him in the name of Asian science, BSC.

**172**  Wolfgang Pauli commented ... that there were no theoretical physicists in Sweden: Letter, Wolfgang Pauli to Oskar Klein, December 12, 1930, reproduced in Karl von Meyen, ed., *Wolfgang Pauli: Wissenschaftlicher Briefwechsel mit Bohr, Einstein, Heisenberg u.a.*, Vol. 2 (Heidelberg: Springer Verlag, 1985), p. 43. Pauli insisted on the need for a representative for modern theoretical physics, referring to Klein.

**172**  Still, in the following year, 1932, nominators began registering impatience: KVA/N, 1932.

**173**  Oskar Klein, David Enskog, and Hulthén: On Klein during this period, see Karl Grandin, *Ett slags modernism i vetenskapen: Teoretisk fysik i Sverige under 1920-talet*, Institutionen för idé- och lärdomshitoria, Uppsala universitet, Skrifter, Nr. 22 (Uppsala, 1999), pp. 101–144; on Enskog, see Mats Fridlund, "International acclaim and Swedish obscurity: The fall and rise of David Enskog," in Svante Lindqvist, ed., *Center on the Periphery: Historical Aspects of Swedish Twentieth-Century Physics*, Uppsala Studies in the History of Science 17 (Canton, MA: Science History Publications, 1993), pp. 238–268.

**174**  The committee majority voted with Oseen; Hulthén reserved his vote: *Protokoll*, NKF, September 29, 1932, Bil. A, B, C.

**174**  considerable defection from the committee majority: Henrik Söderbaum's tally from the vote written in *Protokollsbilagor till KVA:s protoll i Nobelärenden 1932–1934*, p. 420

**174** Actually, only two nominators added Dirac: W. L. Bragg and the Polish physicist Czeslaw Bialobrzeski proposed Dirac.

**174** Overwhelmingly, nominators indicated a clear desire to reward the two pioneers: Einstein and Maurice de Broglie proposed Schrödinger alone; Louis de Broglie proposed a division among Schrödinger, C. J. Davisson, and L. H. Germer; Ludwig Plate called for Schrödinger as a second choice after Sommerfeld; the brilliant young Italian Enrico Fermi and his patron Orso Corbino both proposed Heisenberg alone. Those calling for rewarding both Heisenberg and Schrödinger included Bohr, W. L. Bragg, and Franck.

**175** achieve something truly great: *Protokoll*, NKF, June 2, 1933, Bil. D, on Oseen's concern with Dirac's achievements and on the initial proposal to split between Heisenberg and Schrödinger. One caveat, seemingly originating with Oseen: The proposal for the division had to be contingent on further discussion as to whether this proposal violated the Nobel Foundation's basic bylaws.

**175** He now linked the three as standing head and shoulders above others: *Protokoll*, NKF, September 9, 1933, Bil. E; September 25, 1933, Bil. B.

**176** Some, such as Millikan, emerged as overbearing boosters: Millikan repeatedly (1934–36) supported C. D. Anderson for the discovery of the positron, calling this the greatest achievement of the century in experimental physics. In the late 1930s he then began pressing for Anderson's and Seth Neddermeyer's claims for the mesotron. In both these cases, Millikan himself would be elevated through a prize; the nominations also totally ignored the work of others who also had a claim in the honor of discoveries.

## TEN  It Can Happen That Pure Pettiness Enters

**179** National differences sharpened: Robert E. Kohler, Jr., "The Lewis-Langmuir theory of valence and the chemical community, 1920–1928," *Historical Studies in the Physical Sciences*, **6** (1975), pp. 431–468.

**179** More than ever, chemistry was understood as a key: William H. Brock, *The Fontana History of Chemistry* (London: Fontana Press, 1992), Chapters 13 to 16, provides much insight into chemistry after World War I, although with a British-American emphasis. More technical, but necessarily schematic, relevant chapters appear in the classic Aaron J. Ihde, *The Development of Modern Chemistry* (New York: Harper & Row, 1964).

**180** "Don't shoot the piano player, he's doing the best he can!": Letter, Henrik Söderbaum to Oskar Widman, May 5, 1919, OW.

**180** Swedish chemists held diverging ideas about scientific accomplishment: Letter, Bror Holmberg to Oseen, May 29, 1929, CWO-KVA. Holmberg was discussing the impossibility of gaining a consensus among Swedish chemists because they held diverging views on appropriate methods and research priorities.

**180** In the 1890s, Arrhenius and Nernst were both young rising stars: Erwin N. Hiebert, "Developments in physical chemistry at the turn of the century," in Carl-Gustaf Bernhard, Elisabeth Crawford, and Per Sörbom, eds., *Science, Technology and Society in the Time of Alfred Nobel*, Nobel Symposium, Björkborn, Karlskoga, Sweden, August 17–22, 1981 (Oxford: Pergamon Press for the Nobel Foundation, 1982), pp. 97–114; and "Walther Hermann Nernst," *DSB*, **15**, pp. 432–453; D. K. Barkan, *Walther Nernst and the Transition to Modern Physical Chemistry* (Cambridge: Cambridge University Press, 1999). Barkan treats Arrhenius's opposition to awarding Nernst a prize in "Simply a matter of chemistry? The Nobel prize for 1920," *Perspectives on Science*, **2** (1994), pp. 357–395. The technical details of Arrhenius's evaluations of Nernst's contributions are well covered here.

**180** Rivalry came quickly to the two researchers: In a letter from Arrhenius to Arthur Rindell, September 29, 1896, AR, Arrhenius enviously compares his and Nernst's working conditions.

**180** Arrhenius began a frenzied, defensive correspondence: Related by Otto Pettersson, who witnessed the correspondence in the 1890s, in a letter to Carl Bendicks, November 8–9, [1937], CB.

**180** Nernst hurled abusive comments: Letter, Hans von Euler-Chelpin to Arrhenius, May 22, 1901, SA.

**181** Arrhenius became enraged upon learning: Letter, Arrhenius to Gustaf Tamman, December 22, 1910, copy, SA.

**181** began to nominate Nernst themselves: In 1916, Bror Holmberg, Carl Kullgren, Wilhelm Palmær, The Svedberg, and K. A. Vesterberg all nominated Nernst.

**181** Arrhenius proved a formidable opponent: KU, NKK, 1916. Letter, Arrhenius to Wilhelm Palmær, April 26, 1916, WP. Arrhenius's broad network of correspondents helped him to keep in touch with research and attitudes related to Nernst's work, as in letters from Princeton's Hugh S. Taylor to Arrhenius, February 1 and December 8, 1916, SA.

**181** An autocratic professor was forcing assistants: Letter, Arrhenius to O. Aschan, February 20, 1916, OA. Arrhenius expresses his opposition to Nernst and his theory in letters to Arthur Rindell, January 15, 1916, copy, SA, and to Jacques Loeb, March 4, 1916, JL.

**181** "A[rrhenius] couldn't stomach and couldn't forget a disagreement": A letter from Pettersson to Benedicks, November 7–8, [1937], discusses Arrhenius's longstanding feud with Nernst.

**182** to counter Arrhenius's influence: Indications of Arrhenius's softening of his feelings toward Nernst appear in letters to Edward Hjelt, October 23, 1919, EH, and Ernst Riesenfeld, October 6, 1919, ER. Nernst had a strong position among the candidates. In 1920, again, relatively few (twenty-eight) nominators responded, proposing eighteen candidates. The eight votes for Nernst dwarfed the remaining candidates; these included Fritz Haber's letter of nomination that entailed a strongly argued evaluation of Nernst's achievements (probably pitched with insight into Arrhenius's opposition in mind). Other candidates in 1920 included P. A. Guye, W. H. Perkins, Jr., Georges Urbain, and Henri Le Châtelier, all of whom received no more than a couple of nominations. Carl Bosch, however, was advanced by five persons in protest to his being left out in the 1919 prize to Haber.

**182** But Arrhenius dashed these hopes again: KU, NKK, 1920.

**182** The debate brought swift results: *Protokoll*, IV klassen, October 30, 1920.

**182** But in the Academy: *Protokoll*, KVA/N, November 11, 1920.

**182** they organized a campaign ... to swamp the committee with nominations: Letter, Ludwig Ramberg to Wilhelm Palmær, January 24, 1921, WP.

**182** the most important in German physical chemistry, was being strangled by economic difficulties: Hans von Euler, "Walther Nernst," *Svensk kemisk tidskrift* (1921), pp. 194–198, noted how the prize would help the experimental work that otherwise was threatened by economic problems. Von Euler was in close contact with Nernst, who had suggested that von Euler accept his Berlin professorship. Nernst was planning on leading the national laboratory of physical and technical research [von Euler, *Minnen* (unpublished autobiography), pp. 91-92].

**183** The committee voted to award the reserved 1920 prize to Nernst: Letter, Arrhenius to Richards, May 21 and October 29, 1921, TWR, on expectations that the chemists and the Academy were prepared to give a prize to Nernst. A letter from Söderbaum to Widman, August 3, 1921, OW, discusses Arrhenius's unexpected new report on Nernst. Widman denies Nernst's significance for chemistry proper in a letter to Söderbaum, August 7, 1921, HS-KVA, and finds a rationale to vote for him in KU, NKK, 1921, Bil. 11.

**183** "an energetic opposition" was expected: Letter, von Euler to Svedberg, October 27, 1921, TS.

**183** "for the Academy's sake, luckily such a scene was rare": Letter, Richards to Arrhenius, October 10, 1921, discusses strategy on how to justify a prize for Nernst without accepting the German's priority for the discovery of the heat theorem. Richards began complaining about Nernst's alleged plagiarism in a letter to Arrhenius, October 16, 1913, SA. Although Richards provided important ideas, he had not attained the insight and fullness of Nernst's work. On Arrhenius's behavior at the Academy's meeting, see letter, Åke Ekstrand to Wilhelm Palmær, November 11, 1921, WP.

**183** He now praised the imposing thermodynamical work of Nernst's laboratory: Letter, Arrhenius to Richards, January 3, 1922, TWR.

**184** first-rate research but without spectacular discoveries: Candidates such as Theodor Curtius, Albin Haller, P. A. Guye, Gustav Tammann, and W. H. Perkin, Jr., were thought worthy by respective national nominators, but without an advocate on the committee their frequent appearances on the lists of candidates had little effect.

**184** on several occasions to dismiss the sparse nominations for Soddy: KU, NKK, 1918. Widman had proposed in 1908 a division between collaborators Rutherford and Soddy as an alternative to Rutherford alone. Subsequently, Soddy received two nominations in 1918 (from Rutherford and Wilhelm Schlenck) and one each in 1919 and 1922 (both from Rutherford). Soddy had noted that a radioactive element, when emitting an alpha particle (helium nucleus), changes its chemical properties to that belonging to an element two places to the left on the periodic table. Using the Greek word meaning "same place," he dubbed the term "isotope" for radioactive elements that in transforming themselves through emitting particles eventually occupy the same place in the periodic table but have different origins. Not until Henry Moseley's work did the meaning become clear: The atomic number of an element defined its chemical properties; isotopes of the same element could vary with respect to their atomic weight and yet be chemically inseparable. Rutherford's discovery of the proton as the key constituent of the atomic nucleus gave further understanding of the atomic number, which is the number of protons in the nucleus. Not until the discovery of the neutron in 1932 was there a fuller understanding of why the same chemical element could be found with differing atomic weights: Isotopes of the same element have different numbers of neutrons but the same number of protons. Prior to the discovery of the neutron, models of the atom generally included electrons both in the nucleus and orbiting it.

**185** did these innovations really have an impact on chemistry?: Discussions, *Protokoll*, NKK, 1922.

**185** Söderbaum was one of the original committee members: Gunnar Nilson, "Henrik Gustaf Söderbaum: Minnesord i Svenska Läkaresällskapet den September 26,

1933," *Hygiea*, **95** (1933), pp. 721–725; Christian Barthel, "Henrik Gustaf Söderbaum," *Kungl. Landbruks-akademiens handlingar och tidskrift* (1933), pp. 793–795, emphasizes Söderbaum's ability to transcend pettiness, achieve impartiality, and embody the characteristics of an intellectual aristocrat ["*En andens aristokrat*"].

**185**  it was even more surprising to Soddy, who could not believe the news: Letters, Frederick Soddy to Sir Ernest and Lady Rutherford, November 10, 1922, and Soddy to Rutherford (second letter), November 10, 1922, ER, microfilm copy, AIP.

**185**  too narrow in its overall significance for chemistry: KU, NKK, 1917.

**185**  For better or worse, scientists were not always the most perfect of God's creatures: Letter, Widman to Bror Holmberg, February 17, 1919, BH-LUB.

**186**  Pregl had shown that the traditional methods: Ihde, *Development of Modern Chemistry*, pp. 577–578. Mimliang Xu, "Fritz Pregl," in Laylin James, ed., *Nobel Laureates in Chemistry, 1901–1992* (Washington, DC: American Chemical Society and Chemical Heritage Foundation, 1994), pp. 146–150.

**186**  Widman became a devotee: Arne Tiselius et al., eds., *The Svedberg 1884 30/08 1944* (Uppsala: Almqvist & Wiksells, 1944), p. 716.

**186**  Pregl subsequently disappeared from the list of candidates: In 1917, Eduard Buchner and A.O.R. Windaus proposed Pregl (KVA/N-1917). Widman's praise and other committee members' negative evaluations appear in KU, NKK, 1917.

**186**  Pregl was well received in Stockholm and Uppsala: Pregl seems to have come more than once. In 1921, he helped organize a section for microanalysis in von Euler's Stockholm laboratory [von Euler, *Minnen*, p. 88]. Another trip in 1922 is mentioned in a letter from Pregl to Arrhenius, April 16 and November 23, 1922. On "Gallic imperialism," see letter from Pregl to Ludwig Ramberg, January 15, 1923, LR. Additional correspondence with Ramberg and Oskar Widman also refers to Pregl's correspondence with Hammarsten (LR and OW).

**186**  Nobody else nominated Pregl: K. A.Vesterberg and J. Sjöqvist were the other Swedish chemists to nominate. Von Euler discusses his plans to move to Vienna in *Minnen*, p. 92.

**186**  Haber wrote a letter to Arrhenius: Letter, Haber to Arrhenius, January 5, 1923, SA. Haber and Willstätter hoped the committee could, instead, look favorably on the German chemist Carl Auer von Welsbach, who was also proposed by one of the Swedes, for discovering two new rare earth elements.

**187**  Pregl received a prize in 1923: *Protokoll*, IV klassen, October 27, 1923. The other member to oppose, Christian Barthel, actually worked with microbiological aspects of physiological chemistry.

**187**  the Danish-based results were incontestable: See Hevesy's letters to Fritz Paneth, 1922–1923, Hevesy Collection, microfilm, NBA; letter, Söderbaum to Widman, April 15, 1928, OW. Helge Kragh, "Anatomy of a priority conflict: The case of element 72," *Centaurus*, **23** (1980), pp. 275–301.

**187**  He worked constantly to ease the boycott against German chemists: Arrhenius acted as middleman between the American Chemical Society and Nernst and Ostwald for reinstating membership. See letters, Charles Parsons to Arrhenius, September 10, 1926; Parsons to Wilhelm Ostwald, April 22, 1927, copy; Parsons to Arrhenius, April 26, 1927, SA. His efforts to include Germans in formal and informal international chemistry meetings is a recurring theme in correspondence during the 1920s.

**187** Hammarsten noted in his 1924 evaluation that Wieland's painstaking organic analyses: KU, NKK, April 4, 1924, Hammarsten's special report. Widman subsequently took over responsibility for evaluating Wieland and Windaus.

**188** Conferring closely with Söderbaum, Widman devised a strategy: Letter, Söderbaum to Widman, April 15, 1928, OW.

**188** must be rewarded with a prize: *Protokoll*, NKK, June 12, 1928, Bil. K and L. In the Chemistry Section meeting, Otto Pettersson objected, voting for metallurgist Le Châtelier. Klason repeated his call for giving a prize to von Euler, but in the end he joined the majority in voting for Windaus and Wieland.

**188** Arrhenius concluded, although Lewis's overall contributions were great, they did not satisfy Nobel's conditions for a prize: KU, NKF, 1924. Those mentioned were Gustav Kirchhoff, C. M. Guldberg, J. W. Gibbs, Hermann Helmholtz, Rudolf Clausius, and Max Planck.

**189** he felt threatened that physical chemistry had reached levels of sophistication beyond his grasp: Letter, Arrhenius to "My very dear friend" [T. W. Richards], December 19, 1924, TWR. On Arrhenius's opposition to modifications of his theory and the creation of a theory of strong solutions, see Arne Ölander, "Arrhenius och den elektrolytiska dissociationsteorien," *Svante Arrhenius till 100-årsminnet av hans födelse: Bilaga till Vetenskapsakademiens årsbok 1959* (Stockholm: Almqvist & Wicksells, 1959), pp. 5–33. In particular, in 1908 Lewis had held a lecture on the accumulated evidence against the original theory; although he claimed Arrhenius's theory would not be replaced—only extended—that alone would create long-lasting ill will. Although Arrhenius did not engage in public rebuttals in the 1920s, when a number of theoretical extensions of dissociation theory had been published (by Lewis, Niels Bjerrum, J. N. Brønsted, Peter Debye, and Erich Hückel), many of his disciples and supporters at home disparaged these works, without fully understanding them (p. 25).

**189** Such was the case with the Dane Niels Bjerrum: Oskar Klein related how Arrhenius seemingly accepted Bjerrum's solution theory (letter to Niels Bohr, October 30, 1918, BSC-AIP), but by 1920 Arrhenius was again refusing to accept change; letter Oseen to Arrhenius, December 3, 1920, SA, approves Arrhenius's rejection of the latest theories.

**189** It seemed that little distinguished winners from losers other than luck and committee taste: That German organic chemist Otto Wallach received a prize in 1910 but others of his generation, such as Haller, Perkin, Jr., and Curtius, did not subsequently receive prizes surely left local champions of these and other significant chemists frustrated and perplexed.

**189** Four new members: The four new members of the committee were Theodor Svedberg (1925), Wilhelm Palmær (1926), Ludwig Ramberg (1927), and Hans von Euler-Chelpin (1929). They all figure in the following chapter.

## ELEVEN  One Ought to Think the Matter Over Twice

**190** When The(odor) Svedberg studied at Uppsala: Historical articles in two Festschrift collections, Arne Tiselius et al., eds., *The Svedberg 188430/81944* (Uppsala: Almqvist & Wiksells, 1944); and Bengt Rånby, ed., *Physical Chemistry of Colloids and Macromolecules*, Svedberg Symposium—1984 (Oxford: Blackwell, 1987); Stig Claesson and Kai O. Pedersen, "The Svedberg, 1884–1971," *Biographical Memoirs of Fellows of the Royal Society*, **18** (1972), pp. 595–627; Anders Lundgren, "The ideological use of instrumentation: The Svedberg, atoms, and the structure of matter," in Svante Lindqvist, ed.,

*Center on the Periphery: Historical Aspects of Twentieth-Century Swedish Physics* (Canton, MA: Science History Publications, 1993), pp. 327–346. Svedberg's unpublished autobiography in typescript, "Fragment," [1960] TS.

**190** Svedberg was soon included among the conquering heroes of atomism: Milton Kerker, "The Svedberg and molecular reality," *Isis*, **67** (1976), pp. 190–216. On Jean Perrin's work on atomism and the criticism of Svedberg's claims, see Mary Jo Nye, *Molecular Reality: A Perspective on the Scientific Work of Jean Perrin* (New York: Elsevier, 1972).

**191** much to the dismay of Widman and his successor: Letters, Oskar Widman to Ossian Aschan, June 8, 1920, OA; Ludwig Ramberg to Bror Homberg, March 3 and 10, 1919, BH-LUB.

**191** Widman feared losing a culture of chemistry: Letter, Widman to Edward Hjelt, January 9, 1920, EH.

**191** Svedberg might draw upon resources and students in physics rather than chemistry: Letter, Widman to Aschan, June 8, 1920.

**191** "Americanization" of science: Letter, Vilhelm Bjerknes to C. W. Oseen, February 12, 1925, CWO-KVA.

**191** "Well, we can easily do that!": Svedberg, "Fragment," p. 130ff.

**192** began to devise a new instrument for studying these molecules: On Svedberg's ultracentrifuge, see standard accounts from Svedberg's assistants and students in 1944 and 1984 Svedberg anniversary volumes and in Kai O. Pedersen, "The development of Svedberg's ultracentrifuge," *Biophysical Chemistry*, **5** (1976), pp. 3–18; greater analysis and use of unpublished documents appear in Boelie Elzen, "Scientists and rotors: The development of biochemical ultracentrifuges," Ph. D. dissertation, University of Twente, Netherlands, 1988, Chapter "Theodor Svedberg," pp. 13–142.

**192** "The, I see a dawn": Kai Oluf Pedersen, "The Svedberg and Arne Tiselius. The early development of modern protein chemistry at Uppsala," *Selected Topics in the History of Biochemistry: Personal Recollections I*, vol. 35 of *Comprehensive Biochemistry* (New York: Elsevier, 1983), p. 241.

**192** "That sure ought to make the chemists 'sit up and take notice' . . .": Svedberg quoting J. B. Nichols in a letter to Herman Rinde, July 24, 1924, copy, TS.

**193** When the Chemistry Section voted: Widman admitted that all three were qualified, but because Svedberg and von Euler could be expected to continue to be nominated for a prize and therefore might not be able to participate in the annual evaluation work, it would be best for the committee to elect Palmær. *Protokoll*, KVA, IV klassen, January 10, 1925; letter, Arrhenius to Ernst Riesenfeld, January 28, 1925, ER.

**193** In the Academy Svedberg edged out Palmær by two votes: Letter, Hammarsten to Palmær, January 11, 1925, WP-KVA; *Protokoll*, KVA/N, January 31, 1925, Bil. B. There were seventeen votes for Svedberg, fifteen for Palmær, and one for von Euler.

**193** Svedberg convinced his colleagues: KU, NKK, 1926. A few of the earlier nominations proposed splitting a prize between Zsigmondy and Henry Siedentopf, who had co-invented the ultramicroscope. See letter, Olof Hammarsten to The Svedberg, January 19, 1926, TS, on strategy to convince the committee.

**193** How could anybody expect the government to fund such a huge request?: Letters, Ramberg to Holmberg, April 24, 1926, BH-LUB; Widman to Aschan, November 1, 1926, OA.

**193** "Professor Must Work in Dark Coal-Cellar . . .": *DN*, August 6, 1925 [*"Professorn får arbeta i mörk kolkällare och hans lärjungar på W.C."*]. See also *DN*, October 16 and November 16, 1926.

**194** When members of the Academy had proposed Svedberg in 1918 and 1919: A letter from Widman to Hjelt, January 9, 1920, EH, relates how Söderbaum and Arrhenius fought for Svedberg, while Widman and Hammarsten opposed him. Statements in letter from Hammarsten to Arrhenius, May 28, 1918, SA; Otto Pettersson to IV klassen, October 21, 1918, and *Protokoll*, KVA, IV klassen, October 26, 1918, accused the committee of being too narrow in its understanding of chemistry and thereby allowing important work on the boundaries between physics and chemistry to fall between the chairs. See also letter, Axel Hamberg to Wilhelm Palmær, November 12, 1919, WP.

**194** They held that Max von Laue and the Braggs': Arrhenius had been angling to consider the colloid studies of Perrin, Svedberg, and Einstein physics, but the physics committee's majority refused to accept their consideration, claiming that the experiments of Perrin and Svedberg at best showed a probability of atoms existing. They held to an experimentalist perspective that after 1912 atoms were first proven to exist through the X-ray diffraction experiments. KU, NKF, 1918, 1919.

**194** the Academy had disregarded the principles by which its committees had worked for a quarter century: Letter, Widman to Palmær, November 19, 1826 [sic], WP.

**194** Svedberg . . . experienced both depression and exhilaration: The relevant paragraphs from Svedberg's autobiography, "Fragment," on postprize thoughts are reproduced in translation in Milton Kerker, "The Svedberg and molecular reality: An autobiographical postscript," *Isis*, **77** (1986), pp. 278–282.

**195** Through Svedberg, Swedes could celebrate their own cultural eminence: "Tjugufemte Nobelfesten en vacker högtid . . . Minutlång applåd hälsade den svenske Nobelpristagaren," *Stockholms Tidningen*, December 11, 1926. All the Stockholm papers carried extensive coverage, each emphasizing slightly different aspects, but in all accounts Svedberg was clearly the darling of all in attendance. *Aftonbladet*, December 13, 1926, describes Uppsala's welcome to Svedberg upon his return, including fireworks, a parade, and songs.

**195** enable Svedberg to continue his important research at home, in Sweden: H. G. Söderbaum, speech at presentation of Nobel Prize to Svedberg, *Nobel Lectures in Chemistry, 1922–1941* (New York: Elsevier, 1966), p. 66.

**195** Svedberg's "extraordinary contributions": Meeting with Cabinet Minister Börje Knös described in "Fragment." On the government budget request, see "Våra Nobelpristagare få bättre resurser för forskningarna," *Nya Dagliga Allhända*, January 11, 1927. The expenditures were to cover a few years. On the approval, see *DN*, April 2, 1927. Documentary evidence offers conflicting views on when the university actually accepted Svedberg's request in its budget.

**195** Svedberg was scarcely noted in the diary of visiting officials: Rockefeller Foundation officials forgot his name, Travel log, Augustus Trowbridge, 2/–10/25, 1/–6/26, "Memorandum of visit to Uppsala," October 2, 1925, in General Education Board 12, File 905, RF.

**195** Svedberg, sitting in Uppsala, became a major player in international science: On Rockefeller Foundation support of natural science and especially efforts to stimulate the application of physical methods and theories in biomedical science, see Kohler, *Partners in Science: Foundations and Natural Scientists, 1900–1945* (Chicago: University of

Chicago Press, 1991); Pnina Abir-Am, "The discourse of physical power and biological knowledge in the 1930s: A reappraisal of the Rockefeller Foundation's 'policy' in molecular biology," *Social Studies of Science*, **12** (1982), pp. 341–382. Ulf von Euler quotes a foundation official on wanting more physics in physiology in a letter to Göran Liljestrand, May 28, 1934, GL.

**196** Svedberg proposed dividing a prize between Urey and Washburn: In early drafts of his evaluation, Svedberg considered dividing the prize between Harold Urey and G. N. Lewis, who had also contributed to work on heavy hydrogen and who claimed priority to the discovery. Svedberg subsequently learned, secondhand, of conflicts and accusations over priority, and he decided to change his evaluation to reward Urey alone. Drafts of Svedberg's evaluations for 1934 are in Nobel file in TS. Also, *Protokoll*, NKK, Spring–Autumn 1934.

**196** the best-equipped chemistry laboratory in the world: *SvD*, February 20, 1935.

**196** Svedberg applied to the chemistry committee's special fund: *Protokoll*, NKK, January 1935.

**197** Svedberg wrote to Perrin: Letter, Svedberg to Jean Perrin, February 8, 1935, copy, TS. Perrin repeated his earlier nomination of the couple for the physics prize. Another person with a strong interest in the outcome of the 1935 competition, Rutherford, acted like Svedberg. He hoped to see his protégé, James Chadwick, collect an undivided physics prize for the discovery of the neutron; therefore, he took the unusual step of sending a nomination also for the chemistry prize, for the French couple. (Rutherford had not submitted a chemistry proposal since 1922.) The only other nominator to suggest a chemistry prize for Joliot and Joliot-Curie was Svedberg's committee colleague and tactical ally, von Euler. KVA/N, 1935.

**197** he would have a "battle" to get the chemistry prize awarded to the two physicists: Letter, Arne Westgren to Svedberg, August 29, 1935, TS, referring to Svedberg's own comment.

**197** Langmuir was drawn to consider the physics and chemistry . . . of this artificial little world: Leonard S. Reich, "Irving Langmuir and the pursuit of science and technology in the corporate environment," *Technology and Culture* (1983), pp. 199–221.

**198** Svedberg prevailed, but not because of any suggestion of intrinsic superiority: James Kendall proposed Langmuir in 1932. KU, NKK, 1932, Bil. 10 and 11.

**198** he divorced her and married into a Swedish aristocratic family: Von Euler's version in *Minnen*. Other perspectives: Olof Hammarsten to Gustav Retzius, March 27, 1914, GR, refers to details in Retzius's letter; and witnessed testimony from Astrid Cleve, March 19, 1914, GR.

**199** joining the German war effort: *Minnen*, pp. 82–85; letters, Arrhenius to Gustaf Ekman, March 16, 1916, Gustaf Ekman papers, Landsarkivet i Göteborg; Arrhenius to Jacques Loeb, October 12, 1916, JL.

**199** seemed ready to accept a chair in Vienna: *Minnen*, p. 140; letters, Söderbaum to Widman, April 19, 1922, OW; von Euler to Aschan, October 29, 1922, OA; on his last-minute decision to reject, letters between Arrhenius and Riesenfeld in 1922, SA and ER. Von Euler was also entertaining a proposal for him to take over Nernst's chair in physical chemistry in Berlin.

**199** Arrhenius feared that von Euler would disparage the prize: Letters, Arrhenius to Georg Bredig, November 26, 1926, copy, SA; Arrhenius to Riesenfeld, December 6, 1926, ER.

**199** Similarly, local friends, who were certainly aware of his plans, regularly nominated him: Interview with Willstätter, reported in letters from Augustus Trowbridge to Wickliffe Rose, October 27 and November 2, 1927, RF/IEB, 588, on von Euler and his proposed institute. Planck used his right to nominate during the interwar years to support individuals who the German scientific establishment wanted to help. His choices in chemistry were no doubt made in consultation with colleagues such as Nernst, Willstätter, and Haber. Peter Klason and Finnish chemist Ossian Aschan nominated von Euler regularly.

**199** foreign colleagues to help sell "biochemistry": An example of the way he used foreign colleagues to help biochemistry was the well-respected Finnish chemist Aschan, who was nominating von Euler for a prize, contributing an article on biochemistry's scientific and practical significance in *Svensk kemisk tidskrift* (1927), pp. 131–132.

**200** "an amateur in biochemistry": Letters, Svedberg to Einar Hammarsten, April 3 and 18, 1929, copy, TS. The physics and chemistry committees formally cooperated when a candidate was nominated for both prizes, as they both belonged to the Academy of Sciences and therefore were not subject to the statutory requirement of secrecy. At the time, nobody knew for sure whether committee members from differing prize-awarding institutions (such as the Academy and Caroline Institute) could share information. Hammarsten related some of the problems he was having convincing the medical committee to reward biochemist Otto Warburg. He felt that the predominance of clinicians on the committee favored prizes for work that they could understand without having to strain themselves on sophisticated theory (letter, Hammarsten to Svedberg, April 16, 1929, TS).

**200** To help Svedberg, von Euler sent reprints: Letters, Hans von Euler to Svedberg, April 8 and 17, 1929; Svedberg to von Euler, April 9, 1929, copy, TS. It was von Euler who provided information and arguments to convince Svedberg and others to consider Harden's achievements as being worthy of a prize [*Minnen*, p. 164].

**200** Svedberg admitted that Neuberg's contributions merited a prize: Svedberg, draft of special report on von Euler, Harden, and Neuberg, April 25, 1929, copy, TS.

**201** encouraged the Cambridge biochemist Sir Frederick Gowland Hopkins: Letters, F. G. Hopkins to Svedberg, October 18, 1926; Svedberg to Hopkins, October 22, 1926, copy, TS. Possibly Svedberg had hoped that Hopkins would nominate von Euler after the Englishman had praised him at an international physiological congress in Stockholm earlier that year.

**201** "one ought to think the matter over twice...": Letter, Ramberg to Palmær, June 1, 1929, WP ["*att man bör betänka sig två ganger innan man låter rent biokemiska arbeten rycka in i de kemiska Nobeprisens sfär*"].

**201** "sighing and whining [*suckan och kvidan*]": Related by Westgren to Svedberg, September 24, 1929, TS.

**201** it should be divided between von Euler and Neuberg: *Protokoll*, IV klassen, October 30, 1929.

**202** "Thanks for the prize and the roses, Your friend, Euler": Telegram, von Euler to Svedberg, December 11, 1929. TS

**202** an institute for organic chemistry under his leadership: Letter, W. E. Tisdale to von Euler, October 24, 1928. Although the Rockefeller Foundation subsequently provided funds for new projects and ultimately an organic chemistry institute, correspondence reveals the frustrations of dealing with von Euler and his insistent demands. From the late 1920s, see IEB, Folder 588, RF; from the 1930s, RF, Record 1.1, Series 800, Box 6, Folder 63ff.

**202** he and his laboratory produced results, massive results, some of high quality and some less so: A fellowship report, T. R. Krogness, August 16, 1937, RF, Record 1.1, Series 800, Box 6, Folder 65, contains much criticism, including comments that the American fellow heard in Danish and German laboratories about von Euler. In the 1930s, Warren Weaver and other Rockefeller Foundation officials heard about considerable dissatisfaction with von Euler from other Swedish and German biochemists; some referred to him as a dictator. Of course, von Euler was also able to achieve and lead others toward significant results. Fritz Schlenk, "The dawn of nicotinamide coenzyme research," *Trends in Biochemical Sciences*, **9** (1984), pp. 383–386, shows the positive aspects of von Euler's leadership, but in an unpublished memoir Schlenk reveals the less flattering aspects of life in von Euler's Stockholm laboratory. (I am indebted to Professor Anders Lundgren in Uppsala for sharing this manuscript with me.)

**202** All was set for von Euler's move to Berlin: Some sources imply that he would be head of all German research planning; others suggest that he would be in charge of organizing the Reich's chemistry research. In *Minnen*, p. 131, he implies that he was to be part of a ruling research council; at the time, he referred to the proposed position as that of science dictator. Apparently, Johannes Stark advanced von Euler for the position, but upon seeing von Euler's all-too-good relations with one of Stark's rivals in the Nazi hierarchy, he might well have helped sabotage the nomination. Von Euler's plans for bringing back Jewish notables was described in an interview with von Euler by W. E. Tisdale to Warren Weaver, November 20, 1934, RF, R.G. 1.1, Series 717, Box 2, Folder 10. The fate of his efforts to be named to the post was related in an interview with Warren Weaver, RF, R.G. 12.1, Warren Weaver Diaries 1935, May 28, 1935; and in *Minnen*. He claimed that in his laboratory all that mattered was talent. As related in his unpublished autobiography, von Euler was unable to fathom the evil of the Nazi regime both during and after the war, and he seems to have been able to repress unpleasant facts.

**202** Hitler ordered a German boycott: Göran Liljestrand, "The prizes in physiology," in Nobel Foundation, ed., *Alfred Nobel: The Man and His Prizes*, 2nd ed. (New York: Elsevier, 1962), pp. 334–337, relates details of Hitler's ban and the impact on those who were subsequently awarded prizes. Von Euler turned to Hermann Göring, with whom he was acquainted, in an effort to try to prevent the ban from becoming law. Von Euler saw the Nobel Prize as a means of supporting and celebrating German science [*Minnen*, p. 140].

**203** "detested by most German scientists": As reported by Otto Warburg, who "cannot say enough hard things" about von Euler. W. E. Tisdale Diary, September 30, 1937. RF, R.G. 1.1, Series 717, Box 2, Folder 10. Warburg himself was a difficult person who enraged many colleagues; he also competed with von Euler on a number of biochemical problems. His comments were certainly colored to some extent.

**203** Von Euler openly disapproved of Svedberg's overreliance on physical methods: Related in log on trip of W. E. Tisdale, April 16, 1934, RF, R.G. 1.1, Series 800D, Box 6, Folder 63. In *Minnen*, p. 73, von Euler regrets that he has not been able to keep up his earlier intimate contact with Svedberg.

**203** he did not hesitate to dissent from the committee majority to vote for his own candidates: For example, in 1933 von Euler voted for Swiss colleague Paul Karrer, while the rest of the committee voted to reserve the prize.

**204** "a minor prophet": The student—and later professor—Holger Erdtman had difficulties with both von Euler and organic chemist Bror Holmberg when he chose to work with Robinson and then return as a disciple. Details are related in a memoir sent to Robinson [n.d.], in Robert Robinson papers, Royal Society, London. Erdtman also relates details of the anti-Robinson atmosphere in Sweden in a letter to Anders Lundgren, July 11, 1985. (I thank Professor Lundgren for sharing this letter with me.) Both von Euler and Bror Holmberg, who joined the committee in 1934, disapproved of the development of physical organic chemistry in Britain and the United States.

**204** at the door of the Palace of Life itself: In Harden's Nobel lecture, quoted in T. W. Goodwin, "Arthur Harden," in Laylin James, ed., *Nobel Laureates in Chemistry, 1901–1992* (Washington, DC: American Chemical Society and Chemical Heritage Foundation, 1993), p. 84.

**204** The term biochemistry meant different things: A sizable scholarly literature exists on the subject; for the Swedish case, see Anders Lundgren, "Från fysikalisk kemi till biokemi: Bakgrunden till Arne Tiselius' professur i biokemi," *Forskningsprogrammet Stella: Modern vetenskapshistoria, 1850–2000*, Arbetsrapport Nr. 4 (Uppsala: Avdeling för Vetenskapshistoria, 1995).

**205** Rockefeller Foundation's comprehensive evaluation: An evaluation of American and European chemistry in 1927 placed Copenhagen as the third most significant European center for physical chemistry. Brønsted and Bjerrum were ranked as leaders in their field. IEB, RF, R.G. 1, Box 10, Folder 146.

**205** It justified its earlier decision by claiming that Haber alone had been nominated in 1919: KU, NKK, 1926.

**206** Benedicks' campaigns in the late 1920s: Correspondence in CB.

**206** he drew a sharp line between chemistry and the biological sciences: Arne Fredga, "Ludwig Ramberg. 21.2.1874–12.25.1940. Minnesteckning," *Kungliga. Svenska Vetenskapsakademiens Årsbok för År 1945* (Stockholm: Alonqvist & Wicksells, 1945), pp. 315–340.

**207** "With time I hope . . . to be able to teach the lads: Letters, Ramberg to Holmberg, March 10, 1922, BH-LUB; on Uppsala chemistry, Ramberg to Holmberg, December 30, 1921, BH [*"Fy f-n vad Uppsala-kemisternas skrifter äro trista!"*]; on advocating theory in chemical training, Ramberg to Bror Holmberg, February 10, 1919, BH.

**207** "at present scarcely can be said to maintain [Sweden's] proud traditions": Letter, Ramberg to Hermann Nilsson-Ehle, April 1, 1926, copy, LR. On need for physical theory, Ludwig Ramberg, "Moderna strömningar inom valensläran," *Svensk kemisk tidskrift* (1926), p. 126.

**207** He saw no reason to give prizes to chemical works: KU, NKK, 1931. Comment in Ramberg's unpublished manuscripts, Q10:6:1, LR. In this case he was arguing against von Euler's push for Werner Kuhn.

**207** Ramberg found that Arrhenius's loyal followers kept vigil even after his death in 1927: Letter, Ramberg to Bjerrum, September 15, 1932, on opposition from Widman to Brønsted's acid-base theory, copy, LR. See letter, Ramberg to Erik Larsson, September 15, 1935, copy, LR, on need for a new textbook to introduce and disseminate the newer theories of electrolytic solutions into Swedish chemistry.

**207** Svedberg—with von Euler's support—wanted to give the prize to Irving Langmuir: Letters, Svedberg to Westgren, May 10, 1932, copy; Westgren to Svedberg, May 14, 1932, TS.

**208** "epoch-making contribution to dissociation theory . . .": *Kungliga. Svenska Vetenskapsakademien 200 årsjubileum År 1939* (Uppsala, 1940), pp. 26–27.

**208** Shall the Nobel Prize in chemistry be given to those: Untitled ms. [1936], papers related to Nobel Prize, LR.

**209** Using bogus reasoning: KU, NKK, 1932. Goldschmidt's important transformation of crystal chemistry was similarly downplayed by the committee. KU, NKK, 1934.

## TWELVE  Dazzling Dialects

**213** He urged the Academy to stop the practice of withholding prizes: "Vetenskapsakademien och nobelprisen," *Göteborgs Handels- och Sjöfarts- Tidning* [hereafter GHT], November 10, 1936.

**214** discussed the fate of the building housing the Nobel Institute for Physical Chemistry: Arrhenius's institute was understood as personal. Although he had hoped in 1904 to head the Royal Swedish Academy of Sciences' Nobel Institute for Physics and Chemistry, lack of both money and support for him led to a compromise of a "department" based on physical chemistry and financed from both committees. The Nobel committee established subcommissions after Arrhenius's death to explore alternative options for the future of the building [Protocol of the committees, 1927–1933].

**214** A new, costly institute could siphon from this limited source of grants: *Protokoll*, NKK, February 24, 1930, Bil. A, memorandum, Ludwig Ramberg to NKK. Ramberg considers the problem to be not a need for new laboratories in chemistry but a need for funds to conduct experiments and purchase costly equipment. If a Nobel Institute were to be created for chemistry, it would be best to have it as a depository for expensive instruments that could be made available as needed to researchers.

**214** State-sponsored research councils: Memorandum, Ida Marcovich to Warren Weaver, June 13, 1932, RF, Record 1.1, Series 800D, Box 6, Folder 63, relates the situation in Sweden for funding research. The difficulties in finding support for research are clear from von Euler's and Svedberg's frequent proposals to the Rockefeller Foundation. Newly appointed professor of experimental physics Erik Hulthén described the wretched economic situation for setting up a laboratory in a letter to Niels Bohr, March 19, 1930, BSC.

**214** until a better solution could be found: Correspondence among several committee members shows efforts to find a satisfactory temporary solution that could make use of the building and the limited resources. Siegbahn's efforts in the late 1920s to improve his situation in Uppsala are reviewed in a letter from Siegbahn to Akademiräntmästaren, April 20, 1933, copy, in Ludwig Ramberg's Administrative Papers, LR. Oseen and others had hoped chemist Sven Odén might be willing to head the institute. Oseen might well have been waiting for better economic times to build a new complex for both experimental and theoretical physics—for Siegbahn and himself (letters, Sven Odén to Oseen, October 2, 1929, January 3 and 8 and February 3, 1930, CWO-KVA; and Sven Odén, "Förtrolig promemoria till Professor Svedberg angående Nobelinstitutet och dess omorganisation," December 23, 1929, copy, CWO-KVA). A letter from The Svedberg to Odén, January 2, 1930, copy, CWO-KVA, implies that Oseen was considering a new proposal for the institute. See letters from Bror Holmberg to Oseen, May 29, 1929, on problems facing a chemistry Nobel Institute; Otto Pettersson to Carl Benedicks, October 7 and 10, 1930, CB, on plans for the institute.

**215** Even this modest institute, based in an existing building: *Protokoll*, III and IV klasserna, January 31, February 25, and May 3, 1933. A budget proposal of Oseen, Siegbahn, and Pleijel circulated on January 31. Although most of the funding for the actual building would come from other funds connected to the Nobel Foundation, the special fund, generated from withholding prizes, could be used for aiding the Nobel Institute. Arrhenius received grants from the special fund in the 1920s for renovation and improvement of the institute. A report from 1924 drawn up by members of both prize committees underscores that the special fund could be used to support the Nobel Institute [memo by Allvar Gullstrand and Wilhelm Palmær, *P. M. Angående anvisande av medel från de fysiska och kemiska prisgruppernas s.k. särskilda fonder*, April 1924, copy, WP, Vol. 94], which is repeated in the proposal for converting what was Arrhenius's Institute for Physical Chemistry to one for theoretical physics. Actual experiments and research costs for a Nobel Institute can be supported by a committee's special fund. [*P. M. Angående nobelinstitutet och dess eventuella framtida använding*, January 31, 1930]. A letter from Oseen to Palmær, July 5, 1931, WP, makes the claim for the importance of using the existing building as the basis for an institute for theoretical physics: Eventually Siegbahn will get a Nobel Institute for Experimental Physics, so it would be best to wait to see where this institute will be erected before building a new structure for theoretical physics.

**215** Siegbahn's prize did not have the same effect: Svedberg got a major new institute; Siegbahn got funds for a full-time instrument maker for his laboratory. "Våra Nobelpristagare få bättre resurser för forskningarna," *NDA*, January 11, 1927.

**215** In 1929, Siegbahn's institute had sixteen advanced graduate students: Siegbahn and the laboratory's overtaxed state as well as the fifty-year history of the state's refusal to increase the teaching staff in the university's Physics Institute are included in Uppsala University's Division of Mathematics and Natural Science's petition to the Academic Collegium [*Större akademiska konsistoriet*]; all included in the university's petition to the government to keep Siegbahn and his new laboratory connected with the university as part of the general government proposal on behalf of Siegbahn: "Nr. 64. Utlåtande i anledning av Kungl Maj:ts under riksstatens åttonde huvudtitel gjorde framtällning angående bidrag till vetenskapsakademien," *Bihang till riksdagens protokoll 1936, 6 saml. Nr. 64–65*.

**216** and these matching funds must be pledged by the end of 1930: Letter, Knut Wallenberg to Manne Siegbahn, June 4, 1930, MS. I thank Dr. Thomas Kaiserfeld for providing copies of the Wallenberg–Siegbahn correspondence.

**216** In addition, the Academy would be asked to contribute 20,000 crowns: Letters, Siegbahn to Wallenberg, December 20, 1930, copy; Wallenberg to Siegbahn, February 17, 1930, MS.

**216** assured Siegbahn of Nobel funds for a new building and inventory: The committee's so-called organization fund would be used. This money was put aside in 1900 for a Nobel Institute; part of it went to Arrhenius's institute, and the rest continued to grow from accrued interest. In 1932, the Nobel Foundation channeled an additional 200,000 crowns to each committee's organization fund.

**216** Success seemed almost in reach: Letters, Siegbahn to Wallenberg, February 16, 1934, copy; Gösta Forssell to Siegbahn, February 20, 1934, MS, on the efforts to petition for a professorship.

**216** At best, Wallenberg promised to examine the situation next year: Letter, Wallenberg to Siegbahn, February 17, 1934, MS.

**216** The Nobel Foundation promised 800,000 crowns: Letter, Siegbahn to Wallenberg, February 18, 1935, copy, MS.

**216** He and his supporters still hoped to get the state to pay for a professorship: The physics committee's savings fund, which was to provide money for the institute leader's salary and annual operational budget, was not large enough. Oseen, Pleijel, and Forssell sought to have the state create a personal professorship for Siegbahn and cover for a limited time some of the operating costs; this would allow the Nobel committee's savings fund to continue growing so that when the personal professorship and state subsidy ended, the fund could generate sufficient dividends to cover these expenses [letter,

Oseen, Forssell, and Pleijel to KVA, March 13, 1935, *Protokoll*, KVA, March 13, 1936; *Protokoll*, *KVA:s forvaltnings utskott*, March 13 and May 31, 1935.

**216** Wallenberg guaranteed his foundation's support (55,000 crowns a year): Letter, Wallenberg to Siegbahn, March 31, 1935, MS.

**217** to commission an architect (at a cost of 5,000 crowns): *Protokoll*, NKF, May 22, 1935; *Protokoll*, KVA/N, June 5, 1935. The planned institution was referred to as a department of experimental physics in the Academy's Nobel Institute for Physics.

**217** As it turned out: Government and Parliament's proceedings, "Nr. 64. Utlåtande i anledning av Kungl Maj:ts under riksstatens åttonde huvudtitel gjorde framtällning angående bidrag till vetenskapsakademien," *Bihang till riksdagens protokoll 1936, 6 saml. Nr. 64–65*. The government approved his professorship in the summer of 1936, and from July 1937 Siegbahn assumed the position of the professor of experimental physics at the Royal Academy of Sciences [*Protokoll*, KVA/N, April 22, 1936, on Academy and Nobel Foundation actions]. The name of the institute went through various changes after the war: sometimes a Nobel Institute, sometimes an Academy Institute.

**217** the physics committee's various funds were "drained [*bottenskrapat*]": Letter, Otto Pettersson to Benedicks, February 23, 1936, CB.

**217** "the Firm Siegbahn, Oseen, & Co. from making an economic coup": Letter, Pettersson to Benedicks, April 6, 1936, CB.

**217** When Pettersson looked into where the grants were going: From 1931 to 1935, the chemistry committee's special fund generated 118,150 crowns for grants; 86 percent (101,700 crowns) went to members of the Chemistry Section. This amount included annual grants to Svedberg and von Euler ranging from 5,000 to 10,000 crowns each, coming to a total of 75,400 crowns for the two. Pettersson urged that all members of the Royal Swedish Academy of Sciences should have a right to apply for grants as a means of loosening control. [Letter, Otto Pettersson to Hjalmar Hammarskjöld, March 20, 1936, and copy of petition, "Anförande och förslag . . . för behandling av Nobelärenden . . . den 26 februari 1936," HH. See also *Protokoll*, KVA/N, February 26, 1936.]

**217** ". . . downright scandalous traffic . . . the economic gang that our Nobel Prize winners have established": Letter, Pettersson to Benedicks, February 28, 1936, CB. ["*I virkeligheten tror jag mig ha draget fram en rent av skandalös trafik vilken om den blev allmänt bekant skulle göra den ekonomiska liga som våra Nobelpristagare bildt omöjlig för framtiden.*"]

**217** Oseen shot back a categorical denial of any impropriety: "Blott vart femte år behöver nobelpris ges," *GHT*, November 11, 1936; see also *NDA*, November 11, 1936. Those defending the committees and Academy included Gothenburg physicist Arvid Hedvall, Henning Pleijel, and Wilhelm Palmær.

**217** How, Pettersson wondered: Pettersson's response "Nobelprisens indragning," *GHT*, November 12, 1936.

**218** broke ranks and announced that Pettersson was right: "Kritiken torde inverka på nobelprisutdeningen," *GHT*, November 13, 1936. In supporting Pettersson, Forssman claimed that the Academy's decision in 1935 came after a long debate; he believed it was wrong to reserve (and thereby withhold) the 1934 prize rather than the one for 1935. Enström was quoted as agreeing in principle with Pettersson's position.

**218** It was not until World War I, when prizes began to be reserved, that an ambiguity appeared: Fearing misuse, several barriers were introduced between the committees and the prize money. The money from a withheld prize would normally be returned to the Nobel Foundation's main fund—that is, the fund that provides for the annual prizes and accompanying expenses. But if a significant majority, four-fifths, of the responsible prize-awarding institution approved, a committee could instead vote to place the money, or a large part of it, in its special fund. Fear of misuse of the special funds—withholding prizes to increase their size—was present from the start. Some members of the committee drafting bylaws felt that the idea of the special fund went against Alfred Nobel's intentions; others understood both the fund and the Nobel Institute as necessary compromises to gain the support of the institutions charged with awarding prizes. Extracts of the proposals and debates in 1899 are provided in the protocol of the Nobel Foundation, June 1, 1918, when members deliberated about the myriad motions and countermotions surrounding the question of reserving prizes and the use of withheld prize money; a copy of the extract of this protocol is found in HH. See also letter, Carl Lindhagen to Hjalmar Hammarsköld, April 25, 1899, concerning revisions to the final bylaws against misuse of the special funds, HH, L175/035.

**218** Disagreement plagued the Academy whenever it tried to resolve this problem: An eleven-page summary of deliberations from 1915 through 1933 is provided in a joint petition from the physics and chemistry committees to the Academy, March 6, 1934 [*Protokoll, gemensamma samanträde* NKF and NKK, March 6, 1934, *Protokollsbilagor till KVA:s protokoll i Nobelärenden 1932–1934*]. *Protokoll*, KVA/N, November 13, 1917, refers to a debate on the physics committee's desire to vote first on 1917 prize and then the reserved 1916 prize. As no prizewinner was named that year, the debate was left unresolved. In 1918, when only one winner in physics was proposed (Barkla), the Academy voted thirty-eight to twenty-four to give him the reserved 1917 physics prize, and reserved the 1918 prize.

**218** He ruled that prize-awarding institutions were compelled first to vote on the reserved prize: He also rebuked the committees for having diverted prize money to their respective funds rather than returning it on occasion to the main fund. This practice suggested to the attorney general an unhealthy desire to use the prize institution to aid Swedish science. The bylaws allowed for this practice, but the main thrust of Nobel's testament and the bylaws aimed at awarding prizes. At least by returning the money from a withheld prize to the main fund, the prizes themselves could grow as the interest on this fund would increase. The government therefore proposed that every third time a prize was withheld the prize money would go to the main fund [*P. M.*, Emil Sjöberg, October 1, 1919, in *Protokoll, Styrelsen för Nobelstiftelsen*, December 4, 1919, copy of extract in HH].

**218** This ruling held for a decade; then the committees petitioned: Mentioned in petitions and correspondence among Nobel committees, the Academy of Sciences, and the Nobel Foundation during 1934, in *Protokollsbilagor till KVA:s protokoll i Nobelärenden 1932–1934*. As recently as 1930, Ragnar Sohlman, who more than any other person embodied the Nobel legacy's memory, declared his support for the practice of first deliberating the previously reserved prize and then voting on the current year's prize [letter, Sohlman to Ragnvald Moe (Director, Norwegian Nobel Institute), November 18, 1930, copy, HH]. Actually this statement is not accurate; Sohlman voted with the rest of the board of directors of the Nobel Foundation on May 19, 1916, to back the Academy of Sciences' and the Caroline Institute's desire to award first the current year's prize and then vote on the reserved prize.

**218** the prospect of finding two worthy candidates the following year looked doubtful: The petition claimed that this situation could result in lowering the standards for the prize as the committees might act precipitously to avoid a long series of reserved prizes. See petition to KVA, March 6, 1934. A draft of the petition with annotations, February 17, 1934, is found in Vol. 121 of WP.

**218** The real objective for wanting change was transparent: Some of the correspondence points to the recognition that the funds from withheld prizes would benefit

Swedish research, which was a fair way of recognizing the great time and effort committee members gave to annual evaluations. Other issues also arose in the petitions, such as changing the ruling from 1919 that every third withheld prize must go to the main fund—to help maintain the monetary value of the prizes—rather than be repeatedly directed toward a committee's special fund. The revised practice called for placing into the main fund at least one-third of the money from every withheld prize rather than the entire amount from every third withheld prize. The reasoning was that, especially during a period when the value of the Swedish crown plummeted, it would be valuable to keep a stable trickle of replenishment into the fund that generates money for the prizes. In this way, those who were critical of the idea of the special funds and the use of withheld prize money to aid local research at the expense of Nobel's prizes could be wooed to support the proposed change with respect to withholding prizes [letter, Gösta Forssell to Hjalmar Hammarskjöld, April 10, 1934, HH].

**219** Professor Oseen's dazzling dialectics can offer another justification for the change than just this: Letter, Hans Pettersson to Benedicks, April 7, 1937, CB. ["*Åtskilliga akademiledamöter, som jag talade med, syntes överraskade av paragrafändringen första gången den tillämpades. En av dem, en man vars opartiskhet och rättrådighet stå över varje diskussion, meddelade mig, att enlight hans mening paragrafen i fråga måtte ha ändrats just för att underlätta framtida indragningar i ännu större omfattning än hittils. Och onekligen har ändringen borttagit en allvarlig hämsko på indragningsåtgärdens obefogade bruk. Jag har svårt tro att ens Prof Oseens bländande dialektik kan prestera en annan motivering för ändringen än just denna.*"]

**219** the very act of writing them was the act of erasing any and all extra-scientific criteria: R. M. Friedman, "Text, context, and quicksand: Method and understanding in studying the Nobel science prizes," *Historical Studies in the Physical and Biological Sciences*, **20** (1989), pp. 63–77.

**219** "... which is what the IVth [Chemistry] Section did not do this time": Letter, Otto Pettersson to Benedicks, November 18, 1934, CB. [*"det stora grälet i K.V.A. ... såsom en affär beslutad på förhand i samband med det nyss fattade beslutet att ett reserverat prisbelopp får disponeras till 2/3 för andra ändamål (vilka få vi väl snart se). Nobelpriset 33 skulle delas, icke utdelas ... men; man skall pröva de foreslagna på ett rättvist och anständigt sätt; vilket IVde klassen icke gjort denna gång."*]

**220** "cyclical permutation": The expression was first used by Benedicks in a flurry of letters between him and Pettersson during February 1936, CB, on the question of the special funds.

**221** "who can write more disparagingly than anyone": Letter, J. W. Sandström to Bjerknes, January 4, 1939, VB.

**222** Stern urged fellow refugee James Franck never to forget to mention his own Nobel Prize while in the United States: Letter, Otto Stern to James Franck, January 23, 1934, JF.

**222** languished at Brooklyn Polytechnic Institute: Much remains to be researched on Neuberg's career. In actuality, a vital chemistry department emerged at this time at Brooklyn Polytechnic Institute. Insight on Neuberg's career is gleaned from correspondence with various chemists and should not be considered here as definitive.

**223** Max von Laue asked, to propose giving a prize to ... relief organizations?: Von Laue's letter of August 18, 1933, was included as an appendix to the physics committee meeting, September 9, 1933 [*Protokoll*, NKF, September 9, 1933, Bil. B]. Von Laue acknowledged that many worthy candidates could be considered for the prize, but stressed the extraordinary political circumstances requiring assistance to worthy refugees. He asked that his letter not be answered, obviously fearing reprisals at home. Formally, the committee's response was correct, in that the prize should be given to documented accomplishments, but any number of options were available, if they chose, to find a means to generate money for relief organizations. The bylaw related to the committee's special fund requires that the money not be used for the same purpose as the prize but to further Nobel's general intentions, generally meaning to aid research. Grants to relief organizations could, of course, benefit research and the ideals embodied in Nobel's testament.

**224** "to adulterate the spiritual coinage of the world": Letter, Felix Frankfurter to Raymond B. Fosdick, November 24, 1936, RF, R.G. 1.1, Series 717, Box 2, Folder 10.

**224** who challenge everything that we hold dear: Letter, Frankfurter to Fosdick, December 9, 1936. Frankfurter was soon appointed to the U.S. Supreme Court.

**224** The amounts generated by withholding a prize: The point was repeated in an article in *GHT*, March 1, 1939, after a report of large grants from the special funds going to committee members. The amounts of these grants were compared with rivulets of modest monies irrigating the rice field of Swedish research but which together marked a diversion from Alfred Nobel's intentions [*Protokoll*, NKK, June 1, 1939, Bil. O]. Indeed, prior to the late 1930s, the relatively small amounts, seen from today's perspective, were nevertheless essential components of the expansive plans of von Euler, Svedberg, and Siegbahn, as these allowed them to woo funding from other sources (namely, the Wallenberg and Rockefeller foundations) for equipment. Subsequently, the costs of instrumentation for cutting-edge research and the start of massive state support for research, coupled with accusations, ended this two-decade-long episode of committee members' hunt for funds to fill the holes in their research budgets.

## THIRTEEN   Completely Lacking an Unambiguous, Objective Standard

**226** Cavendish Laboratory in Cambridge made many of the experimental breakthroughs: John Hendry, ed., *Cambridge Physics in the Thirties* (Bristol: Adam Hilger, 1984).

**226** Lawrence devised the machine that made him and Berkeley physics famous: Unless otherwise stated, references to Lawrence's plans and accomplishments are from the masterful and massive first volume of the history of this saga in the history of physics, J. L. Heilbron and Robert W. Seidel, *Lawrence and His Laboratory: A History of the Lawrence Berkeley Laboratory*, Vol. 1 (Berkeley: University of California Press, 1984); also of great value is the shorter and well-illustrated *Lawrence and His Laboratory: Nuclear Science at Berkeley, 1931–1961*, by J. L. Heilbron, Robert W. Seidel, and Bruce R. Wheaton (Berkeley: Office for History of Science and Technology, University of California, 1981).

**228** Prior to 1939, Lawrence had received a mere sprinkling of nominations: In 1938 only, KVA/N, 1938; on the evaluation of Lawrence and others who had worked on particle accelerators, KU, NKF, 1938.

**228** theirs was not the definitive solution: Siegbahn's negative evaluations appear in KU, NKF, 1937 and 1938.

**229** Richardson believed it only fair to let Lawrence wait: Letter, O. W. Richardson to Niels Bohr, January 14, 1939, BSC.

**229** Having spent so much time and energy raising funds to bring his Copenhagen institute into the nuclear age: Finn Aaserud, "Niels Bohr as fund raiser," *Physics Today*, October 1985, pp. 38–46; Finn Aaserud, *Redirecting Science: Niels Bohr, Philanthropy, and the Rise of Nuclear Physics* (Cambridge: Cambridge University Press, 1990).

**229** "originated a new branch of science": Letter, W. L. Bragg to NKF, January 13, 1939, KVA/N, 1939. W. H. Bragg also nominated Appleton.

**229** Lawrence suddenly stood out, with a surprising eleven nominations: Nominations were from E. F. W. Alexanderson, Edoardo Amaldi, Niels Bohr, Karl Compton, H. Cushing, C. J. Davisson, Enrico Fermi, Irving Langmuir, S. Loria, C. V. Raman, and Franco Rasetti.

**229** He failed, however, to make a firm case for Lawrence: KU, NKF, 1939.

**230** "has proven to be by far the most effective tool for the experimental study . . .": KVA/N-1939, Karl Compton to NKF.

**230** Again, the committee rejected them, saying that because of the intense ongoing developments: KU, NKF, 1939.

**230** meteorology was not physics: In 1919, Oseen had concluded his comprehensive lectures on theoretical physics by noting that in the future the frontier for this science might no longer be in the subatomic domain but in the even more complex processes of atmospheric changes. His opposition to Bjerknes had as much if not more to do with local science politics (opposition to a Swedish disciple of Bjerknes, among other issues) as principled separation of the atmosphere from the realm of physics. See R. M. Friedman, "Nobel physics prize in perspective," *Nature*, August 27, 1981, pp. 793–798; and correspondence between V. W. Ekman and Carl Benedicks in the mid- and late 1930s, CB, on Oseen's opposition to Bjerknes's candidacy.

**231** a minority of the Physics Section refused Lawrence: *Protokoll*, III klassen, October, 1939. See letters to and from Benedicks with Ekman and Otto Pettersson in CB.

**231** Lawrence, fifty, Bjerknes, nineteen, and several for not awarding the prize: Letter, Benedicks to Otto Pettersson, November 10, 1939, copy, BC.

**231** I think it would be splendid in every way for the Foundation to make a grant to Professor Siegbahn: Letters, E. O. Lawrence to F. L. Hansen, January 15, 1943; Hansen to Lawrence, January 12, 1943; copy, RF, Record 1.1, Series 800, Box 6, Folder 54. Siegbahn sent a radiogram to the Rockefeller Foundation, November 29, 1944, saying that he needed an even larger cyclotron and requested yet more money.

**233** Lise Meitner: Ruth Lewin Sime, *Lise Meitner: A Life in Physics* (Berkeley: University of California Press, 1996). Sime's masterful biography includes the results of her numerous articles on various aspects of Meitner's career; relevant ones are cited below.

**234** nobody could quite make sense of the complex and confusing results of bombarding heavy elements . . . with neutrons: Spencer R. Weart and Fritz Krafft each discuss how preexisting theoretical and conceptual paradigms as well as experimental practices shaped the interpretations of nuclear research that precluded considering fission a possibility. See, respectively, "The discovery of fission and a nuclear physics paradigm," and "Internal and external conditions for the discovery of fission by the Berlin team," in William R. Shea, ed., *Otto Hahn and the Rise of Nuclear Physics* (Dordrecht, Netherlands: Reidel, 1983), pp. 91–134 and 135–166.

**234** The Fermi team was propelled into the limelight: Gerald Holton, "Fermi's group and the recapture of Italy's place in physics," in *The Scientific Imagination: Case Studies* (Cambridge: Cambridge University Press, 1978), pp. 155–198.

**234** In Berlin, Meitner led the team: Ruth Lewin Sime, "Lise Meitner and the discovery of fission," *Journal of Chemical Education*, **66** (1989), pp. 373–376.

**234** Meitner had to flee Germany: Ruth Lewin Sime, "Lise Meitner's escape from Germany," *American Journal of Physics*, **58** (1990), pp. 262–267.

**235** Careful historical research: In addition to Sime's articles and biography, already cited, the pioneering work that began to break through legend and myth was by Fritz Krafft, "Lise Meitner: Her life and times—On the centenary of the great scientist's birth," *Angewandte Chemie: International Edition in English*, **17** (1978), pp. 826–842, and *Im Schatten der Sensation: Leben und Wirken von Fritz Strassmann* (Weinheim: Verlag Chemie, 1981).

**235** From Hahn's perspective as a chemist: Sime, "Meitner and the discovery of fission," p. 373, quoting Otto Hahn in *Chemische Berichte*, 1937.

**235** Strassmann noted many years later that it was fortunate her opinion: Krafft, *Im Schatten der Sensation*, pp. 84, 208, 211; Sime, "Meitner and the discovery of fission," p. 374.

**236** "... There could still perhaps be a series of unusual coincidences that has given us deceptive results": Otto Hahn and Fritz Strassmann, "Über den Nachweis und das Verhalten der bei der Bestrahlung des Urans mittels Neutronen entstehenden Erdalkalimetalle," *Naturwissenschaften*, **27** (1939), pp. 11–15; quotation in Sime, "Meitner and the discovery of fission," p. 374. Krafft also notes Hahn's reluctance to interpret the data.

**236** "... one cannot unconditionally say, 'It is impossible' ": Letter, Meitner to Hahn, December 21, 1938, quoted in Sime, "Meitner and the discovery of fission," p. 374.

**237** "Oh, what fools we have been! We ought to have seen that before": Otto Robert Frisch, "How it all began," *Physics Today*, November, 1967, p. 47. Various forms of Bohr's reaction have been noted, but all indicate surprise over an unexpected insight; see Roger H. Stuewer, "The origin of the liquid-drop model and the interpretation of nuclear fission," *Perspectives on Science*, **2** (1994), pp. 76–129, which analyzes the relation of Meitner and Frisch's work to Bohr's.

**237** Before the paper was in print, word spread like a chain reaction: Through a number of mishaps, the publication was delayed over a month; Bohr was expecting that it would appear momentarily. Once word leaked out, Bohr tried to keep Frisch and Meitner's contribution in focus, but in the avalanche of research suddenly precipitated by Hahn and Strassmann's article and Bohr's leak, this was not easily achieved. See, among many relevant letters, George B. Pegram to Niels Bohr, February 3, 1939; Bohr to Otto Stern, February 2, 1939, copy; Bohr to Ebbe Rasmussen, February 4 and 14, 1939, copies; and Rasmussen to Bohr, February 20, 1939, BSC-Suppl. See also Roger H. Stuewer, "Bringing the news of fission to America," *Physics Today*, October 1985, pp. 49–56; the history of nuclear research and the work leading to controlled fission and then an atomic bomb is presented in Richard Rhodes, *The Making of the Atomic Bomb* (New York: Simon & Schuster, 1986).

**237** Again, though, he did not provide any hint of his communications with Meitner: Ruth Lewin Sime, "Lise Meitner in Sweden, 1938–1960: Exile from physics," *American Journal of Physics*, **62** (1994), pp. 665–701.

**238** Meitner feared that her new Swedish colleagues would interpret: This point comes out clearly in the correspondence from Meitner to colleagues; in particular, letters from Meitner's Swedish longtime friend, the physicist Eva von Bahr-Bergius, indicate that Meitner was clearly obsessed with the need to prove herself in her new home. (Copies of Meitner's typed letters to Bahr-Bergius are preserved in LM, but most of Meitner's handwritten letters to her have not been located; Bahr-Bergius's responses provide insight into what Meitner had written.) Krafft quotes excerpts from Meitner's letters to Otto Hahn, and Sime refers to these as well as other correspondence in LM.

**238** Oseen was actually one of the few in Sweden who extended hospitality: Meitner describes lunches and other meetings with Oseen in letters to Bahr-Bergius.

Meitner's difficulties in learning Swedish and adjusting to a new culture added to the problem. A cycle of negative reinforcement began early, in which her frustrations with herself and with her new home led to a negative attitude, which in turn was compounded by the insensitivity of some Swedish scientists. Having come from Berlin, Meitner experienced much of Swedish culture and Swedish academics' behavior toward a refugee (at a time when refugees were not common in Sweden, in contrast to the present-day situation) as decidedly provincial.

**239** "She can do nothing with her hands": Letter, C. W. Oseen to Vilhelm Bjerknes, November 17, 1938, VB.

**239** She had to ask him for even the smallest of appropriations: This pattern began early and persisted. At first, Meitner was told to be patient, as Siegbahn's budget was severely strained, but the sense of being treated as a dependent and inexperienced inferior continued. Meitner's many letters to Bahr-Bergius, Hahn, and Max von Laue during the period from autumn of 1938 through 1945 contain reference to minor and major problems arising from Siegbahn's treatment. Many letters to and from von Laue appear in Jost Lemmerich, *Lise Meitner–Max von Laue. Briefwechsel 1938–1948, Berliner Beiträge zur Geschichte der Naturwissenschaften und der Technick* (Berlin: ERS, 1998); excerpts from Meitner's letters and postcards to Hahn in Krafft, "Lise Meitner: Her life and times," pp. 837–839.

**239** But this plan also faltered. The sudden demand for physicists: Letter, Oseen to Bjerknes, December 17, 1938, VB. Oseen hoped that Bjerknes could recommend a young Norwegian physicist who might be willing to be an assistant for a few years.

**239** ". . . Perhaps Professor Meitner must seek another home and another institute": Letter, Oseen to Bjerknes, November 17, 1938, VB.

**240** Svedberg nominated Hahn and Meitner for the chemistry prize: Svedberg proposed either Hahn alone or possibly a division with Meitner. He also submitted a nomination suggesting a shared prize between Wendell Stanley on the one hand and F. C. Bawden and N. W. Pirie on the other, for virus research. These nominations enabled Svedberg to study recent developments in both these fields and to begin considering how a prize for either fission or virus research might be awarded. These nominations did not necessarily reflect Svedberg's choice for a prize, but rather, as is often the case, a committee member submitted a proposal for individuals who had not been named by nominators but who the committee might want to begin assessing.

**241** he claimed that the role of elucidating the theory of fission belonged largely to Bohr: KU, NKK, 1939; Svedberg's evaluation was dated March 31, 1939. See KVA/N, 1939, for Svedberg's letter of nomination.

**241** In 1940, nominators provided a corrective: KVA/N, 1940.

**241** Compton considered the "discovery" of fission to consist of both Hahn's chemical and Meitner's physical results: Letter, A. H. Compton to NKF, December 12, 1939, KVA/N, 1940. The British physicist C. G. Darwin also considered fission to be the most important recent discovery and nominated "Professor Hahn and Miss Meitner" together, but he wondered, based on the list of recent prizes, whether this work was more suitable for a chemistry prize.

**241** "I do not need to emphasize the importance of this discovery . . .": Letter, James Franck to NKF, December 13, 1940, KVA/N, 1942. Franck's letter arrived too late to be eligible for 1941, but it was included as a nomination for 1942 prize.

**242** Svedberg again concluded that Hahn alone deserved to be rewarded for fission: KU, NKK, 1941. Those who proposed Hahn for a chemistry prize were M. Jaeger

(Groeningen, Netherlands), who nominated Hahn and Meitner (the second choice after Wendell Stanley), and committee secretary Arne Westgren, who nominated a prize for either Hahn or for Hahn and Meitner. (The latter nomination may well have been a formal proposal to allow the committee to continue evaluating important candidates who otherwise might not be nominated that particular year.) The committee did not register the nomination from physicist C. D. Anderson for dividing a chemistry prize between Hahn and Meitner that was sent to the physics committee.

**242** In 1942, Wilhelm Palmær openly expressed his uncertainty: Palmær nominated Hahn and Meitner both for their earlier work on radioactivity and for fission. He wanted the committee to reexamine the case for a split prize. Letters, Palmær to NKK, January 31, 1942, KVA/N, 1942, and Palmær to NKK, February 4, 1942, *Protokoll*, NKK, February 4, 1942.

**242** Westgren claimed that Hahn and Meitner's transuranic "mistakes" should be kept separate: KU, NKK, 1942, Bil. 4.

**243** In 1943, the situation remained the same with respect to the chemistry prize: KU, NKK, 1943. Westgren's nomination for Hahn alone, KVA/N, 1943.

**243** the physics committee again refused to evaluate: KU, NKF, 1943. See also Siegbahn's and Franck's letters, KVA/N, 1943.

**243** The committee added that if, when the Academy voted, the political situation still prohibited Hahn: *Protokoll*, NKK, September 11, 1944, Bil. N. See KU, NKK, 1944, on the evaluation. The prize was reserved. Meitner's diary traces Hahn's socializing and lecturing during the visit to Stockholm in October 1943; fragments are reproduced in Lemmerich, *Meitner–Max von Laue*, p. 315. Hahn was well known to Stockholm scientists; he had visited more than once, as reported in Hans von Euler, *Minnen* (unpublished autobiography).

**243** At its meeting of June 4: *Protokoll*, NKK, June 4, 1945.

**243** The committee agreed on September 10 to propose reserving the 1945 prize: *Protokoll*, NKK, September 10, 1945, Bil. G; NKK, September 20, 1945, Bil. A.

**244** But when the Academy deliberated, a member of its Medical Section, Göran Liljestrand, objected: Related by Oskar Klein to Niels Bohr, November 17, 1945, BSC-Suppl. Slightly over half of the votes went to Hahn and just under half for postponement. Liljestrand's action was allegedly motivated by the fear that the Chemistry Section's vote seemed to be prompted by an "opportunistic political basis" to appease the Americans. Klein felt that there was widespread sympathy to award the next physics prize to Meitner and Frisch or to Meitner alone.

**244** She made no secret of her feelings: Her annoyance with Siegbahn's treatment of her was most clearly expressed in letters to Hahn and Bahr-Bergius; she focused on his deficiencies as a nuclear physicist and the leader of the institute in these letters as well as letters to von Laue.

**244** Central among these were Lund University physicist Torsten Gustafson and Tage Erlander: The immediate postwar nuclear developments in Sweden are laid out in Stefan Lindström's doctoral dissertation, "Hela Nationens Tacksamhet: Svensk Forskningspolitik på Atomenergiområdet, 1945–1956," Statsvetenskapeliga Insitutionen, Stockholms Universitetet, 1991. (I thank adviser Professor Björn Wittrock for a copy.) Further insight can be found in Karl Grandin, "En neutral atompolitik? Tage Erlander, Torsten Gustafson och folkhemsfysiken, 1945–1953," *Arbetsrapport Nr. 18 Forskningsprogrammet Stella: Modern vetenskapshistoria, 1850-2000* (Uppsala Universitet, Avdeling för Vetenskapshistoria, 1997). In addition to Tage Erlander's diaries and correspondence between him and Gustafson, which are treated in these publications, Oskar Klein's correspondence with Niels and Margarethe Bohr (Bohr Personal Correspondence, NBA) from the mid-1940s provides further insight.

**245** for research related to atomic weapons: Lindström, "Hela Nationens Tacksamhet," p. 55; based on revelations by Legal Chief of the Department of Defense Olof Forssberg in *Svensk Kärnvapenforskning* (Stockholm: Försvardept., 1987).

**245** nuclear research will be "monopolized" by Siegbahn and his institute: Lindström, "Hela Nationens Tacksamhet," p. 61; letter, Gustafson to Erlander, n.d. [November], 1945. Niels Bohr also expressed dismay over Siegbahn's control of nuclear research as well as Siegbahn's willingness to declare to the media that he was interested in building an atomic bomb.

**245** Oskar Klein hoped to create a nuclear physics research unit: Klein was interested both in improving the resources available to theoretical physics in general and in assisting Meitner. Many plans were being discussed in 1945 and 1946; some called for a smaller unit at the Högskola and a more comprehensive nuclear physics unit at the Royal Technical College. Klein and Gustafson hoped that Meitner could find a single or multiple institutional basis for working and for helping to develop Swedish nuclear research. (Part or full time in Lund was a possibility, especially when plans for a Nordic institution were being considered for Lund; a primary affiliation with the more academically oriented Högskola but with direct access to the engineering college's facilities was considered most probable.) A letter from Klein to Bohr, October 22, 1945, BSC, also notes the paucity of innovative nuclear physicists in Sweden. Bohr's enthusiasm for bringing Meitner and Frisch to the Högskola is expressed in a letter from Bohr to Klein, November 7, 1945, copy, BSC.

**245** Here, then, was a challenge to Siegbahn's dominance: It was clear to Klein that the plans for creating a new center for nuclear physics where Meitner and also possibly Frisch could work might run into problems from Siegbahn, who was less competent in this field and who would feel threatened. See letters, Klein to Bohr, November 17, 1945, BSC; Meitner to Margarethe Bohr, November 25, 1945, Bohr Personal Correspondence, on Siegbahn's opposition to creating a nuclear physics unit at the Högskola.

**246** Klein was finally elected to the Academy's Physics Section early in 1945: KVA/N, December 11, 1944. Klein, however, was narrowly defeated for committee membership in 1944 after Oseen's death. Siegbahn replaced Oseen as official leader of the Nobel Institute, and with that he occupied a permanent place on the committee; the Physics Section voted three for Ivar Waller and three for Klein to fill Siegbahn's position on the committee; in the Academy, the vote was twenty-five for Waller and twenty-two for Klein.

**246** reluctance to provide the extra funding Siegbahn requested for his massive cyclotron: Lindström, "Hela Nationens Tacksamhet," p. 65.

**246** Klein understood that the task of achieving fairness would not be easy: Klein followed the developments and reported to Bohr, in part via Niel's wife, Margarethe. Both Bohr and Klein feared that the physics committee would overlook Meitner's and Frisch's roles in the discovery of nuclear fission. Bohr felt he had a responsibility to inform the committee that the record of publication did not give an accurate picture of the pivotal role of both Meitner and Meitner and Frisch. [Letters, Klein to Margarethe Bohr, June 24, 1945, Bohr Private Correspondence, NBA; Niels Bohr to Klein, September 11, 1945, OK; Klein to Bohr, September 16, 1945, BSC; Klein to Bohr, November 17, 1945, BSC; Bohr to Klein, December 28, 1945, OK; Bohr to Klein, January 21, 1946, OK.]

**246** Hulthén made an agreement with Siegbahn: Letter, Meitner to Margarethe Bohr, November 25, 1945, NBP.

**246** A number of physicists discussed possible scenarios: The Klein–Bohr correspondence in the autumn of 1945 and winter of 1946 contains references. See also Lindström "Hela Nationens Tacksamhet," and Grandin, "En neutral atompolitik?" for Gustafson and Erlander's discussions.

**246** his institute could build a Swedish atomic bomb within weeks: Siegbahn's proclamations were particularly disturbing for Gustafson and Erlander, who were trying to stake out a neutral path for Sweden while Americans were pressuring to have access to massive Swedish low-grade uranium ore and fretting over the possibility of a Soviet grab for the same ore. Sweden needed to avoid stirring further suspicion by appearing to be preparing for military nuclear capability (see Lindström, "Hela Nationens Tacksamhet," and Grandin, "En neutral atompolitik?"). Bohr, Klein, and Gustafson were aware of Meitner's antimilitaristic feelings; she had refused to move to Great Britain or the United States during the war to help develop an atomic bomb.

**247** Meitner and Frisch were again among the candidates: KVA/N, 1946.

**247** Bohr and Klein tried to set the record straight: Their letters of nomination reflect their private correspondence during the winter of 1945–1946 on how to open the committee's eyes to the importance of Meitner's and Frisch's contributions.

**247** "it would be striking [*påfallende*]" if it was not rewarded with a Nobel Prize in physics: E. A. Hylleraas to NKF, January 29, 1946, KVA/N, 1946.

**248** Hulthén and Siegbahn both had connections with the military's research efforts: Lindström, "Hela Nationens Tacksamhet," p. 60. Hulthén was on the steering committee for the Defense Research Institute.

**248** And, of course, Siegbahn was not only nervously waiting the outcome of his petition: The stakes in 1946 grew tremendously as the Rockefeller Foundation rejected Siegbahn's request for additional funds to build an even larger cyclotron than that which he had proposed during the war. A radiotelegram, Warren Weaver to Siegbahn, March 6, 1946, relayed the rejection [RF, Record 1.1, Series 800, Box 6, Folder 54].

**248** Instead, Hulthén dug in: Hulthén's report in KU, NKF, 1938. At the same time that Hulthén submitted his report, the committee considered Meitner's application for a grant from the committee's special fund that emphasized research on the transuranic elements [*Protokoll*, NKF, September 13, 1938, Bil. E; Hulthén's subsequent report, KU, NKF, 1946].

**248** But when the Physics Section took up the measure, Klein was waiting: *Protokoll*, III klassen, October 25, 1946.

**248** Siegbahn and others preferred to recruit less quarrelsome and less intellectually gifted colleagues: Pleijel asked to retire April 1947; Erik Hulthén expressed his decision to vote against Klein, allegedly to keep peace within the committee, using the possibility of Klein repeating his strong advocacy for Meitner and Frisch as a reason (letter to Ivar Waller, May 25, 1947, IW). The Physics Section voted five for geophysicist Gustaf Ising, three for Klein; on June 4, 1947, the KVA voted eighteen for Ising, six for Klein.

**249** The chemistry committee members washed their hands of the matter: Letters, Bohr to Klein, January 21 and 27, 1947, OK, on Bohr's thinking with respect to circumventing the physics committee's opposition; KU, NKK, 1946 and 1947, on the refusal to consider Meitner and Frisch.

**249** "Sweden Aids Atom Study": *New York Times*, December 31, 1946.

## FOURTEEN   The Knights Templar

**251**   "Knights Templar": Interview with Svedberg, *Upsala nya tidning*, March 2, 1946.

**251**   "Nobel festivities" at the Waldorf-Astoria hotel: Letters from American organizers to Hjalmar Hammarskjöld and Nobel Foundation, 1944, HH. Account of festivities for the fiftieth anniversary of Alfred Nobel's death at New York's American Museum of Natural History, *Christian Science Monitor*, October 23, 1946.

**252**   the prize became an icon: Postwar American enthusiasm converged with the fiftieth anniversary of awarding prizes and led to numerous publications, especially books of mini-biographies and histories of the prizewinners and their contributions. Such works were, of course, the occasion for grinding many an ideological ax. See, for example, Herbert Dingle's foreword to *Nobel Prize Winners in Physics, 1901–1950* (New York: Henry Schuman, 1953), on physics as the search for truth in the abstract divorced from any social duties.

**252**   During the first five years after the war, nominators . . . provided no clear mandate for any candidate: KVA/N, 1940–1950.

**252**   The war did not mark a dividing line in the history of awarding the prize: Several persons, including members of the Nobel establishment in Sweden, have informally expressed to me a belief that, although the awarding of the science prizes prior to the war might well have been marred by personal influences, after the war, somehow, the process of selecting winners managed to achieve an impartial objectivity, free from intrigues and interests.

**252**   Waller was much more in contact with foreign researchers: As evidenced by his scientific correspondence [IW] and scientific production, Waller achieved considerably greater integration into international networks than his predecessor, Oseen. Although Waller did not suffer from Oseen's hypercritical temperament, he did develop his own sense of how, in choosing among several prizeworthy candidates, particular scientific agendas could be furthered. On Waller's early career, see Karl Grandin, *Ett slags modernism i vetenskapen: Teoretisk fysik i Sverige under 1920-talet*, Institutionen för Idé- och Lärdomhistoria, Uppsala Universitet, Skrifter, Nr. 22 (Uppsala, 1999), pp. 145–174ff.

**252**   but merely logical postulates, philosophical constructions: KU, NKF, 1943; *Protokoll*, NKF, June 11, 1943. Oseen provided an additional review of Pauli's work but did not prepare a formal supplemental report.

**252**   in a short telegram from Einstein: KVA/N, 1945. Nominations for Pauli for the 1944 prize from Harvard physicist H. J. Van Vleck and Dutch physicist H. A. Kramers that had arrived too late were included in the 1945 proposals. Six nominations for 1945 that arrived too late to be formally included were nevertheless used in Waller's advocacy for Pauli as proof of growing insight into the fundamental role of Pauli's exclusion principle in contemporary physics. KU, NKF, 1945, Bil. 5.

**253**   Euler again dismissed Robinson as "not worthy": KU, NKK, 1942, Bil. 6. Euler praised Robinson's great energy, massive scientific production, and experimental rigor, but he continued to find grounds to claim the Englishman as not worthy of a prize. Compare letters of nomination, KVA/N, 1942: "There is no man alive who has done so much as Robinson, both by theory and by practice, to help on the future progress of organic chemistry" [N.V. Sidgwick to NKK, November 10, 1941].

**253**   also included work on penicillin: Robinson collaborated with two Oxford colleagues, Ernst Chain and Howard Florey, who shared the 1945 Nobel Prize in medicine in 1945 with the discoverer of penicillin, Alexander Fleming. Von Euler claimed that Robinson had a right to share in the honor of elaborating the penicillin's chemistry, along with Chain and Robinson's student E. P. Abraham, but because this work was not yet finished, the committee should wait in deciding on a prize [KU, NKK, 1946, Bil. 7]. Von Euler claims in his autobiography, *Minnen*, that he championed a prize to Chain for work on penicillin by bringing his accomplishments to the attention of the medical committee.

**253** other committee members made sure that this accomplishment did not enter the formal justification for the prize: In 1946, Bror Holmberg was ready to back Robinson for the chemistry prize. He was a traditional organic chemist who previously, like von Euler, adamantly refused to acknowledge Robinson's theoretical contributions. Although he was willing to find other reasons for rewarding Robinson, among which was the Englishman's enormous production, Holmberg clearly had his specific likes and dislikes, as related in a letter from Arne Westgren to The Svedberg, August 30, 1946, TS.

**253** Appleton was repeatedly passed over: In KU, NKF, 1935, the committee noted that several others, including Breit and Tuve, had contributed in the development of this field [KU, NKF, 1939]. KU, NKF, 1941, repeated the claim from 1939 that this work was not of great enough significance to warrant a prize and that nothing new had been added since the last evaluation. In 1942, laureates O. W. Richardson and W. L. Bragg again nominated Appleton, but the committee felt that no reason existed to change the earlier evaluations [KU, NKF, 1942].

**253** Pleijel decided to back Appleton: KU, NKF, 1947, Bil. 7. Appleton never enjoyed strong support from nominators; at best, there were two or three nominations, and these were not persistent. Over the years, Appleton received support almost solely from British-based scientists O. W. Richardson, W. H. and W. L. Bragg, E. N. da Andrade, and J. T. Macgregor-Norris; in 1935, he received support from the Swiss Hans Zickendraht. Pleijel notified the committee on April 17, 1947, that he wanted to be released from duties as of June 1, 1947 [*Protokoll*, KVA/N, June 4, 1947]. For the 1947 physics prize, twenty-seven nominators proposed twenty-two candidates.

**254** In 1944, when Allied victory was assured, and the prizes again were given out, Americans dominated: In physics, the reserved 1943 prize went to Otto Stern and the one for 1944 went to I. I. Rabi; three Americans and a Danish researcher received the reserved 1943 and 1944 prizes in medicine.

**254** Svedberg gave a lecture extolling American science: Mss., TS. The lecture seems to have been held at an unofficial celebration of the Nobel prizes announced in 1944.

**254** he was always wary of the prize losing its claim to international impartiality: Svedberg clearly had no sympathy for the Nazi regime and, in contrast to some of his colleagues, little special affection for German science. He understood well that in the United States and other nontotalitarian nations, disgust and grief over Nazi treatment of Jewish and dissident academics were very strong, but in spite of his own tendencies to share these feelings, Svedberg tried to steer the Nobel Prize clear of any political coloring. [Interview with Svedberg, *Göteborgs Handels- og Sjöfarts-Tidning*, July 27, 1933; letters, C. L. Lange to Svedberg, January 15, 1936, and Svedberg to Lange, January 20, 1936, copy, TS.]

**255** notwithstanding public proclamations to the contrary, he did not ignore nationality: Semonov and Kapitzka received prizes many years later (1956 and 1978, respectively). Over and beyond questions of scientific worthiness for a prize, some committee members might not have wanted to give a prize to a Russian or in any other way provide the Soviet Union with an accolade. In 1947, the cold war was becoming a reality; continued Soviet pressure on Finland and fears for Soviet interest in Swedish uranium ore surely reinforced traditional antagonism, especially among older conservative scientists.

**255** He remained in close touch with several Swedish chemists: Virtanen was often invited to Sweden by friendly colleagues. Von Euler urged his friend to attend the 1929 opening of his Institute for Biochemistry and the Nobel Prize celebration. Among others, von Euler's disciple and later successor at the Stockholm Högskola and on the Nobel committee, Karl Myrbäck, was close socially and professionally. Von Euler devoted special attention to Virtanen in his autobiography, *Minnen*.

**255** Virtanen had few nominations: Virtanen's main support came from colleagues Oskari Routala and Niilo Toivonen, who nominated him regularly beginning in 1933. During the war, von Euler took over the task of nominating Virtanen; in 1945, after not sending in nominations during most of the war, Finnish chemists returned with proposals for Virtanen, as did von Euler. In addition to von Euler, Christian Barthel also contributed special reports on Virtanen's contributions in the 1930s.

**255** the silage method had been so important in Finland and abroad that it was itself worthy of a prize: In the Chemistry Section, if any questions were forthcoming, Virtanen had additional close friends to argue for his case such as von Euler's protégé, Karl Myrbäck. In addition to von Euler's special report, the committee co-opted agricultural chemist Ragnar Nilsson to report on the silage methods. In general, the reports consistently seem to exaggerate Virtanen's claims for the international success of his method [KU, NKK, 1945, Bil. 6 and 7].

**255** Members of the Nobel Committee for Chemistry initiated a donation, and the physicists followed suit: Letter, Arne Westgren to The Svedberg, February 23, 1940, TS; and letters with Olof Lamm, TS.

**256** "for we will . . . just be compounding the illusion that the Nobel Committee has created": RF, Log, Warren Weaver Interviews, May 31, 1948. Chibnall actually recommended that the foundation continue support; his comments were not motivated by competition for Rockefeller Foundation research funds.

**256** a theory advanced by his mentor, Robert Millikan: Robert H. Kargon, *The Rise of Robert Millikan: Portrait of a Life in American Science* (Ithaca, NY: Cornell University Press, 1982), pp. 151–162.

**256** Then, however, it made a complete about-face: Blackett had made no new major discovery; he had just recently advanced a speculative theory on the earth's magnetism, which was noted in the evaluation but not considered in assessing his worthiness. It is possible that new committee member Gustaf Ising was interested in the theory and might even have wanted to give Blackett greater authority to promote the theory, but it is unlikely that other committee members shared Ising's interest. Other cosmic-ray researchers nominated to share with him included Bruno Rossi and Pierre Auger. KU, NKF, 1940, argues that Anderson's prize precluded one for Blackett, which was similarly claimed in KU, NKF, 1947, while also rejecting Blackett's work on cosmic rays on the grounds that others were also engaged in important work in this field.

**257** give the greatest possible authority and prestige to Blackett: In his evaluation, Gustaf Ising reasoned that, because none of the Englishmen who could be expected to have "inside information" on the work at Cambridge in the 1930s included collaborator Guiseppe Occhialini in their nominations, the Italian's role must have been minor compared with that of Blackett. Maybe; but the comment reveals a naïveté about the motives of nominators in choosing whom they propose. Either Ising simply needed to find a rationale to eliminate other cosmic-ray researchers or he failed to appreciate that nominations sometimes involve trying to bestow authority and prestige on a particular scientist, even at the expense of others who might also have a right to share in the honor of discovery. Nominations, especially in the immediate postwar years, were problematic sources for judging worthiness of accomplishment. British nominators such as Bernal certainly wanted to boost Blackett politically; a divided prize would have less impact. Other examples included Robert Millikan, who aggressively promoted his own protégés from Cal Tech while ignoring others who deserved consideration for contributions to the particular achievement. Beginning in the late 1930s and in the 1940s, he pushed for C. D. Anderson and S. H. Neddermeyer for the discovery of the mesotron while ignoring and even deprecating Hideki Yukawa's claims for the same. (Even though it turned out to be two different subatomic particles, the degree of partial advocacy was clear.)

**258** "threaten our belief that Stanley crystallized truly pure virus": *Protokoll*, NKK, September 23, 1946, Bil. 8 to Bil. A [draft of the general report based on the September 4 meeting of the committee].

**258** Svedberg did not attend that meeting: Arne Westgren had urged Svedberg to come, noting the many differences of opinion in the committee. He hoped that if the medicine prize could go to Stanley, then it would be best if, at the same time, the chemistry prize could be divided between Sumner and Northrop. He suggested that the best scenario would be to give Robinson the prize that year, but in any case he called on Svedberg to help the committee out of an impasse [letter, Westgren to Svedberg, August 30, 1946, TS]. But Svedberg did not attend the meeting; the committee proposed dividing the prize between Sumner and Northrop [*Protokoll*, NKK, September 23, 1946, Bil. A, distributed in advance of the meeting on September 23]. But, clearly, Svedberg wanted Stanley to receive a chemistry prize. His actions and arguments at the meeting on September 23 are clear in the changes of the general report; some paragraphs related to Stanley were deleted, others added. In the end, the prize was split, with one half going to Sumner and the other half divided between Northrop and Stanley [*Protokoll*, NKK, September 23, 1946, Bil. B. (draft of general report agreed upon during the September 23 meeting)].

**259** This was soon understood as the actual key for understanding virus self-replication and not, as Stanley maintained, enzyme action and crystal growth: Lily E. Key, "W. M. Stanley's crystallization of the tobacco mosaic virus, 1930–1940," *Isis*, **77** (1986), pp. 450–472.

**259** "I have had to swallow a lot": Quotation cited in Albert B. Costa, "Wendell Stanley," in Laylin K. James, ed., *Nobel Laureates in Chemistry 1901–1992* (Washington, DC: American Chemical Society and Chemical Heritage Foundation, 1993), p. 303.

**259** In 1950, Svedberg contacted Lawrence: Letters, Svedberg to Ernest O. Lawrence, April 5 and October 7, 1950, copies; Lawrence to Svedberg, October 3 and November 1, 1950, TS. Svedberg's first cyclotron and his nuclear institute opened officially in 1949; a much more powerful particle accelerator was being constructed and was planned to be inaugurated in the autumn of 1951.

**259** Just in case, Svedberg sent in his own proposal: Letter, Svedberg to NKK, January 31, 1951, copy, TS.

**260** Having repeatedly rejected John Cockcroft and E. T. S. Walton . . . the physics committee suddenly gave them the 1951 prize: The committee's rejection of their work in 1938 and 1939 has been discussed with respect to Lawrence's candidacy. KU, NKF, 1943, rejected their work as being limited to smashing lighter elements and precluded by Lawrence's prize. In 1946, the committee concluded that no reason could be found to reexamine Cockcroft.

**260** This time, the physics committee did comment: KU, NKF, 1947, KVA/N, 1947 for nominations, especially W. L. Bragg.

**260** Cockcroft, Lawrence, and other distinguished guests sat around candlelit tables: Cockcroft's Nordic travels and the Lucia ceremony are related in Guy Hartcup and T. E. Allibone, *Cockcroft and the Atom* (Bristol: Hilger, 1984), pp. 163–167.

**261** approximately three hundred invitations to nominate: Letter, Ivar Waller to Egil Hylleraas, April 9, 1949, copy, IW.

**261** (Currently, more than two thousand invitations to nominate are sent out: Carl Nordling, "Hur man får Nobpelpriset i fysik," *Kosmos* (1995), pp. 32–33.

**261** who had been passed over by the committees: Some of these efforts to make the record complete succeeded, such as those including all major contributors to quantum mechanics. Late in life, Max Born and Walther Bothe, in physics, and Kurt Alder, Otto Diels, and Hermann Staudinger, in chemistry, received recognition. Others did not. Laureates such as James Franck, Enrico Fermi, Harold Urey, and Otto Hahn repeatedly advanced candidates such as Samuel Goudschmidt and George Ullenbeck (for electron spin), Hans Kramers, Lise Meitner, and Fritz Paneth [letters, Franck to NKF and NKK, copies, JF].

**262** The committee's choice was based on its own priorities: KU, NKK, 1948, 1949; KVA/N, 1948, 1949. Those nominating Giauque in 1949 were Tiselius, Svedberg, Gunnar Hägg, and G. A. Ölander.

**262** which does not necessarily mean that the practice has continued: It is not clear whether this practice has continued. At least one recent committee member has resolutely refused to nominate, aiming to be as impartial as possible: Interview with Bo G. Malmström, in Kenneth Levebeck, "För många 70-åringar får nobelpriset i kemi," *Kemisk tidskrift*, Nr. 12 (1987), pp. 7–8.

**262** In 1962, the Swedish organic chemist Holger Erdtmann turned to Robert Robinson: Letter, Holger Edrtmann to Robert Robinson, November 14, 1962, Robert Robinson papers, Royal Society, London (quotation by permission of the President and Council of the Royal Society). Erdtmann was right. Fredga's candidate, Giulio Natta, shared the 1963 prize with the German Karl Ziegler for high polymer chemistry; in 1964, Crowfoot Hodgkin received a full prize; Woodward got the 1965 prize.

**264** Some considered such tactics the introduction of "politics" into science: Letter, Werner Heisenberg to Ivar Waller, September 10, 1968, IW. A letter from Nicholas Kemmer to Waller, November 17, 1971, IW, discusses negative reactions to his efforts to support Salam. Both Kemmer and Waller favored Salam.

**264** Some advanced reasons why it was important to reward a great Asian physicist: See, for example, letter, I. H. Usmani to Waller, May 18, 1967, IW; the slight was in reference to Salam's priority claim for the theory of parity violation, in which Yang and Lee made the primary breakthrough.

**264** had not as yet led to a significant discovery: Letter, T. D. Lee to Waller, January 13, 1967, IW.

**264** the most deserving contributors in theoretical high-energy physics: Paul Dirac to Waller, July 25, 1968, IW.

**264** but he was not giving up on Salam: Letter, Waller to Dirac, January 15, 1974, January 7, 1975, January 14, 1976, copies, IW. Waller retired from the committee in 1971 but remained a member of the Physics Section.

**264** A major problem entailed the makeup of the committee: Letter, Stig Lundqvist to Waller, December 13, 1976, IW.

**265** He had predicted the existence of significant processes: Letter, Hideki Yukawa to Ivar Waller, January 27, 1970, IW. Yukawa mentions Sakata's 1956 model for hadrons as well as earlier contributions to meson theory, including the possibility of spontaneous decay of the neutral pion into photons. Also mentioned was the two-meson hypothesis Sakata formulated with Yasukata Tanikawa (1942), five years before Powell's discovery of pions and muons in cosmic rays.

**265** he interviewed persons present at critical seminars and who were in contact with participants in the developments: Letter, Jan S. Nilsson to Waller, April 22, 1969, IW.

**265** it simply could not make a "just selection": Letter, Waller to Yukawa, February 21, 1970, copy, IW.

**265** asked to know what the Nobel committee thought of Sakata's merits, for that would perhaps bring them consolation: Letters, Yukawa to Waller, September 26 and October 19, 1970, IW.

## Further Reflections

**269** Berlin professors filed a complaint to the Academy: SvD, June 28, 1912.

**270** The emperor was determined to have Prussia compensate: R. Steven Turner, "The growth of professorial research in Prussia, 1818 to 1848: Causes and context," *Historical Studies in the Physical Sciences*, **3** (1971), pp. 137–182.

**270** the prestige of the prize seemed to increase geometrically with the distance from Sweden: Related in a series of letters in 1924, Vilhelm Bjerknes to Svante Arrhenius (SA), C. W. Oseen (CWO-KVA), and Sem Sæland (UBO).

**270** A culture of academic competition gathered strength in the United States: Roger L. Geiger, *To Advance Knowledge: The Growth of American Research Universities, 1900–1940* (New York: Oxford University Press, 1986).

**271** The annual rituals in Stockholm provided opportunities . . . to comment on the state of national research: The topic of how the annual announcements and ceremonies provided a forum for advocating and educating, in Sweden and other nations, is immense. During the immediate postwar years it seems that the scientific elite used the occasion of the prize to preach new visions for social relations and social justification of Big Science and international cooperation. As scientists became increasingly dependent upon government for research, the rhetoric of the freedom of research became more prominent. See, for example, Svedberg's speech on the occasion of Arne Tiselius's Nobel Prize in 1948 [Mss., D5:1, TS].

**271** West German laureates used their status to assume leadership roles in rebuilding their national scientific community: Otto Hahn's correspondence with Georg Hevesy in the late 1940s and 1950s reveals some efforts. Hahn and other German prizewinners were very keen to advance the candidacy for a prize of fellow Germans who they hoped could join the ranks of an elite with political clout [Hevesy correspondence, microfilm, NBA].

**272** Rutherford . . . roared approvingly that he and his wife had the time of their lives: Letter, Ernest Rutherford to Arrhenius, January 1, 1909, SA.

**272** to have his democratic principles tested by the Stockholm ceremony: Letters, Rutherford to Harold Urey, November 27, 1934, copy, and Urey to Rutherford, December 5, 1934, Rutherford correspondence, microfilm, AIP. Urey noted that his democratic principles had already received shocks in the United States.

**272** freshmen at the California Institute of Technology: Informal communication to author by students in undergraduate seminar on the Nobel Prize at the University of California, San Diego, Spring 1995.

**273** bemoaned that "athletics" was turning into "sport": Fridtjof Nansen, "Idealitet og karakter," in *Nansens Røst: Artikler og Taler, 1897–1915*, II (Oslo: Dybwalds, 1945), p. 275.

**273** Vince Lombardi: Whether Lombardi actually said this can be discussed and researched, but this philosophy of winning is widely attributed to him—and seems to be part of popular American culture and mythology.

**274** *The Double Helix* caused a great stir: An edition of the original text and commentaries: Gunther S. Stent, ed., and James D. Watson, *The Double Helix: A Personal Account of the Discovery of the Structure of DNA. A New Critical Edition Including Text, Commentary, Reviews, Original Papers* (London: Weidenfeld & Nicolson, 1981).

**274** that the Frenchman's work had not "benefited mankind": Protokoll, KVA, III klassen, October 26, 1926. Committee members never articulated criteria for assessing how basic research benefited mankind or how to judge degree of benefit on mankind of such contributions to knowledge. Occasionally a committee member would go to great lengths to differentiate between useful for science and useful for daily practical needs: Opening new possibilities for research was hailed as the proper understanding of "benefit on mankind" and not the social beneifts that might arise [KU, NKK, 1943, with reference to Hevesy, whose use of isotopic tracers was a great practical as well as scientific development].

**274** As Oseen confessed: "The interpretation of words changes: Letter, C. W. Oseen to Vilhelm Bjerknes, February 4, 1938, VB.

**275** a moral vision of the professor insulated from commercial values: Charles E. McClelland, *State, Society, and University in Germany 1700–1814* (Cambridge: Cambridge University Press, 1980), and Anthony J. La Vopa, *Grace, Talent, and Merit: Poor Students, Clerical Careers, and Professional Ideology in Eighteenth-Century Germany* (Cambridge: Cambridge University Press, 1988); R. Steven Turner, "Justus Liebig versus Prussian chemistry: Reflections on early institute-building in Germany," *Historical Studies in the Physical Sciences*, **13** (1982), pp. 129–162; David A. Hollinger, "Inquiry and uplift: Late 19th-century American academics and the moral efficacy of science," in Thomas L. Haskell, ed., *The Authority of Experts: Studies in History and Theory* (Bloomington: Indiana University Press, 1984); Michael Aaron Dennis, "Accounting for research: New histories of corporate laboratories and the social history of American science," *Social Studies of Science*, **17** (1987), pp. 479–518.

**275** Gilman did not separate the intellectual growth of the disciplines from the development of moral character: See Gilman's vision for the university in Dennis, "Accounting for research"; Owen Hannaway, "The German model of chemical education in America: Ira Remsen at Johns Hopkins University (1876–1913)," *Ambix*, **23** (1976), pp. 145–164; Larry Owens, "Pure and sound government: Laboratories, playing fields, and gymnasia in the nineteenth-century search for order," *Isis*, **76** (1985), pp. 182–194.

**276** postwar era of Big Science, big bucks, and competition for prestige and resources: The literature is vast; for an excellent overview and a provocative essay, see, respectively, Roger L. Geiger, *Research and Relevant Knowledge: American Research Universities Since World War II* (New York: Oxford University Press, 1993); and Paul Forman, "Behind quantum electronics: National security as basis for physical research in the United States, 1940–1960," *Historical Studies in the Physical Sciences*, **18** (1987), pp. 149–229.

**276** "This game of improving an educational operation is great fun to play": Stuart W. Leslie, "Playing the education game to win: The military and interdisciplinary research at Stanford," *Historical Studies in the Physical Sciences*, **18** (1987), pp. 55–88; Stuart W. Leslie, "Profit and loss: The military and MIT in the postwar era," *Historical Studies in the Physical Sciences*, **21** (1990), pp. 59–85; Rebecca S. Lowen, "Exploiting a wonderful opportunity: The patronage of scientific research at Stanford University, 1937–1965," *Minerva*, **30** (1992), pp. 391–421. On the drift of American physics toward amoral "fun," see Paul Forman, "Social niche and self-image of the American physicist," in Mario Michelangelo de Maria and Fabio Sebastiani Grilli, eds., *The Restructuring of Physical Science in Europe and the United States, 1945–1960. Proceedings of the International Conference, Università 'La Sapienza', Rome, Italy, September 19–23, 1988* (Singapore: World Scientific, 1989).

**277** As the fury of competition increases, a buzz of discontent is heard: Columbia University's Robert Pollack, for example, expressed dismay over the lack of a social structure of collegiality that is required for high morale in science. He describes from one researcher's experiences how today's culture of hypercompetition demoralizes professors and students alike: "The crisis of morale in scientific research," *Science and Government Report*, October 1, 1996, p. 6, from a symposium, "Science in Crisis at the Millennium," George Washington University, September 19, 1996. For a fine survey of literature and themes related to competition in science, see David Edge, "Competition in modern science," in Tore Frängsmyr, ed., *Solomon's House Revisited: The Organization and Institutionalization of Science*, Nobel Symposium 75 (Canton, MA: Science History Publications and Nobel Foundation, 1990), pp. 208–232. Pollack, Edge, and many others might look on with envy at the poem that greeted guests at the fifteenth meeting of the Scandinavian Natural Scientists in 1898 in Stockholm. In welcoming the guests and extolling the value of such congresses, the poem calls attention to the importance of friendship, and even the role of personality, for the advance of science; these help maintain the "free republic of science" [*Helsning till Gästerna vid Det 15 de Skandinaviska Naturforskaremötet i Stockholm* (Stockholm: Central, 1898)].

**277** "Is this all there is?": *The Days Are Just Packed. A Calvin and Hobbes Collection by Bill Watterson* (Kansas City, MO: Andrews and MacMeel, 1993), p. 34. (Permission to quote granted by Andrews and MacMeel.)

# Index

Abbott, C. G., 148
Academy. *See* Royal Swedish Academy of Sciences
Accelerators, linear, 226
Acid-base chemistry, 205
Adams, W. S., 146
af Wirsén, Carl David
   and establishment of Nobel prizes, 16–17
   and prize ceremonies, 64
   as social conservative, 60, 95
*Aftonbladet* [newspaper], 72–73, 75
Alpha particles, 150, 164, 225
Amino acids, 111
Anderson, Carl David
   nomination and evaluation of, 220
   physics prize for (1936), 256
Ångström, Anders, 41, 44
Ångström, Knut
   career of, 46, 48
   election to physics committee of, 24
   experimentalist bias of, 43–44, 50
   nomination and evaluation of, 50–51, 57, 145
*Annalen der Physik* [periodical], 125
Anti-Semitism
   in hiring practices, 94, 209
   in Nazi Germany, state-sanctioned, 121, 202, 221, 235
   in nominee evaluations, 138–139, 209
   protests against, 273
   in Sweden, 130, 238
   in the United States, 241
Appleton, Edward Victor, 229–230, 253–254, 256
Argon, discovery of, 42
Arrhenius, Svante
   as advocate for international science, 99–100, 120
   and breaking chemistry committee consensus, 33, 55
   career of, 29–30, 41, 45–47, 108–109, 198, 214–215
   and changing the bylaws, 93
   chemistry prize for (1903), 29–31, 45, 67
   election to physics committee of, 24, 45, 144
   and establishment of chemistry prizes, 16–18, 22
   and evaluation of Einstein, 127, 129, 132, 135, 137–140
   and evaluation of Nernst, 180–184, 191
   and feud with Hammarsten, 56
   and feud with Hasselberg, 20, 43–52
   and feud with Mittag-Leffler, 20, 34, 48–53
   and feud with Nernst, 36–39, 180–181
   fund raising by, 152
   influence of on chemistry committee, 193, 199, 207
   and internationalism in science, 83, 86–92, 107
   money-saving maneuvers by, 144–149, 154–157, 160–161
   nominations and evaluations by, 35–36, 52, 74, 145, 187–188
   and opposing chemistry candidates, 34–38, 186, 188–189
   and opposing physics candidates, 43–48
   personal attacks on and by, 29–30, 33–34, 43–49
   physics committee work by, 100–102, 110–111, 114–115
   posthumously honoring wishes of, 165–168, 188, 196
   and prize ceremonies, 65, 67
   and promoting chemistry candidates, 37, 99
   and wartime candidates, 95, 97–98
Arrhenius Medal, 208

Artistic movements. *See* Movements, social
Aschan, Ossian, 38
Association of German Chemical Societies, 188
Aston, Francis William
    chemistry prize for (1922), 185, 189
    nomination and evaluation of, 184–185
Astronomy, 148–149
Astrophysics, 45–46. *See also* Solar physics
    elimination of, from Nobel physics, 144–151, 230
Atmospheric physics, 120, 145, 150–151, 157, 229–231, 253, 274
Atomic Committee (Sweden), 245–246
Atomic Energy Research Establishment (Britain), 260
Atomic physics, 44, 95, 102, 111, 119–120, 152–153, 158. *See also* Nuclear physics; Quantum theory; Spectroscopy
    and relativity, 172, 174
Atomic weights of elements, 74
Atoms, reality of, 35, 153, 159, 190
Aurivillius, Christopher, 66, 73, 89, 142
Austrian Peace League, 17
Awards ceremonies. *See* Ceremonies

Baeyer, Adolf von. *See* von Baeyer, Adolf
Bahr-Bergius, Eva von. *See* von Bahr-Bergius, Eva
Balkan wars, 71
Bárány, Robert
    medicine prize for (1914), 90, 94
Barbier, P. A., 39
Barkla, Charles Glover
    nomination and evaluation of, 101–102, 106, 129
    physics prize for (1917), 103, 109, 115
Barthel, Christian, 193, 255
BASF Corporation, 104, 112
Bawden, Frederick, 258
Becquerel, Henri
    nominations and evaluations by, 47
    physics prize for (1903), 52, 66
Beethoven, Ludwig van, 76

Behring, Emil von. *See* von Behring, Emil
Benedicks, Carl
    career of, 159, 194, 217
    challenge to physics committee authority by, 159–167, 169, 206
    nominations and evaluations by, 206
Benoît, René, 128
Bergius, Friedrich
    chemistry prize for (1931), 206, 274
Bernal, J. D., 257
Berthelot, Marcelin, 29, 31, 33–34
Berzelius, Jacob, 6
Beta particles, 234
Bethe, Hans
    physics prize for (1967), 151
Big Science, 176, 213, 225–250, 276
    fund raising for, 227
    in the United States, 232–233
Biochemistry, 179, 187–188, 199, 204, 209, 262–263. *See also* Medicine; Organic chemistry
Biomedicine, 227
Birkeland, Kristian, 104, 147
Bjerknes, Vilhelm
    career of, 157, 270
    election to physics committee of, 47–48
    nomination and evaluation of, 52–53, 146, 151, 229–231, 274
Bjerrum, Niels, 189, 207–209
Blackbody radiation, 47–49
Blackett, Patrick Maynard Stuart
    career of, 164
    nomination and evaluation of, 220, 256–257
    nominations and evaluations by, 260
    physics prize for (1948), 221, 257
Bohr, Niels
    career of, 103, 120, 135, 143, 170–171, 174, 187, 235–237, 241–242, 245–246
    nomination and evaluation of, 111, 123, 132–134, 136–137, 163
    nomination to physics committee of, 168, 218
    nominations and evaluations by, 127, 153, 166, 169–170, 172–173, 197, 221, 228–229, 238, 244, 246–247, 249

physics prize for (1922), 138, 146, 168
Boltzmann, Ludwig
   career of, 52
   nomination and evaluation of, 47
Bonding, chemical, 188–189, 197, 207
Born, Max
   career of, 171
   nomination and evaluation of, 171, 174–175, 220
   physics prize for (1954), 221
Bosch, Carl
   career of, 104–105
   chemistry prize for (1931), 112, 206, 274
   nomination and evaluation of, 104–106, 112, 205–206
Boström, Gustaf, 58, 66
Bowen, Ira, 151
Boycotts
   of English physics by Germans, 83, 153
   of German physics after World War I, 120–121, 138, 153, 182, 187–188, 269
   of Nobel institutions by Hitler, 202–203, 229, 243
Bragg, William H.
   career of, 75, 101, 143, 154, 194
   nomination and evaluation of, 88, 146
   physics prize for (1915), 91–92, 115
Bragg, William L.
   career of, 75, 143, 154, 194
   nomination and evaluation of, 88, 146
   nominations and evaluations by, 160, 229, 253, 260
   physics prize for (1915), 91–92, 115
Branting, Hjalmar, 72, 97, 113–114
   peace prize for (1921), 113
Braun, Ferdinand
   physics prize for (1909), 52–53, 274
Bridgeman, P. W., 229, 248
Broglie, Maurice de. *See* de Broglie, Maurice
Broglie, Victor Louis de. *See* de Broglie, Victor Louis
Brønsted, J. N., 205, 207, 209
Brownian movement, 35, 135, 153, 159, 190, 194

Buchner, Eduard
   career of, 200
   chemistry prize for (1907), 55–56
Butenandt, Adolph Friedrich
   chemistry prize for (1939), 203, 209
Bylaws, 2, 23–24. *See also* Committees; Nominators; Reserving prizes; Special funds; Withholding prizes
   and deceased nominees, 89
   and frequency of awards, 94, 154
   and interpretation of "benefit on mankind," 31, 53, 160, 274–277
   and interpretation of "discoveries or innovations," 160, 173
   and interpretation of "recent," 27, 31–34, 50, 274
   and nationality, 76–77, 93–95, 254–255
   reform of, 93–94, 261–262
   and "secrecy," 31, 53, 162, 217
   and split awards, 200

California Institute of Technology, 149–150, 228, 264, 270, 272
Campbell, W. W., 146
Cannizzaro, Stanislao, 31, 34
Carlheim-Gyllensköld, Vilhelm
   career of, 122, 175, 218
   election to physics committee of, 47
   and feuding, 51
   grant from special funds to, 148
   on nature of physics, 48
   nominations and evaluations by, 146–147
   physics committee work by, 100–103, 111, 114, 129, 132, 148–149, 154, 156–157, 161, 166–169
   on science in wartime, 98
Carnegie-Mellon University, 222
Caro, Heinrich, 38, 104
Caroline Institute of Medicine
   and awarding of physiology/medicine prizes, 2, 14
   and establishment of physiology/medicine prize, 17, 19, 22
   and nominating rights, 23
   and wartime prizes, 90
Catalysis, 36–37
Causality, 171, 173
Celsius, Anders, 6
Ceremonies, 34, 58–68, 267–269, 272
   changing date of, 65, 72, 115

Chadwick, James
  nomination and evaluation of, 196–197, 220
  nominations and evaluations by, 244
Charles XII (King of Sweden), 4, 59
Châtelier, Henri Le. *See* Le Châtelier, Henri
Chemistry. *See also* Acid-base chemistry; Biochemistry; Electrochemistry; Geochemistry; Industrial chemistry; Nuclear chemistry; Organic chemistry; Physical chemistry; Structural chemistry; Surface chemistry; Thermodynamics
  establishment of prizes for, 14–15, 19
  history of, 26–27, 179–180, 191, 204–207, 209–210
Chemistry committee. *See also* Committees; Physics committee
  and age limits, 189
  changes in membership of, 37, 193, 253–254
  initial membership of, 24–25
Chibnall, A. C., 256
Christiansen, Christian, 45
Ciamician, Giacomo, 34
Civilization
  hierarchy and progression of, 60–62, 268–269
  and "Manifesto of the Ninety-Three," 76–78
  new physics and decline of, 125–126
Cleve, Astrid, 198
Cleve, P. T.
  and Arrhenius's work, 29–30
  career of, 198
  election to chemistry committee of, 24
Cloud chambers, 153–154, 164–165, 226
Cockcroft, John
  career of, 226, 260–261
  nomination and evaluation of, 228, 230
  physics prize for (1951), 228, 260
Cofactors, 200
Cold War, 255, 271, 276–277
Collège de France, 8, 137
Colloids, 190–194
Committees. *See also* Chemistry committee; Physics committee
  and nominating rights, 1, 23, 261
  role of, 2, 19–21, 23–25, 30
Competition, peaceful
  and nationalism, 58–63, 68, 71, 115, 122, 268–272, 277
Compton, Arthur Holly
  nomination and evaluation of, 160–161
  nominations and evaluations by, 241, 249, 256
  physics prize for (1926), 164–165
Convents, Hugo, 84
Copenhagen University, 23
Cosmic rays, 226, 256
Cosmical physics. *See* Astrophysics
Coster, Dirk, 187, 235, 238
Cryogenics, 50, 57
Crystallography, 74–75, 152, 175
Cultural movements. *See* Movements, social
Culture of science, 273–274
  and deference, 55–56
  and Nobel fever, 251–252, 267–268, 272
Curie, Marie
  chemistry prize for (1911), 37, 51–52, 67
  physics prize for (1903), 52, 66
Curie, Pierre
  physics prize for (1903), 52, 66
Curtius, Theodor, 34
Cyclotrons, 197, 225–232, 239, 244–246, 259–261

*Dagens Nyheter* [newspaper], 79, 194
Dalén, Nils Gustaf
  nomination and evaluation of, 57
  physics prize for (1912), 206, 274
Darboux, Gaston, 52
Davisson, Clinton Joseph
  nomination and evaluation of, 169, 220
Davy, Humphry, 81
de Broglie, Maurice
  nominations and evaluations by, 249
de Broglie, Victor Louis
  career of, 143, 154, 166, 172
  nomination and evaluation of, 169
  nominations and evaluations by, 172, 197, 249

physics prize for (1929), 169
de Laval, Gustaf, 57
Debye, Peter
　career of, 235
　chemistry prize for (1936), 209
　nomination and evaluation of,
　　207–208
Defense Research Institute (Sweden),
　245–246
Dehmel, Richard, 84
Delbrück, Hans, 84
Descartes, René, 86
Deslandres, Henri
　career of, 51
　nomination and evaluation of, 75, 95,
　　128, 145–149, 151
Diels, Otto, 220
Diffraction, 134. *See also* Spectroscopy
　of electrons, 169, 220
　through crystals, 74–75, 143
Diffusion of gases, 174
Dirac, Paul
　career of, 174–175, 221, 256
　nomination and evaluation of, 174
　nominations and evaluations by, 264
　physics prize for (1933), 175
Dissociation theory, 34–35, 208
DNA (nucleic acid), 273
Domagk, Gerhard
　medicine prize for (1939), 203
*Double Helix, The* (Watson), 273–274
Durkheim, Emile, 85
Dynamite, 13

Eclipses, 123, 126–127, 131. *See also*
　Astrophysics; Relativity theory
Economics, prizes in honor of Nobel in,
　271
Eddington, Arthur, 123, 127, 149,
　151
Edén, Nils, 97
Edison, Thomas, 57
Ehrlich, Paul, 76
Einstein, Albert
　career of, 40–41, 120, 123–126, 149,
　　170–174, 233, 252–253
　as celebrity, 124, 133
　and "Manifesto of the Ninety-Three,"
　　83, 124
　nomination and evaluation of,
　　100–101, 111, 123, 126–138, 163,
　　269

　nominations and evaluations by, 153,
　　160, 172–173, 190, 206
　physics prize for (1921), 137–141,
　　146
Einthoven, Willem
　medicine prize for (1924), 159
Ekstrand, Åke, 38, 89, 106, 111, 113,
　193
Ekström, Anders, 61
Electrochemistry, 29–30, 181, 189, 205,
　207
Electrodynamics, 40–41
Electromagnetism, 40, 52, 74, 110
　and quantum theory, 119–120,
　　134
Electrons
　discovery of, 40
　dual nature of, 169, 220
Electrophoresis, 192, 258, 262
Elements
　atomic weights of, 74
　periodic table of, 33, 88
　transuranium, 234–236, 240, 242,
　　248, 259
Embden, Gustav, 203
Emeleus, K. G., 253
Emich, Friedrich, 186–187
Energy
　conservation of, 40, 190
　and equivalence principle,
　　125
Enskog, David, 173–175
Enström, Axel, 218
Enzymes, 37, 55, 186, 198, 200,
　257–259
Equilibrium, 36
Erdtmann, Holger, 262–263
Eriksson, Jakob, 66, 89
Erlander, Tage, 245–246
Ether, 41–44, 124–125, 131, 134,
　170
Eugenics, 87, 115. *See also* Race
Euler-Chelpin, Hans von. *See*
　von Euler-Chelpin, Hans
Exhibitions, 61, 268
Explosives, 13, 107
Eyde, Sam, 104

Fåhraeus, Robin, 192
Faraday, Michael, 83, 254
Fermentation, cell-free, 55–56,
　200–201

Fermi, Enrico
  career of, 226, 234, 259
  nomination and evaluation of, 240–241, 248
  physics prize for (1938), 238
Fertilizers, 38, 104–107, 112
First Colloid Symposium, 191
Fischer, Emil
  career of, 55
  chemistry prize for (1902), 28–31
  and "Manifesto of the Ninety-Three," 76, 109, 111
  nomination and evaluation of, 111, 201, 206
  nominations and evaluations by, 32–33, 182
Fission, nuclear, 230, 233–242, 246
Fleming, J. A., 165–166
Fluorescence, 153
Fluorine, isolation of, 32
Forman, Paul, 101
Forssell, Gösta, 156, 216, 231
Forssell, Hans, 16
Forssman, Johan, 218
France, Anatole
  literature prize for (1921), 183, 270
Franck, James
  career of, 202, 222, 233
  nomination and evaluation of, 149, 153–154, 158, 161
  nominations and evaluations by, 169, 172, 221, 241, 243, 247
  physics prize for (1925), 161
Frank, Adolph, 38, 104
Frankfurter, Felix, 224
Franklin, Benjamin, 270
Fredga, Arne, 253, 263
Fredholm, Ivar, 130
Frisch, Otto Robert
  career of, 236–238, 245, 250
  nomination and evaluation of, 241–243, 246–249
Funding of science. *See also* Big Science; Nobel Foundation; Rockefeller Foundation; Special funds; Wallenberg Foundation
  history of, 3, 121, 251, 269, 275
  and Swedish physics, 141–147, 176, 194–195, 199, 215–219, 239, 244–246
  using special funds, 94, 100, 151–152, 162, 213–214, 238

Gamow, George, 228
Gases, 42, 50
Gehrcke, Ernst, 129
Gell-Mann, Murray
  physics prize for (1969), 264–265
General Electric Corporation, 166, 197
Geochemistry, 209–210
Geophysics, 45, 230, 254
Gerlach, Walther, 221
German Chemical Society, 90, 183, 187, 234–235
German Cultural Association, 77
German National Science Association, 127
Germany. *See also* Nazi Germany
  academic culture of, 6, 76–77, 268, 270
  and chemical weapons, 105–107, 112, 114
  links to, of Swedish science and culture, 6, 32, 60–62, 72, 80, 224
  militarism of, 76–77, 87
  organic chemistry in, 27, 31, 39, 185–187, 202–203
  postwar science in, 249–250, 271
  science in, 40, 81–82, 275
  and Weimar Republic, 124, 138–139, 370
  and World War I, 71–72, 76–78, 96–97, 102, 105–107
Germer, L. H., 220
Giauque, William, 262
Gilman, Daniel Coit, 275–276
Gjellerup, Karl, 96
Glashow, Sheldon
  physics prize for (1979), 265
Goethe, Johann Wolfgang von, 76, 84
Goldschmidt, Victor Moritz, 209
Gomberg, Moses, 205
Gosse, Edmund, 86
*Göteborgs Handels- och Sjöfarts- Tidning* [newspaper], 213, 217
Grandqvist, Gustaf
  career of, 142, 144, 146–147
  election to physics committee of, 50, 52
  grant from special funds to, 148

physics committee work by, 50–53,
111, 131–132, 137–138
Grignard, Victor
chemistry prize for (1912), 36, 38–39
and nationalism, 85–86
nominations and evaluations by,
206
Guillaume, Charles-Edouard
nominations and evaluations by, 153,
160, 206, 220
physics prize for (1920), 127–129
Gullstrand, Allvar
career of, 142
election to physics committee of, 51
medicine prize for (1911), 57
money-saving maneuvers by, 145,
148, 154–157
opposition to Einstein of, 129–135,
137
physics committee work by, 75, 109,
111, 146, 149, 160–161, 166–167,
169, 274
and prize ceremonies, 108
as "small pope in Uppsala," 122
Gustaf II Adolph (King of Sweden),
62
Gustaf V Adolph (King of Sweden),
140
Gustaf Vasa (King of Sweden), 59
Gustafson, Torsten, 245–246, 250
Guye, C. E., 127
Guye, P.-A., 109

Haas, Arthur, 100
Haber, Fritz
career of, 104–105, 112, 114, 188
chemistry prize for (1919), 112–114,
139, 179–180, 184, 205, 254
nomination and evaluation of,
104–106, 111, 205–206, 255
nominations and evaluations by, 182,
186, 209
Haber-Bosch process, 104–107, 206,
240
Hague conventions, 72, 79–80, 114
Hahn, Otto
career of, 233–239
nomination and evaluation of,
240–244, 246–247, 253
physics prize for (1945), 244, 249,
254

Hale, George Ellery
nomination and evaluation of, 75, 88,
95, 128, 145–149, 151
nominations and evaluations by,
42–43, 46, 150
Hammarskjöld, Hjalmar, 72, 78–80,
90–91, 97, 240
Hammarsten, Olof
and awarding prizes during wartime,
73
and building chemistry committee
consensus, 242
chemistry committee work by, 38–39,
56, 105–107, 111–114, 179,
182–183, 185, 187–188, 194,
198–199, 201, 208
election to chemistry committee of,
37, 193
and feud with Arrhenius, 56
and prize ceremonies, 67
Hann, Julius von. *See* von Hann, Julius
Harden, Arthur
chemistry prize for (1929), 204
nomination and evaluation of,
200–201
Hasselberg, Bernhard
career of, 46, 128–129, 134, 144,
147
election to physics committee of, 24,
144
and establishment of physics prizes,
20–21
experimentalist bias of, 43–48
and feud with Arrhenius, 20,
43–52
nomination and evaluation of, 42
nominations and evaluations by, 128,
146
physics committee work by, 111,
131–132, 138, 145
and prize ceremonies, 64
Haworth, Walter
chemistry prize for (1937), 209
Heat
radiation of, 74
specific, 137
Heaviside, Oliver, 47
Hedin, Sven G., 93, 201
Heidenstam, Werner von. *See*
von Heidenstam, Werner
Heine, Heinrich, 140

Heisenberg, Werner
  career of, 166, 170, 263–264
  nomination and evaluation of,
    171–175, 220
  nominations and evaluations by, 197,
    221
  physics prize for (1932), 174–175
Helium, liquefaction of, 50, 57
Helsinki University, 23
Heredity, 273
Hertz, Gustav
  physics prize for (1925), 161
Hertz, Heinrich, 52, 134
Hevesy, Georg von. *See* von Hevesy,
    Georg
Hildebrand, Emil, 62, 64
Hildebrandsson, Hugo Hildebrand
  election to physics committee of,
    24, 144
  experimentalist bias of, 46
  and feuding, 48, 50–51, 98
  physics committee work by, 150
Hinshelwood, Cyril, 255
Hitler, Adolf, 202–203, 224, 229
Hjärne, Harald, 78, 93, 95
Hjelt, Edward, 28–29, 31, 54–55
Hodgkin, Dorothy Crowfoot,
    262–263
Högskola. *See* Stockholm Högskola
Holton, Gerald, 41
Hopkins, Frederick Gowland, 201
Hormones, 208–209
Hubble, Edwin, 151
Hulthén, Erik
  election to physics committee of,
    167
  physics committee work by, 169,
    172–174, 220, 246–249
Hydrodynamics, 142, 152, 157
Hydrogen, allotropic, 171, 174
Hylleraas, Egil, 247

Ibsen, Henrik, 16, 249
IEB. *See* International Education Board
    (IEB)
IG Farben Corporation, 206
Indigo, synthesis of, 32
Industrial chemistry, 104–107,
    205–206
Industrialization
  in Germany, 81–82, 112
  in Sweden, 5, 62

Ingold, Christopher, 205
Institute for Advanced Studies
    (Princeton), 237, 253
Institute for Theoretical Physics
    (Copenhagen), 120, 229
Interferometers, 42–44, 50
International Bureau of Weights and
    Measures, 42, 127–129
International Congress of Physics,
    144
International Education Board (IEB),
    155–156, 199. *See also* Rockefeller
    Foundation
International law, 79
Internationalism
  interwar, 107, 137
  nationalist threats to, 75–78, 83–86
  post–World War II, 244, 257
  pre–World War I, 73
  and World War I, 81–83, 86–92,
    97–98, 101, 126
Ionosphere, 229
Isotopes
  biochemical uses of, 196, 230–231,
    244
  discovery of, 184, 190
  radioactive, 196, 220, 226, 230–231,
    234
*Italia* [airship], 166, 170

Japan, science in, 265–266
Johns Hopkins University, 241, 275
Joliot, Frédéric
  career of, 226–228
  chemistry prize for (1935), 196, 228
  nomination and evaluation of, 197,
    220
Joliot-Curie, Irène
  career of, 226–228, 234, 240
  chemistry prize for (1935), 197, 228
  nomination and evaluation of, 197,
    220
Jordan, David Starr, 154

Kaiser Wilhelm Institute for Physical
    Chemistry (Berlin), 105–106, 222,
    233, 247
Kaiser Wilhelm Institute for Theoretical
    Physics (Berlin), 125
Kaiser Wilhelm Institutes for the
    Advancement of Science (Berlin),
    100

Kamerlingh Onnes, Heike
  nomination and evaluation of, 50, 57
  nominations and evaluations by, 127, 153
  physics prize for (1913), 57
Kant, Immanuel, 76
Kapitzka, Peter, 254
Karrer, Paul
  chemistry prize for (1937), 209
  nomination and evaluation of, 203, 208, 220
Kelvin, William Thomson, Lord, 47
Kendrew, J. C., 263
Kennedy, J. W., 259
Key, Axel, 17
Khvol'son, Orest, 154
Kinetic theory, 52
Kinetics, chemical, 28
King-in-council, 22, 58, 73, 78
Kjellin, Fredrick, 57
Klason, Peter
  career of, 103
  chemistry committee work by, 33, 103–107, 112–114, 187
  election to chemistry committee of, 24, 193
  nominations and evaluations by, 38, 111
Klein, Felix, 76
Klein, Oskar
  career of, 173–175, 245–246, 249
  nominations and evaluations by, 173, 246–248, 250
Knudsen, Jakob, 96
Kögl, Fritz, 203, 208
Kohler, Robert, 204
Kohlrausch, Friedrich, 30
Kowalevski, Sonja, 8
Krafft, Fritz, 235, 238
Kuhlmann, W. H., 185–187
Kuhn, Richard
  chemistry prize for (1938), 203, 209

Landsteiner, Karl
  medicine prize for (1930), 197
  nominations and evaluations by, 198
Langevin, Paul, 169
Langmuir, Irving
  chemistry prize for (1932), 198

  nomination and evaluation of, 166, 170, 197–198, 203, 207
Lapworth, Arthur, 205
Laue, Max von. *See* von Laue, Max
Laval, Gustaf de. *See* de Laval, Gustaf
Lavoisier, Antoine, 86, 162
Lawrence, Ernest Orlando
  career of, 226–227, 259–261
  nomination and evaluation of, 225, 228–231, 252
  physics prize for (1939), 231
Le Châtelier, Henri, 206, 274
Le Ronsignal, Robert, 106
Lee, Tsung Dao
  physics prize for (1957), 264
Lenard, Philipp
  anti-Semitism of, 138–139
  career of, 137
  and "Manifesto of the Ninety-Three," 76
  and nationalism, 83, 126–127
  nomination and evaluation of, 101
  nominations and evaluations by, 47, 109, 129
*Les Atoms* (Perrin), 159
*Les Prix Nobel* [periodical], 58
Lewis, Gilbert Newton
  Arrhenius Medal for, 208
  career of, 207
  nomination and evaluation of, 188–189, 197–198, 209, 220
Liebig, Justus, 185
Light
  dual nature of, 119–120, 134
  speed of, 125
Liljestrand, Göran, 244
Lilljeqvist, Rudolf
  and establishment of Nobel prizes, 14, 16, 22
Lindh, Axel, 175
Lindhagen, Carl
  and establishment of Nobel prizes, 14–17, 19, 22
  and prize ceremonies, 58, 63–64
Lindström, Stefan, 245
Linnaeus, Carl, 6
Lippmann, Gabriel
  physics prize for (1908), 49, 57
Literary movements. *See* Movements, social

Literature
  establishment of Nobel prize for, 14–15
  prizes for, 159, 183, 269–270
  and wartime prizes, 78, 89–91, 95–96, 98
Livingston, M. Stanley, 227
Ljungberg, Erik Johan, 57
Lombardi, Vince, 273
Lorentz, Hendrik Antoon
  career of, 52, 124–125
  nomination and evaluation of, 101, 126
  nominations and evaluations by, 49, 127, 153, 160
  physics prize for (1902), 47, 49
Lowry, T. M., 205
Ludendorff, Erich, 98, 102, 106–107
Lund University, 8, 23

MacArthur Foundation, 271–272
McMillan, Edwin, 259
Macromolecules, 192
Magnetism, 169
  and quantum mechanics, 221
  and secular variation, 148
  and sunspots, 145
  and Zeeman effect, 47
"Manifesto of the Ninety-Three," 76, 84
  opposition to, 76–77, 101, 124
  regret over, 83, 109, 111
Marconi, Guglielmo
  physics prize for (1909), 52–53, 274
Mass media. *See* Media
Mass spectrometers, 184
Massachusetts Institute of Technology, 229
Masson, Frédéric, 85
Mathematics
  in chemistry, 206
  in physics, 50–51, 145, 170
Max Planck Institutes (Berlin), 100
Maxwell, James Clerk, 134, 254
Measurement, precision, 41–44, 128, 149, 215
Media. *See also* Aftonbladet; Dagens Nyheter; Göteborgs Handels- och Sjöfarts- Tidning; New York Times; Nya Dagligt Allehanda; Svenska Dagbladet
  and cult of prizewinners, 122, 162, 252, 268, 271
  history of, 5–6, 268–269, 271
  manipulation of, for financial gain, 194–195
  and prize ceremonies, 64–67
  reaction of, to new physics, 124–126
  reaction of, to Nobel's testament, 62–63
  reaction of, to postwar prizes to Germans, 113–114
  reaction of, to wartime reservation of prizes, 79, 95
  and scandal over use of special funds, 213, 217–218
  and use of propaganda, 234, 268
Medicine. *See also* Biochemistry; Caroline Institute of Medicine; Organic chemistry
  establishment of Nobel prize for, 14–15, 19
  and medicine committee work, 90, 94, 196, 200
  prizes for, 57, 90, 94, 159, 197, 203
Meitner, Lise
  career of, 232–250
  grant from special funds to, 238–239
  nomination and evaluation of, 225, 240–244, 246–249, 252
*Memoirs of a Minor Prophet* (Robinson), 253
Mendeleyev, Dmitry, 31, 33–35, 50
Meteorology. *See* Atmospheric physics
Metrology, 42–43, 128
Meyer, Edgar, 150
Meyer, Stefan, 154
Michelson, Albert Abraham
  career of, 149
  nominations and evaluations by, 49–50
  physics prize for (1907), 41–44, 47, 50, 128
Michelson-Morley experiment, 44, 124
Microanalysis, 185–187
Milk, 255
Millikan, Robert
  career of, 134, 136, 149, 228, 256
  nomination and evaluation of, 111, 149–150, 161
  nominations and evaluations by, 153, 160, 176
  physics prize for (1923), 270

Mineralogy, 209
Mittag-Leffler, Gösta
    on the bourgeoisie, 269
    and feud with Arrhenius, 20, 34,
        48–53, 56
    and lack of mathematics prize, 17
    nominations and evaluations by, 145
    and opposition to experimentalist
        physics, 48–53, 130
    and prize ceremonies, 66–67
    and prize deliberations, 56–57, 90–91
    and women in science, 8
Moissan, Henri
    chemistry prize for (1906), 32–33
Montelius, Oscar, 84, 93
Morley, Edward W., 41
Mörner, Karl, 66, 78, 94
Moseley, Henry
    career of, 101, 143, 154
    nomination and evaluation of, 88–89,
        103
Movements, social, 59–60, 60, 124, 190
Mussolini, Benito, 121, 234

Nansen, Fridtjof, 272
Napoleonic wars, 72, 81, 270
Nationalism
    in British science, 77, 86, 90–91
    in French science, 85–86
    in German science, 77–78, 83–85, 88,
        126–127, 139
    and peaceful competition, 58–63, 68,
        71, 115, 122, 268–272, 277
    in Sweden, 5
    and World War I, 71
*Nature* [periodical], 77, 237
Nazi Germany
    boycott of Nobel institutions by,
        202–203, 229, 243
    and Meitner, 234–235, 237–238
    racial laws in, 223–224, 235, 237
    science in, 121, 126, 176, 221–224,
        253
Nernst, Walther
    career of, 30
    and chemical weapons, 105
    chemistry prize for (1920), 183,
        185
    and feud with Arrhenius, 36–39,
        180–181
    and "Manifesto of the Ninety-Three,"
        76

nomination and evaluation of,
    180–184, 189, 191
nominations and evaluations by, 194
Neuberg, Carl, 200–201, 203, 220,
    222–223
Neutrality. *See also* Hague conventions
    and awarding prizes, 87–92
    of Sweden during World War I,
        71–73, 80, 199
    of Sweden during World War II, 240
Neutrons, 196, 220, 225
*New York Times* [newspaper], 249
Newspapers. *See* Media
Newton, Isaac, 83, 129, 133, 254
Nichols, J. B., 192
Nitrogen fixation, 38, 104–107, 255
Nitroglycerine, 13
Nobel, Alfred, 13–14
    bequest of, 2, 13–19
    determining intent of, 58, 275,
        276–277
    failure of dream of, 115, 222
Nobel, Emmanuel, 18–19
Nobel Brothers Naphtha Company, 19
Nobel Day, 65–66
Nobel Foundation, 22, 90, 115, 216–217
Nobel institutes
    for chemistry, 147
    for cosmical physics, 147
    establishment of, 21–23, 115, 147,
        244
    for experimental physics, 217
    failed plans for a single, all-
        encompassing, 18–21
    for medicine, 94
    for physical chemistry, 144, 214
    for physics, 147
    for theoretical physics, 142, 215
Nobel laureates
    nominating rights of, 1, 23
    veneration of, 1, 158, 247–248,
        251–268, 271
Nobel Palace, 65
Nobel prizes. *See also* Bylaws;
        Ceremonies; Nominators;
        Standards; *under specific discipline*
    establishment of, 13–25
    financial value of, 63
    selection process for, 1–3, 19–25
Nobile, Umberto, 166
Noble gases, 29, 33
Noddack, Ida, 234

Nominators
  eligibility for, 1–2, 19–21, 23, 205, 261
  and pressure tactics, 263–264
Nordensköld, Adolf Erik, 7, 17
Northrop, John Howard
  chemistry prize for (1946), 258
  nomination and evaluation of, 257–258
Norway
  history of, 4, 59, 63
  nuclear power in, 260
Noyes, A. A., 150
Nuclear chemistry, 197, 231
Nuclear physics, 119–120, 176, 220, 232–233
  and alpha particles, 150, 225
  and astrophysics, 151
  and cloud chambers, 164
  and cyclotrons, 197, 225–232, 244–246
  and fission, 233–242, 246
  and neutrons, 225
  and radioactivity, 190
Nucleic acids, 259, 273
*Nya Dagligt Allehanda* [newspaper], 138–139

Occhialini, Giuseppe, 256–257
Oil-drop experiment, 149
Olympic Games, 63, 122, 269, 272–273
  prizes as cultural, 268
Optics, 57, 130, 153
Organic chemistry. *See also* Biochemistry; Medicine
  French, 39
  German, 27, 31, 39, 185–187, 202–203
  reactions in, 253
  structural, 28
  Swedish, 206–207
Osborn, Henry Fairfield, 150
Oscar II (King of Sweden), 5, 60–61, 64
Oscarian era, 63–65
Oseen, Carl Wilhelm
  fund raising by, for Swedish theoretical physics, 141–144, 214–215, 217–220
  money-saving maneuvers by, 145–146, 148, 238–239
  physics committee work by, 131, 133–138, 150–158, 160–162, 208, 220–221, 224, 230–231, 246, 252, 274
  and reassertion of physics committee authority, 163–176
  as "small pope in Uppsala," 122
Oslo University, 23, 209
Osmotic pressure, 30
Ossietzky, Carl von. *See* von Ossietzky, Carl
Ostwald, Wilhelm
  career of, 30, 107
  chemistry prize for (1909), 34–37, 181
  and nationalism, 77–78, 88
  nominations and evaluations by, 126

Palmær, Wilhelm
  career of, 193
  chemistry committee work by, 113, 182, 198, 201, 204–207, 242
  nominations and evaluations by, 206
Paris Academy of Sciences, 7, 51–52, 77, 81, 85
Parity, nonconservation of, 264
Parliament, Norwegian [Storting], 14, 63
Parliament, Swedish [Riksdagen], 5
Partial differential equations, 50
Particle physics, 176, 220, 241, 248–249, 263–265
  and cloud chambers, 164
Paschen, Friedrich, 129, 133, 150, 153, 157
Pasteur, Louis, 55, 60
Patriotism. *See* Nationalism
Pauli, Wolfgang
  career of, 171
  nomination and evaluation of, 174–175, 220, 252
  nominations and evaluations by, 172
  physics prize for (1945), 221, 246, 253
Pauling, Linus
  chemistry prize for (1954), 259
Peace
  establishment of Nobel prize for, 14–15
  prizes for, 113, 202
  and wartime prizes, 91
Pedagogy, 8, 275
Penicillin, 253

Pérez Galdós, Benito, 96
Periodic table, 33, 88
Perrin, Jean
　career of, 159
　nomination and evaluation of, 153, 158–161, 194, 274
　nominations and evaluations by, 164, 171, 190, 197, 206
　physics prize for (1926), 161–162, 164, 270
Perutz, Max, 263
Pettersson, Hans
　career of, 167, 246
　charges of impropriety by, 213, 217–219, 224
　grant from special funds to, 154, 156
Pettersson, Otto
　on Arrhenius, 30, 181
　chemistry committee work by, 107, 213, 217, 219–220
　election to chemistry committee of, 24
　and establishment of chemistry prizes, 16, 18
　and feuding, 52
　on Hasselberg, 46
　nominations and evaluations by, 33–34
　on Uppsala "popes," 168
Photoelectric effect
　law of, 133–137, 149
　theory of, 135–136, 138
Photography, color, 49, 57
Physical chemistry, 30, 35–36, 45, 179–183, 207
Physics. *See also* Astrophysics; Atmospheric physics; Atomic physics; Geophysics; Nuclear physics; Particle physics; Quantum theory; Relativity theory; Solar physics; Thermodynamics
　early experimentalist bias in, 41–48, 52–53, 101, 175
　establishment of Nobel prize for, 14–15, 19, 144–145
　history of, 40–41, 119–123, 143, 191, 251, 263–264
　and interplay of theory and experiment, 141–144, 173
　interwar rise of theoretical, 110–111, 119–123

Physics committee. *See also* Chemistry committee; Committees
　challenge to authority of, 159–162
　changes in membership of, 37, 45, 47–48, 50–52, 144, 167–168, 175, 193, 199, 207, 252–253
　experimentalist bias in, 43–44, 175
　initial membership of, 24–25, 144
　reassertion of authority of, 163–176
Physiology. *See* Biochemistry; Medicine
Pirie, Norman, 258
Planck, Max
　career of, 30, 40–41, 47, 52, 110, 125–126, 135, 171, 174, 233
　and internationalism in science, 83, 89
　and "Manifesto of the Ninety-Three," 76, 109
　nomination and evaluation of, 47, 49, 57, 74–75, 88, 95, 100, 110, 126, 146, 153
　nominations and evaluations by, 129, 171, 182, 199–200, 209, 221, 249
　physics prize for (1918), 111–112, 122–123, 137, 139, 146
Pleijel, Henning
　election to physics committee of, 167–168
　physics committee work of, 216–217, 253–254
Plutonium, 37, 259
Poincaré, Henri
　career of, 124–125
　nomination and evaluation of, 47, 49–52, 57, 145
Politics. *See also* Nationalism; Neutrality
　of awarding prizes during wartime, 73, 79, 252
　in science, 1, 97–98, 203, 249
Positrons, 175, 220, 256
Postponing prizes. *See* Reserving prizes
Power
　hydroelectric, 104
　nuclear, 232, 245, 260–261
Poynting, J. H., 47
Prandtl, Ludwig, 142
Prandtl, Wilhelm, 105–106, 111
Precision. *See* Measurement, precision
Pregl, Fritz
　chemistry prize for (1923), 187, 189
　nomination and evaluation of, 185–187
Probability, 119–120, 170–171, 173

"Proclamation to the Civilized World." *See* "Manifesto of the Ninety-Three"
Propaganda. *See* Media
Proteins, 192, 194, 196, 198, 257–259
Prussian Academy of Sciences, 100, 125
Prytz, Kristian, 42, 128

Quantum theory
  of the atom, 40, 49, 74, 103, 110–111, 119–122, 133–137, 143, 153–154
  attempts to resist, 149, 170–176, 252
  in chemistry, 206
  of energy, 134, 136, 153
  of light, 119, 134, 136, 160
  need for physics committee expertise in, 168–169
  of the nucleus, 120, 228
  relativistic, 172, 174, 220

Race. *See also* Anti-Semitism; Eugenics; Nazi Germany
  and Social Darwinism, 6, 59–63, 68
  and World War I, 76–77
Radar, 232, 260
Radiation
  electromagnetic, 110, 136, 156
  of heat, 74
  solar, 41, 148
Radio, 53, 165–166, 170, 229, 253
Radioactivity
  artificially induced, 196, 220, 226, 230–231, 234, 244
  and cloud chambers, 164
  discovery of, 27, 40
  research on, 35, 120, 184, 190
Radium, 52
Raman, Chandrasekhara Venkata
  nominations and evaluations by, 221
  physics prize for (1930), 171–172
Ramberg, Ludwig
  career of, 214
  chemistry committee work by, 193, 201, 204–208
  election to chemistry committee of, 199, 207
Ramsay, William
  career of, 100
  chemistry prize for (1904), 29–31, 33, 185

and nationalism, 77, 86, 90–91
  nominations and evaluations by, 74
Rates of reactions, 36
Rathenau, Walther, 139
Rayleigh, John William Strutt, Lord
  physics prize for (1904), 42, 64
Reactions, 36
Relativity theory
  general, 120–126, 130–132, 134–136, 139–140
  influence on other disciplines of, 123–126, 130, 170
  in quantum mechanics, 172, 174
  special, 40–41, 100–101, 103, 120–126, 130–132, 135, 139–140
Remsen, Ira, 276
Research, 8, 274–276. *See also* Big Science; Funding of science
Reserving prizes, 38, 56, 217–219
  between the wars, 132, 154, 156, 158–161, 167, 174, 182
  during World War I, 78–81, 93, 105, 107, 113
  during World War II, 240
Retzius, Gustaf, 33, 73
Reymont, Wladyslaw Stanislaw
  literature prize for (1924), 159
Richards, Theodore William
  career of, 109, 183
  chemistry prize for (1914), 37–38, 91–92, 115, 197
  nomination and evaluation of, 50, 73–74, 78, 88
  nominations and evaluations by, 150, 182
Richardson, O. W.
  nomination and evaluation of, 165–167
  nominations and evaluations by, 228–229, 253
  physics prize for (1928), 169–170, 274
Richet, Charles, 86
Riesenfeld, Ernst, 238
Riksdagen. *See* Parliament, Swedish
RNA (nucleic acid), 259
Robinson, Robert
  career of, 204
  nomination and evaluation of, 220, 243, 253, 256
  nominations and evaluations by, 262–263

Rockefeller Foundation
  and biochemistry, 199, 202, 258
  and chemistry, 205, 215–216, 256–258
  and Nazi Germany, 224
  and nuclear physics, 228–229, 231
  and physical chemistry, 194–197
  and physics, 155–156, 158
Rolland, Romain, 90–91, 96
Röntgen, Wilhelm
  career of, 74
  and "Manifesto of the Ninety-Three," 76
  nominations and evaluations by, 47
Royal Astronomical Society (London), 123
Royal Institute of Technology (Stockholm), 23, 33
Royal Institution (London), 8
Royal Jubilee (Sweden), 61, 63
Royal Society (London), 6–7, 45, 123, 126
Royal Swedish Academy of Sciences
  and establishment of physics and chemistry prizes, 14–17, 24
  feuding in, 48–53
  history of, 6–8
  and nominating rights, 1, 20–23
  and prize selection, 54
  royal patronage of, 60
Russell, H. N., 151
Russia. *See also* Soviet Union
  and Bolshevik revolution, 97
  as traditional Swedish enemy, 72
  and World War I, 97–98
Russo-Japanese War, 71
Rutherford, Ernest
  career of, 75, 87–88, 164, 196, 225–226
  chemistry prize for (1908), 35, 66, 184–185, 272
  nomination and evaluation of, 148, 150
  nominations and evaluations by, 101–103, 145, 153, 164, 169, 172
Ruzicka, Leopold
  chemistry prize for (1939), 203, 209
Rydberg, Johannes, 44–45

Sabatier, Paul
  chemistry prize for (1912), 39
  and nationalism, 85–86
Saha, Meghnad, 151
Sakata, Soichi, 265–266
Salam, Abdus
  nomination and evaluation of, 263–265
  physics prize for (1979), 265
Scandinavian Art and Industry Exhibition, 61
Scheele, Carl, 6
Schlieffen Plan, 76
Schrödinger, Erwin
  career of, 166
  nomination and evaluation of, 169, 171–173, 220
  physics prize for (1933), 174–175
Schück, Henrick, 115
Science. *See also* Big Science; Chemistry; Funding of science; Internationalism; Nationalism; Physics
  Americanization of, 191–192, 251
  and the Cold War, 255, 271, 276–277
  and cultural view of scientists, 60, 251–252
  history of, 3–4, 267–272
  links of, to industry and defense, 232, 276
  as a national resource, 81, 98–100, 119
  nature of, 3–4, 161, 272–274
  societal role of, 8, 152, 274–277
  specialization of, 262
  women in, 8
*Scientia* [periodical], 82
Seaborg, Glenn, 259
Segrè, Emilio, 259
Semonov, N. N., 255
Senderens, J.-B., 39
Shaw, W. N., 146
Siegbahn, Manne
  career of, 102, 154–155, 167, 175–176, 231, 257
  fund raising by, for nuclear research, 225–232, 232–233, 235, 248–249, 259
  fund raising by, for Swedish experimental physics, 141–144, 214–217, 220, 224
  and Meitner, 235, 238–239, 244–249

Siegbahn, Manne (*continued*)
  money-saving maneuvers by,
    145–146, 148, 151–152, 154–158,
    161–162
  nomination and evaluation of,
    154–157, 163, 218
  nominations and evaluations by, 243
  physics committee work by, 150, 164,
    168, 220, 224, 229–232, 254, 257,
    259
  physics prize for (1925), 157
  as "small pope in Uppsala," 122, 155,
    157–158
Silage, 255–256
Sime, Ruth, 235, 238
Social Darwinism, 6, 59–63, 68
Social Democrats, 5, 72, 97, 113–114,
    244, 259
Social movements, 59–60, 60, 124, 190
Society, role of science in, 8, 152,
    274–277
Soddy, Frederick
  career of, 190
  chemistry prize for (1921), 185, 189
  nomination and evaluation of,
    184–185
  nominations and evaluations by, 206
Söderbaum, Henrik
  career of, 166
  chemistry committee work by, 31,
    106, 112–113, 179–180, 182,
    184–185, 187–188, 194–195, 207
  election to chemistry committee of,
    24, 193
  nominations and evaluations by, 38
  and reserving prizes, 89
Sohlman, Ragnar
  and establishment of Nobel prizes,
    14–19, 22
  and prize ceremonies, 58, 63–64, 66
  and Swedish neutrality, 240
Solar constant, 50, 145
Solar physics, 46, 75, 145–146, 151
Solar radiation, 41, 148
Solvay conferences, 74, 248–249
Sommerfeld, Arnold
  career of, 102, 143, 157, 221
  grant from special funds to, 109, 148,
    223
  nomination and evaluation of, 123,
    133, 153–154
Sörensen, Soren Peter Lauritz, 205

Soviet Union. *See also* Russia
  and Bolshevik revolution, 97
  and the Cold War, 255
  Stalinist, 224
Special funds
  establishment of, 21, 217
  grants from, 109, 148, 154, 159, 214,
    223, 238–239
  and interwar prizes, 156, 158, 160,
    174
  as matching funds, 156, 162,
    213–214
  scandal over use of, 213, 217–218
  and wartime prizes, 78, 98, 107
Spectroheliographs, 145
Spectroscopy. *See also* Mass
    spectrometers
  and astrophysics, 127
  and atomic structure, 41–45, 129,
    133, 136, 143
  and electric fields, 95
  and isotopes, 184
  and magnetism, 47
  and spectrometers, 156–157
  X-ray, 88, 101–102
Spengler, Ostwald, 125
Spitteler, Carl, 78, 96
Standards
  lowering of, 54–58, 103, 201
  raising of, 94–95, 149, 174–175,
    220
Stanley, Wendell Meredith
  chemistry prize for (1946), 258
  nomination and evaluation of,
    257–259
Stark, Johannes
  career of, 109, 111, 134, 136–137
  and nationalism, 126, 139
  nomination and evaluation of, 95,
    101–102, 109–111, 114
  nominations and evaluations by, 172,
    221
  physics prize for (1919), 111–112,
    137
Statistical mechanics, 52
Staudinger, Hermann, 192
Stern, Otto
  career of, 233
  nomination and evaluation of,
    220–223
  physics prize for (1943), 221–222
Stockholm Exhibition, 61, 64, 180

Stockholm Högskola
  and establishment of physics and
    chemistry prizes, 17, 22–23
  history of, 8–9, 15
  style of chemistry at, 30
  style of physics at, 45–47, 154
Stockholm Hospital, 17
Størmer, Carl, 147
Storting. *See* Parliament, Norwegian
Strassmann, Fritz
  career of, 233–238
  nomination and evaluation of, 241, 243, 246, 249
Strikes, 71
Strindberg, August, 16, 60, 67, 190
Strömholm, Daniel, 190
Structural chemistry, 28
Sugars, chemistry of, 28–29
Sumner, James Batcheller
  chemistry prize for (1946), 258
  nomination and evaluation of, 257–258
Surface chemistry, 197–198, 207
Svedberg, The(odor)
  career of, 183, 189–198, 231, 251, 257
  chemistry committee work by, 200–204, 207–209, 220, 222, 224, 240–244
  chemistry prize for (1926), 161–162, 194, 199, 247
  election to chemistry committee of, 193
  fund raising by, for Swedish chemistry, 215–217, 243, 245, 259
  grant from special funds to, 214
  nomination and evaluation of, 159, 161, 194
  nominations and evaluations by, 171, 174, 254, 257–260, 262
  rise to prominence of, 190–198
*Svenska Dagbladet* [newspaper]
  on Einstein's prize, 138
  on nationalism, 62, 83–85, 96
  on wartime reservation of prizes, 79–80, 89–90
Sweden
  anti-Semitism in, 130, 238
  conservatism in, 7–8, 16, 59–61, 79
  government of, 113–114
  history of, 4–9, 59–63, 72–73, 97, 108, 254
  links to Germany of, 6, 32, 60–62, 72, 80, 224
  and "Manifesto of the Ninety-Three," 77
  national identity of, 59–61, 80, 272
  physics in, 41, 44, 141–144, 232
  progressivism in, 8–9, 59–61, 79
  role of Royal Academy in science of, 54, 122
Swedish Academy of Sciences. *See* Royal Swedish Academy of Sciences
Swedish Chemistry Society, 207
Swedish Inventors' Association, 57
Swedish-America Foundation, 100

Terman, Frederick E., 276
Terpenes, 38
Thalén, Robert
  and Arrhenius's work, 29
  election to physics committee of, 24
  experimentalist bias of, 44, 46, 50
Thermiotic effect, 170
Thermodynamics, 28, 41, 52, 181, 183, 207, 262
Thomson, George Paget, 220, 244
Thomson, Joseph John
  career of, 48, 100, 123, 127, 184
  nominations and evaluations by, 47, 153, 164
  physics prize for (1906), 47
Tiselius, Arne
  career of, 192, 257–258
  chemistry prize for (1948), 262
  nominations and evaluations by, 262
Todd, Alexander R., 263
Trowbridge, Augustus, 155–156

Ultracentrifuges, 192, 194–195
Ultramicroscopes, 190, 193
Uncertainty principle, 252
United States. *See also* Big Science
  anti-Semitism in, 241
  and the Cold War, 271
  and competition, 272–274, 277
  history of, 108, 120, 270–271
  as maturing scientific nation, 41, 120–121, 155, 191–192, 230–231
  post–World War II, 251–252, 254, 270–271
  in World War I, 96, 98

Universities. *See also specific university*
  and German academic culture, 6, 76–77, 268, 270
  nineteenth-century laboratories in, 18
  and nominating rights, 1–2, 23
  Nordic, 20, 23
University of California, Berkeley, 226, 270
University of Chicago, 241
University of Wisconsin, 191–192
Uppsala University
  history of, 7–8
  and nominating rights, 23
  style of chemistry at, 29–30, 206
  style of physics at, 43–47, 52, 130, 142–143, 154–155, 175, 238
Uranium, 234–236, 242, 247
Urbain, Georges, 187
Urey, Harold
  chemistry prize for (1934), 196, 219, 272
  nomination and evaluation of, 219–220
  nominations and evaluations by, 244

Vacuum tubes, 166, 197
van der Waals, Johannes Diderik
  nomination and evaluation of, 146
  physics prize for (1910), 50–51
  and prize ceremonies, 68
van't Hoff, Jacobus Hendricus
  chemistry prize for (1901), 28–31
  nominations and evaluations by, 33–35
*Vega* [ship], 60
Victoria (Queen of Great Britain), 61
Virtanen, Artturi Ilmari
  chemistry prize for (1945), 243, 255
  nomination and evaluation of, 253, 255–256
Viruses, 257–259
Vitamins, 199, 203, 208–209, 263
von Baeyer, Adolf
  career of, 32, 38, 55
  chemistry prize for (1905), 31–32, 247
  and "Manifesto of the Ninety-Three," 76
  nominations and evaluations by, 34, 75
von Bahr-Bergius, Eva, 236, 239
von Behring, Emil, 76
von Euler-Chelpin, Hans
  career of, 193, 197–198, 271, 273
  chemistry committee work by, 182–183, 189, 198–204, 207–209, 224, 253, 255–258
  chemistry prize for (1929), 202
  grant from special funds to, 214, 217
  nomination and evaluation of, 200–201
  nominations and evaluations by, 55, 186, 198, 222, 254
von Hann, Julius, 146
von Heidenstam, Werner, 67, 80, 96
von Hevesy, Georg, 187, 253
von Laue, Max
  career of, 126, 133, 171, 194, 223, 233
  chemistry prize for (1914), 90–92
  nomination and evaluation of, 74–75, 78, 88, 146
  nominations and evaluations by, 153–154, 221, 247
von Ossietzky, Carl
  peace prize for (1935), 202

Wachtmeister, Fredrik, 65, 68
Wallach, Otto
  chemistry prize for (1910), 38–39
Wallenberg, Knut, 215–216, 225
Wallenberg Foundation, 215–216, 231
Waller, Ivar
  career of, 215
  election to physics committee of, 252
  nominations and evaluations by, 174–175
  physics committee work by, 264–266
Walton, Ernest Thomas Sinton
  career of, 226, 261
  nomination and evaluation of, 228, 230
  physics prize for (1951), 228, 260
Warburg, Emil, 75, 101, 127, 129
Warburg, Otto, 200
Warner, Gustaf, 259
Washburn, Edward, 196
Watson, James, 273–274
Weapons
  atomic, 231–232, 237, 243, 245–246, 248, 259
  chemical, 105–107, 112, 114

Weather. *See* Atmospheric physics
Weaver, Warren, 256
Weinberg, Steven
  physics prize for (1979), 265
Weiss, Pierre, 100, 169
Wells, Herbert George, 86, 269
Werner, Alfred
  chemistry prize for (1913), 37
Westgren, Arne, 242–243, 249
Weyland, Paul, 126, 139
Wideröe, Rolf, 226
Widman, Oskar
  career of, 190–191, 206, 262
  chemistry committee work by,
    28–29, 31–33, 37–39, 54, 74, 106,
    161, 182–183, 185–188, 194, 201,
    208
  election to chemistry committee of,
    24, 193
  and establishment of chemistry
    prizes, 15–16
  nominations and evaluations by,
    35–36, 39, 186
Wieland, Heinrich
  chemistry prize for (1927), 188
  nomination and evaluation of, 187,
    196
  nominations and evaluations by,
    222
Wien, Wilhelm
  and "Manifesto of the Ninety-Three,"
    76
  and nationalism, 83–85
  nomination and evaluation of, 47, 49,
    57
  nominations and evaluations by, 111,
    126, 160, 169
  physics prize for (1911), 74
Willstätter, Richard
  career of, 55
  chemistry prize for (1915), 39,
    90–92, 273
  and "Manifesto of the Ninety-Three,"
    76
  nomination and evaluation of, 73–74,
    88
  nominations and evaluations by, 186,
    199–200, 222
Wilson, Charles Thomson Rees
  nomination and evaluation of,
    153–154

nominations and evaluations by,
  172
physics prize for (1927), 164–165,
  170
Wilson, Woodrow, 98, 107–108
Windaus, Adolf
  chemistry prize for (1928), 188
  nomination and evaluation of, 187,
    196
Withholding prizes
  bases for, 21
  and funding Swedish science,
    144–149, 151–152, 154–158,
    161–162, 213
  between the wars, 154, 156, 158,
    163, 182, 219–221
  during World War I, 73, 78, 93–95,
    146
  during World War II, 240
Women in science, 8. *See also*
  5Curie, Marie; Kowalevski, Sonja;
  Meitner, Lise; von Bahr-Bergius,
  Eva
Wood, Robert, 153, 158, 160, 169, 172
Woodward, R. B., 263
World War I, 71, 75–76, 96–98, 102,
  105–107, 240, 269
World War II, 240, 243, 256–257
  and Big Science, 232–233
World's fairs, 61, 268

X-rays, 27, 40, 101. *See also* Spectroscopy
  and crystallography, 74–75, 154–157,
    209
  diffraction of, through crystals,
    74–75, 143

Year of Miracles, 196, 225
Yukawa, Hideki
  nominations and evaluations by,
    265–266
  physics prize for (1949), 265

Zeeman, Pieter
  nomination and evaluation of,
    146
  nominations and evaluations by,
    47, 50, 127, 153, 160
  physics prize for (1902), 47
Zsigmondy, Richard, 190, 193–194,
  196